AF388656

Carbapenemase-Producing Enterobacterales

Carbapenemase-Producing Enterobacterales

Editor

Francesca Andreoni

MDPI • Basel • Beijing • Wuhan • Barcelona • Belgrade • Manchester • Tokyo • Cluj • Tianjin

Editor
Francesca Andreoni
Department of Biomolecular
Sciences
University of Urbino
Fano
Italy

Editorial Office
MDPI
St. Alban-Anlage 66
4052 Basel, Switzerland

This is a reprint of articles from the Special Issue published online in the open access journal *Antibiotics* (ISSN 2079-6382) (available at: www.mdpi.com/journal/antibiotics/special_issues/ Carbapenemase).

For citation purposes, cite each article independently as indicated on the article page online and as indicated below:

LastName, A.A.; LastName, B.B.; LastName, C.C. Article Title. *Journal Name* **Year**, *Volume Number*, Page Range.

ISBN 978-3-0365-4030-6 (Hbk)
ISBN 978-3-0365-4029-0 (PDF)

Contents

About the Editor . . . . **ix**

Preface to "Carbapenemase-Producing Enterobacterales" . . . . **xi**

Gurleen Taggar, Muhammad Attiq Rehman, Patrick Boerlin and Moussa Sory Diarra
Molecular Epidemiology of Carbapenemases in *Enterobacteriales* from Humans, Animals, Food and the Environment
Reprinted from: *Antibiotics* **2020**, *9*, 693, doi:10.3390/antibiotics9100693 . . . . **1**

Marilena Agosta, Daniela Bencardino, Marta Argentieri, Laura Pansani, Annamaria Sisto and Marta Luisa Ciofi Degli Atti et al.
Prevalence and Molecular Typing of Carbapenemase-Producing Enterobacterales among Newborn Patients in Italy
Reprinted from: *Antibiotics* **2022**, *11*, 431, doi:10.3390/antibiotics11040431 . . . . **23**

Efthymia Protonotariou, Georgios Meletis, Dimitrios Pilalas, Paraskevi Mantzana, Areti Tychala and Charalampos Kotzamanidis et al.
Polyclonal Endemicity of Carbapenemase-Producing *Klebsiella pneumoniae* in ICUs of a Greek Tertiary Care Hospital
Reprinted from: *Antibiotics* **2022**, *11*, 149, doi:10.3390/antibiotics11020149 . . . . **45**

Ângela Novais, Rita Veiga Ferraz, Mariana Viana, Paula Martins da Costa and Luísa Peixe
NDM-1 Introduction in Portugal through a ST11 KL105 *Klebsiella pneumoniae* Widespread in Europe
Reprinted from: *Antibiotics* **2022**, *11*, 92, doi:10.3390/antibiotics11010092 . . . . **67**

Reo Onishi, Katsumi Shigemura, Kayo Osawa, Young-Min Yang, Koki Maeda and Shiuh-Bin Fang et al.
The Antimicrobial Resistance Characteristics of Imipenem-Non-Susceptible, Imipenemase-6-Producing *Escherichia coli*
Reprinted from: *Antibiotics* **2021**, *11*, 32, doi:10.3390/antibiotics11010032 . . . . **79**

Yoshiro Sakai, Kenji Gotoh, Ryuichi Nakano, Jun Iwahashi, Miho Miura and Rie Horita et al.
Infection Control for a Carbapenem Resistant Enterobacteriaceae Outbreak in an Advanced Emergency Medical Services Center
Reprinted from: *Antibiotics* **2021**, *10*, 1537, doi:10.3390/antibiotics10121537 . . . . **91**

Jehane Y. Abed, Maxime Déraspe, Ève Bérubé, Matthew D'Iorio, Ken Dewar and Maurice Boissinot et al.
Complete Genome Sequences of *Klebsiella michiganensis* and *Citrobacter farmeri*, KPC-2-Producers Serially Isolated from a Single Patient
Reprinted from: *Antibiotics* **2021**, *10*, 1408, doi:10.3390/antibiotics10111408 . . . . **101**

Emanuela Roscetto, Rosa Bellavita, Rossella Paolillo, Francesco Merlino, Nicola Molfetta and Paolo Grieco et al.
Antimicrobial Activity of a Lipidated Temporin L Analogue against Carbapenemase-Producing *Klebsiella pneumoniae* Clinical Isolates
Reprinted from: *Antibiotics* **2021**, *10*, 1312, doi:10.3390/antibiotics10111312 . . . . **109**

Estefanía Cano-Martín, Inés Portillo-Calderón, Patricia Pérez-Palacios, José María Navarro-Marí, María Amelia Fernández-Sierra and José Gutiérrez-Fernández
A Study in a Regional Hospital of a Mid-Sized Spanish City Indicates a Major Increase in Infection/Colonization by Carbapenem-Resistant Bacteria, Coinciding with the COVID-19 Pandemic
Reprinted from: *Antibiotics* **2021**, *10*, 1127, doi:10.3390/antibiotics10091127 . . . . **119**

Pilar Lumbreras-Iglesias, María Rosario Rodicio, Pablo Valledor, Tomás Suárez-Zarracina and Javier Fernández
High-Level Carbapenem Resistance among OXA-48-Producing *Klebsiella pneumoniae* with Functional OmpK36 Alterations: Maintenance of Ceftazidime/Avibactam Susceptibility
Reprinted from: *Antibiotics* **2021**, *10*, 1174, doi:10.3390/antibiotics10101174 . . . . **133**

Hassan Al Mana, Sathyavathi Sundararaju, Clement K. M. Tsui, Andres Perez-Lopez, Hadi Yassine and Asmaa Al Thani et al.
Whole-Genome Sequencing for Molecular Characterization of Carbapenem-Resistant Enterobacteriaceae Causing Lower Urinary Tract Infection among Pediatric Patients
Reprinted from: *Antibiotics* **2021**, *10*, 972, doi:10.3390/antibiotics10080972 . . . . **143**

Pamela Barbadoro, Daniela Bencardino, Elisa Carloni, Enrica Omiccioli, Elisa Ponzio and Rebecca Micheletti et al.
Carriage of Carbapenem-Resistant Enterobacterales in Adult Patients Admitted to a University Hospital in Italy
Reprinted from: *Antibiotics* **2021**, *10*, 61, doi:10.3390/antibiotics10010061 . . . . **153**

Suzan Mohammed Ragheb, Mahmoud Mohamed Tawfick, Amani Ali El-Kholy and Abeer Khairy Abdulall
Phenotypic and Genotypic Features of *Klebsiella pneumoniae* Harboring Carbapenemases in Egypt: OXA-48-Like Carbapenemases as an Investigated Model
Reprinted from: *Antibiotics* **2020**, *9*, 852, doi:10.3390/antibiotics9120852 . . . . **165**

Ana Margarida Guerra, Agostinho Lira, Angelina Lameirão, Aurélia Selaru, Gabriela Abreu and Paulo Lopes et al.
Multiplicity of Carbapenemase-Producers Three Years after a KPC-3-Producing *K. pneumoniae* ST147-K64 Hospital Outbreak
Reprinted from: *Antibiotics* **2020**, *9*, 806, doi:10.3390/antibiotics9110806 . . . . **181**

Kyriaki Xanthopoulou, Alessandra Carattoli, Julia Wille, Lena M. Biehl, Holger Rohde and Fedja Farowski et al.
Antibiotic Resistance and Mobile Genetic Elements in Extensively Drug-Resistant *Klebsiella pneumoniae* Sequence Type 147 Recovered from Germany
Reprinted from: *Antibiotics* **2020**, *9*, 675, doi:10.3390/antibiotics9100675 . . . . **193**

Mykhailo Savin, Gabriele Bierbaum, Nico T. Mutters, Ricarda Maria Schmithausen, Judith Kreyenschmidt and Isidro García-Meniño et al.
Genetic Characterization of Carbapenem-Resistant *Klebsiella* spp. from Municipal and Slaughterhouse Wastewater
Reprinted from: *Antibiotics* **2022**, *11*, 435, doi:10.3390/antibiotics11040435 . . . . **207**

Florence Hammer-Dedet, Fabien Aujoulat, Estelle Jumas-Bilak and Patricia Licznar-Fajardo
Persistence and Dissemination Capacities of a bla_{NDM-5}-Harboring IncX-3 Plasmid in *Escherichia coli* Isolated from an Urban River in Montpellier, France
Reprinted from: *Antibiotics* **2022**, *11*, 196, doi:10.3390/antibiotics11020196 . . . . **221**

Lungisile Tshitshi, Madira Coutlyne Manganyi, Peter Kotsoana Montso, Moses Mbewe and Collins Njie Ateba
Extended Spectrum Beta-Lactamase-Resistant Determinants among Carbapenem-Resistant *Enterobacteriaceae* from Beef Cattle in the North West Province, South Africa: A Critical Assessment of Their Possible Public Health Implications
Reprinted from: *Antibiotics* **2020**, *9*, 820, doi:10.3390/antibiotics9110820 . . . . **235**

Annamária Földes, Edit Székely, Septimiu Toader Voidăzan and Minodora Dobreanu
Comparison of Six Phenotypic Assays with Reference Methods for Assessing Colistin Resistance in Clinical Isolates of Carbapenemase-Producing Enterobacterales: Challenges and Opportunities
Reprinted from: *Antibiotics* **2022**, *11*, 377, doi:10.3390/antibiotics11030377 . . . . **255**

Alicja Sekowska and Tomasz Bogiel
The Evaluation of Eazyplex® SuperBug CRE Assay Usefulness for the Detection of ESBLs and Carbapenemases Genes Directly from Urine Samples and Positive Blood Cultures
Reprinted from: *Antibiotics* **2022**, *11*, 138, doi:10.3390/antibiotics11020138 . . . . **273**

Katja Probst, Dennis Nurjadi, Klaus Heeg, Anne-Marie Frede, Alexander H. Dalpke and Sébastien Boutin
Molecular Detection of Carbapenemases in Enterobacterales: A Comparison of Real-Time Multiplex PCR and Whole-Genome Sequencing
Reprinted from: *Antibiotics* **2021**, *10*, 726, doi:10.3390/antibiotics10060726 . . . . **283**

About the Editor

Francesca Andreoni

Francesca Andreoni received a BSc in Biological Sciences and her Ph.D. in Biochemical and Pharmacological Methodologies from the University of Urbino and a Specialization in Clinical Biochemistry from the University of Camerino, Italy. She is an Adjunct Professor of Cell Biology and Genetics at the Department of Biomolecular Sciences of the University of Urbino. Her research interests are antimicrobial resistance, carbapenemase, Enterobacterales, clinical microbiology, molecular typing, plasmid characterization, and the identification of vaccine candidates. Her main research areas are the identification of vaccine candidates for the development of a vaccine against *Photobacterium damselae* subsp. *piscicida* fish pathogen and the molecular characterization of microbial species of clinical interest and the development of procedures to trace and control the antimicrobial resistance spread. She has supervised and participated in several research projects in these areas, authoring more than 50 peer-reviewed publications (articles, abstracts, and book chapters) and 60 scientific communications in congresses (oral and poster communications). She also has experience in student supervision at several levels (PhD, MSc and BSc). She was responsible for the diagnostic laboratory of Molecular Biology and Medical Genetics Cante di Montevecchio from 2016 to 2018.

Preface to "Carbapenemase-Producing Enterobacterales"

The large family of Enterobacteriaceae includes such species as *Klebsiella pneumoniae* and *Escherichia coli*, which are commonly responsible for healthcare infections. The increasing prevalence of resistance to antibiotics used to treat severe infections and diseases, in particular to carbapenems, is due to a rise in multidrug-resistant pathogens which pose an urgent threat to public health. Carbapenem resistance is mainly associated with the production of carbapenemase—encoded by mobile genetic elements, which are usually plasmids that are horizontally acquired and highly transmissible. Carbapenem-resistant Enterobacterales (CRE) are a common cause of infections in both community and healthcare settings. For this reason, the implementation of control measures and screening programs on CRE carriage is an important practical application toward limiting the dissemination of these strains between clinical wards. The focus of this Special Issue includes any aspects concerning plasmid-mediated antimicrobial resistance along with other carbapenem resistance mechanisms. Understanding the prevalence and routes of transmission of CRE is important in developing specific interventions for healthcare facilities. No less important is the general impact of CRE circulation on the environment. It is known that residues of antimicrobials that are widely used in clinical settings and also entering water and soil during intensive breeding create a selective pressure contributing to the increasing antibiotic resistance of microorganisms. In light of this, attention must be focused on carbapenemase testing in order to provide advanced phenotypic and molecular assays for the identification of CRE. Furthermore, the optimization of protocols could be a valid tool for active global surveillance, and from this perspective, the study of resistance mechanisms can provide significant support for the development of new and appropriate antimicrobial molecules. For all of these reasons, the phenomenon of carbapenem resistance deserves more attention, for the sake of public health.

Francesca Andreoni
Editor

Review

Molecular Epidemiology of Carbapenemases in *Enterobacteriales* from Humans, Animals, Food and the Environment

Gurleen Taggar [1,2], **Muhammad Attiq Rheman** [1], **Patrick Boerlin** [2] and **Moussa Sory Diarra** [1,*]

1. Guelph Research and Development Center, Agriculture and Agri-Food Canada (AAFC), 93, Stone Road West, Guelph, ON N1G 5C6, Canada; gtaggar7@gmail.com (G.T.); attiq.muhammad@canada.ca (M.A.R.)
2. Department of Pathobiology, Ontario Veterinary College, University of Guelph, Guelph, ON N1G 2W1, Canada; pboerlin@uoguelph.ca
* Correspondence: moussa.diarra@canada.ca

Received: 14 September 2020; Accepted: 7 October 2020; Published: 13 October 2020

Abstract: The Enterobacteriales order consists of seven families including *Enterobacteriaceae*, Erwiniaceae, Pectobacteriaceae, Yersiniaceae, Hafniaceae, Morganellaceae, and Budviciaceae and 60 genera encompassing over 250 species. The *Enterobacteriaceae* is currently considered as the most taxonomically diverse among all seven recognized families. The emergence of carbapenem resistance (CR) in *Enterobacteriaceae* caused by hydrolytic enzymes called carbapenemases has become a major concern worldwide. Carbapenem-resistant *Enterobacteriaceae* (CRE) isolates have been reported not only in nosocomial and community-acquired pathogens but also in food-producing animals, companion animals, and the environment. The reported carbapenemases in *Enterobacteriaceae* from different sources belong to the Ambler class A (bla_{KPC}), class B (bla_{IMP}, bla_{VIM}, bla_{NDM}), and class D ($bla_{\mathrm{OXA\text{-}48}}$) β-lactamases. The carbapenem encoding genes are often located on plasmids or associated with various mobile genetic elements (MGEs) like transposons and integrons, which contribute significantly to their spread. These genes are most of the time associated with other antimicrobial resistance genes such as other β-lactamases, as well as aminoglycosides and fluoroquinolones resistance genes leading to multidrug resistance phenotypes. Control strategies to prevent infections due to CRE and their dissemination in human, animal and food have become necessary. Several factors involved in the emergence of CRE have been described. This review mainly focuses on the molecular epidemiology of carbapenemases in members of *Enterobacteriaceae* family from humans, animals, food and the environment.

Keywords: carbapenemases; *Enterobacteriales*; human; animal; food; environment

---

1. Introduction

The actual pandemic outbreak of the COVID-19 killing several thousands of people along with its serious negative global economic impacts worldwide is a clear indication that a lot of efforts need to be deployed to fight against infectious diseases and the increased global issue of antimicrobial resistance. The World Health Organization (WHO) published a global priority list of antimicrobial resistant pathogenic bacteria including some *Enterobacteriales* for which new antibiotics are urgently needed [1]. The genera within the order *Enterobacteriales* are composed of Gram-negative bacteria of class *Gammaproteobacteria*, which encompasses many harmless symbiotic and pathogenic strains, including members of the genera *Dickeya*, *Pectobacterium*, *Brenneria*, *Erwinia* and *Pantoea* [2]. The pathogenic strains mainly *Klebsiella pneumoniae*, *Yersinia pestis*, *Escherichia* spp., *Salmonella enterica* serovars and *Enterobacter* spp. cause a broad range of intestinal and extra intestinal diseases in humans and animals [3]. The expended spectrum cephalosporins (ESC) and cephamycins are frequently used against infectious diseases caused by *Enterobacteriaceae*. Due to the emergence of multidrug resistance, carbapenems in addition to tigecycline and colistin are among the last line of defense against *Enterobacteriaceae*, because co-resistance to both colistin and tigecycline among the carbapenem-resistant *Enterobacteriaceae* (CRE) has been rarely reported [4,5]. Carbapenems are a powerful group of broad-spectrum antibiotics which, in many cases, are the last line of defense against multi-resistant bacterial infections. Carbapenems are classified under β-lactams antibiotics, slightly different from the penicillin by substitution of a carbon atom for a sulfur atom and addition of a double bond to the five-membered ring of the penicillin nucleus (Figure 1).

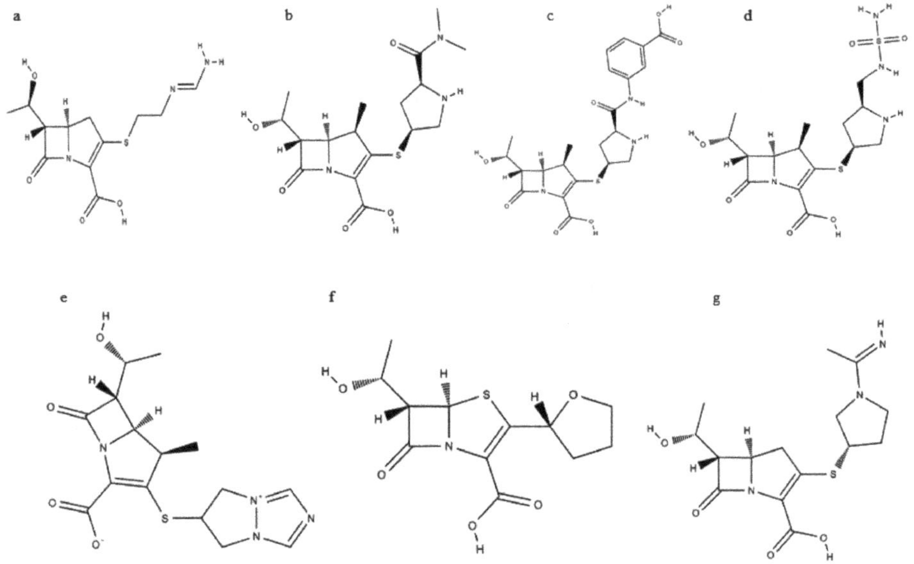

Figure 1. Chemical Structures of various carbapenems: (**a**) Imipenem; (**b**) Meropenem; (**c**) Ertapenem; (**d**) Doripenem; (**e**) Biapenem; (**f**) Faropenem; (**g**) Panipenem, obtained from the National Center for Biotechnology Information (NCBI) PubChem database.

Carbapenems bind tightly to the bacterial penicillin-binding proteins (PBPs), which are vital for elongation and cross-linkage of the cell wall peptidoglycan, leading to bacterial lysis [6] (Figure 2).

Currently four of carbapenems (imipenem, meropenem, ertapenem and doripenem) are approved for clinical use in the United States of America [7] and additional three of them (biapenem, faropenem, panipenem) in Canada. During the last decade, several monitoring studies have reported the emergence of carbapenem resistant *Enterobacteriaceae* (CRE) [8]. The three major mechanisms of carbapenem resistance in these bacteria include: (i) the presence of β-lactamase enzymes called carbapenemases, and (ii) the combined effect of other β-lactamases with bacterial cell membrane permeability due to alteration or mutations in the porins and/or (iii) increased efflux pump-action (Figure 3). The modification of penicillin binding proteins (PBPs) has been reported as the forth mechanism of resistance to carbapenems in Gram-negative bacteria [9].

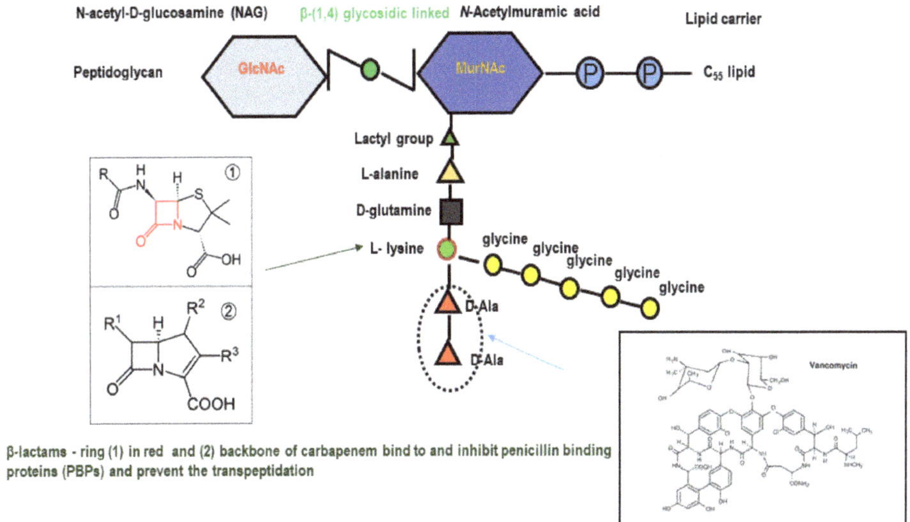

Figure 2. Mechanism of action of β-lactam antibiotics compared to that of vancomycin on the bacterial cell wall. Beta-lactams bind to and inhibit enzymes (PBPs: transpeptidases) which catalyse the final crosslinking (transpeptidation) of the nascent peptidoglycan layer which disrupt cell wall synthesis. Updated from Neu and Gootz, 1996, Ch. 11. Antimicrobial Chemotherapy in Medical Microbiology. 4th edition. Baron, editor. Galveston (TX): University of Texas Medical Branch at Galveston, USA.

Carbapenemases induce resistance essentially by hydrolysis of carbapenem using active catalytic substrates either serine or zinc [10] as indicated in Table 1. The non-metallo-carbapenemase-A (Nmc-A) was first described to cause carbapenem resistance in *Enterobacteriaceae* in 1993 [11]. According to the Ambler's molecular classification, class B metallo β-carbapenemases (MBL types), class A (*Klebsiella pneumoniae* carbapenemases (KPC) types), and class D oxacillinases (OXA types) are epidemiologically important in *Enterobacteriaceae* [12–14]. The genes encoding these carbapenemases can be located either on the chromosome or on mobile genetic elements (MGEs) like plasmids, integrons, and transposons [15,16]. The carbapenemases KPC and NDM (New Delhi metallo β-carbapenemases) producing bacteria have shown resistance against most of the β-lactams, fluoroquinolones, and aminoglycosides [17]. However, OXA type in particular OXA-48-like carbapenemases are less active against carbapenems and can induce a high resistance level only when associated with extended-spectrum β-lactamases (ESBLs) [18,19]. The dissemination of carbapenemases genes by MGEs among clinical isolates is a source of serious public health and food safety concerns. *Enterobacteriaceae* with acquired carbapenem resistance genes have been isolated from humans, animals, food, and the environment. The aim of this review is to discuss the molecular epidemiology of carbapenemases in *Enterobacteriaceae*.

Table 1. Characteristics of the three most common classes of carbapenemases in *Enterobacteriales*.

Ambler Class.	Functional Class [a]	Representative Gene [b]	No. of Variants [c]	Active Site [d]	Substrate	Inhibitor(s)	Genetic Location	Species of Origin
A	2F	KPC	22	Serine	carbapenems, cephalosporins, Penicillins	Clavulanic acid	Chromosomally encoded; *Inc*FIIK2, *Inc*F1A, *Inc*I2, multiple types; Tn*4401*	*Klebsiella pneumoniae*
		IMI	9				Chromosomally encoded, *Inc*F	*Enterobacter cloacae*
		SME	5				Chromosomally encoded, SmarGI1 novel genomic island	*Serratia marcescens*
		NMC-A	1				Chromosomally encoded	*Enterobacter cloacae*
		GES	27				Class I integrons	*Pseudomonas aeruginosa*
B	3	NDM	16	Zinc	Most β-lactams including carbapenems	EDTA	*Inc*A/C, multiple; IS*Aba125*; Tn125	*Klebsiella pneumoniae*
		IMP	56				*Inc*L/M, *Inc*A/C, multiple types; class I integrons	*Serratia marcescens*
		VIM	48				*Inc*N, *Inc*I1, multiple types; class I integrons	*Pseudomonas aeruginosa*
		GIM	2				class I integrons	*Pseudomonas aeruginosa*
		SPM	1				Plasmid-mediated	*Pseudomonas aeruginosa*
		KHM	1		Most β-lactams except monobactams		Plasmid-mediated	*Citrobacter freundii*
		CcrA	1		Most β-lactams including carbapenems		Chromosomally encoded	*Bacteroides fragilis*
		BcII	1				Plasmid-mediated	*Bacillus cereus*
		CphA	8				Plasmid-mediated	*Aeromonas hydrophilia*
		L1					not determined	*Stenotrophomonas maltophilia*
D	2	OXA	489	Serine	Most β-lactams including carbapenems	Clavulanic acid	*Inc*L/M, Tn1999, IS1999, ColE plasmids, Tn2013, IS*Ecp1*, IS*Aba125*	*Klebsiella pneumoniae*

[a] cited in [12], [b] Most common carbapenemases identified, [c] Based on the Comprehensive Antibiotic Resistance Database (CARD v 3.0.2), [d] Carbapenemases induce resistance essentially by hydrolysis of carbapenem using active catalytic substrates either serine or zinc.

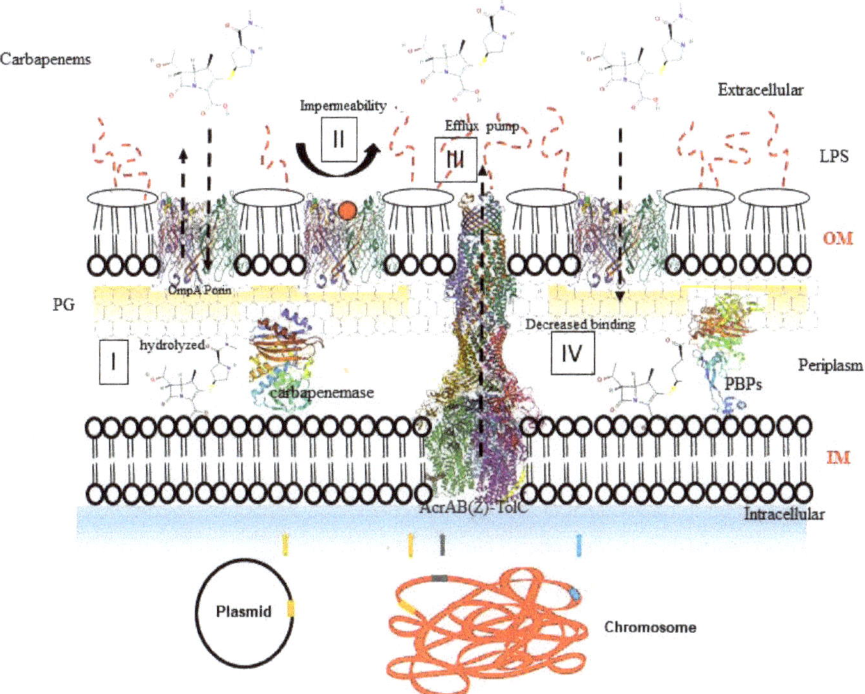

Figure 3. Major mechanisms associated with carbapenem-resistance in Gram-negative bacteria. (I) Production of carbapenemases enzyme from gene located on chromosome or plasmid that hydrolyze carbapenem antibiotics shown in golden rectangles. (II) Decreased permeability of the outer membrane due to structural mutations in porins (modified porins shown as red circle). (III) Drug efflux pumps. The blue and grey rectangles represent the chromosomal loci that encode various membrane associated proteins. Abbreviations: LPS, lipopolysaccharides; OM, outer membrane; IM, inner membrane; PG, peptidoglycan.

2. Carbapenemases-Producing *Enterobacteriaceae* from Humans

The β-lactams antibiotics such as penicillins and cephalosporins have been wildly used against pathogenic *Enterobacteriaceae* because of their broad-spectrum activity [20]. However, this practice contributed to the emergence and spread of several types of β-lactamases including ESBLs. To overcome the resistance against extended spectrum cephalosporins (ESC) and cephamycins in *Enterobacteriaceae*, carbapenems have been introduced in human medicines. According to the recent 2020 Canadian Antimicrobial Resistance Surveillance System (CARSS) annual report, the carbapenems use in human medicine has been increased from 3.0 to 6.8 defined daily doses (DDDs) per 1000 inhabitants between 2014 and 2018 in Canada. Consequently, a concomitant nine-fold increase in the number of patients colonized by carbapenemases-producing organisms (CROs) without signs of infection has been reported in Canada [20]. Globally, the population of CRE is increasing dramatically [21,22]. There could be several factors leading to the spread of carbapenemase-producing pathogens in humans. These factors include continuous exposure to antibiotics, usage of different concentrations of antibiotics, and contamination of surgical equipment used [12,23,24].

The prevalence of carbapenem-producing *Enterobacteriaceae* (CPE) is high in humans with advanced age, primarily, due to their frequent visits to hospitals, long-term stay in health care facilities, tertiary care hospitals, and teaching hospitals [25–30]. Hospital stay represents a particularly high risk to be colonize or in developing an infection with a CRE. Recent studies found that CRE colonization among

ICU patients showing new emerging mechanisms of resistance continue to rise in the United States of America [31,32]. There seems to have no different in other developed countries. For instance, Canada's Nosocomial Infection Surveillance Program data suggested that incidences of bla_{NDM}, bla_{KPC}, bla_{OXA-48}, and bla_{VIM} producing *K. pneumoniae* and *Enterobacter* spp. increased significantly in Canadian health care facilities since 2007 [33]. These health care facilities could be a reservoir for patient spreading CPE to multiple regions. For instance, bla_{KPC}-producing *K. pneumonia* have been reported in hospital outbreaks in many European countries such as Greece, Italy, Spain, France and Germany [3,34–40]. A recent study reported cases of bla_{NDM-1} and bla_{KPC-2} -producing *K. pneumonia* among transplanted patients in Brazil [41]. Although, little is known about the spread and clinical relevance of CRE in Africa, two studies reported their prevalence in hospital and community settings among several African countries [42,43]. Issues about carbapenemases include their potential link with multidrug-resistance genes on the same MGEs. For instance, the bla_{KPC} gene encoding KPC enzyme to hydrolyze all β-lactams was found on plasmids carrying multiple other antimicrobial resistance determinants [44,45]. Outbreaks caused by multidrug resistant and bla_{KPC}-positive *K. pneumoniae* opportunistic pathogenic strains have been reported in North America, Europe, Asia and South America [3,46–48]. A recent study from China reported a *Morganella morganii* isolate, an opportunistic pathogen, harboring bla_{NDM-5} gene on a self-transmissible *IncX3* plasmid from a stool sample of a cancer patient [49].

Another important factor responsible for worldwide dissemination of CRE is the international travelling and medical tourism. There are several reports demonstrating the role of travelling to affected developed countries in the epidemiology of CRE. Pathogenic strains of *K. pneumoniae* and *Enterobacter cloacae* containing bla_{KPCs} have been isolated from patients from France and Greece hospitalized in New York [50,51]. Overcrowding coupled with poor sanitation conditions including inappropriate waste management system and misuse of antibiotics could play roles in the spread of antimicrobial resistance genes in general and those for carbapenemases in particular. Furthermore, urbanization and globalization are greatly involved in spreading antimicrobial resistance pathogens all over the world. Accordingly, class B MBL, bla_{NDM-1}-producing *K. pneumoniae*, *Escherichia coli*, *E. cloacae*, *Citrobacter* spp., *Proteus* spp., *and Klebsiella oxytoca* strains were originally isolated in India. These same bla_{NDM-1} producing *K. pneumoniae* and *E. coli* strains were subsequently isolated from Sweden and UK patients who had travelled to India recently. The patients in Sweden and UK may either have been hospitalized or underwent any medical intervention in India which could led to their infection or colonization by these pathogenic strains producing bla_{NDM-1} [52–54]. A similar scenario has been reported in Canada and the United States where bla_{NDM-1}-producing *Enterobacteriaceae* were isolated from patients who visited and received medical cares in the Indian subcontinent [55,56]. Furthermore, a recent study reported two hyper-virulent *K. pneumoniae* clones of ST86 harboring plasmid mediated bla_{KPC-2}, isolated from a Canadian patient who visited Greece [57]. Likewise, a novel bla_{KPC-3} variant (bla_{KPC-50}) was recently identified in multi-drug resistant *K. pneumoniae* clinical isolate conferring resistance to ceftazidime-avibactam in Switzerland that was most likely acquired in Greece [58]. Several studies have demonstrated that medical tourism is another way to introduce CREs from an endemic country to a non-endemic country [59–61]. For instance, a case has been reported in an Israelis hospital where four non-Israelis patients, were positive for bla_{OXA-48}-producing *Enterobacteriaceae* [61]. The *OXA*-48 positive bacteria were absent from this hospital before these patients' admission in the Israel, two and one of them were respectively hospitalized in Jordan and Georgia where bla_{OXA-48} producing *Enterobacteriaceae* were prevalent, demonstrating thus the role of medical tourism in the epidemiology of CRE [60].

3. Companion Animals

Carbapenems are not licensed for the treatment of infectious diseases in companion animals in most of countries. As a result, pathogenic strains of *Enterobacteriaceae* causing infections in companion animals are not usually screened for carbapenemase resistance genes in veterinary laboratories. The possible way companion animals may get infected with CRE is through direct contact with colonized hosts and contaminated environment. Eventually, companion animal may become a reservoir for CPE [62–65]. For instance, a bla_{OXA-48} carbapenemase-producing *K. pneumoniae* has been transmitted from human to companion animals (dogs) through contaminated hands [66]. In 2015, the transmission of an *IncX3* plasmid bearing bla_{NDM-5} in *E. coli* ST167 was detected in dogs and humans in Finland [67]. A recent study found *IncX3* plasmid mediated bla_{NDM-5} in *E. coli* ST410 in four Korean dogs implying the HGT an important mechanism of their spread among companion animals. Carbapenemase-producing *E. coli* ST131 has also been reported repeatedly in both household members and their companion animals [68–70]. An Australian study also reported the *IncHI2* type plasmid bearing bla_{IMP-4} in *Salmonella* Typhimurium isolates from infected cats [71]. Importantly this variant showed sequence similarity to two bla_{IMP-8} carrying *IncHI2* plasmids from *Enterobacter* spp. from humans with an indication of nosocomial spread and broader risk to humans, animals and the environment. *Enterobacteriaceae* producing carbapenemase (bla_{NDM-1} and bla_{OXA-48}) have been isolated from companion animals like dogs, cats, and horses in the USA, Germany, Greece and UK [66,72–75]. Another recent study in Switzerland reported the isolation of various carbapenemase-producing *Enterobacteriales* in companion animal clinics mainly associated with poor clinical practices [76]. The isolation of CRE in companion animals brings the attention to reconsider the use of any off-label use of carbapenems in the veterinary medicines. Even carbapenems are not registered to use in companion animals, these antibiotic are used as off-label for the treatment of urinary tract infections in dogs and horses and to treat after surgical procedure infections caused by multidrug resistance *E. coli* in the UK and some other European union countries [77]. Identification of CPE in companion animals could become significant for public health due to not only host-to-host transmission but also possible gene transfers between commensals and pathogens. Due to selection pressure, treated animals (pets) may become colonized with CPE that could be transmitted to human through fecal – oral contaminations.

4. Carbapenemases-Producing *Enterobacteriaceae* in Other Animals

Carbapenemases-producing *Enterobacteriaceae* is not only a threat to humans; but animals may also get colonized and/or affected by CPE. The CPE could be isolated from food producing animals such as chicken, swine and cattle. A geographical distribution of CPE isolates from animal clinical samples has been recently reported [9].

5. Food Producing Animals

Various studies reported the presence of CPE in livestock which could constitute a food safety issue. Based on these studies, bla_{VIM-1} and bla_{NDM-1} were the most prevalent carbapenemase enzymes among *Enterobacteriaceae* in food producing animals [78–82]. In Germany, bla_{VIM-1} producing *S. enterica* and *E. coli* have been isolated from swine and chicken farms. In both cases, bla_{VIM-1} gene was located on an *IncHI2* type plasmid. These isolates showed multidrug resistance due to the presence of other resistance genes including bla_{ACC-1}, bla_{ADD-1} and $strA/B$ in addition to the bla_{VIM-1} on same *IncHI2* type plasmids [80,81]. Apart from bla_{VIM-1}, three *E. coli* isolates were found to carry the bla_{NDM-5} gene on a *IncX3* transferrable plasmid, including one co-harboring the colistin resistance *mcr-1* gene on *IncHI2* plasmid of ST446 and the other two belonged to ST2, isolated from three dairy cows in China [83]. Moreover, a novel *IncX3*-type plasmid harboring a bla_{NDM} variant (bla_{NDM-20}) due to three point mutations compared to bla_{NDM-1}, was recovered recently in an *E. coli* ST1114 from swine in China thus, suggesting that food-animal could be a source of new carbapenemase genes [84].

6. Carbapenemases Producing *Enterobacteriaceae* in Food

CREs from food-producing animals could find their way into the food chain, leading to an alarming food safety issue. For public health concerns with respect to risk of transfer to humans via the food source, several resistance surveillance systems for retail meat are in place which include carbapenem susceptibility monitoring in many countries. Interestingly, CRO could be isolated from the food that escape resistance surveillance programs such as seafood [85]. Besides live animals, several reports demonstrating the spread of CPE in retail meat in Egypt, China and Pakistan [86–88].

As reported in 2016 in China, the increasing spread of carbapenem and colistin-resistant *E. coli* clone ST167 from chicken meat harboring bla_{NDM-9} and *mcr-1* co-located on a plasmids has raised concerns worldwide because of potential transfer of this resistance plasmids to other Gram-negative pathogens and to other countries [88]. Moreover, VIM-1 producing *S. enterica* serovar Infantis have been isolated from minced pork meats in Germany [89] and NDM-1 producing *S. enterica* serovar Indiana have been isolated from chicken carcasses in China [90]. Additionally, a recent study reported the contamination of retail meat samples from pork, chicken and beef with *E. coli* and *K. pneumonia* containing bla_{NDM} genes located on *Inc*X3 plasmid [91]. Between 2016 to 2018, a substantial increase of such CRE from 9.4% to 22. 2% in the retail meat samples was reported.

As stated above CRE could be found in seafood as reported as VIM-1 producing *E. coli* from retail squid, sea squirt, clams and seafood medley in Germany, China and Korea [92]. Trading of seafood from endemic countries to non-endemic countries could result in the spread of carbapenem resistance genes. For instance, the Canadian Integrated Program for Antimicrobial Resistance Surveillance (CIPARS) reported carbapenem-resistant *Enterobacter* spp. in fresh and frozen raw shrimp collected imported from Southeast Asia to Canada [93]. A recent study from Netherlands reported the isolation of *E. cloacae* ST813 bearing plasmid with bla_{IMI-2} and a novel Ambler class A carbapenemases bla_{FLC-1} from a frozen vannamei white shrimp (*Litopenaeus vannamei*) originating in India. This new bla_{FLC-1} carbapenemases was found to be related to previously known French imipenemase (FRI), with 82% amino acid identity to bla_{FRI-1} and 87% to bla_{FRI-5} [94]. Although rare but raw milk containing OXA-48-producing *K. pneumoniae* ST530, an epidemic clone, was reported in Lebanon [95].

A role for fresh produces in spreading antimicrobial-resistant has been suggested by several studies. In China, fresh lettuce was found to be contaminated with *E. coli* ST877 co-producing NDM-1 and KPC-2, while also carrying *fosA3* and *floR* genes on a transferrable *Inc*A/C2 type plasmid [96]. Various species of CRE were recovered from leek, radish, basil, spinach, lettuce, traditional and commercial salads in Iran and OXA-48 producing *K. pneumonia* were detected in fresh vegetables from Algeria [97]. It has also been suggested that international trade of fresh vegetables and spices could be the possible route for the spread of CRE. Coriander imported from Asia to Switzerland and many other countries was positive to $bla_{OXA-181}$-producing *Klebsiella variicola* [98,99]. Consequently, there is a need for resistance surveillance programs for both carbapenemase-producing pathogenic and non-pathogenic organisms in the food chain to find the potential reservoirs of carbapenemase genes and to prevent their spread from food to humans.

7. Carbapenemases-Producing *Enterobacteriaceae* from the Environment

As discussed above, several studies demonstrate the spread of CPE all around the world among humans and animals. However, there are very few reports on the role of environmental contamination in the spread of CPE. The environment, surrounded by the CPE carriers, may be contaminated with these bacteria and further act as a vector for their dissemination. Dissemination of environmental CPE (eCPE) can negatively impact human health [100]. The prevalence rate of eCPE is high especially around intensive care units, acute and long-term health care facilities. Exposure of health care personnel to infected patients and cleaning methods used in the health facilities could potentially be responsible for dissemination of eCPE [101]. A study reported the presence of eCPE on the bed surface, personal table and infusion pump used by one CPE-infected patient [101]. The potential presence of eCPE on health care personnel brings attention toward the importance of adopting cleaning methods used in hospitals to disinfect the surfaces and material used by CPE carriers [102]. In hospitals, carbapenems are frequently used to treat infectious diseases caused by extended spectrum cephalosporinase (ESC) producing bacteria. Carbapenems are not entirely metabolized in the body and some residues present in human excreta can get into hospital sewage. Due to selection pressure, there is a chance that pathogens present in hospital effluent may become resistant against carbapenems. It has been reported that hospital sewage may act as a reservoir for resistance genes and a point where organisms likely acquire resistance through horizontal gene transfer events [103]. For instance, NDM-4 producing *E. coli* with downstream bleomycin resistance gene, ble_{MBL} on a plasmid type associated with complete IS *Aba*125 were isolated from hospital sewage in India, suggesting hospital wastes as major reservoir of resistance genes [104]. Likewise, antibiotic residues released into the municipal wastewater along with human excreta could contribute to selection of CPE the ground and surface water sources. A VIM-1-producing resistant *K. pneumoniae* strain has been isolated from rivers in Spain, Switzerland and Sweden [105–107]. The presence of CRE in wastewater is a potential concern because this environment may serve as major reservoir leading to HGT events and an increased risk of carbapenem resistance spreading into the environment. The data collected from a recent US survey of seven wastewater treatment plants reported the detection of 20% carbapenem-resistant *E. coli* isolates of sequence types associated with extra-intestinal infections in humans harboring predominantly bla_{VIM} and bla_{KPC} genes [108]. Similarly, a study from the United States documented the presence of CRE including *E. coli* and *P. mirabilis* isolates harboring bla_{IMP-27} gene on *IncQ1* plasmids in both environmental and fecal samples of swine production system [109]. The presence of bla_{KPC-2} gene has been reported in the United States in the metagenome from the feces of beef cattle regardless of antibiotic use in the farm [110]. Moreover, two most recent studies reported the presence and survival of carbapenem-resistant organisms harboring a plasmid-borne bla_{OXA-23} gene from swine manure environment from a Croatian pig farm and NDM-5 producing *E. coli* ST156 from a poultry farm in China. These suggest the possible dissemination of CRE in the food chain through animal manure fertilization [111,112]. On the other hand, irrigation water and various water matrices are also considered as major source of fresh produce contamination with resistant bacteria including CRE that can potentially be transferred to the consumer especially when consumed raw [113]. Likewise, a recent study reported carbapenem resistant *K. pneumoniae* isolates from Austrian rivers showing genetic similarities to clinical isolates from hospitals raises concerns regarding the role of surface water in the of dissemination of CRE [114]. Surprisingly, tap water could serve as reservoir of community exposure to CRE in high income countries. A recent multi-state study form United States screened drinking water samples from both public and private water systems in six states and detected 6.4% CRE harboring bla_{OXA-48}-type carbapenemase gene [115]. Another Recent study from Egypt, investigated the occurrence of β-lactamase and CRE in the integrated agriculture–aquaculture environment and isolated several *Enterobacteriaceae* strains resistant carrying predominantly the carbapenemase resistant gene bla_{KPC} either alone or with the β-lactamase genes ($bla_{CTX-M-15}$, bla_{SHV}, bla_{TEM}, and bla_{PER-1}). This study suggests transmission of the resistance genes among *Enterobacteriaceae* strains in integrated agriculture–aquaculture system with serious public health implication [116].

8. Molecular Epidemiology of Carbapenem Resistance (CR) Genes

Spread of CRE: The most frequently identified carbapenem genes are the ambler class A including bla_{KPC} followed by class B metallo- beta lactamases (MLBs) such as bla_{NDM}, and the class D OXA-type gene like bla_{OXA-48} (Table 1). Since its identification in the North Carolina, USA in 1996, in *K. pneumonia* patient, the epidemiology of bla_{KPC} producing isolates has expanded mostly globally in Africa, America, Asia, Australia, and Europe especially the clonal group (CG) 258, which includes the lineages ST258 and ST11). Since 2015, several outbreaks related to ST258 have been documented in hospital of Israel closely related to strain found in New York [117]. Likewise, the class B metallo- beta lactamases genes, including bla_{NDM}, bla_{SME}, bla_{GES}, bla_{VIM}, and bla_{IMP}, have also been disseminated worldwide, bla_{NDM-1} being the most prevalent worldwide [118]. The gene bla_{NDM-1} was first detected in a *K. pneumonia* isolate from a Swedish patient of Indian origin in 2008 in a multidrug-resistant *K. pneumoniae*, suffering from urinary tract infection acquired in India [119]. Subsequently, the bla_{NDM-1} was also found in *K. pneumonia* isolate from Croatia, from a patient arrived from Bosnia and Herzegovina. The second geographical origin of bla_{NDM-1} considered to be eastern Balkans. In May 2010, a case of infection with bla_{NDM-1} producing *E. coli* was reported in Coventry in the United Kingdom. Moreover, the bla_{NDM-1} gene was also reported in Australia, Austria, Belgium, China, Canada, France, Germany, Hong Kong, Japan, Netherland, Norway, and Sweden. Likewise, the bla_{SME} were reported in England with sporadic reports of such gene in USA. Since 2004 and it has also been reported in Argentina, Australia, Brazil, Canada and Switzerland but the increasing number of this gene have been reported in America and UK [120]. The bla_{VIM} gene was originally described in Italian *Pseudomonas aeruginosa* in the mid-1990s and afterwards CRE carrying bla_{VIM} were predominantly reported in Europe and large number of cases were also reported in other countries including Africa, Taiwan, Mexico, Saudi Arabia and the United States. However, the bla_{VIM} gene was found in Greece with more than 48 variants of this gene showing global dissemination [121]. Furthermore, the bla_{IMP} gene was first originated in south pacific Asia predominantly among isolates of *E. coli* and *Enterobacter* on class 1 integrons and genomic evidences are now emerging of its dissemination globally [122]. Finally, among the various variants known of Class D today, the $bla_{OXA\text{-}48}$ gene was first identified in Turkey in 2001. Although endemic during that time mostly in Turkey and Malta but since 2015, it has been spread to multiple countries including Singapore, Algeria, Korea, south Africa and have recently emerged in Canada and other western countries.

CRE carriage: Several studies particularly in China and also around the globe have recently attempted to investigate the carriage of CRE in animals and humans that possibly contributes to its dissemination in societies and healthcare facilities [123,124]. For instance, a recent case study in Singapore determined the duration of CRE carriage among various hospital workers by investigating 21 CRE carriers for more than 1 year. The authors of this study reported a mean carriage duration of 86 days among hospital workers and further suggested that the prolonged carriage could be associated with use of antimicrobial drugs and the probability of decolonization in a year was 98.5% [124]. Despite reports in healthcare facilities, a recent study by Zhai, et al. [125] monitored the carriage of CRE in a Chinese poultry farm for over a year (January 2017 to April 2018). During this period these authors collected 350 cloacal samples from four broiler farms in additional to 582 environmental samples and found that CRE negative 1-day-old broilers acquired bla_{NDM} within 24 h of transfer. Furthermore, the same study also analyzed the persistence and transmission of bla_{NDM}-producing bacteria in a Chinese poultry farm and found that the *IncX3* plasmid accounts for 71% of bla_{NDM} carriage that persisted in farm over 16 months and about 20% also carried either the colistin resistance *mcr-1* or *mcr-8* gene. This study further suggests that the contaminated in-house environment contributes to the persistence and transmission of bla_{NDM} producing bacteria in Chinese poultry farms.

Food chain transmission: The rise of CRE in food-producing animals and food supply is of growing concern globally and, given the risks of CRE to human health, there have been a zero-tolerance policy and an international ban on the sale of food items contaminated with CRE in several countries [126]. Therefore, prevention of CRE occurrence and spread in food-producing, wildlife and companion animals is a major public health priority to protect both persons with direct exposure and consumers. Authorities in European countries have already reacted by establishing active carbapenem resistance surveillance programs targeting food-producing animals mainly poultry, pork, cattle, and retail meat products [127]. There is concern that without these programs the presence of carbapenem genes in bacteria has the potential to enter the food supply undetected and subsequently transmission to humans [19]. For example, in 2014 a study by Morrison and Rubin [85], reported a 3.94% prevalence of CPOs in seafood imported in Canada that could pose a potential risk for transfer to clinically relevant bacteria and eventually to humans which is certainly of concerns. Strong evidences exist that the transmission of resistance might have taken place from animals to humans. Although many of these evidences were not direct and were based on the similarities between the carbapenem resistance genes of bacteria isolated from food-producing animals including poultry, pigs, cattle and from humans having close contact with these animals such as farm workers, animal caretakers and their family members [128]. To support risk assessment for zoonotic CRE, a comprehensive systematic review on CRE that carried bla_{VIM}, bla_{KPC}, bla_{NDM}, blaIMP and bla_{OXA}, from wildlife, food-producing, and companion animals was conducted recently by Köck, el al [72]. This study was primarily focused on the dissemination of CRE in livestock, food items including seafood, companion animals and their potential of transmission to exposed humans. The most prevalent carbapenem genes were bla_{VIM}, bla_{KPC},

bla_{NDM}, bla_{OXA}, and *blaIMP* in *E. coli* and *K. pneumoniae* isolates. Interestingly, only two independent studies reported that 33–67% of exposed humans on poultry farms carried CRE closely related to isolates from the poultry farm environment suggesting fecal-oral transmission or transmission via the food sources has the higher potential to spread CREs to healthier population. Wang, et al. tested six fecal specimens from farmers and workers from a Chinese commercial chicken farm, of which three (50%) were positive for plasmid-borne bla_{NDM-5} producing *E. coli* and demonstrated that two of the farmers isolates shared sequence types (ST10, ST746) with isolates from local dogs, flies, and chickens and clustered closely together, suggesting exposure and transmission to human on the farm environment. In their follow up study, Wang et al. [129] demonstrated clonal commonality between bla_{NDM}-positive *E. coli* isolates from chicken farms, slaughterhouses, supermarkets, and humans, belonging to sequence types ST10 and ST156. Recently, another study in Egypt by Elshafiee et al. [130] provided strong evidences of direct transmission of CRE to humans from farm animals. Another study by Li et al. [124], investigated the prevalence, risk factors, and drivers of CRE transmission between humans and their backyard animals in rural China that provided direct evidence of inter-host transmission of bla_{NDM} producing *E. coli* between humans and backyard animals. Moreover, the incidence of carbapenem resistant genes in food-producing animals has also been reported from several other countries with chickens and pig being the most investigated species, in which carbapenem resistant genes has been most frequently detected [128].

Detection of CRE: In the recent past, several molecular genotyping methods such as microarrays, single and multiplex polymerase chain reaction (PCR) assays have been used to detect the common carbapenemases, including bla_{KPC}, bla_{NDM}, bla_{IMP}, bla_{VIM} and bla_{OXA-48}, in bacterial isolates or directly from clinical specimens [131]. However, these molecular methods can accurately detect specific carbapenemase genes but cannot detect novel carbapenemase genes. On the other hand, whole genome sequencing (WGS) is a powerful new method with vastly improved resolution over current gold-standard techniques that not only identifying CRE but also providing detailed insights into their evolution and dissemination [132]. WGS potentially represent the ultimate molecular detection by probing the complete genomic content, chromosomal and extrachromosomal, of bacteria for the detection of carbapenem resistant genes. Moreover, WGS provide an opportunity to extrapolate additional information, including strain relatedness, molecular epidemiology, plasmids of replicon types harboring the carbapenemase, prediction of factors influencing carbapenem resistance for example point mutation and presence of other resistance factors and data can be analyzed in real-time or stored for future analysis. There is no doubt that WGS enhances our ability to characterize and resolve outbreaks of carbapenem resistant bacterial populations, understand and predict epidemiological trends, and create new machine learning tools for rapid detection of novel variants. As new algorithms are developing for tracking of transmission of carbapenem resistant bacteria in various fields including health care facilities and agricultural products (poultry, beef, and pork), likely to become increasing efficient and interpreting with the level of precision necessary to guide the modification of infection control procedures and food safety measures to limit the spread of CPO, allowing WGS to gain broader acceptance [133].

Role of insertion sequences, transposons and integrons: Dissemination of CPE among humans and animals has been enhanced by the horizontal transfer of CR genes on MGEs such as insertion sequences, transposons and integrons and their associated plasmids. [16,50,134–137]. The genetic analysis of bla_{IMP} and bla_{IVIM} from endemic areas in Southern Europe and Southeast Asia showed an association between both genes with class 1 integrons [138]. Molecular typing method such as PCR based replicon typing and plasmid multi-locus sequence typing (pMLST) methods have been used to identify a range of plasmid incompatibility groups responsible for the epidemiology of CR genes [139–141]. Most of the CR genes associated with the Ambler class A, class B, and Class D β-lactamases are located on plasmids, except Ambler class A CR genes ($bla_{SME-1\ to\ 3}$, bla_{IMI} and bla_{NMC-A}), which are usually located on the chromosome (Table 1). Some bla_{KPC}, bla_{VIM}, bla_{IMP}, and bla_{OXA} type carbapenemases genes have been located within transposons on plasmids which are responsible for the spread of such CR genes among *Enterobacteriaceae*. In addition, clonal expansion also acts as a way to spread the CR genes as described for bla_{KPC} in *K. pneumoniae* ST258 [142] and bla_{NDM-5} in *E. coli* on the *IncX3* type plasmid [143]. The insertion sequences and transposons play a role in the genetic variability. Accordingly, IS*Aba*1/3 and IS*Aba*125 have been reported to induce expression of several carbapenem resistance genes such as $bla_{OXA-23-51-66}$ -like and bla_{NDM-1} in *A. baumannii* and *K. pneumoniae* [144–148]. Similarly, a novel Tn*4401* transposon variant (Tn*4401*h, a188-bp deletion) on conjugative plasmids has been associated with enhanced bla_{KPC} expression and the degree of phenotypic resistance to carbapenemases in *K. pneumoniae* [149].

Plasmid: Plasmids belonging to different incompatibility groups play an important role in the epidemiology of CR genes. Plasmid-mediated spread of bla_{NDM-1} gene in *Enterobacteriaceae* has attracted attention, primarily, due to its predominant presence in nosocomial isolates (*K. pneumoniae*), as well as in community-acquired isolates (*E. coli*). Another characteristic is the propensity of conjugative plasmids to transfer resistant genes among other *Enterobacteriaceae* [150,151]. The bla_{NDM-1} gene was first detected in New Delhi, and in the Indian subcontinents [52]. Since then it has been isolated in UK, Pakistan, Bangladesh, Central and South America, the US and Canada [151]. In this case, urbanization and international travelling were described to contribute in the propagation of the bacterial clones and plasmids with bla_{NDM-1} gene from Indian subcontinent to other countries and continents [151]. Genetic studies demonstrated that the bla_{NDM-1} gene was associated with different bacterial clones and has spread among *Enterobacteriaceae* and *non-Enterobacteriaceae* species. This bla_{NDM-1} gene is carried by a broad range of multidrug plasmids belonging to the *IncA/C*, *IncL/M*, *IncN* and *IncF* incompatibility groups [45]. These plasmids with bla_{NDM-1} gene, also harbor other genes conferring resistance to aminoglycosides, quinolones, trimethoprim, sulphonamides, tetracyclines, colistin and heavy metal [3,44,152,153]. Moreover, bla_{NDM-1} has been identified in *E. coli* ST90, ST131 and *E. cloacae* ST231 whereas bla_{NDM-5} gene in ST101, ST167, and ST405 clones bearing diverse transmitting vectors were responsible for community acquired infections [154,155]. Thus, infections caused by NDM pathogens could be difficult to be treated [52,156,157].

Some CR genes including bla_{OXA-48} are associated with a single type of plasmid incompatibility group in *Enterobacteriaceae* [39,158]. The dissemination of bla_{OXA-48} gene among *K. pneumoniae*, *E. coli* and *E. cloacae* is due to the presence of this gene on mono-drug resistance plasmids belonging to the *IncL/M* incompatibility group sharing common characteristics such as self-transferability, no additional antibiotic resistance genes and size of plasmid range between 60–70 kbp [45]. In companion animals, the bla_{OXA-48} gene was found to be located on a multidrug-resistance plasmids belong to same *IncL/M* incompatibility group also carrying and extended spectrum β-lactamases and Amp C β-lactamases [45,66,159].

The bla_{KPC} genes have usually been located on self-transferable and multidrug-resistant large size plasmids, which are varying in size as well as phage-like plasmids [160,161]. The transferability of the bla_{KPC} plasmids has been reported between *C. freundii* and *K. oxytoca* belonging to two different bacterial genera [162,163]. These plasmids also carry additional antibiotic resistance genes for aminoglycosides, ESBLs and fluoroquinolones [164]. Different *K. pneumoniae* clones and different strains may carry KPC β-lactamase. In all reported cases, the bla_{KPC} gene has been associated with the Tn3 type transposon named Tn*4401*. This transposon may be a major source for transmission of bla_{KPC} gene [160].

9. Conclusions

The driving factor for the continuous increase in antibiotic resistance is antimicrobial usage in human and veterinary medicine as well as in agriculture. The European Surveillance of Antibiotic Consumption Network data in 2017 reported that in 10 out of 25 countries, consumption of carbapenems among humans increased since 2012, while only one country (Portugal) showed a decreasing trend during the same period (https://ecdc.europa.eu/en/publications-data/summary-latest-data-antibiotic-consumption-eu-2017). The carbapenems resistance in addition to ESC and cephamycins resistance in *Enterobacteriaceae* represent an important threat for public health. Out of the three classes, the class A carbapenemases KPC has spread globally and become more prevalent in the USA and Greece. The class B carbapenemase VIM and IMP has also been reported worldwide, but seem more endemic in Taiwan, Japan, and some European countries. The carbapenemase KPC, IMP, VIM, NDM, and OXA types have been mainly reported in nosocomial *K. pneumonia* strains. Pathogenic *E. coli* strains carrying bla_{NDM-1} and bla_{OXA-48} genes also have been found in community acquired infections. Therefore, proper identification and surveillance programs of carbapenem resistance pathogens and non-pathogenic strains have become necessary to support the control of CRE infections in both animals and humans. Source attribution studies along with developing alternative infection control strategies are warranted.

Author Contributions: M.S.D. (principle investigator) and P.B. conceptualized the study design; G.T. and M.A.R. wrote the review with major contributions from M.S.D. and P.B.; M.S.D. provided overall guidance, mentorship, and resources throughout the scope of this review. All authors have read and agreed to the published version of the manuscript.

Funding: This work was supported by Agriculture and Agri-Food Canada to M.S.D. through A-Base project #2411 and the Genomic Research and Development Initiative on AMR (GRDI-AMR) mitigation project.

Conflicts of Interest: The authors declared no conflict of interest.

References

1. WHO. *World Health Organization. Gloval Priority List of Antiviotic-Resistand Bacteria to Guide Research, Discovery, and Development of New Antibiotics*; WHO: Geneva, Switzerland, 2017.

2. Adeolu, M.; Alnajar, S.; Naushad, S.; Gupta, R.S. Genome-based phylogeny and taxonomy of the 'Enterobacteriales': Proposal for Enterobacterales ord. nov. divided into the families *Enterobacteriaceae*, *Erwiniaceae* fam. nov., *Pectobacteriaceae* fam. nov., *Yersiniaceae* fam. nov., *Hafniaceae* fam. nov., *Morganellaceae* fam. nov., and *Budviciaceae* fam. nov. *Int. J. Syst. Evol. Microbiol.* **2016**, *66*, 5575–5599. [CrossRef] [PubMed]

3. Nordmann, P.; Naas, T.; Poirel, L. Global spread of Carbapenemase-producing *Enterobacteriaceae*. *Emerg. Infect. Dis.* **2011**, *17*, 1791–1798. [CrossRef] [PubMed]

4. Kumar, M. Colistin and tigecycline resistance in Carbapenem-resistant *Enterobacteriaceae*: Checkmate to our last line of defense. *Infect. Control. Hosp. Epidemiol.* **2016**, *37*, 624–625. [CrossRef] [PubMed]

5. Sheu, C.-C.; Chang, Y.-T.; Lin, S.-Y.; Chen, Y.-H.; Hsueh, P.-R. Infections caused by Carbapenem-resistant *Enterobacteriaceae*: An update on therapeutic options. *Front. Microbiol.* **2019**, *10*, 80. [CrossRef]

6. Papp-Wallace, K.M.; Endimiani, A.; Taracila, M.A.; Bonomo, R.A. Carbapenems: Past, present, and future. *Antimicrob. Agents Chemother.* **2011**, *55*, 4943–4960. [CrossRef]

7. Kumar, P. Pharmacology of specific drug groups. In *Pharmacology and Therapeutics for Dentistry*; Dowd, F.J., Johnson, B.S., Mariotti, A.J., Eds.; Elsevier: Milton, ON, Canada, 2017; pp. 457–487.

8. Djahmi, N.; Dunyach-Rémy, C.; Pantel, A.; Dekhil, M.; Sotto, A.; Lavigne, J.-P. Epidemiology of Carbapenemase-producing *Enterobacteriaceae* and *Acinetobacter baumannii* in Mediterranean countries. *BioMed Res. Int.* **2014**, *2014*, 1–11. [CrossRef]

9. Anderson, R.E.V.; Boerlin, P. Carbapenemase-producing *Enterobacteriaceae* in animals and methodologies for their detection. *Can. J Vet Res* **2020**, *84*, 3–17.

10. Cheng, Z.; Thomas, P.W.; Ju, L.; Bergstrom, A.; Mason, K.; Clayton, D.; Miller, C.; Bethel, C.R.; Vanpelt, J.; Tierney, D.L.; et al. Evolution of New Delhi metallo-β-lactamase (NDM) in the clinic: Effects of NDM mutations on stability, zinc affinity, and mono-zinc activity. *J. Biol. Chem.* **2018**, *293*, 12606–12618. [CrossRef]

11. Naas, T.; Nordmann, P. Analysis of a Carbapenem-hydrolyzing class A beta-lactamase from *Enterobacter cloacae* and of its LysR-type regulatory protein. *Proc. Natl. Acad. Sci. USA* **1994**, *91*, 7693–7697. [CrossRef]

12. Queenan, A.M.; Bush, K. Carbapenemases: The versatile β-Lactamases. *Clin. Microbiol. Rev.* **2007**, *20*, 440–458. [CrossRef]

13. Bush, K.; Jacoby, G.A. Updated functional classification of β-Lactamases. *Antimicrob. Agents Chemother.* **2009**, *54*, 969–976. [CrossRef] [PubMed]

14. Boyd, S.E.; Livermore, D.M.; Hooper, D.C.; Hope, W.W. Metallo-β-lactamases: Structure, function, epidemiology, treatment options, and the development pipeline. *Antimicrob. Agents Chemother.* **2020**, *10*, e00397-20. [CrossRef] [PubMed]

15. Mathers, A.J.; Stoesser, N.; Chai, W.; Carroll, J.; Barry, K.; Cherunvanky, A.; Sebra, R.; Kasarskis, A.; Peto, T.E.; Walker, A.S.; et al. Chromosomal integration of the *Klebsiella pneumoniae* Carbapenemase gene, blaKPC, in Klebsiella species is elusive but not rare. *Antimicrob. Agents Chemother.* **2016**, *61*, e01823-16. [CrossRef] [PubMed]

16. Botelho, J.; Roberts, A.P.; León-Sampedro, R.; Grosso, F.; Peixe, L. Carbapenemases on the move: It's good to be on ICEs. *Mob. DNA* **2018**, *9*, 37. [CrossRef] [PubMed]

17. Souli, M.; Galani, I.; Giamarellou, H. Emergence of extensively drug-resistant and pandrug-resistant Gram-negative bacilli in Europe. *Euro Surveill. Eur. Commun. Dis. Bull.* **2008**, *13*, 19045.

18. Evans, B.A.; Amyes, S. OXA-Lactamases. *Clin. Microbiol. Rev.* **2014**, *27*, 241–263. [CrossRef]

19. Codjoe, F.S.; Donkor, E.S. Carbapenem resistance: A review. *Med. Sci.* **2017**, *6*, 1. [CrossRef]

20. Moxon, C.A.; Paulus, S. Beta-lactamases in *Enterobacteriaceae* infections in children. *J. Infect.* **2016**, *72*, S41–S49. [CrossRef]

21. Yigit, H.; Queenan, A.M.; Anderson, G.J.; Domenech-Sanchez, A.; Biddle, J.W.; Steward, C.D.; Alberti, S.; Bush, K.; Tenover, F.C. Novel Carbapenem-hydrolyzing β-Lactamase, KPC-1, from a Carbapenem-resistant strain of *Klebsiella pneumoniae*. *Antimicrob. Agents Chemother.* **2001**, *45*, 1151–1161. [CrossRef]

22. Drew, R.J.; Turton, J.; Hill, R.; Livermore, D.; Woodford, N.; Paulus, S.; Cunliffe, N.A. Emergence of Carbapenem-resistant *Enterobacteriaceae* in a UK paediatric hospital. *J. Hosp. Infect.* **2013**, *84*, 300–304. [CrossRef]

23. Livermore, D.M.; Woodford, N. The β-lactamase threat in *Enterobacteriaceae*, Pseudomonas and Acinetobacter. *Trends Microbiol.* **2006**, *14*, 413–420. [CrossRef] [PubMed]

24. European Centre for Disease Prevention and Control (ECDC). *Risk Assessment on the Spread of Carbapenemase-Producing Enterobacteriaceae (CPE) through Patient Transfer between Healthcare Facilities, with Special Emphasis on Cross-Border Transfer*; ECDC Technical Report; ECDC: Stockholm, Sweden, 2011.

25. Azap, O.; Otlu, B.; Yesilkaya, A.; Yakupoğulları, Y. Detection of OXA-48-like Carbapenemase-producing *Klebsiella pneumoniae* in a tertiary care center in Turkey: Molecular characterization and epidemiology. *Balk. Med. J.* **2013**, *30*, 259–260. [CrossRef] [PubMed]

26. Peleg, A.Y.; Hooper, D.C. Hospital-acquired infections due to gram-negative bacteria. *N. Engl. J. Med.* **2010**, *362*, 1804–1813. [CrossRef] [PubMed]

27. Choi, S.-H.; Ahn, M.Y.; Chung, J.-W.; Lee, M.-K. In vitro antibacterial activity of Doripenem against gram-negative blood isolates in a Korean tertiary care center. *Infect. Chemother.* **2015**, *47*, 175–180. [CrossRef]

28. Aschbacher, R.; Pagani, L.; Doumith, M.; Pike, R.; Woodford, N.; Spoladore, G.; Larcher, C.; Livermore, D.M. Metallo-β-lactamases among *Enterobacteriaceae* from routine samples in an Italian tertiary care hospital and long-term care facilities during 2008. *Clin. Microbiol. Infect.* **2010**, *17*, 181–189. [CrossRef]

29. Ahn, C.; Syed, A.; Hu, F.; O'Hara, J.A.; Rivera, J.I.; Doi, Y. Microbiological features of KPC-producing Enterobacter isolates identified in a U.S. hospital system. *Diagn. Microbiol. Infect. Dis.* **2014**, *80*, 154–158. [CrossRef]

30. Chopra, T.; Rivard, C.; Awali, R.A.; Krishna, A.; Bonomo, R.A.; Perez, F.; Kaye, K.S. Epidemiology of Carbapenem-resistant *Enterobacteriaceae* at a long-term acute care hospital. *Open Forum Infect. Dis.* **2018**, *5*, 224. [CrossRef]

31. Tamma, P.D.; Kazmi, A.; Bergman, Y.; Goodman, K.E.; Ekunseitan, E.; Amoah, J.; Simner, P.J. The likelihood of developing a Carbapenem-resistant infection during the hospital stay. *Antimicrob Agents Chemother* **2019**. [CrossRef]

32. Senchyna, F.; Gaur, R.L.; Sandlund, J.; Truong, C.; Tremintin, G.; Kültz, D.; Gomez, C.A.; Tamburini, F.B.; Andermann, T.M.; Bhatt, A.; et al. Diversity of resistance mechanisms in Carbapenem-resistant *Enterobacteriaceae* at a health care system in Northern California, from 2013 to 2016. *Diagn. Microbiol. Infect. Dis.* **2019**, *93*, 250–257. [CrossRef]

33. Kohler, P.P.; Melano, R.G.; Patel, S.N.; Shafinaz, S.; Faheem, A.; Coleman, B.L.; Green, K.; Armstrong, I.; Almohri, H.; Borgia, S.; et al. Emergence of Carbapenemase-producing *Enterobacteriaceae*, South-Central Ontario, Canada1. *Emerg. Infect. Dis.* **2018**, *24*, 1674–1682. [CrossRef]

34. Souli, M.; Galani, I.; Antoniadou, A.; Papadomichelakis, E.; Poulakou, G.; Panagea, T.; Vourli, S.; Zerva, L.; Armaganidis, A.; Kanellakopoulou, K.; et al. An outbreak of infection due to β-Lactamase *Klebsiella pneumoniae* Carbapenemase 2–producing *K. pneumoniae* in a Greek University Hospital: Molecular characterization, epidemiology, and outcomes. *Clin. Infect. Dis.* **2010**, *50*, 364–373. [CrossRef] [PubMed]

35. Giakkoupi, P.; Papagiannitsis, C.C.; Miriagou, V.; Pappa, O.; Polemis, M.; Tryfinopoulou, K.; Tzouvelekis, L.S.; Vatopoulos, A. An update of the evolving epidemic of blaKPC-2-carrying *Klebsiella pneumoniae* in Greece (2009–10). *J. Antimicrob. Chemother.* **2011**, *66*, 1510–1513. [CrossRef] [PubMed]

36. Grundmann, H.; Livermore, D.M.; Giske, C.G.; Rossolini, G.M.; Campos, J.; Vatopoulos, A.; Gniadkowski, M.; Toth, A.; Pfeifer, Y.; Jarlier, V.; et al. Carbapenem-non-susceptible *Enterobacteriaceae* in Europe: Conclusions from a meeting of national experts. *Eurosurveillance* **2010**, *15*, 15. [CrossRef] [PubMed]

37. Carbonne, A.; Thiolet, J.M.; Fournier, S.; Fortineau, N.; Kassis-Chikhani, N.; Boytchev, I.; Aggoune, M.; Séguier, J.C.; Sénéchal, H.; Tavolacci, M.P.; et al. Control of a multi-hospital outbreak of KPC-producing *Klebsiella pneumoniae* type 2 in France, September to October 2009. *Eurosurveillance* **2010**, *15*, 19734. [CrossRef] [PubMed]

38. Oteo, J.; Saez, D.; Bautista, V.; Fernández-Romero, S.; Hernández-Molina, J.M.; Pérez-Vázquez, M.; Aracil, B.; Campos, J. Carbapenemase-producing *Enterobacteriaceae* in Spain in 2012. *Antimicrob. Agents Chemother.* **2013**, *57*, 6344–6347. [CrossRef] [PubMed]

39. Nordmann, P.; Cuzon, G.; Naas, T. The real threat of *Klebsiella pneumoniae* Carbapenemase-producing bacteria. *Lancet Infect. Dis.* **2009**, *9*, 228–236. [CrossRef]

40. Ducomble, T.; Faucheux, S.; Helbig, U.; Kaisers, U.; Konig, B.; Knaust, A.; Lübbert, C.; Möller, I.; Rodloff, A.; Schweickert, B.; et al. Large hospital outbreak of KPC-2-producing *Klebsiella pneumoniae*: Investigating mortality and the impact of screening for KPC-2 with polymerase chain reaction. *J. Hosp. Infect.* **2015**, *89*, 179–185. [CrossRef]

41. Raro, O.H.F.; Da Silva, R.M.C.; Filho, E.M.R.; Sukiennik, T.C.T.; Stadnik, C.; Dias, C.A.G.; Iglesias, J.O.; Pérez-Vázquez, M. Carbapenemase-producing *Klebsiella pneumoniae* from transplanted patients in Brazil: Phylogeny, resistome, virulome and mobile genetic elements harboring blaKPC-2 or blaNDM-1. *Front. Microbiol.* **2020**, *11*, 1563. [CrossRef]

42. Manenzhe, R.I.; Zar, H.J.; Nicol, M.P.; Kaba, M. The spread of Carbapenemase-producing bacteria in Africa: A systematic review. *J. Antimicrob. Chemother.* **2014**, *70*, 23–40. [CrossRef]

43. Singh-Moodley, A.; Perovic, O. Antimicrobial susceptibility testing in predicting the presence of carbapenemase genes in *Enterobacteriaceae* in South Africa. *BMC Infect. Dis.* **2016**, *16*, 536. [CrossRef]

44. Poirel, L.; Pitout, J.D.; Nordmann, P. Carbapenemases: Molecular diversity and clinical consequences. *Futur. Microbiol.* **2007**, *2*, 501–512. [CrossRef] [PubMed]

45. Poirel, L.; Bonnin, R.A.; Nordmann, P. Genetic features of the widespread plasmid coding for the Carbapenemase OXA-48. *Antimicrob. Agents Chemother.* **2011**, *56*, 559–562. [CrossRef] [PubMed]

46. Lee, C.-R.; Lee, J.H.; Park, K.S.; Kim, Y.B.; Jeong, B.C.; Lee, H.S. Global dissemination of Carbapenemase-producing *Klebsiella pneumoniae*: Epidemiology, genetic context, treatment options, and detection methods. *Front. Microbiol.* **2016**, *7*, 895. [CrossRef] [PubMed]

47. French, C.; Coope, C.; Conway, L.; Higgins, J.P.; McCulloch, J.; Okoli, G.; Patel, B.; Oliver, I. Control of Carbapenemase-producing *Enterobacteriaceae* outbreaks in acute settings: An evidence review. *J. Hosp. Infect.* **2017**, *95*, 3–45. [CrossRef] [PubMed]

48. Senchyna, F.; Gaur, R.; Sandlund, J.; Truong, C.; Tremintin, G.; Küeltz, D.; Gomez, C.A.; Tamburini, F.B.; Andermann, T.M.; Bhatt, A.S.; et al. Diverse mechanisms of resistance in Carbapenem-resistant *Enterobacteriaceae* at a health care system in Silicon Valley, California. *bioRxiv* **2018**. [CrossRef]

49. Guo, X.; Rao, Y.; Guo, L.; Xu, H.; Lv, T.; Yu, X.; Chen, Y.; Liu, N.; Han, H.; Zheng, B. Detection and genomic characterization of a *Morganella morganii* isolate from China that produces NDM-5. *Front. Microbiol.* **2019**, *10*, 1156. [CrossRef] [PubMed]

50. Cuzon, G.; Demachy, M.C.; Nordmann, P.; Naas, T. Plasmid-mediated Carbapenem-hydrolyzing β-Lactamase KPC-2 in *Klebsiella pneumoniae* isolate from Greece. *Antimicrob. Agents Chemother.* **2007**, *52*, 796–797. [CrossRef] [PubMed]

51. Naas, T.; Vedel, G.; Nordmann, P.; Poyart, C. Plasmid-mediated Carbapenem-hydrolyzing β-Lactamase KPC in a *Klebsiella pneumoniae* isolate from France. *Antimicrob. Agents Chemother.* **2005**, *49*, 4423–4424. [CrossRef]

52. Kumarasamy, K.K.; Toleman, M.A.; Walsh, T.R.; Bagaria, J.; Butt, F.; Balakrishnan, R.; Chaudhary, U.; Doumith, M.; Giske, C.G.; Irfan, S.; et al. Emergence of a new antibiotic resistance mechanism in India, Pakistan, and the UK: A molecular, biological, and epidemiological study. *Lancet Infect. Dis.* **2010**, *10*, 597–602. [CrossRef]

53. Nahid, F.; Khan, A.A.; Rehman, S.; Zahra, R. Prevalence of metallo-β-lactamase NDM-1-producing multi-drug resistant bacteria at two Pakistani hospitals and implications for public health. *J. Infect. Public Heal.* **2013**, *6*, 487–493. [CrossRef]

54. Van Der Bij, A.K.; Pitout, J.D.D. The role of international travel in the worldwide spread of multiresistant *Enterobacteriaceae*. *J. Antimicrob. Chemother.* **2012**, *67*, 2090–2100. [CrossRef] [PubMed]

55. Tijet, N.; Alexander, D.C.; Richardson, D.; Lastovetska, O.; Low, N.E.; Patel, S.N.; Melano, R.G. New Delhi Metallo-β-Lactamase, Ontario, Canada. *Emerg. Infect. Dis.* **2011**, *17*, 306–307. [CrossRef] [PubMed]

56. Bernick, J.; Beliavsky, A.; Bogoch, I.I. Endometritis and bacteremia with a New Delhi Metallo-Beta-Lactamase 1 (NDM-1)-containing organism in a remote traveler. *J. Obstet. Gynaecol. Can.* **2019**, *41*, 753–754. [CrossRef] [PubMed]

57. Mataseje, L.F.; Boyd, D.A.; Mulvey, M.R.; Longtin, Y. Report on two hypervirulent *Klebsiella pneumoniae* producing a blaKPC-2 Carbapenemase from a Canadian patient. *Antimicrob Agents Chemother* **2019**. [CrossRef]

58. Poirel, L.; Vuillemin, X.; Juhas, M.; Masseron, A.; Bechtel-Grosch, U.; Tiziani, S.; Mancini, S.; Nordmann, P. KPC-50 confers resistance to Ceftazidime-Avibactam associated with reduced Carbapenemase activity. *Antimicrob. Agents Chemother.* **2020**, *64*, e00321-20. [CrossRef]

59. López, J.; Correa, A.; Navon-Venezia, S.; Torres, J.; Briceno, D.; Montealegre, M.C.; Quinn, J.; Carmeli, Y.; Villegasz, M. Intercontinental spread from Israel to Colombia of a KPC-3-producing klebsiella pneumoniae strain. *Clin. Microbiol. Infect.* **2011**, *17*, 52–56. [CrossRef]

60. Adler, A.; Shklyar, M.; Schwaber, M.J.; Navon-Venezia, S.; Dhaher, Y.; Edgar, R.; Solter, E.; Benenson, S.; Masarwa, S.; Carmeli, Y. Introduction of OXA-48-producing *Enterobacteriaceae* to Israeli hospitals by medical tourism. *J. Antimicrob. Chemother.* **2011**, *66*, 2763–2766. [CrossRef]

61. ECDC. *European Centre for Disease Prevention and Control. Outbreak of VIM-Producing Carbapenem Resistant Pseudomonas aeruginosa Linked to medical Tourism to Mexico*; ECDC: Stockholm, Sweden, 2019.

62. Cheng, V.C.; Wong, S.-C.; Wong, S.C.; Ho, P.-L.; Yuen, K.-Y. Control of Carbapenemase-producing *Enterobacteriaceae*: Beyond the hospital. *EClinicalMedicine* **2019**, *6*, 3–4. [CrossRef]

63. Gentilini, F.; Turba, M.E.; Pasquali, F.; Mion, D.; Romagnoli, N.; Zambon, E.; Terni, D.; Peirano, G.; Pitout, J.D.D.; Parisi, A.; et al. Hospitalized pets as a source of Carbapenem-resistance. *Front. Microbiol.* **2018**, *9*, 2872. [CrossRef]

64. Hong, J.S.; Song, W.; Park, H.-M.; Oh, J.-Y.; Chae, J.-C.; Han, J.-I.; Jeong, S.H. First detection of New Delhi Metallo-β-Lactamase-5-producing *Escherichia coli* from companion animals in Korea. *Microb. Drug Resist.* **2019**, *25*, 344–349. [CrossRef]

65. Adams, R.J.; Kim, S.S.; Mollenkopf, D.F.; Mathys, D.A.; Schuenemann, G.M.; Daniels, J.B.; Wittum, T. Antimicrobial-resistant *Enterobacteriaceae* recovered from companion animal and livestock environments. *Zoonoses Public Health* **2018**, *65*, 519–527. [CrossRef] [PubMed]

66. Stolle, I.; Prenger-Berninghoff, E.; Stamm, I.; Scheufen, S.; Hassdenteufel, E.; Guenther, S.; Bethe, A.; Pfeifer, Y.; Ewers, C. Emergence of OXA-48 Carbapenemase-producing *Escherichia coli* and *Klebsiella pneumoniae* in dogs. *J. Antimicrob. Chemother.* **2013**, *68*, 2802–2808. [CrossRef] [PubMed]

67. Grönthal, T.; Österblad, M.; Eklund, M.; Jalava, J.; Nykäsenoja, S.; Pekkanen, K.; Rantala, M. Sharing more than friendship-transmission of NDM-5 ST167 and CTX-M-9 ST69 *Escherichia coli* between dogs and humans in a family, Finland, 2015. *Eurosurveillance* **2018**, *23*, 1700497. [CrossRef] [PubMed]

68. Guo, S.; Brouwers, H.J.M.; Cobbold, R.N.; Platell, J.L.; Chapman, T.A.; Barrs, V.R.; Johnson, J.R.; Trott, D.J. Fluoroquinolone-resistant extraintestinal pathogenic *Escherichia coli*, including O25b-ST131, isolated from faeces of hospitalized dogs in an Australian veterinary referral centre. *J. Antimicrob. Chemother.* **2013**, *68*, 1025–1031. [CrossRef]

69. Johnson, J.R.; Clabots, C.; Kuskowski, M.A. Multiple-host sharing, long-term persistence, and virulence of *Escherichia coli* clones from human and animal household members. *J. Clin. Microbiol.* **2008**, *46*, 4078–4082. [CrossRef]

70. Nicolas-Chanoine, M.-H.; Bertrand, X.; Madec, J.-Y. *Escherichia coli* ST131, an intriguing clonal group. *Clin. Microbiol. Rev.* **2014**, *27*, 543–574. [CrossRef]

71. Abraham, S.; O'Dea, M.; Trott, D.J.; Abraham, R.J.; Hughes, D.; Pang, S.; McKew, G.; Cheong, E.Y.L.; Merlino, J.; Saputra, S.; et al. Isolation and plasmid characterization of Carbapenemase (IMP-4) producing *Salmonella enterica Typhimurium* from cats. *Sci. Rep.* **2016**, *6*, 35527. [CrossRef]

72. Köck, R.; Daniels-Haardt, I.; Becker, K.; Mellmann, A.; Friedrich, A.W.; Mevius, D.; Schwarz, S.; Jurke, A. Carbapenem-resistant *Enterobacteriaceae* in wildlife, food-producing, and companion animals: A systematic review. *Clin. Microbiol. Infect.* **2018**, *24*, 1241–1250. [CrossRef]

73. Wang, Y.; Wang, X.; Schwarz, S.; Zhang, R.; Lei, L.; Liu, X.; Lin, D.; Shen, J. IMP-45-producing multidrug-resistant *Pseudomonas aeruginosa* of canine origin. *J. Antimicrob. Chemother.* **2014**, *69*, 2579–2581. [CrossRef]

74. Rubin, J.; Pitout, J.D. Extended-spectrum β-lactamase, Carbapenemase and AmpC producing *Enterobacteriaceae* in companion animals. *Veter Microbiol.* **2014**, *170*, 10–18. [CrossRef]

75. Reynolds, M.E.; Phan, H.T.T.; George, S.; Hubbard, A.; Stoesser, N.; Maciuca, E.I.; Crook, D.W.; Timofte, D. Occurrence and characterization of *Escherichia coli* ST410 co-harbouring blaNDM-5, blaCMY-42 and blaTEM-190 in a dog from the UK. *J. Antimicrob. Chemother.* **2019**, *74*, 1207–1211. [CrossRef] [PubMed]

76. Schmidt, J.S.; Kuster, S.P.; Nigg, A.; Dazio, V.; Brilhante, M.; Rohrbach, H.; Bernasconi, O.J.; Büdel, T.; Campos-Madueno, E.I.; Brawand, S.G.; et al. Poor infection prevention and control standards are associated with environmental contamination with Carbapenemase-producing Enterobacterales and other multidrug-resistant bacteria in Swiss companion animal clinics. *Antimicrob. Resist. Infect. Control.* **2020**, *9*, 1–13. [CrossRef] [PubMed]

77. Gibson, J.; Morton, J.; Cobbold, R.; Sidjabat, H.; Filippich, L.; Trott, D. Multidrug-resistant *E. coli* and Enterobacter extraintestinal infection in 37 dogs. *J. Veter. Intern. Med.* **2008**, *22*, 844–850. [CrossRef] [PubMed]

78. Fischer, J.; Rodríguez, I.; Schmoger, S.; Friese, A.; Roesler, U.; Helmuth, R.; Guerra, B. *Escherichia coli* producing VIM-1 Carbapenemase isolated on a pig farm. *J. Antimicrob. Chemother.* **2012**, *67*, 1793–1795. [CrossRef]

79. Poirel, L.; Berçot, B.; Millemann, Y.; Bonnin, R.A.; Pannaux, G.; Nordmann, P. Carbapenemase-producing Acinetobacter spp. in Cattle, France. *Emerg. Infect. Dis.* **2012**, *18*, 523–525. [CrossRef]

80. Fischer, J.; Rodríguez, I.; Schmoger, S.; Friese, A.; Roesler, U.; Helmuth, R.; Guerra, B. *Salmonella enterica* subsp. enterica producing VIM-1 Carbapenemase isolated from livestock farms. *J. Antimicrob. Chemother.* **2012**, *68*, 478–480. [CrossRef]

81. Fischer, J.; José, M.S.; Roschanski, N.; Schmoger, S.; Baumann, B.; Irrgang, A.; Friese, A.; Roesler, U.; Helmuth, R.; Guerra, B. Spread and persistence of VIM-1 Carbapenemase-producing *Enterobacteriaceae* in three German swine farms in 2011 and 2012. *Veter. Microbiol.* **2017**, *200*, 118–123. [CrossRef]

82. Wang, Y.; Wu, C.; Zhang, Q.; Qi, J.; Liu, H.; Wang, Y.; He, T.; Ma, L.; Lai, J.; Shen, Z.; et al. Identification of New Delhi Metallo-β-lactamase 1 in *Acinetobacter lwoffii* of food animal origin. *PLoS ONE* **2012**, *7*, e37152. [CrossRef]

83. He, T.; Wei, R.; Zhang, L.; Sun, L.; Pang, M.; Wang, R.; Wang, Y. Characterization of NDM-5-positive extensively resistant *Escherichia coli* isolates from dairy cows. *Veter. Microbiol.* **2017**, *207*, 153–158. [CrossRef]

84. Liu, Z.; Li, J.; Wang, X.; Liu, D.; Ke, Y.; Wang, Y.; Shen, J. Novel variant of New Delhi Metallo-β-lactamase, NDM-20, in *Escherichia coli. Front. Microbiol.* **2018**, *9*, 248. [CrossRef]

85. Morrison, B.J.; Rubin, J. Carbapenemase producing bacteria in the food supply escaping detection. *PLoS ONE* **2015**, *10*, e0126717. [CrossRef]

86. Abdallah, H.; Reuland, E.A.; Wintermans, B.B.; Al Naiemi, N.; Koek, A.; Abdelwahab, A.M.; Ammar, A.M.; Mohamed, A.A.; Vandenbroucke-Grauls, C.M.J.E. Extended-spectrum β-Lactamases and/or Carbapenemases-producing *Enterobacteriaceae* isolated from retail chicken meat in Zagazig, Egypt. *PLoS ONE* **2015**, *10*, e0136052. [CrossRef] [PubMed]

87. Ahmad, K.; Khattak, F.; Ali, A.; Rahat, S.; Noor, S.; Mahsood, N.; Somayya, R. Carbapenemases and extended-spectrum β-Lactamase-producing multidrug-resistant *Escherichia coli* isolated from retail chicken in Peshawar: First report from Pakistan. *J. Food Prot.* **2018**, *81*, 1339–1345. [CrossRef] [PubMed]

88. Yao, X.; Doi, Y.; Zeng, L.; Lv, L.; Liu, J.-H. Carbapenem-resistant and colistin-resistant *Escherichia coli* co-producing NDM-9 and MCR-1. *Lancet Infect. Dis.* **2016**, *16*, 288–289. [CrossRef]

89. VIM-1-producing *Salmonella infantis* isolated from swine and minced pork meat in Germany. *J. Antimicrob. Chemother.* **2017**, *72*, 2131–2133. [CrossRef]

90. Wang, W.; Baloch, Z.; Peng, Z.; Hu, Y.; Xu, J.; Fanning, S.; Li, F. Genomic characterization of a large plasmid containing a bla NDM-1 gene carried on *Salmonella enterica* serovar Indiana C629 isolate from China. *BMC Infect. Dis.* **2017**, *17*, 479. [CrossRef]

91. Zhang, Q.; Lv, L.; Huang, X.; Huang, Y.; Zhuang, Z.; Lu, J.; Liu, E.; Wan, M.; Xun, H.; Zhang, Z.; et al. Rapid increase in Carbapenemase-producing *Enterobacteriaceae* in retail meat driven by the spread of the blaNDM-5-carrying IncX3 plasmid in China from 2016 to 2018. *Antimicrob. Agents Chemother.* **2019**, *63*, e00573-19. [CrossRef]

92. Roschanski, N.; Guenther, S.; Vu, T.T.T.; Fischer, J.; Semmler, T.; Huehn, S.; Alter, T.; Roesler, U. VIM-1 Carbapenemase-producing *Escherichia coli* isolated from retail seafood, Germany 2016. *Eurosurveillance* **2017**, *22*, 17–32. [CrossRef]

93. Janecko, N.; Martz, S.-L.; Avery, B.P.; Daignault, D.; Desruisseau, A.; Boyd, D.; Irwin, R.J.; Mulvey, M.R.; Reid-Smith, R.J. Carbapenem-resistant Enterobacter spp. in retail seafood imported from Southeast Asia to Canada. *Emerg. Infect. Dis.* **2016**, *22*, 1675–1677. [CrossRef]

94. Brouwer, M.S.; Tehrani, K.H.M.E.; Rapallini, M.; Geurts, Y.; Kant, A.; Harders, F.; Mashayekhi, V.; Martin, N.I.; Bossers, A.; Mevius, D.J.; et al. Novel Carbapenemases FLC-1 and IMI-2 encoded by an Enterobacter cloacae complex isolated from food products. *Antimicrob. Agents Chemother.* **2019**, *63*, e02338-18. [CrossRef]

95. Diab, M.; Hamze, M.; Bonnet, R.; Saras, E.; Madec, J.-Y.; Haenni, M. OXA-48 and CTX-M-15 extended-spectrum beta-lactamases in raw milk in Lebanon: Epidemic spread of dominant *Klebsiella pneumoniae* clones. *J. Med. Microbiol.* **2017**, *66*, 1688–1691. [CrossRef] [PubMed]

96. Wang, J.; Yao, X.; Luo, J.; Lv, L.; Zeng, Z.; Liu, J.-H. Emergence of *Escherichia coli* co-producing NDM-1 and KPC-2 Carbapenemases from a retail vegetable, China. *J. Antimicrob. Chemother.* **2017**, *73*, 252–254. [CrossRef]

97. Touati, A.; Mairi, A.; Baloul, Y.; Lalaoui, R.; Bakour, S.; Thighilt, L.; Gharout, A.; Rolain, J.-M. First detection of *Klebsiella pneumoniae* producing OXA-48 in fresh vegetables from Béjaïa city, Algeria. *J. Glob. Antimicrob. Resist.* **2017**, *9*, 17–18. [CrossRef] [PubMed]

98. Zurfluh, K.; Poirel, L.; Nordmann, P.; Klumpp, J.; Stephan, R. First detection of *Klebsiella variicola* producing OXA-181 Carbapenemase in fresh vegetable imported from Asia to Switzerland. *Antimicrob. Resist. Infect. Control.* **2015**, *4*, 1–3. [CrossRef] [PubMed]

99. Hsu, L.Y.; Apisarnthanarak, A.; Khan, E.; Suwantarat, N.; Ghafur, A.; Tambyah, P.A. Carbapenem-resistant *Acinetobacter baumannii* and *Enterobacteriaceae* in South and Southeast Asia. *Clin. Microbiol. Rev.* **2016**, *30*, 1–22. [CrossRef]

100. Walsh, T.R.; Weeks, J.; Livermore, D.M.; Toleman, M.A. Dissemination of NDM-1 positive bacteria in the New Delhi environment and its implications for human health: An environmental point prevalence study. *Lancet Infect. Dis.* **2011**, *11*, 355–362. [CrossRef]

101. Lerner, A.; Adler, A.; Abu-Hanna, J.; Meitus, I.; Navon-Venezia, S.; Carmeli, Y. Environmental contamination by Carbapenem-resistant *Enterobacteriaceae*. *J. Clin. Microbiol.* **2012**, *51*, 177–181. [CrossRef]

102. Chen, W.K.; Yang, Y.; Tan, B.H. Increased mortality among Carbapenemase-producing Carbapenem-resistant *Enterobacteriaceae* carriers who developed clinical isolates of another genotype. *Open Forum Infect. Dis.* **2019**, *6*, 6. [CrossRef]

103. Weingarten, R.A.; Johnson, R.C.; Conlan, S.; Ramsburg, A.M.; Dekker, J.P.; Lau, A.F.; Khil, P.; Odom, R.T.; Deming, C.; Park, M.; et al. Genomic analysis of hospital plumbing reveals diverse reservoir of bacterial plasmids conferring Carbapenem resistance. *mBio* **2018**, *9*, 02011–02017. [CrossRef]

104. Khan, A.U.; Parvez, S. Detection of bla NDM-4 in *Escherichia coli* from hospital sewage. *J. Med. Microbiol.* **2014**, *63*, 1404–1406. [CrossRef]

105. Zurfluh, K.; Hächler, H.; Nüesch-Inderbinen, M.; Stephan, R. Characteristics of extended-spectrum β-Lactamase-and Carbapenemase-producing *Enterobacteriaceae* isolates from rivers and lakes in Switzerland. *Appl. Environ. Microbiol.* **2013**, *79*, 3021–3026. [CrossRef] [PubMed]

106. Khan, F.A.; Hellmark, B.; Ehricht, R.; Söderquist, B.; Jass, J. Related Carbapenemase-producing *Klebsiella* isolates detected in both a hospital and associated aquatic environment in Sweden. *Eur. J. Clin. Microbiol. Infect. Dis.* **2018**, *37*, 2241–2251. [CrossRef] [PubMed]

107. Piedra-Carrasco, N.; Fàbrega, A.; Calero-Cáceres, W.; Cornejo-Sánchez, T.; Brown-Jaque, M.; Mir-Cros, A.; Muniesa, M.; González-López, J.J. Carbapenemase-producing *Enterobacteriaceae* recovered from a Spanish river ecosystem. *PLoS ONE* **2017**, *12*, e0175246. [CrossRef] [PubMed]

108. Hoelle, J.; Johnson, J.R.; Johnston, B.D.; Kinkle, B.; Boczek, L.; Ryu, H.; Hayes, S. Survey of US wastewater for Carbapenem-resistant *Enterobacteriaceae*. *J. Water Health* **2019**, *17*, 219–226. [CrossRef] [PubMed]

109. Mollenkopf, D.F.; Stull, J.W.; Mathys, D.A.; Bowman, A.S.; Feicht, S.M.; Grooters, S.V.; Daniels, J.B.; Wittum, T. Carbapenemase-Producing Enterobacteriaceae Recovered from the Environment of a Swine Farrow-to-finish operation in the United States. *Antimicrob. Agents Chemother.* **2016**, *61*, 61. [CrossRef] [PubMed]

110. Vikram, A.; Schmidt, J.W. Functional blaKPC-2 sequences are present in U.S. beef cattle feces regardless of antibiotic use. *Foodborne Pathog. Dis.* **2018**, *15*, 444–448. [CrossRef] [PubMed]

111. Hrenovic, J.; Music, M.S.; Durn, G.; Dekic, S.; Hunjak, B.; Kisic, I. Carbapenem-resistant *Acinetobacter baumannii* recovered from swine manure. *Microb. Drug Resist.* **2019**, *25*, 725–730. [CrossRef] [PubMed]

112. Tang, B.; Chang, J.; Cao, L.; Luo, Q.-X.; Xu, H.; Lyu, W.; Qian, M.; Ji, X.; Zhang, Q.; Xia, X.; et al. Characterization of an NDM-5 Carbapenemase-producing *Escherichia coli* ST156 isolate from a poultry farm in Zhejiang, China. *BMC Microbiol.* **2019**, *19*, 82. [CrossRef] [PubMed]

113. Adegoke, A.A.; Fatunla, O.K.; Okoh, A.I. Critical threat associated with carbapenem-resistant gram-negative bacteria: Prioritizing water matrices in addressing total antibiotic resistance. *Ann. Microbiol.* **2020**, *70*, 1–13. [CrossRef]

114. Lepuschitz, S.; Schill, S.; Stoeger, A.; Pekard-Amenitsch, S.; Huhulescu, S.; Inreiter, N.; Hartl, R.; Kerschner, H.; Sorschag, S.; Springer, B.; et al. Whole genome sequencing reveals resemblance between ESBL-producing and carbapenem resistant *Klebsiella pneumoniae* isolates from Austrian rivers and clinical isolates from hospitals. *Sci. Total. Environ.* **2019**, *662*, 227–235. [CrossRef]

115. Tanner, W.D.; Vanderslice, J.A.; Goel, R.K.; Leecaster, M.K.; Fisher, M.A.; Olstadt, J.; Gurley, C.M.; Morris, A.G.; Seely, K.A.; Chapman, L.; et al. Multi-state study of *Enterobacteriaceae* harboring extended-spectrum beta-lactamase and Carbapenemase genes in U.S. drinking water. *Sci. Rep.* **2019**, *9*, 3938. [CrossRef]

116. Hamza, D.; Dorgham, S.; Ismael, E.; El-Moez, S.I.A.; ElHariri, M.; Elhelw, R.; Hamza, E. Emergence of β-lactamase-and Carbapenemase-producing *Enterobacteriaceae* at integrated fish farms. *Antimicrob. Resist. Infect. Control.* **2020**, *9*, 1–12. [CrossRef]

117. Bonomo, R.A.; Burd, E.M.; Conly, J.; Limbago, B.M.; Poirel, L.; Segre, J.A.; Westblade, L.F. Carbapenemase-producing organisms: A global scourge. *Clin. Infect. Dis.* **2017**, *66*, 1290–1297. [CrossRef]

118. Nordmann, P.; Dortet, L.; Poirel, L. Carbapenem resistance in *Enterobacteriaceae*: Here is the storm! *Trends Mol. Med.* **2012**, *18*, 263–272. [CrossRef] [PubMed]

119. Yong, D.; Toleman, M.A.; Giske, C.G.; Cho, H.S.; Sundman, K.; Lee, K.; Walsh, T.R. Characterization of a new Metallo-β-Lactamase gene, blaNDM-1, and a novel erythromycin esterase gene carried on a unique genetic structure in *Klebsiella pneumoniae* sequence type 14 from India. *Antimicrob. Agents Chemother.* **2009**, *53*, 5046–5054. [CrossRef] [PubMed]

120. Bush, K.; Bradford, P.A. Epidemiology of β-Lactamase-producing pathogens. *Clin. Microbiol. Rev.* **2020**, *33*, 33. [CrossRef] [PubMed]

121. Pournaras, S.; Tsakris, A.; Maniati, M.; Tzouvelekis, L.S.; Maniatis, A.N. Novel variant (blaVIM-4) of the Metallo-β-Lactamase gene blaVIM-1 in a clinical strain of *Pseudomonas aeruginosa*. *Antimicrob. Agents Chemother.* **2002**, *46*, 4026–4028. [CrossRef] [PubMed]

122. Matsumura, Y.; Peirano, G.; Motyl, M.R.; Adams, M.D.; Chen, L.; Kreiswirth, B.; DeVinney, R.; Pitout, J.D.D. Global molecular epidemiology of IMP-producing *Enterobacteriaceae*. *Antimicrob. Agents Chemother.* **2017**, *61*, e02729-16. [CrossRef] [PubMed]

123. Shen, Z.; Hu, Y.; Sun, Q.; Hu, F.; Zhou, H.; Shu, L.; Ma, T.; Shen, Y.; Wang, Y.; Li, J.; et al. Emerging carriage of NDM-5 and MCR-1 in *Escherichia coli* from healthy people in multiple regions in China: A cross sectional observational study. *EClinicalMedicine* **2018**, *6*, 11–20. [CrossRef]

124. Li, J.; Bi, Z.; Ma, S.; Chen, B.; Cai, C.; He, J.; Schwarz, S.; Sun, C.; Zhou, Y.; Yin, J.; et al. Inter-host transmission of Carbapenemase-producing *Escherichia coli* among humans and backyard animals. *Environ. Health Perspect.* **2019**, *127*, 107009. [CrossRef]

125. Zhai, R.; Fu, B.; Shi, X.; Sun, C.; Liu, Z.; Wang, S.; Shen, Z.; Walsh, T.R.; Cai, C.; Wang, Y.; et al. Contaminated in-house environment contributes to the persistence and transmission of NDM-producing bacteria in a Chinese poultry farm. *Environ. Int.* **2020**, *139*, 105715. [CrossRef] [PubMed]

126. ECDC. *European Centre for Disease Prevention and Control. Carbapenem-Resistant Enterobacteriaceae, second Update*; ECDC: Stockholm, Sweden, 2019.

127. Schrijver, R.; Stijntjes, M.; Rodríguez-Baño, J.; Tacconelli, E.; Rajendran, N.B.; Voss, A. Review of antimicrobial resistance surveillance programmes in livestock and meat in EU with focus on humans. *Clin. Microbiol. Infect.* **2018**, *24*, 577–590. [CrossRef] [PubMed]

128. Bonardi, S.; Pitino, R. Carbapenemase-producing bacteria in food-producing animals, wildlife and environment: A challenge for human health. *Ital. J. Food Saf.* **2019**, *8*, 7956. [CrossRef] [PubMed]

129. Wang, Y.; Zhang, R.; Li, J.; Wu, Z.; Yin, W.; Schwarz, S.; Tyrrell, J.M.; Zheng, Y.; Wang, S.; Shen, Z.; et al. Comprehensive resistome analysis reveals the prevalence of NDM and MCR-1 in Chinese poultry production. *Nat. Microbiol.* **2017**, *2*, 16260. [CrossRef]

130. Elshafiee, E.A.; Nader, S.M.; Dorgham, S.M.; Hamza, D. Carbapenem-resistant *Pseudomonas aeruginosa* originating from farm animals and people in Egypt. *J. Veter.-Res.* **2019**, *63*, 333–337. [CrossRef]

131. Cui, X.; Zhang, H.; Du, H. Carbapenemases in *Enterobacteriaceae*: Detection and antimicrobial therapy. *Front. Microbiol.* **2019**, *10*, 1823. [CrossRef]

132. Balloux, F.; Brynildsrud, O.B.; Van Dorp, L.; Shaw, L.P.; Chen, H.; Harris, K.A.; Wang, H.; Eldholm, V. From theory to practice: Translating whole-genome sequencing (WGS) into the clinic. *Trends Microbiol.* **2018**, *26*, 1035–1048. [CrossRef]

133. Lynch, T.; Petkau, A.; Knox, N.; Graham, M.; Van Domselaar, G. A Primer on infectious disease bacterial genomics. *Clin. Microbiol. Rev.* **2016**, *29*, 881–913. [CrossRef]

134. Partridge, S.R. Analysis of antibiotic resistance regions in Gram-negative bacteria. *FEMS Microbiol. Rev.* **2011**, *35*, 820–855. [CrossRef]

135. Stokes, H.W.; Gillings, M.R. Gene flow, mobile genetic elements and the recruitment of antibiotic resistance genes into Gram-negative pathogens. *FEMS Microbiol. Rev.* **2011**, *35*, 790–819. [CrossRef]

136. Toleman, M.A.; Walsh, T.R. Combinatorial events of insertion sequences and ICE in Gram-negative bacteria. *FEMS Microbiol. Rev.* **2011**, *35*, 912–935. [CrossRef] [PubMed]

137. Hardiman, C.A.; Weingarten, R.A.; Conlan, S.; Khil, P.; Dekker, J.P.; Mathers, A.J.; Sheppard, A.E.; Segre, J.A.; Frank, K.M. Horizontal transfer of Carbapenemase-encoding plasmids and comparison with hospital epidemiology data. *Antimicrob. Agents Chemother.* **2016**, *60*, 4910–4919. [CrossRef]

138. Liapis, E.; Bour, M.; Triponney, P.; Jové, T.; Zahar, J.-R.; Valot, B.; Jeannot, K.; Plésiat, P. Identification of diverse integron and plasmid structures carrying a novel Carbapenemase among *Pseudomonas* species. *Front. Microbiol.* **2019**, *10*, 404. [CrossRef] [PubMed]

139. Sugawara, Y.; Akeda, Y.; Sakamoto, N.; Takeuchi, D.; Motooka, D.; Nakamura, S.; Hagiya, H.; Yamamoto, N.; Nishi, I.; Yoshida, H.; et al. Genetic characterization of blaNDM-harboring plasmids in carbapenem-resistant *Escherichia coli* from Myanmar. *PLoS ONE* **2017**, *12*, e0184720. [CrossRef]

140. Carattoli, A.; Miriagou, V.; Bertini, A.; Loli, A.; Colinon, C.; Villa, L.; Whichard, J.M.; Rossolini, G.M. Replicon typing of plasmids encoding resistance to newer β-Lactams. *Emerg. Infect. Dis.* **2006**, *12*, 1145–1148. [CrossRef] [PubMed]

141. Carattoli, A. Resistance plasmid families in *Enterobacteriaceae*. *Antimicrob. Agents Chemother.* **2009**, *53*, 2227–2238. [CrossRef]

142. Cuzon, G.; Naas, T.; Truong, H.; Villegas, M.-V.; Wisell, K.T.; Carmeli, Y.; Gales, A.C.; Navon-Venezia, S.; Quinn, J.P.; Nordmann, P. Worldwide diversity of *Klebsiella pneumoniae* that produce β-Lactamase blaKPC-2gene1. *Emerg. Infect. Dis.* **2010**, *16*, 1349–1356. [CrossRef]

143. Zhang, X.; Wang, W.; Yu, H.; Wang, M.; Zhang, H.; Lv, J.; Tang, Y.-W.; Kreiswirth, B.N.; Du, H.; Chen, L.; et al. New Delhi Metallo-β-Lactamase 5-producing *Klebsiella pneumoniae* sequence type 258, Southwest China, 2017. *Emerg. Infect. Dis.* **2019**, *25*, 1209–1213. [CrossRef]

144. Lopes, B.S.; Amyes, S.G.B. Role of ISAba1 and ISAba125 in governing the expression of bla ADC in clinically relevant *Acinetobacter baumannii* strains resistant to cephalosporins. *J. Med. Microbiol.* **2012**, *61*, 1103–1108. [CrossRef]

145. Figueiredo, S.; Poirel, L.; Croize, J.; Recule, C.; Nordmann, P. In vivo selection of reduced susceptibility to Carbapenems in *Acinetobacter baumannii* related to ISAba1-mediated overexpression of the natural blaOXA-66 Oxacillinase gene. *Antimicrob. Agents Chemother.* **2009**, *53*, 2657–2659. [CrossRef]

146. Bontron, S.; Nordmann, P.; Poirel, L. Transposition of Tn125encoding the NDM-1 Carbapenemase in *Acinetobacter baumannii*. *Antimicrob. Agents Chemother.* **2016**, *60*, AAC.01755-16. [CrossRef]

147. Mukherjee, S.; Bhattacharjee, A.; Naha, S.; Majumdar, T.; Debbarma, S.K.; Kaur, H.; Dutta, S.; Basu, S. Molecular characterization of NDM-1-producing *Klebsiella pneumoniae* ST29, ST347, ST1224, and ST2558 causing sepsis in neonates in a tertiary care hospital of North-East India. *Infect. Genet. Evol.* **2019**, *69*, 166–175. [CrossRef] [PubMed]

148. Antunes, N.T.; Lamoureaux, T.L.; Toth, M.; Stewart, N.K.; Frase, H.; Vakulenko, S.B. Class D β-Lactamases: Are they all Carbapenemases? *Antimicrob. Agents Chemother.* **2014**, *58*, 2119–2125. [CrossRef] [PubMed]

149. Cheruvanky, A.; Stoesser, N.; Sheppard, A.E.; Crook, D.W.; Hoffman, P.S.; Weddle, E.; Carroll, J.; Sifri, C.D.; Chai, W.; Barry, K.; et al. Enhanced *Klebsiella pneumoniae* Carbapenemase expression from a novel Tn4401 deletion. *Antimicrob. Agents Chemother.* **2017**, *61*, e00025-17. [CrossRef]

150. Moellering, R.C. NDM-1-A cause for worldwide concern. *New Engl. J. Med.* **2010**, *363*, 2377–2379. [CrossRef]

151. Nordmann, P.; Poirel, L.; Toleman, M.A.; Walsh, T.R. Does broad-spectrum beta-lactam resistance due to NDM-1 herald the end of the antibiotic era for treatment of infections caused by Gram-negative bacteria? *J. Antimicrob. Chemother.* **2011**, *66*, 689–692. [CrossRef]

152. Li, X.; Mu, X.; Zhang, P.; Zhao, D.-D.; Ji, J.; Quan, J.; Zhu, Y.; Yu, Y. Detection and characterization of a clinical *Escherichia coli* ST3204 strain coproducing NDM-16 and MCR-1. *Infect. Drug Resist.* **2018**, *11*, 1189–1195. [CrossRef]

153. Yang, Q.E.; Agouri, S.R.; Tyrrell, J.M.; Walsh, T.R. Heavy metal resistance genes are associated with blaNDM-1- and blaCTX-M-15-carrying *Enterobacteriaceae*. *Antimicrob. Agents Chemother.* **2018**, *62*, e02642-17. [CrossRef]

154. Liu, C.M.; Stegger, M.; Aziz, M.; Johnson, T.J.; Waits, K.; Nordstrom, L.; Gauld, L.; Weaver, B.; Rolland, D.; Statham, S.; et al. *Escherichia coli* ST131-H22 as a foodborne uropathogen. *mBio* **2018**, *9*, e00470-18. [CrossRef]

155. Pecora, N.D.; Zhao, X.; Nudel, K.; Hoffmann, M.; Li, N.; Onderdonk, A.B.; Yokoe, D.; Brown, E.; Allard, M.; Bry, L. Diverse vectors and mechanisms spread New Delhi Metallo-β-Lactamases among Carbapenem-resistant *Enterobacteriaceae* in the Greater Boston area. *Antimicrob. Agents Chemother.* **2018**, *63*, e02040-18. [CrossRef]

156. Muir, A.; Weinbren, M. New Delhi metallo-β-lactamase: A cautionary tale. *J. Hosp. Infect.* **2010**, *75*, 239–240. [CrossRef] [PubMed]

157. Wu, W.; Feng, Y.; Tang, G.; Qiao, F.; McNally, A.; Zong, Z. NDM Metallo-β-Lactamases and their bacterial producers in health care settings. *Clin. Microbiol. Rev.* **2019**, *32*, e00115-18. [CrossRef] [PubMed]

158. Carrër, A.; Poirel, L.; Yilmaz, M.; Akan, O.A.; Feriha, C.; Cuzon, G.; Matar, G.; Honderlick, P.; Nordmann, P. Spread of OXA-48-encoding plasmid in Turkey and beyond. *Antimicrob. Agents Chemother.* **2010**, *54*, 1369–1373. [CrossRef]

159. Xie, L.; Dou, Y.; Zhou, K.; Chen, Y.; Han, L.; Guo, X.; Sun, J. Coexistence of blaOXA-48 and truncated blaNDM-1 on different plasmids in a *Klebsiella pneumoniae* isolate in China. *Front. Microbiol.* **2017**, *8*, 133. [CrossRef] [PubMed]

160. Naas, T.; Cuzon, G.; Villegas, M.-V.; Lartigue, M.-F.; Quinn, J.P.; Nordmann, P. Genetic structures at the origin of acquisition of the β-Lactamase blaKPC gene. *Antimicrob. Agents Chemother.* **2008**, *52*, 1257–1263. [CrossRef] [PubMed]

161. Galetti, R.; Andrade, L.N.; Varani, A.M.; Darini, A.L.C. A phage-like plasmid carrying blaKPC-2 gene in Carbapenem-resistant *Pseudomonas aeruginosa*. *Front. Microbiol.* **2019**, *10*, 572. [CrossRef]

162. Rasheed, J.K.; Biddle, J.W.; Anderson, K.F.; Washer, L.; Chenoweth, C.; Perrin, J.; Newton, D.W.; Patel, J.B. Detection of the *Klebsiella pneumoniae* Carbapenemase type 2 Carbapenem-hydrolyzing enzyme in clinical isolates of *Citrobacter freundii* and *K. oxytoca* carrying a common plasmid. *J. Clin. Microbiol.* **2008**, *46*, 2066–2069. [CrossRef]

163. Schweizer, C.; Bischoff, P.; Bender, J.; Kola, A.; Gastmeier, P.; Hummel, M.; Klefisch, F.-R.; Schoenrath, F.; Frühauf, A.; Pfeifer, Y. Plasmid-mediated transmission of KPC-2 Carbapenemase in *Enterobacteriaceae* in critically ill patients. *Front. Microbiol.* **2019**, *10*, 276. [CrossRef]

164. Cai, J.C.; Zhou, H.W.; Zhang, R.; Chen, G.-X. Emergence of *Serratia marcescens*, *Klebsiella pneumoniae*, and *Escherichia coli* isolates possessing the plasmid-mediated Carbapenem-hydrolyzing β-Lactamase KPC-2 in intensive care units of a Chinese hospital. *Antimicrob. Agents Chemother.* **2008**, *52*, 2014–2018. [CrossRef]

Article

Prevalence and Molecular Typing of Carbapenemase-Producing Enterobacterales among Newborn Patients in Italy

Marilena Agosta [1,†], Daniela Bencardino [2,†], Marta Argentieri [1], Laura Pansani [1], Annamaria Sisto [1], Marta Luisa Ciofi Degli Atti [3], Carmen D'Amore [3], Lorenza Putignani [4], Pietro Bagolan [5], Barbara Daniela Iacobelli [5], Andrea Dotta [6], Ludovica Martini [6], Luca Di Chiara [7], Mauro Magnani [2], Carlo Federico Perno [1], Francesca Andreoni [2,‡] and Paola Bernaschi [1,*,‡]

1   Microbiology and Diagnostic Immunology Unit, Department of Diagnostic and Laboratory Medicine, Bambino Gesù Children's Hospital, IRCCS, 00165 Rome, Italy; marilena.agosta@opbg.net (M.A.); marta.argentieri@opbg.net (M.A.); laura.pansani@opbg.net (L.P.); annamaria.sisto@opbg.net (A.S.); carlofederico.perno@opbg.net (C.F.P.)
2   Department of Biomolecular Sciences, University of Urbino "Carlo Bo", 61032 Fano, Italy; daniela.bencardino@uniurb.it (D.B.); mauro.magnani@uniurb.it (M.M.); francesca.andreoni@uniurb.it (F.A.)
3   Clinical Pathways and Epidemiology Unit, Bambino Gesù Children's Hospital, IRCCS, 00165 Rome, Italy; marta.ciofidegliatti@opbg.net (M.L.C.D.A.); carmen.damore@opbg.net (C.D.)
4   Human Microbiome Unit, Department of Diagnostics and Laboratory Medicine, Bambino Gesù Children's Hospital, IRCCS, 00165 Rome, Italy; lorenza.putignani@opbg.net
5   Neonatal Surgery Unit, Medical and Surgical Department of the Fetus-Newborn-Infant, Bambino Gesù Children's Hospital, IRCCS, 00165 Rome, Italy; pietro.bagolan@opbg.net (P.B.); bdaniela.iacobelli@opbg.net (B.D.I.)
6   Neonatal Intensive Care Unit, Medical and Surgical Department of the Fetus-Newborn-Infant, Bambino Gesù Children's Hospital, IRCCS, 00165 Rome, Italy; andrea.dotta@opbg.net (A.D.); ludovica.martini@opbg.net (L.M.)
7   Pediatric Cardiac Intensive Care Unit, Department of Cardiology and Cardiac Surgery, Bambino Gesù Children's Hospital, IRCCS, 00165 Rome, Italy; luca.dichiara@opbg.net
*   Correspondence: paola.bernaschi@opbg.net; Tel.: +39-06-6859-2205
†   Co-first author: These authors contributed equally to this work. Author order was determined alphabetically.
‡   Co-last author: These authors contributed equally to this work. Author order was determined alphabetically.

Citation: Agosta, M.; Bencardino, D.; Argentieri, M.; Pansani, L.; Sisto, A.; Ciofi Degli Atti, M.L.; D'Amore, C.; Putignani, L.; Bagolan, P.; Iacobelli, B.D.; et al. Prevalence and Molecular Typing of Carbapenemase-Producing Enterobacterales among Newborn Patients in Italy. *Antibiotics* **2022**, 11, 431. https://doi.org/10.3390/antibiotics11040431

Academic Editor: Masafumi Seki

Received: 1 March 2022
Accepted: 21 March 2022
Published: 23 March 2022

Publisher's Note: MDPI stays neutral with regard to jurisdictional claims in published maps and institutional affiliations.

Abstract: The spread of carbapenemase-producing Enterobacterales (CPE), especially *Klebsiella pneumoniae* (*K. pneumoniae*) and *Escherichia coli* (*E. coli*), is a serious public health threat in pediatric hospitals. The associated risk in newborns is due to their underdeveloped immune system and limited treatment options. The aim was to estimate the prevalence and circulation of CPE among the neonatal intensive units of a major pediatric hospital in Italy and to investigate their molecular features. A total of 124 CPE were isolated from rectal swabs of 99 newborn patients at Bambino Gesù Children's Hospital between July 2016 and December 2019. All strains were characterized by antimicrobial susceptibility testing, detection of resistance genes, and PCR-based replicon typing (PBRT). One strain for each PBRT profile of *K. pneumoniae* or *E. coli* was characterized by multilocus-sequence typing (MLST). Interestingly, the majority of strains were multidrug-resistant and carried the bla_{NDM} gene. A large part was characterized by a multireplicon status, and FII, A/C, FIA (15%) was the predominant. Despite the limited size of collection, MLST analysis revealed a high number of Sequence Types (STs): 14 STs among 28 *K. pneumoniae* and 8 STs among 11 *E. coli*, with the prevalence of the well-known clones ST307 and ST131, respectively. This issue indicated that some strains shared the same circulating clone. We identified a novel, so far never described, ST named ST10555, found in one *E. coli* strain. Our investigation showed a high heterogeneity of CPE circulating among neonatal units, confirming the need to monitor their dissemination in the hospital also through molecular methods.

Keywords: Enterobacterales; carbapenem resistance; neonates; plasmid-typing; sequence type

1. Introduction

The spread of carbapenem-resistant Enterobacterales (CRE) is an emerging concern worldwide and, noteworthy, carbapenem-resistant gram-negative bacterial infections are counted in the WHO priority pathogen list [1,2]. Due to continuous evolution of molecular mechanisms of resistance, the epidemiology of CRE is changing, and an increasing number of people are being colonized and infected by these organisms worldwide [3,4]. These infections are associated with high mortality ranging from 40% to 50%, and their prevalence continue to increase [4]. Indeed, carbapenemase-producing Enterobacterales (CPE) typically carry genes that confer resistance to other antibiotics and they are often extensively drug-resistant or even pan-drug resistant, limiting treatment options and leading to a high rate of therapeutic failures and subsequently to fatalities [5,6]. Globally, *Klebsiella pneumoniae* carbapenemases (KPCs) are the most common transmissible genes circulating in Enterobacterales. They have the ability to hydrolyze all β-lactams, and strains carrying bla_{KPC} often acquire resistance to several antibiotic agents, resulting in Multidrug-Resistant Organisms (MDROs) [7]. The rapid spread of KPC-Enterobacterales is due to the successful ST258 lineage, multidrug-resistant strains of *K. pneumoniae* endemic in several countries and responsible for many outbreaks [8].

However, in the last decade, the emergence of the most recently described carbapenemase New Delhi metallo-lactamase (NDM-1) was widely described as a matter of concern in clinical settings. This carbapenemase belongs to the metallo-lactamases class (MBLs) hydrolyzing a broad range of β-lactams with the exception of monobactams such as aztreonam. Infections caused by NDM producers include urinary tract infections, peritonitis, septicemia, pulmonary infections, soft tissue infections, and device-associated infections [9]. In contrast with KPC genes, the spread of NDM-type MBLs seems to be not associated with clonal lineages, but it is mediated by different plasmid incompatibility (Inc) groups [4,9]. Indeed, the clonal spread of these microorganisms among different patients develops very easily, and resistance genes to carbapenems can be transmitted between microorganisms of different species through plasmids [10].

In this scenario, the spread of (MDROs) is particularly worrisome in the Neonatal Intensive Care Unit (NICU): in this complex care setting, the emergence and dissemination of CRE and CPE, especially *K. pneumoniae* and *E. coli*, can occur with particular frequency and serious risks for newborns [11–13]. The epidemiological context and, in particular, the level of endemicity in communities and care settings, play a very important role in the risk of importing isolates and/or genetic resistance determinants into a NICU. Community-acquired (CA) pathogens are introduced in the hospital through the patients (including neonates and infants) themselves, their parents, and healthcare professionals. Furthermore, in the reference NICU of a specific territorial area, the epidemiology of microorganisms is strongly influenced by patient movements from and to other NICUs. Alongside the condition for which they are transferred (e.g., malformations, surgical emergencies, etc.), they present additional risk factors for the carriage of MDRO, i.e., previous exposure to another high-risk care setting [14–17].

On the other hand, prolonged hospitalization in such fragile patients is known to predispose to hospital-acquired colonization (HAC) [18,19]. For these reasons, to monitor the spread of carbapenemase-producing organisms (CPO), different international institutions have developed numerous guidelines [20–22], which have also been adopted in Italy.

Current prevention strategies for carriage and infections by CPE include active surveillance using culture and/or molecular methods, adoption of contact precautions, isolation or cohorting and, in selected cases, decolonization. Several studies emphasize the importance of identifying individuals carrying antimicrobial-resistant bacteria in both the patient and healthy population [23,24]. In fact, asymptomatic rectal carriers are considered the main source of spread of multidrug-resistant organisms, especially among fragile individuals, such as pediatric ones, where colonization is a precursor to infections such as bacteraemia, pneumonia, central nervous system (CNS), and urinary tract infections [25,26].

A large amount of information is reported in the literature for adults but, to date, little is known about the epidemiology and screening strategy of CPE in children, and even less in newborns.

The aim of this study was to estimate the distribution of CPE strains isolated during a three-year period from newborn patients admitted to the NICU and cardiac intensive care unit (CICU) of a large, tertiary care pediatric Italian hospital. All CPE strains were characterized in terms of their antimicrobial resistance patterns and molecular features such as determinants and plasmid replicons. Each strain of *K. pneumoniae* and *E. coli*, representative for each PBRT profile, was further characterized by multilocus-sequence typing (MLST).

2. Results

2.1. Epidemiological Features

Between July 2016 and December 2019, 124 CPE strains were consecutively collected from 99 newborn patients (44 females and 55 males) aged from six days to one year. A large part of newborns (72%) was hospitalized in the neonatal surgical unit (NSU), 15% in neonatal intensive therapy (NIT), 6% in the sub-intensive neonatal unit (SNU), and the remaining 5% in the cardiac intensive care unit (CICU). All strains were collected from rectal swabs: 105 (84%) were isolated from Italian patients, whereas 19 (15.3%) from 14 foreign patients: three from Burundi, two from Libya, Russia, and Romania, and one from Iraq, Georgia, Central African Republic, Sierra Leone, and Ukraine. From all patients, at least one strain was isolated (one strain from each swab), with the exception of 25 children subjected to a second swab during the hospitalization stay or after the second admission. In detail, 50 strains were isolated from 25 patients, and the remaining 74 strains were isolated from 74 patients. Moreover, the major part of the strains ($n = 103$; 82%) were isolated from patients colonized during their hospitalization, whereas 21 strains (17%) were community-acquired because they were isolated from patients who were colonized at the time of their admission to the hospital.

2.2. Isolate Identification

Of 124 CPE strains, *K. pneumoniae* ($n = 55$; 44%), *E. coli* ($n = 44$; 35%), *Klebsiella oxytoca* ($n = 7$; 6%), *Enterobacter cloacae* ($n = 7$; 6%), *Citrobacter freundii* ($n = 6$; 5%), *Klebsiella aerogenes* ($n = 2$; 2%), *Citrobacter koseri* ($n = 1$; 1%), *Serratia marcescens* ($n = 1$; 1%), and *Morganella morganii* ($n = 1$; 1%) were identified. The most prevalent species of *K. pneumoniae* and *E. coli* were widely detected both in NICU and CICU wards, with a significant presence in NSU ($n = 41/55$ for *K. pneumoniae*; $n = 35/44$ for *E. coli*). One foreign patient carried *Citrobacter koseri* and another carried *Morganella morganii*.

2.3. Antimicrobial Resistance Characterization

All CPE strains were found to be resistant to nearly all antibiotics tested in this study, including imipenem ($n = 121$; 97%) and meropenem ($n = 119$; 96%). High resistance was detected towards amoxicillin–clavulanic acid (100%), cefotaxime (100%), ceftazidime (100%), and piperacillin–tazobactam (100%). The lowest rate of resistance was recorded for tigecycline ($n = 4$; 3%). All strains of *K. pneumoniae* were resistant to all the antibiotics tested, whereas those of *E. coli* were susceptible only to tigecycline. Among 21 strains related to community-acquired colonization, 12 different antibiotic patterns were recorded, showing very high variability.

All 124 CPE strains were positive for carbapenemase genes, with the prevalence of 73% for the New-Delhi Metallo-β-lactamase gene (bla_{NDM}). In detail, the bla_{NDM} gene was detected in combination with bla_{KPC} and bla_{VIM} in 24 and 2 strains, respectively. By contrast, the remaining bla_{KPC} ($n = 15$), bla_{VIM} ($n = 13$), and bla_{OXA}-48 ($n = 3$) genes were individually detected.

2.4. Replicon Typing

A summary of replicons detected among the 122 CPE strains is given in Figure 1. Overall, only two strains of *E. cloacae* were not typeable by the PBRT, and IncA/C was the most common Inc group (65%), followed by IncFIA, IncFII, and IncFIIK found in 43%, 39%, and 32% of isolates, respectively. In this study, 18 out of 30 replicons were identified, whereas IncHI1, IncI2, IncB0, IncP1, IncW, IncI1γ, IncFIIS, IncN2, IncX1, IncX2, IncK, and IncX4 were not detected.

The majority of isolates ($n = 94/122$; 77%) were characterized by the multireplicon status carrying three or more different Inc groups. In detail, the predominant multireplicon profiles were FII, A/C, FIA ($n = 19$; 15%) and FII, A/C, FIB ($n = 11$; 9%), both found only among *E. coli* strains. Instead, the two multireplicon profiles consisting of A/C, R, FIIK, FIB KQ and FIB KQ, FIIK, R, A/C, and FIA were the most common in *K. pneumoniae*.

In our study, we observed that all forty-nine PBRT patterns were species-specific with few exceptions. In fact, the profile composed by the single replicon A/C was found in *E. coli*, *E. cloacae*, *M. morganii*, *K. aerogenes*, and *K. oxytoca*, whereas the pattern IncFIB KQ, IncFIIK was found in *K. pneumoniae* and *C. freundii*.

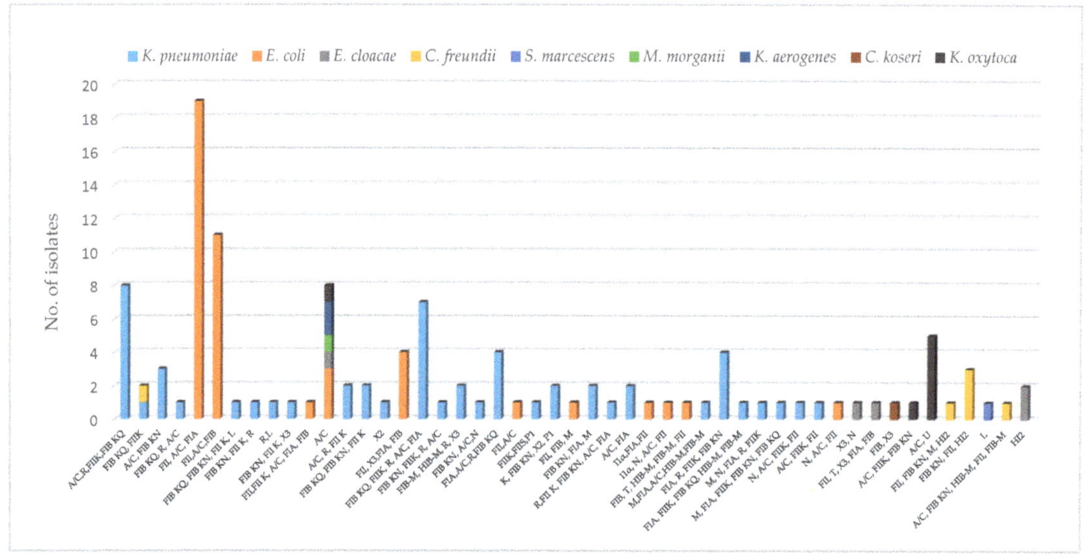

Figure 1. Distribution of PBRT patterns among CPE strains. Forty-nine PBRT profiles were distributed among 122 out of 124 CPE strains. All of these PBRT patterns were associated with a single species with the exception of replicon A/C (*E. coli*, *E. cloacae*, *M. morganii*, *K. aerogenes*, and *K. oxytoca*) and IncFIB KQ, IncFIIK (*K. pneumoniae* and *C. freundii*).

2.5. Correlation between PBRT and Antibiotic Resistance Patterns

The distribution of PBRT and antibiotic resistance within the collected strains is reported in Table 1. Although we detected a consistent heterogeneity, 12 (24%) PBRT profiles seemed to be related with specific antibiotic-resistance patterns and the associated carbapenemase-resistance genes. For instance, the PBRT profile IncFII, IncA/C, IncFIA (indicated with the number 5) was present in 18 *E. coli* strains, and all of them reported the same antibiotic resistance pattern (AMK-AMC-FEP-CTX-CAZ-CIP-GEN-IPM-TZP-MEM) and the same carbapenemase determinant (bla_{NDM}). This type of analysis was performed for all PBRT profiles, which included at least two strains, and numbered as 2, 13, 14, 16, 17, 19, 24, 28, 33, 46, and 49 (Table 1). Moreover, within the PBRT profiles IncFII, IncA/C, IncFIA and IncFIB KQ, IncFIIK, IncR, IncA/C, and IncFIA (numbered 5 and 17, respectively) we observed the highest number of strains that not only shared the same antibiotic-resistance pattern and related determinants, but also the same species (*E. coli* and *K. pneumoniae*, respectively), origin of patients (Italy), and type of colonization. By contrast, the highest variability was found for the PBRT profile IncA/C (number 12) wherein, out of eight strains, five different species (*E. coli*, *K. aerogenes*, *K. oxytoca*, *M. morganii*, and *E. cloacae*), five different antibiotic-resistance patterns, and three different carbapenemase determinant patterns were recorded.

Overall, correlations between replicons and multiple resistance profiles found in this study confirmed that plasmids are an important reservoir for the spread of resistance in Enterobacterales.

2.6. Multilocus-Sequence Typing

MLST analysis performed on 39 strains, each representative of a PBRT profile found in *E. coli* ($n = 11$) or *K. pneumoniae* ($n = 28$), revealed 22 STs. As described in Table 1, 14 different STs in 28 carbapenemase-producing *K. pneumoniae* isolates were found, with the prevalence of ST307 ($n = 5$; 18%), followed by ST17 ($n = 4$; 14%) and ST395 (n= 4; 14%). Among four *K. pneumoniae* strains isolated from foreign-born neonates, four different STs were detected (ST307, ST11, ST323, and ST1412). All strains were isolated from patients colonized upon their admission (CAC), with the exception of the strain belonging to ST323 related to hospital-acquired colonization. Moreover, these patients were hospitalized in NIT and NSU wards. Despite the modest size of collection, these findings suggest a certain degree of heterogenicity, due to the high number of circulating clones independently by origin of patient and hospitalization ward.

In light of this, we constructed phylogenetic trees in order to investigate the relationship among all STs identified in our analysis, and clonal relatedness of strains is shown in Figure 2. This analysis involved 28 nucleotide sequences for *K. pneumoniae* with 3018 positions in the final dataset.

Only the two strains belonging to ST323 and ST1164 seemed to be less related with the others, whereas strains typed with the most common identified ST307 showed higher phylogenetic relation with strains belonging to ST35 and ST466. Thus, all *K. pneumoniae* strains of our collection showed a clonal relatedness with the exception of two strains that were not genetically related with the main cluster. All strains grouped in this main cluster are well-known to be responsible for carbapenem resistance spread.

On the other hand, eight different STs were observed among 11 carbapenemase-producing *E. coli*, and ST131 was the main, found in three strains (two isolated from foreign patients). Instead, two strains belonged to ST69, and the remaining STs (ST617, ST80, ST74, ST162, ST167, and the novel ST10555) were individually found. Moreover, all foreign patients were hospitalized in the NSU ward, and four out of five strains isolated from them belonged to different STs (ST131, ST617, ST167, and ST69), reporting community-acquired colonization. By contrast, all *E. coli* strains isolated from Italian patients were associated with hospital-acquired colonization concerning not only the NSU ward but also the other wards.

Notwithstanding the low number of strains typed by MLST, we detected a wide diversity of STs in our *E. coli* collection as confirmed by the phylogenetic analysis. It involved 11 nucleotide sequences with a total of 3423 positions in the final dataset and revealed a higher relation among strains belonging to ST131, the novel ST10555, ST80, and ST74 (Figure 2). Of note, a main cluster was not identified, but we observed an interesting similarity of the novel ST10555 with the most common ST131. The latter is known to be an emerging carbapenemase-producing clone, suggesting the ability of the novel ST10555 to disseminate carbapenemase genes.

Table 1. Genotypic and phenotypic profile of CPE strains isolated from rectal swabs.

Strains [a]	Age [b]	Sex	Origin [c]	CAC or HAC [d]	Clinical Ward [e]	Date	Species	Antibiotic Resistance Pattern [f]	Genes	PBRT [g]	ST
84-A	5 mths	M	Russia	CAC	NSU	6/9/17					
116	24 days	M			NIT	8/7/17					395
127	6 days	F			NSU	8/14/17		AMK-AMC-FEP-CTX-CAZ-CIP-GEN-IPM-TZP-SXT-MEM			
123	14 days	M			NSU	8/14/17		AMK-AMC-FEP-CTX-CAZ-CIP-GEN-IPM-TZP-SXT-MEM			
145	2 mths	M	Italy	HAC	SNU	10/2/17	*K. pneumoniae*		bla_{KPC} + bla_{NDM}	(1) A/C, R, FIIK, FIB KQ	
172	19 days	F			SNU	11/6/17					
122	11 days	F			NIT	8/14/17		AMK-AMC-FEP-CTX-CAZ-CIP-GEN-IPM-TZP-MEM			
126-D	1 mth	M			NSU	8/14/17		AMK-AMC-FEP-CTX-CAZ-CIP-GEN-IPM-TZP-TGC-SXT-MEM			
158	2 mths	F	Italy	CAC	NIT	10/14/17	*K. pneumoniae*	AMC-CTX-CAZ-IPM-TZP-MEM	bla_{KPC}	(2) FIB KQ, FIIK	1164
155	4 mths			HAC	NSU	10/9/17	*C. freundii*				
159	2 mths				SNU			AMK-AMC-FEP-CTX-CAZ-GEN-IPM-TZP-MEM			17
168a-L	13 days	M	Italy	HAC	NSU	10/16/17	*K. pneumoniae*		bla_{NDM}	(3) A/C, FIB KN	
160	20 days				NSU			AMK-AMC-FEP-CTX-CAZ-GEN-IPM-TZP-SXT-MEM			
189	3 mths	F	Italy	HAC	NIT	12/11/17	*K. pneumoniae*	AMK-AMC-FEP-CTX-CAZ-GEN-IPM-TZP-MEM	bla_{NDM}	(4) FIB KQ, R, A/C	54
167-G	4 mths	F			NSU	10/16/17		AMK-AMC-FEP-CTX-CAZ-CIP-GEN-IPM-TZP-SXT-MEM			
138-D	2 mths	M			NSU	9/18/17					
187-O	2 mths	F			NSU	12/4/17					
188	15 days	F			SNU	12/11/17					131
191	4 mths	M	Italy	HAC	SNU	12/11/17	*E. coli*	AMK-AMC-FEP-CTX-CAZ-CIP-GEN-IPM-TZP-MEM	bla_{NDM}	(5) FII, A/C, FIA	
204-R	1 mth	M			NSU	2/5/18					
207-Q	2 mths	M			NSU	2/9/18					
211-T	17 days	F			NSU	2/19/18					
214	18 days	F			NSU	2/26/18					

Table 1. *Cont.*

Strains [a]	Age [b]	Sex	Origin [c]	CAC or HAC [d]	Clinical Ward [e]	Date	Species	Antibiotic Resistance Pattern [f]	Genes	PBRT [g]	ST
215	1 mth	M				2/26/18					
216	2 mths	M				2/26/18					
218-U	2 mths	F				2/26/18					
222	3 mths	M			NIT	3/5/18					
227	1 mth	M			SNU	3/12/18					
226-Z	2 mths	F				3/12/18					
228-S	1 year	M				3/12/18					
232	6 days	M			NSU	3/19/18					
240-Y	1 mth	M				4/16/18					
274-Z	7 mths	F				8/8/18					
163-M	1 mth	M	Italy	HAC	NSU	10/16/17	*E. coli*	AMK-AMC-FEP-CTX-CAZ-GEN-IPM-TZP-MEM	*bla*$_{NDM}$		
168b-L	13 days	M			NSU	10/16/17					
169-I	1 mth	F			NSU	10/16/17					
210	1 mths	M			NIT	2/19/18					
245	6 mths	F			NIT	4/23/18					74
250	29 days	M			NSU	5/7/18				(6) FII, A/C, FIB	
261	2 mths	F		CAC	NIT	6/30/18		AMC-FEP-CTX-CAZ-IPM-TZP-SXT	*bla*$_{VIM}$		
275	4 mths	M			NSU	8/20/18					
266	14 days	M		HAC	SNU	7/25/18		AMC-FEP-CTX-CAZ-IPM-TZP-SXT-MEM	*bla*$_{VIM}$		
336	1 year	M	S. Leone		NSU	7/15/19			*bla*$_{NDM}$		
199-P	3 mths	M	Italy		NIT	1/10/18		AMK-AMC-FEP-CTX-CAZ-TZP	*bla*$_{NDM}$		
247	5 mths	F	Romania	CAC	NIT	4/28/18	*K. pneumoniae*	AMC-FEP-CTX-CAZ-CIP-GEN-IPM-TZP-SXT	*bla*$_{OXA-48}$	(7) FIB KQ, FIB KN, FII K, L	307
285	1 mth	M	Italy	CAC	NIT	9/28/18	*K. pneumoniae*	AMK-AMC-FEP-CTX-CAZ-CIP-GEN-IPM-TZP-MEM	*bla*$_{KPC}$	(8) FIB KN, FII K, R	101
16	1 mth	F	Romania	CAC	NSU	9/22/16	*K. pneumoniae*	AMK-AMC-FEP-CTX-CAZ-CIP-GEN-IPM-TZP-TGC-MEM	*bla*$_{OXA-48}$	(9) R, L	11

Table 1. *Cont.*

Strains [a]	Age [b]	Sex	Origin [c]	CAC or HAC [d]	Clinical Ward [e]	Date	Species	Antibiotic Resistance Pattern [f]	Genes	PBRT [g]	ST
83	1 mths	M	Italy	HAC	NSU	6/5/17	*K. pneumoniae*	AMK-AMC-FEP-CTX-CAZ-CIP-IPM-TZP-MEM	bla_{KPC}	(10) FIB KN, FII K, X3	512
90-A	5 mths	M	Russia	HAC	NSU	6/19/17	*E. coli*	AMK-AMC-FEP-CTX-CAZ-CIP-GEN-IPM-TZP-SXT-MEM	bla_{NDM}	(11) FII, FII K, A/C, FIA, FIB	131
104-B	5 mths	M	Italy	HAC	NSU	7/18/17	*E. coli*	AMK-AMC-FEP-CTX-CAZ-CIP-GEN-IPM-TZP-SXT-MEM	bla_{NDM}	(12) A/C	10555
129	1 mth	M	Italy	HAC	NSU	8/14/17	*K. aerogenes*	AMK-AMC-FEP-CTX-CAZ-GEN-IPM-TZP-MEM	bla_{NDM}	(12) A/C	
130	15 days	M	Italy	HAC	NSU	8/18/17	*E. coli*	AMK-AMC-FEP-CTX-CAZ-GEN-IPM-TZP-MEM	bla_{NDM}	(12) A/C	
131-C	1 mth	F	Italy	HAC	NSU	8/21/17	*E. coli*	AMK-AMC-FEP-CTX-CAZ-GEN-IPM-TZP-MEM	bla_{NDM}	(12) A/C	
183	6 mths	M	Italy	HAC	NSU	11/27/17	*K. aerogenes*	AMK-AMC-FEP-CTX-CAZ-GEN-IPM-TZP-MEM	$bla_{KPC}+bla_{NDM}$	(12) A/C	
229	7 mths	M	Italy	HAC	NSU	3/12/18	*K. oxytoca*	AMC-CTX-CAZ-IPM-TZP-MEM	bla_{NDM}	(12) A/C	
221-K	1 year	M	Russia	HAC	NSU	3/2/18	*M. morganii*	AMK-AMC-CTX-CAZ-GEN-IPM-MEM	$bla_{NDM}+bla_{VIM}$	(12) A/C	
246	4 mths	M	Italy	HAC	NSU	4/26/18	*E. cloacae*	AMK-AMC-FEP-CTX-CAZ-GEN-TZP-MEM	bla_{NDM}	(12) A/C	
106	1 mth	M	Italy	HAC	NSU	7/24/17	*K. pneumoniae*	AMK-AMC-FEP-CTX-CAZ-CIP-GEN-IPM-TZP-SXT-MEM	$bla_{KPC}+bla_{NDM}$	(13) A/C, R, FII K	395
117-C	1 mth	F	Italy	HAC	NSU	8/7/17	*K. pneumoniae*	AMK-AMC-FEP-CTX-CAZ-CIP-GEN-IPM-TZP-SXT-MEM	$bla_{KPC}+bla_{NDM}$	(13) A/C, R, FII K	
124	11 days	F	Italy	HAC	NSU	8/14/17	*K. pneumoniae*	AMC-CTX-CAZ-IPM-TZP-MEM	bla_{KPC}	(14) FIB KQ, FIB KN, FII K	17
125	5 days	F	Italy	HAC	NSU	8/14/17	*K. pneumoniae*	AMC-CTX-CAZ-IPM-TZP-MEM	bla_{KPC}	(14) FIB KQ, FIB KN, FII K	
103-B	5 mths	M	Italy	HAC	NSU	7/17/17	*K. pneumoniae*	AMC-FEP-CTX-CAZ-CIP-GEN-IPM-TZP-SXT	bla_{KPC}	(15) X2	307
134-E	1 mth	F	Burundi	CAC	NSU	8/30/17	*E. coli*	AMC-FEP-CTX-CAZ-CIP-GEN-IPM-TZP-SXT-MEM	bla_{NDM}	(16) FII, X3, FIA, FIB	617
137-F	1 mth	F	Burundi	CAC	NSU	9/4/17	*E. coli*	AMC-FEP-CTX-CAZ-CIP-GEN-IPM-TZP-SXT-MEM	bla_{NDM}	(16) FII, X3, FIA, FIB	
166-N	1 mth	F	Italy	HAC	NSU	10/16/17	*E. coli*	AMC-FEP-CTX-CAZ-CIP-GEN-IPM-TZP-SXT-MEM	bla_{NDM}	(16) FII, X3, FIA, FIB	
174	5 mths	F	Libya	HAC	NSU	11/13/17	*E. coli*	AMC-FEP-CTX-CAZ-CIP-GEN-IPM-TZP-SXT-MEM	bla_{NDM}	(16) FII, X3, FIA, FIB	
139	3 mths	M	Italy	HAC	NSU	9/18/17	*K. pneumoniae*	AMK-AMC-FEP-CTX-CAZ-CIP-GEN-IPM-TZP-SXT-MEM	$bla_{KPC}+bla_{NDM}$	(17) FIB KQ, FIIK, R, A/C, FIA	395
144-H	4 mths	F	Italy	HAC	NSU	10/2/17	*K. pneumoniae*	AMK-AMC-FEP-CTX-CAZ-CIP-GEN-IPM-TZP-SXT-MEM	$bla_{KPC}+bla_{NDM}$	(17) FIB KQ, FIIK, R, A/C, FIA	
141	13 days	F	Italy	HAC	NSU	10/2/17	*K. pneumoniae*	AMK-AMC-FEP-CTX-CAZ-CIP-GEN-IPM-TZP-SXT-MEM	$bla_{KPC}+bla_{NDM}$	(17) FIB KQ, FIIK, R, A/C, FIA	
143-G	4 mths	F	Italy	HAC	NSU	10/2/17	*K. pneumoniae*	AMK-AMC-FEP-CTX-CAZ-CIP-GEN-IPM-TZP-SXT-MEM	$bla_{KPC}+bla_{NDM}$	(17) FIB KQ, FIIK, R, A/C, FIA	

Table 1. *Cont.*

Strains [a]	Age [b]	Sex	Origin [c]	CAC or HAC [d]	Clinical Ward [e]	Date	Species	Antibiotic Resistance Pattern [f]	Genes	PBRT [g]	ST
162-M	1 mth	M				10/16/17					
171-N	1 mth	F				10/23/17					
209-S	11mths	M				2/12/18					
146-I	16 days	F	Italy	HAC	NSU	10/2/17	*K. pneumoniae*	AMK-AMC-FEP-CTX-CAZ-GEN-IPM-TZP-MEM	bla_{KPC}+bla_{NDM}	(18) FIB KN, FIIK, R, A/C	17
153-F	3 mths	F	Burundi	HAC	NSU	10/9/17	*K. pneumoniae*	AMC-FEP-CTX-CAZ-CIP-IPM-TZP-SXT-MEM	bla_{NDM}	(19) FIB-M, HIB-M, R, X3	323
154-E	3 mths					10/9/17					
161-O	17 days	F	Italy	HAC	NSU	10/16/17	*K. pneumoniae*	AMK-AMC-FEP-CTX-CAZ-GEN-IPM-TZP-TGC-MEM	bla_{NDM}	(20) FIB KN, A/C, N	17
177-P	2 mths					11/20/17			bla_{NDM}		54
178	26 days	M	Italy	HAC	NSU	11/20/17	*K. pneumoniae*	AMK-AMC-FEP-CTX-CAZ-GEN-IPM-TZP-MEM	bla_{KPC}+bla_{NDM}	(21) FIA, A/C, R, FIB KQ	
181-Q	13 days					11/27/17			bla_{NDM}		
182	19 days					11/27/17			bla_{NDM}		
190-H	6 mths	F	Italy	HAC	NSU	12/11/17	*E. coli*	AMK-AMC-FEP-CTX-CAZ-GEN-TZP-SXT-MEM	bla_{NDM}	(22) FII, A/C	80
195-R	16 days	M	Italy	HAC	NSU	1/3/18	*K. pneumoniae*	AMK-AMC-FEP-CTX-CAZ-CIP-GEN-IPM-TZP-TGC-SXT-MEM	bla_{KPC} + bla_{NDM}	(23) FIIK, FIIS, P1	395
200	9 mths	M	Italy	CAC	NSU	1/17/18	*K. pneumoniae*	AMC-FEP-CTX-CAZ-CIP-GEN-IPM-TZP-SXT-MEM	bla_{KPC}	(24) K, FIB KN, X2, P1	307
289	2 mths			HAC	CICU	10/15/18					
224-V	6 mths	M	Italy	HAC	CICU	3/11/18	*E. coli*	AMC-FEP-CTX-CAZ-IPM-TZP-SXT-MEM	bla_{VIM}	(25) FII, FIB, M	162
225-V	6 mths	M	Italy	HAC	CICU	3/11/18	*K. pneumoniae*	AMC-FEP-CTX-CAZ-IPM-TZP-SXT-MEM	bla_{NDM} + bla_{VIM}	(26) FIB KN, FIA, M	466
257	20 days	F			CICU	6/13/18		AMC-FEP-CTX-CAZ-GEN-IPM-TZP-SXT-MEM	bla_{NDM}		
230-U	3 mths	F	Italy	HAC	NSU	3/12/18	*K. pneumoniae*	AMK-AMC-FEP-CTX-CAZ-CIP-GEN-IPM-TZP-SXT-MEM	bla_{NDM}	(27) R, FII K, FIB KN, A/C, FIA	231
231-T	1 mth	F	Italy	CAC	NSU	3/18/18	*K. pneumoniae*	AMK-AMC-FEP-CTX-CAZ-GEN-IPM-TZP-MEM	bla_{NDM}	(28) A/C, FIA	35
237-Y	23 days	M		HAC		4/9/18					

Table 1. *Cont.*

Strains [a]	Age [b]	Sex	Origin [c]	CAC or HAC [d]	Clinical Ward [e]	Date	Species	Antibiotic Resistance Pattern [f]	Genes	PBRT [g]	ST
278	2 mths	F	C.A.R.	CAC	NSU	9/11/18	*E. coli*	AMK-AMC-FEP-CTX-CAZ-CIP-GEN-IPM-TZP-SXT-MEM	bla_{NDM}	(29) I1α, FIA, FII	167
341-X	3 mths	F	Ukraine	CAC	NSU	8/4/19	*E. coli*	AMK-AMC-FEP-CTX-CAZ-CIP-GEN-IPM-TZP-SXT-MEM	bla_{NDM}	(30) I1α, N, A/C, FII	69
312	7 mths	F	Georgia	CAC	NSU	3/9/19	*E. coli*	AMC-FEP-CTX-CAZ-IPM-TZP-SXT-MEM	bla_{NDM}	(31) FIB, T, HIB-M, FIB-M, FII	131
369-W	4 mths	F	Italy	HAC	NSU	10/29/19	*K. pneumoniae*	AMK-AMC-FEP-CTX-CAZ-GEN-IPM-TZP-MEM	bla_{NDM}	(32) M, FIA, A/C, HIB-M, FIB-M	160
288	3 mths	F				10/11/18					101
298	2 mths	M	Italy	CAC	NSU	12/3/18	*K. pneumoniae*	AMK-AMC-FEP-CTX-CAZ-CIP-GEN-IPM-TZP-MEM	bla_{KPC}	(33) FIA, R, FIIK, FIB KN	
308	6 mths	F				2/21/19					
393	3 mths	M	Iraq	HAC	CICU	1/28/20					
235	3 mths	F	Italy	CAC	NSU	4/4/18	*K. pneumoniae*	AMC-CTX-CAZ-CIP-IPM-TZP-MEM	bla_{KPC}	(34) FIA, FIIK, FIB KQ, HIB-M, FIB-M	307
304	6 days	F	Italy	CAC	NSU	1/29/19	*K. pneumoniae*	AMK-AMC-FEP-CTX-CAZ-CIP-GEN-IPM-TZP-MEM	bla_{KPC}	(35) M, N, FIA, R, FIIK	101
333	2 mths	F	Italy	HAC	NIT	7/1/19	*K. pneumoniae*	AMK-AMC-FEP-CTX-CAZ-CIP-GEN-IPM-TZP-SXT-MEM	bla_{KPC} + bla_{NDM}	(36) M, FIA, FIIK, FIB KN, FIB KQ	307
342-X	3 mths	F	Ukraine	CAC	NSU	8/4/19	*K. pneumoniae*	AMC-FEP-CTX-CAZ-CIP-GEN-IPM-TZP-SXT-MEM	bla_{NDM}	(37) N, A/C, FIIK, FII	1412
345	22 days	M	Italy	HAC	NSU	8/19/19	*K. pneumoniae*	AMC-FEP-CTX-CAZ-CIP-GEN-IPM-TZP-SXT-MEM	bla_{NDM}	(38) A/C, FIIK, FII	1412
350-W	3 mths	F	Italy	HAC	NSU	9/2/19	*E. coli*	AMK-AMC-FEP-CTX-CAZ-CIP-GEN-IPM-TZP-SXT-MEM	bla_{NDM}	(39) N, A/C, FII	69
142	2 mths	F	Burundi	HAC	NSU	10/2/17	*E. cloacae*	AMC-FEP-CTX-CAZ-GEN-IPM-TZP-MEM	bla_{KPC} + bla_{NDM}	(40) X3, N	
165	1 mth	F	Italy	HAC	NSU	10/16/17	*E. cloacae*	AMC-FEP-CTX-CAZ-IPM-TZP-MEM	bla_{NDM}	(41) FII, T, X3, FIA, FIB	
179	5 mths	F	Libya	HAC	NSU	11/20/17	*C. koseri*	AMC-FEP-CTX-CAZ-IPM-TZP-MEM	bla_{NDM}	(42) FIB, X3	

Table 1. *Cont.*

Strains [a]	Age [b]	Sex	Origin [c]	CAC or HAC [d]	Clinical Ward [e]	Date	Species	Antibiotic Resistance Pattern [f]	Genes	PBRT [g]	ST
208-K	1 year	M	Russia	HAC	NSU	2/11/18	*K. oxytoca*	AMK-AMC-CTX-CAZ-GEN-IPM-TZP-MEM	bla_{NDM}	(43) A/C, FIIK, FIB KN	
234	15 days					4/3/18			bla_{KPC} + bla_{NDM}		
241	4 mths					4/16/18					
242	2 mths	M	Italy	HAC	NSU	4/16/18	*K. oxytoca*	AMK-AMC-FEP-CTX-CAZ-GEN-IPM-TZP-MEM	bla_{NDM}	(44) A/C, U	
243	1 mth					4/16/18					
248	19 days					4/30/18					
374	1 mth	F	Italy	HAC	NSU	11/11/19	*C. freundii*	AMC-FEP-CTX-CAZ-GEN-IPM-TZP-MEM	bla_{VIM}	(45) FII, FIB KN, M, HI2	
370	1 mth	M				11/4/19					
375	2 mths	F	Italy	HAC	NSU	11/11/19	*C. freundii*	AMC-FEP-CTX-CAZ-IPM-TZP-MEM	bla_{VIM}	(46) FIB KN, FII, HI2	
388	3 mths	M				12/10/19					
377	18 days	F	Italy	CAC	NIT	11/12/19	*S. marcescens*	AMK-AMC-FEP-CTX-IPM-TZP-MEM	bla_{OXA-48}	(47) L	
381	1 mth	M	Italy	HAC	NSU	11/18/19	*C. freundii*	AMK-AMC-FEP-CTX-CAZ-GEN-IPM-TZP-MEM	bla_{NDM}	(48) A/C, FIB KN, HIB-M, FII, FIB-M	
337	1 year	M	Italy	HAC	NIT	7/15/19	*E. cloacae*	AMC-FEP-CTX-CAZ-CIP-GEN-IPM-TZP-SXT-MEM	bla_{VIM}	(49) HI2	
389	28 days				NSU	12/16/19					
373	2 days	M	Italy	CAC	SNU	11/7/19	*E. cloacae*	AMC-FEP-CTX-CAZ-GEN-IPM-TZP-SXT-MEM	bla_{VIM}	UT	
4	8 mths	F	Italy	HAC	NIT	7/25/16	*E. cloacae*	AMC-FEP-CTX-CAZ-GEN-IPM-TZP-MEM	bla_{VIM}	UT	

[a] Strains isolated from the same patient were indicated with the same letter (e.g., 137-F and 153-F); [b] mths (months); [c] CAR (Central African Republic), S. Leone (Sierra Leone); [d] HAC (hospital-acquired colonization), CAC (community-acquired colonization); [e] NIT (neonatal intensive unit), NSU (neonatal surgery unit), SNU (sub-intensive neonatal unit), CICU (cardiac intensive care unit); [f] amikacin (AMK), amoxicillin–clavulanic acid (AMC), cefepime (FEP), cefotaxime (CTX), ceftazidime (CAZ), ciprofloxacin (CIP), gentamicin (GEN), imipenem (IPM), meropenem (MEM), piperacillin–tazobactam (TZP), tigecycline (TGC), trimethoprim-sulfamethoxazole (SXT); [g] PBRT (profiles were sorted numerically from 1 to 49); UT = untypeable.

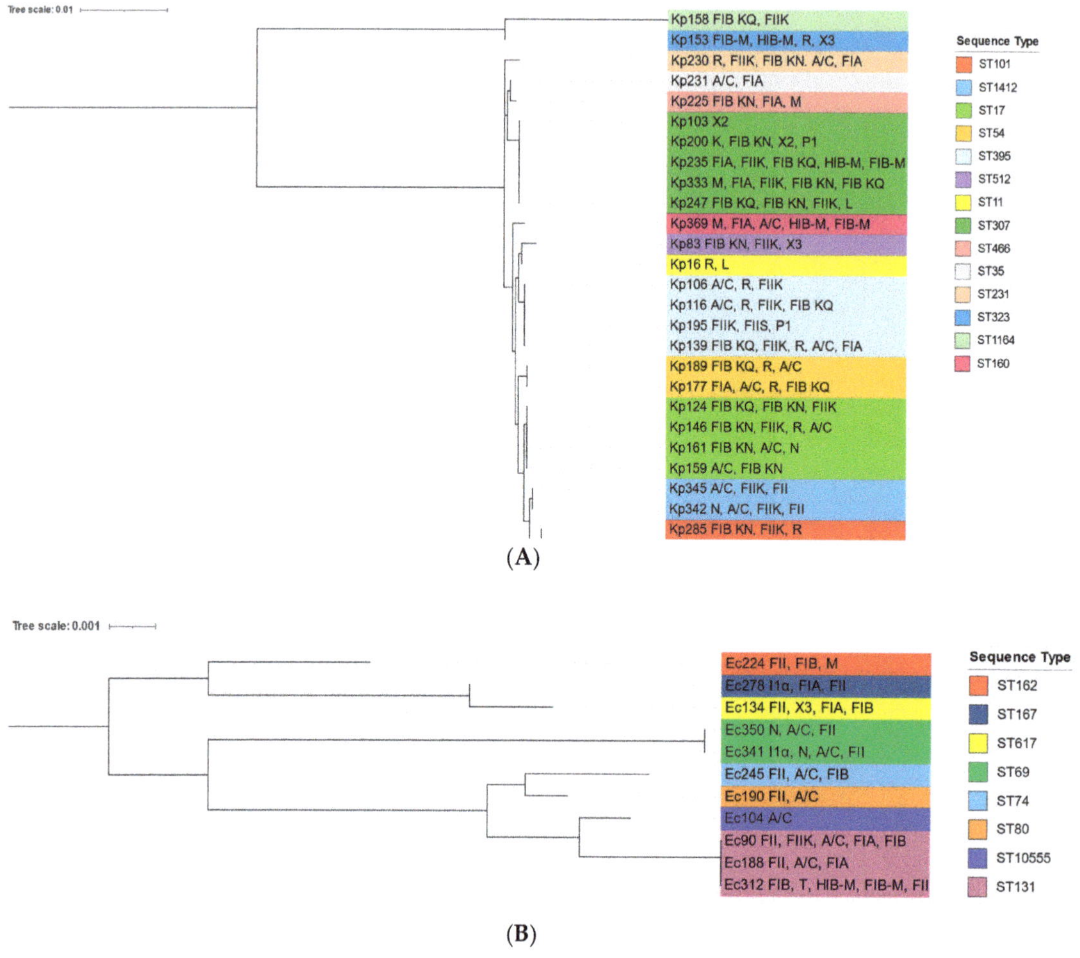

(A)

(B)

Figure 2. MLST-based phylogenetic trees of *K. pneumoniae* and *E. coli* strains. Trees of concatenated nucleotide sequences of seven housekeeping genes of *K. pneumoniae* (**A**) and *E. coli* (**B**) were obtained using the Maximum Likelihood method and the General Time Reversible model by MEGA software (version 10.0) with bootstrap percentages retrieved in 1000 replications. Graphical representation of both were obtained using Interactive Tree of Lifes (iTOL v.6) software. Scale bars indicate the number of nucleotide substitutions per site. For each strain, the name was reported as Kp for *K. pneumoniae* and Ec for *E. coli*. STs and replicon profiles were also indicated for each strain.

2.7. The Novel ST10555

Interestingly, a new sequence type was identified among *E. coli* clinical isolates (named *E. coli*_104) as shown in Table 1 and Figure 2. This strain was isolated from an Italian male patient (five months of age) in 2017. The isolate exhibited the A/C replicon and the following antibiotic-resistance profile AMK-AMC-FEP-CTX-CAZ-CIP-GEN-IPM-TZP-SXT-MEM; carbapenem resistance was confirmed by the detection of bla_{NDM} gene. It revealed a novel allelic variant for the *recA* gene (recorded as 772) confirmed through whole-genome sequencing. Successively, the entire genome was submitted to EnteroBase database at https://enterobase.warwick.ac.uk/species/index/ecoli (accessed on 21 March 2022), resulting in the associated ST10555 (*adk* 13, *fumC* 52, *gyrB* 10, *icd* 14, *mdh* 17, *purA* 25, *recA* 772).

Moreover, we used genomic analysis tools to explore the whole-genome sequencing data. Through ResFinder 4.1 we confirmed the resistance observed phenotypically and found additional resistance towards tobramycin, cefoxitin, ampicillin, and ertapenem. The same analysis revealed the following resistance genes: *aac(6′)-Ib3* (amikacin, tobramycin), *aph(3′)-VI* (amikacin), *rmtC* (amikacin, gentamicin, tobramycin, kanamycin), *bla*$_{CMY-6}$ (amoxicillin, amoxicillin–clavulanic acid, ampicillin, cefotaxime, ceftazidime, piperacillin–tazobactam), *bla*$_{NDM-1}$ (amoxicillin, amoxicillin + clavulanic acid, ampicillin, cefepime, cefotaxime, cefoxitin, ceftazidime, ertapenem, imipenem, meropenem), and *bla*$_{CTX-M-15}$ (amoxicillin, ampicillin, cefepime, cefotaxime, ceftazidime, piperacillin). Moreover, through PlasmidFinder 2.0 and pMLST-2.0 Server, we confirmed that this isolate belonged to IncC group and to the profile IncA/C PMLST, respectively.

Finally, SerotypeFinder 2.0 Server revealed the serotype H6 encoded by the *flic* gene (99.94% of identity) as well as the serotype O2 encoded by *wzy* and *wzx* genes, with the identities of 98.03% and 98.65%, respectively. The genetic features of this newly emerged strain type highlight the risk associated with its circulation, especially in terms of antibiotic resistance spread that is a matter of concern in the neonatal population.

3. Discussion

Although CPE infections are endemic in Italy and neonates and infants are considered patients at high risk, few studies have been carried out to evaluate their dissemination among neonatal intensive units [12,13,27–29]. Retrospective studies carried out among the NICUs of different pediatric Italian hospitals reported high rates of colonization by strains with important molecular features, highlighting the role of risk factors on infection incidence. For instance, Montagnani and colleagues confirmed the immunosuppressive state of children with hematologic/oncologic conditions, invasive devices, history of surgery or hospitalization, and prior use of antibiotics as important risk factors associated with CPE infections [27]. Instead, Mammina and colleagues showed that feeding by formula was significantly associated with colonization due to cross-contamination and poor infection control procedures [30]. On the other hand, a more recent study demonstrated that the efficacy of screening programs associated with proactive measures to control cross-contamination and hospital-acquired colonization are effective to reduce the spread of CPE also in the pediatric setting [13]. Moreover, these studies were useful also to collect information about the emerging clones circulating among neonatal wards, tracing the evolution of their molecular aspects.

Here, we described CPE distribution in neonatal wards, including NICU, and a CICU of a major tertiary pediatric referral hospital in Italy, where foreign patients are also frequently admitted. The first interesting point was the high variability of CPE detected in terms of isolates and associated molecular features. In accordance with other studies, within the variety of species detected in our collection and widely associated with neonatal colonization, *K. pneumoniae* and *E. coli* were the most common [12,13,31–33]. Most strains (76%) exhibited resistance to at least one agent of three or more antimicrobial categories tested in our panel. Hence, they were classified as multidrug-resistant (MDR) as described by Majorakos et al. [34], increasing the risk of treatment failure. To note, almost all strains were susceptible to tigecycline (97%). This was in agreement with results reported by the SENTRY surveillance program where a good activity of tigecycline against these microorganisms was indicated [35]. Conversely, other studies reported higher levels of resistance towards this antimicrobial agent [27], and recent investigations suggest an improved efficacy of tigecycline against carbapenem-resistant Enterobacterales in combination with colistin [36] or with ceftazidime–avibactam [37]. These findings suggest the need to monitor the use of this or other last-resort antibiotics in order to limit the impairment of treatment due to increased levels of resistance.

Moreover, CPE strains have the potential to spread widely due to the localization of carbapenemase genes on mobile genetic elements, which further complicate the treatment, increasing mortality and morbidity rates [38,39]. Overall, 26% of strains coharbored a combination of two resistance genes, thus contributing to increased risk in terms of public health. It was not surprising that the predominant determinant of carbapenem resistance found in our study was bla_{NDM}. Indeed, the rapid dissemination of New Delhi metallo-beta-lactamase (NDM)-producing carbapenem-resistant Enterobacterales have been largely detected in central Italy in the last years [40]. This spread of NDM-producing strains confirms a significant change in CPE epidemiology consisting in replacement of previous endemicity of bla_{KPC} genes.

It is well-known that *K. pneumoniae* and *E. coli* are the main drivers of this rapid and consistent spread, and the dissemination of NDM is generally mediated by plasmids with a variety of replicon types [41]. In accordance with the literature, the predominant IncA/C plasmid (66%) detected in the present study was strongly associated with bla_{NDM} genes [42,43]. This plasmid type, identified in all enterobacterial species, possesses a broad host range of replication and has been identified not only from human but also from animal isolates [44,45]. For all of these reasons, the IncA/C-type plasmid is considered an increasing threat to public health. Moreover, the majority of isolates (77%) carried three or more replicons defining the multireplicon-status, with a high percentage of the IncF family that it is known to be restricted to Enterobacterales. The multireplicon status promotes acquisition of plasmids carrying incompatible replicons, enlarging the host range replication also between different species [46]. A typical multireplicon IncF plasmid carries FII, FIA, and FIB, as also observed in our collection, confirming their large diffusion among clinical Enterobacterales [46,47]. From a functional point of view, FII is silent, whereas the activity of FIA as well as that of FIB is exclusively related to enteric bacteria [46].

Additional interesting information about the diversity of strains investigated in the present study was revealed by multilocus-sequence typing analysis. The identification of ST was performed for a single strain representative for each PBRT pattern in order to obtain preliminary information about the clones circulating in the neonatal intensive unit. Of all the obtained STs, ST307 was the most common in *K. pneumoniae*, confirming the recent and rapid spread of this emerging clone in Italy [48–50]. Noteworthy, the diffusion of ST307 in Italy traces the evolution of the current epidemiological change because it seems to replace the global ST258 and its endemic Italian variant ST512. Indeed, after the first outbreak occurred in Sicily (Italy) in 2008 [51], where KPC-3-producing *K. pneumoniae* ST258 was mainly responsible, a second surveillance carried out in 2014 revealed the emerging of KPC-3-producing strains ST307 [29,52]. As also detected in our investigation, ST307 is a novel distinctive lineage carrying bla_{KPC} determinants [49], even if in our study we found it in association with bla_{NDM} or bla_{OXA-48}, suggesting the potential to acquire advantageous features for the adaption to clinical niches. Furthermore, the increased detection of ST307 also in other countries in the last years suggests not only its pivotal role in the spread of antibiotic resistance but also the potential to become one of the most clinically relevant clones. Our data confirmed the increasing CPE prevalence in neonates and infants and showed that the molecular characteristics of strains are evolving. Moreover, CPE colonization represents an important reservoir for nosocomial infections due to their stronger virulence and transmission. Hence, an active screening also through molecular typing is very useful to classify genotypes and is a priority in vulnerable patients like newborns, where multiple-antibiotic resistance can result in failed infection treatment.

On the other hand, *E. coli* strains characterized by MLST revealed several STs, but this collection was too limited, and the frequency of these clones could be underestimated. Among them, ST131, known to be responsible for hospital- and community-acquired urinary tract infections as well as bloodstream infections, was most commonly detected. Most strains belonging to ST131 are MDR, therefore limiting therapeutic options that cause recurrent infections [53].

The identification of the novel ST10555 highlights the variability observed among *E. coli* strains despite the low size of collection. It was isolated from an Italian patient and exhibited resistance towards all of antibiotics tested in our study, with the exception for tigecycline. Whole-genome analysis revealed the carriage of other determinants besides bla_{NDM}, confirming the risk associated with its distribution among newborn patients. This novel ST was phylogenetically related with ST131, described as a high-risk clone due to its role in antibiotic resistance spread, confirming the continuous emergence of new clones with potential to colonize clinical niches (Figure 2).

Notwithstanding the limited size of our collection, the richness and variability of genetic repertoire showed by these strains made more evident the dynamic evolution of CPE and, consequently, the importance of monitoring their distribution in healthcare settings.

From a health perspective, the results of this study highlight the urgency of addressing the surveillance of CPE distribution in pediatric hospitals. One important challenge for hospitals consists of collaborative efforts with different and external professional figures of health sciences. Indeed, in addition to the commonly reported guidelines implemented by WHO protocols to manage the prescription of antibiotics in therapeutic plans, strict cooperation among diagnostic laboratories, hospitals, and research institutes is strongly recommended to reduce the spread of these dangerous strains and prevent the emergence of serious infections in newborns. Considering this, we highlight the usefulness of molecular typing to better understand the distribution of resistant strains in clinical settings, especially where patients at risk, like neonates and children, are hospitalized.

4. Materials and Methods

4.1. Ethics Statement

The Institutional Review Board (IRB) of the IRCCS Bambino Gesù Children's Hospital approved the study protocol (n.2156_OPBG_2020). As the data in this study were collected and analyzed retrospectively, the study did not infringe upon the rights or welfare of the patients and did not require their consent.

4.2. Study Design

This study was a retrospective investigation carried out from July 2016 to December 2019. A total of 9914 surveillance rectal swabs were performed for CPE screening from newborn patients hospitalized or admitted to the neonatal surgical unit (NSU), neonatal intensive therapy (NIT), the sub-intensive neonatal unit (SNU), and the cardiac intensive care unit (CICU) of Bambino Gesù Children's Hospital in Rome (Italy), according to the active surveillance protocol issued by the hospital infection control committee. Wards considered in this study were those in which a possible transfer of patients frequently occurred during their hospital stay.

Strains were considered community-acquired when isolated from patients that were colonized at the time of their admission to the hospital. Otherwise, strains isolated from hospitalized patients were classified as hospital-acquired.

We collected demographic and clinical information about the enrolled patients from electronic medical records, including sex, age, geographical origin, the date of rectal swab collection, and department hospitalization (e.g., neonatal surgical unit, neonatal intensive therapy, sub-intensive neonatal unit, and cardiac intensive care unit). After isolation and identification, all CPE strains were characterized by antimicrobial susceptibility testing, detection of carbapenemase-encoding genes, and PBRT. Finally, one single strain of *K. pneumoniae* or *E. coli* representative for each PBRT profile was analyzed by MLST.

4.3. Microbiological Cultures and Antibiotic Susceptibility Testing

For cultural screening, each swab was inoculated on a set of two plates: MacConkey agar plate (bioMérieux, Craponne, France) with a 10 µg meropenem disk (Oxoid, Basingstoke, UK), according to EUCAST guidelines for the detection of resistance mechanisms (https://www.eucast.org/resistance_mechanisms/ (accessed on 21 February 2022)) and a CHROMID® CARBA plate (bioMérieux, Craponne, France), implemented in our routine laboratory practice.

Plates were incubated at 37 °C overnight. All the morphologically different colonies growing into the meropenem disk halo (zone diameter < 28 mm) and on the selective chromogenic medium were picked up and subcultured for purity onto a MacConkey agar plate (bioMérieux, Craponne, France) [54,55].

Isolated colonies were identified by using Matrix-Assisted Laser Desorption Ionization–Time-of-Flight Mass Spectrometry (MALDI-TOF, Bruker Daltonics, Bremen, Germany). Antimicrobial susceptibility testing was performed using the automated Vitek 2 (bioMérieux, Craponne, France) instrument. The following antimicrobial agents were tested with the automated system: amikacin (AMK), amoxicillin–clavulanic acid (AMC), cefepime (FEP), cefotaxime (CTX), ceftazidime (CAZ), ciprofloxacin (CIP), gentamicin (GEN), imipenem (IPM), meropenem (MEM), piperacillin–tazobactam (TZP), tigecycline (TGC), trimethoprim-sulfamethoxazole (SXT).

The MIC value of meropenem was confirmed with gradient test methods by MIC Test Strip (Liofilchem, Roseto degli Abruzzi, Italy) on Muller–Hinton agar (bioMérieux, Craponne, France), incubated at 37 °C overnight. The isolates were identified as resistant to carbapenems according to clinical breakpoints based on the European Committee on Antimicrobial Susceptibility Testing (EUCAST) breakpoints tables. We adopted the updated EUCAST breakpoints tables (version 6.0 to version 9.0, from 2016 to 2019) (https://www.eucast.org/clinical_breakpoints/ (accessed on 21 February 2022)).

4.4. PCR-Based Methods for Carbapenemase Genes

The Xpert Carba-R molecular assay (Cepheid, Sunnyvale, CA, USA) was performed to confirm the presence of resistance genes to carbapenems. Briefly, subcultured isolated colonies were diluted in 0.45% saline to the turbidity of a 0.5 McFarland standard. Ten microliters of the suspension were inoculated into a 5 mL sample reagent vial and vortexed for 30 s. Finally, 1.7 mL of this suspension was transferred into an Xpert Carba-R cartridge using a disposable transfer pipette. The cartridge was loaded onto the GeneXpert system, and the assay was performed. The test, based on automated real-time PCR, is designed for rapid detection and differentiation of the bla_{KPC}, bla_{NDM}, bla_{VIM}, bla_{OXA-48}, and bla_{IMP-1} gene sequences associated with carbapenem-non-susceptible Gram-negative bacteria. The results were interpreted by the GeneXpert System from measured fluorescent signals.

4.5. Bacterial DNA Extraction and Plasmid Typing

Total DNA of isolated colonies that were positive for carbapenemase genes was extracted using an EZ1 DNA Tissue Kit (Qiagen, Hilden, Germany). Briefly, isolated colonies were diluted in 0.45% saline to the turbidity of 0.5 McFarland standard, and 200 μL of the suspension was transferred into a 2 mL vial. The vial and prefilled reagent cartridges were loaded onto the EZ1 Advanced XL (Qiagen, Hilden, Germany) instrument, and the protocol for automated purification of bacterial DNA, using magnetic particle technology, was started. DNA was eluted in a final volume of 100 μL and stored at −20 °C until use.

One μL of the extracted DNA was used for plasmid typing using a PCR-based replicon typing (PBRT) kit 2.0 (Diatheva, Fano, Italy). This novel PBRT assay, consisting of eight multiplex PCRs, is able to detect 30 different replicons of the main plasmid families in Enterobacterales [56]. This kit was used following the manufacturer's instructions, including positive controls. Amplification products were resolved and visualized directly on a closed ready-to-use 2.2% agarose gel-cassette system (FlashGel-Lonza, Basel, Switzerland) using the 100 bp FlashGel DNA marker. The obtained fragments were compared to positive controls of each multiplex PCRs.

4.6. Multilocus Sequence Typing

Multilocus sequence typing (MLST) was performed to subtype *K. pneumoniae* and *E. coli* strains using housekeeping genes. For the former, the seven gene fragments (*gapA, infB, mdh, pgi, phoE, rpoB,* and *tonB*) were amplified by PCR and sequenced as described by protocol 2 of the Institute Pasteur Klebsiella MLST database (https://bigsdb.pasteur.fr/klebsiella/primers_used.html (accessed on 21 February 2022)). Instead, the seven housekeeping genes (*adk, fumC, gyrB, icd, mdh, purA,* and *recA*) to type *E. coli* strains were selected from the Enterobase MLST database (http://enterobase.warwick.ac.uk/species/index/ecoli (accessed on 28 February 2022). Primer sequences and reaction conditions were previously described by Wirth and colleagues (2006) [57].

All amplicons were sequenced using a BigDye Terminator v. 1.1 Cycle Sequencing kit on an ABI PRISM® 310 Genetic Analyzer (Thermo Fisher Scientific, Waltham, MA, USA). The alignment between sequences and the related reference was carried out using Unipro UGene version 38.0 software [58]. The allele numbers and the sequence types (STs) were determined through the corresponding MLST database. Finally, phylogenetic analyses of concatenated allelic variants were performed using Molecular Evolutionary Genetics Analysis (MEGA) software version 10.0. The evolutionary history was inferred using the Maximum Likelihood method and the General Time Reversible model. Initial trees for the heuristic search were obtained applying the Neighbor-Joining method to a matrix of pairwise distances estimated using the Maximum Composite Likelihood (MCL) approach. Branch quality was evaluated using a bootstrap test with 1000 replicates [59]. Graphical representation of both were obtained using Interactive Tree Of Lifes (iTOL v.6) software [60].

4.7. Whole-Genome Sequencing (WGS)

Whole-genome sequencing was performed in accordance with Lindsted et al. (2018) [61] for *E. coli*_104 strain, which revealed a new *recA* allele by MLST analysis. Briefly, genomic DNA concentration was determined using a Qubit dsDNA BR Assay Kit and a Qubit 2.0 Fluorometer (Thermo Fisher Scientific, Waltham, USA) in order to obtain a DNA input concentration between 100 and 500 ng. Library preparation was performed using an Illumina DNA Prep Library Kit (Illumina, Berlin, Germany) according to the manufacturer's instructions. The reference strain of *E. coli* ATCC 25922 was used as sequencing quality control. Sequencing was performed using a MiSeq Reagent Micro Kit on an Illumina MiSeq desktop platform (Illumina Inc., San Diego, CA, USA) for ~20 h to produce paired-end sequences (2 × 300 base pair).

Raw Illumina reads were paired and assessed for sufficient coverage ($\geq$Q30), and bases with low quality (<Q30) were discarded. Finally, the reads in FASTQ format were uploaded and submitted to the EnteroBase database (https://enterobase.warwick.ac.uk/species/index/ecoli (accessed on 21 February 2022)). High-throughput sequencing data were submitted to the Sequence Read Archive (SRA) (GeneBank accession number SUB10115582).

Futhermore, raw reads were analyzed using a bioinformatics tool of the Center for Genomic Epidemiology such as ResFinder 4.1 [62], PlasmidFinder 2.0 [63], pMLST-2.0 Server [63], and SerotypeFinder 2.0 Server [64].

5. Conclusions

Great attention should be given to high prevalence of CPE in NICUs and CICU due to the vulnerability of newborn patients. The molecular features of investigated strains showed their potential to be a serious threat to the health of neonates and infants. Our study contributed to increasing the knowledge concerning CPE circulation in Italian pediatric hospitals, and it highlighted the need to strengthen control measures to avoid their spread among wards. This is crucial to avoid complications for the treatment of neonatal infections. Moreover, introducing molecular typing methods in clinical routine procedures is becoming more and more necessary to monitor the spread of high-risk clones, to track their circulation, and to promptly identify emerging clones with advantageous ability to colonize clinical niches. We provided the evidence that PBRT can be commonly integrated within the diagnostic routine of a hospital in order to obtain important information about the epidemiology of circulating strains. Data concerning replicons harbored by MDROs in general, and CPE in our case study, represent an added value for a successful surveillance program. PBRT allows clinicians to prevent or monitor the spread of strains with potential ability to acquire and disseminate resistance genes through plasmids among wards. On the other hand, we used MLST because it enabled us to complete the description of the molecular profile of circulating clones, and the phylogenetic tree was also useful for comparing the novel identified ST with the well-known others. Moreover, typing housekeeping genes can be a very useful way to identify the dominance of a specific lineage responsible for an important clinical outbreak.

Author Contributions: Conceptualization, M.A. (Marilena Agosta), D.B., M.L.C.D.A., C.F.P., F.A., and P.B. (Paola Bernaschi); methodology, M.A. (Marilena Agosta), D.B., M.A. (Marta Argentieri), L.P. (Laura Pansani), L.M., A.S., C.D., L.P. (Lorenza Putignani), P.B. (Pietro Bagolan), B.D.I., L.D.C., A.D., F.A., and P.B. (Paola Bernaschi); formal analysis, M.A. (Marilena Agosta), D.B., M.A. (Marta Argentieri), L.P. (Laura Pansani), A.S., C.F.P., F.A., and P.B. (Paola Bernaschi); investigation, M.A. (Marilena Agosta), D.B., M.A. (Marta Argentieri), L.P. (Laura Pansani), A.S., M.L.C.D.A., C.D., L.P. (Lorenza Putignani), L.M., L.D.C., P.B. (Pietro Bagolan), B.D.I., A.D., M.M., C.F.P., F.A., and P.B. (Paola Bernaschi); writing—original draft preparation, M.A. (Marilena Agosta), D.B., M.L.C.D.A., F.A., and P.B. (Paola Bernaschi); writing—review and editing, M.A. (Marilena Agosta), D.B., M.L.C.D.A., P.B. (Pietro Bagolan), B.D.I., A.D., L.M., L.D.C., M.M., C.F.P., F.A., and P.B. (Paola Bernaschi); supervision, M.M., C.F.P., F.A., and P.B. (Paola Bernaschi); funding acquisition, M.M. and C.F.P. All authors have read and agreed to the published version of the manuscript.

Funding: This research was funded by Fano Ateneo (University of Urbino, Italy) and the Department of Diagnostic and Laboratory Medicine, Microbiology and Diagnostic Immunology Unit, Bambino Gesù Children's Hospital (IRCCS, Rome, Italy).

Institutional Review Board Statement: The study was conducted in accordance with the Declaration of Helsinki and approved by the Institutional Review Board of the IRCCS Bambino Gesù Children's Hospital, study protocol (n.2156_OPBG_2020).

Informed Consent Statement: As the data in this study were collected and analyzed retrospectively, the study did not infringe upon the rights or welfare of the patients and did not require their consent.

Data Availability Statement: All data are described within the text. The reads in FASTQ format of ST10555 were uploaded and submitted to the EnteroBase database (https://enterobase.warwick.ac.uk/species/index/ecoli (accessed on 21 February 2022)) and deposited in the Sequence Read Archive (SRA) (GeneBank accession number SUB10115582).

Conflicts of Interest: The authors declare no conflict of interests.

References

1. Elshamy, A.A.; Aboshanab, K.M. A Review on Bacterial Resistance to Carbapenems: Epidemiology, Detection and Treatment Options. *Futur. Sci. OA* **2020**, *6*, FSO438. [CrossRef] [PubMed]
2. Tacconelli, E.; Carrara, E.; Savoldi, A.; Harbarth, S.; Mendelson, M.; Monnet, D.L.; Pulcini, C.; Kahlmeter, G.; Kluytmans, J.; Carmeli, Y.; et al. Discovery, Research, and Development of New Antibiotics: The WHO Priority List of Antibiotic-Resistant Bacteria and Tuberculosis. *Lancet Infect. Dis.* **2018**, *18*, 318–327. [CrossRef]
3. Sherry, N.L.; Lane, C.R.; Kwong, J.C.; Schultz, M.; Sait, M.; Stevens, K.; Ballard, S.; Da Silva, A.G.; Seemann, T.; Gorrie, C.L.; et al. Genomics for Molecular Epidemiology and Detecting Transmission of Carbapenemase-Producing Enterobacterales in Victoria, Australia, 2012 to 2016. *J. Clin. Microbiol.* **2019**, *57*, e00573-19. [CrossRef] [PubMed]
4. Logan, L.K.; Weinstein, R.A. The Epidemiology of Carbapenem-Resistant Enterobacteriaceae: The Impact and Evolution of a Global Menace. *J. Infect. Dis.* **2017**, *215*, S28–S36. [CrossRef] [PubMed]
5. Perez, F.; El Chakhtoura, N.G.; Papp-Wallace, K.M.; Wilson, B.M.; Bonomo, R.A. Treatment Options for Infections Caused by Carbapenem-Resistant Enterobacteriaceae. *Expert Opin. Pharmacother.* **2016**, *17*, 761–781. [CrossRef] [PubMed]
6. Räisänen, K.; Lyytikäinen, O.; Kauranen, J.; Tarkka, E.; Forsblom-Helander, B.; Grönroos, J.O.; Vuento, R.; Arifulla, D.; Sarvikivi, E.; Toura, S.; et al. Molecular Epidemiology of Carbapenemase-Producing Enterobacterales in Finland, 2012–2018. *Eur. J. Clin. Microbiol. Infect. Dis.* **2020**, *39*, 1651–1656. [CrossRef] [PubMed]

7.   Patel, G.; Bonomo, R.A. "Stormy Waters Ahead": Global Emergence of Carbapenemases. *Front. Microbiol.* **2013**, *4*, 48. [CrossRef] [PubMed]

8.   Munoz-Price, L.S.; Poirel, L.; Bonomo, R.A.; Schwaber, M.J.; Daikos, G.L.; Cormican, M.; Cornaglia, G.; Garau, J.; Gniadkowski, M.; Hayden, M.K.; et al. Clinical Epidemiology of the Global Expansion of Klebsiella Pneumoniae Carbapenemases. *Lancet Infect. Dis.* **2013**, *13*, 785–796. [CrossRef]

9.   Dortet, L.; Poirel, L.; Nordmann, P. Worldwide Dissemination of the NDM-Type Carbapenemases in Gram-Negative Bacteria. *Biomed Res. Int.* **2014**, *2014*, 1–12. [CrossRef]

10.   Carattoli, A. Plasmids in Gram Negatives: Molecular Typing of Resistance Plasmids. *Int. J. Med. Microbiol.* **2011**, *301*, 654–658. [CrossRef]

11.   Labi, A.K.; Bjerrum, S.; Enweronu-Laryea, C.C.; Ayibor, P.K.; Nielsen, K.L.; Marvig, R.L.; Newman, M.J.; Andersen, L.P.; Kurtzhals, J.A.L. High Carriage Rates of Multidrug-Resistant Gram- Negative Bacteria in Neonatal Intensive Care Units from Ghana. *Open Forum Infect. Dis.* **2020**, *7*, ofaa109. [CrossRef]

12.   Maida, C.M.; Bonura, C.; Geraci, D.M.; Graziano, G.; Carattoli, A.; Rizzo, A.; Torregrossa, M.V.; Vecchio, D.; Giuffrè, M. Outbreak of ST395 KPC-Producing Klebsiella Pneumoniae in a Neonatal Intensive Care Unit in Palermo, Italy. *Infect. Control Hosp. Epidemiol.* **2018**, *39*, 496–498. [CrossRef]

13.   Castagnola, E.; Tatarelli, P.; Mesini, A.; Baldelli, I.; La Masa, D.; Biassoni, R.; Bandettini, R. Epidemiology of Carbapenemase-Producing Enterobacteriaceae in a Pediatric Hospital in a Country with High Endemicity. *J. Infect. Public Health* **2019**, *12*, 270–274. [CrossRef]

14.   Ciccolini, M.; Donker, T.; Grundmann, H.; Bonten, M.J.M.; Woolhouse, M.E.J. Efficient Surveillance for Healthcare-Associated Infections Spreading between Hospitals. *Proc. Natl. Acad. Sci. USA* **2014**, *111*, 2271–2276. [CrossRef]

15.   Donker, T.; Wallinga, J.; Slack, R.; Grundmann, H. Hospital Networks and the Dispersal of Hospital-Acquired Pathogens by Patient Transfer. *PLoS ONE* **2012**, *7*, e35002. [CrossRef]

16.   Smith, D.L.; Dushoff, J.; Perencevich, E.N.; Harris, A.D.; Levin, S.A. Persistent Colonization and the Spread of Antibiotic Resistance in Nosocomial Pathogens: Resistance Is a Regional Problem. *Proc. Natl. Acad. Sci. USA* **2004**, *101*, 3709–3714. [CrossRef]

17.   Aguilera-Alonso, D.; Escosa-García, L.; Saavedra-Lozano, J.; Cercenado, E.; Baquero-Artigao, F. Carbapenem-Resistant Gram-Negative Bacterial Infections in Children. *Antimicrob. Agents Chemother.* **2020**, *64*, 1–19. [CrossRef]

18.   Tfifha, M.; Ferjani, A.; Mallouli, M.; Mlika, N.; Abroug, S.; Boukadida, J. Carriage of Multidrug-Resistant Bacteria among Pediatric Patients before and during Their Hospitalization in a Tertiary Pediatric Unit in Tunisia. *Libyan J. Med.* **2018**, *13*, 1419047. [CrossRef]

19.   Thuy, D.B.; Campbell, J.; Hoang Nhat, L.T.; VanMinh Hoang, N.; Van Hao, N.; Baker, S.; Geskus, R.B.; Thwaites, G.E.; Van Vinh Chau, N.; Louise Thwaites, C. Hospital-Acquired Colonization and Infections in a Vietnamese Intensive Care Unit. *PLoS ONE* **2018**, *13*, e0203600. [CrossRef]

20.   ECDC. *Surveillance of Antimicrobial Resistance in Europe 2018*; ECDC: Solna, Sweden, 2018.

21.   Public Health England Health Protection Report: Latest Surveillance Reports. Available online: https://www.gov.uk/government/collections/health-protection-report-latest-infection-reports (accessed on 2 July 2021).

22.   CDC Carbapenem-Resistant Enterobacterales (CRE). Available online: https://www.cdc.gov/hai/organisms/cre/index.html (accessed on 2 July 2021).

23.   Yin, L.; He, L.; Miao, J.; Yang, W.; Wang, X.; Ma, J.; Wu, N.; Cao, Y.; Wang, L.; Lu, G.; et al. Active Surveillance and Appropriate Patient Placement in Contact Isolation Dramatically Decreased Carbapenem-Resistant Enterobacterales Infection and Colonization in Paediatric Patients in China. *J. Hosp. Infect.* **2020**, *105*, 486–494. [CrossRef]

24.   Simon, A.; Tenenbaum, T. Surveillance of Multidrug-Resistant Gram-Negative Pathogens in High-Risk Neonates-Does It Make a Difference? *Pediatr. Infect. Dis. J.* **2013**, *32*, 407–409. [CrossRef] [PubMed]

25.   Gagliotti, C.; Ciccarese, V.; Sarti, M.; Giordani, S.; Barozzi, A.; Braglia, C.; Gallerani, C.; Gargiulo, R.; Lenzotti, G.; Manzi, O.; et al. Active Surveillance for Asymptomatic Carriers of Carbapenemase-Producing Klebsiella Pneumoniae in a Hospital Setting. *J. Hosp. Infect.* **2013**, *83*, 330–332. [CrossRef] [PubMed]

26.   Meredith, H.R.; Kularatna, S.; Nagaro, K.; Nagahawatte, A.; Bodinayake, C.; Kurukulasooriya, R.; Wijesingha, N.; Harden, L.B.; Piyasiri, B.; Hammouda, A.; et al. Colonization with Multidrug-Resistant Enterobacteriaceae among Infants: An Observational Study in Southern Sri Lanka. *Antimicrob. Resist. Infect. Control* **2021**, *10*, 72. [CrossRef] [PubMed]

27.   Montagnani, C.; Prato, M.; Scolfaro, C.; Colombo, S.; Esposito, S.; Tagliabue, C.; Lo Vecchio, A.; Bruzzese, E.; Loy, A.; Cursi, L.; et al. Carbapenem-Resistant Enterobacteriaceae Infections in Children: An Italian Retrospective Multicenter Study. *Pediatr. Infect. Dis. J.* **2016**, *35*, 862–868. [CrossRef]

28.   Giuffrè, M.; Bonura, C.; Geraci, D.M.; Saporito, L.; Catalano, R.; Di Noto, S.; Nociforo, F.; Corsello, G.; Mammina, C. Successful Control of an Outbreak of Colonization by Klebsiella Pneumoniae Carbapenemase-Producing K. Pneumoniae Sequence Type 258 in a Neonatal Intensive Care Unit, Italy. *J. Hosp. Infect.* **2013**, *85*, 233–236. [CrossRef]

29.   Geraci, D.M.; Bonura, C.; Giuffrè, M.; Saporito, L.; Graziano, G.; Aleo, A.; Fasciana, T.; Di Bernardo, F.; Stampone, T.; Palma, D.M.; et al. Is the Monoclonal Spread of the ST258, KPC-3-Producing Clone Being Replaced in Southern Italy by the Dissemination of Multiple Clones of Carbapenem-Nonsusceptible, KPC-3-Producing Klebsiella Pneumoniae? *Clin. Microbiol. Infect.* **2015**, *21*, e15–e17. [CrossRef]

30. Mammina, C.; Di Carlo, P.; Cipolla, D.; Giuffrè, M.; Casuccio, A.; Di Gaetano, V.; Plano, M.R.A.; D'Angelo, E.; Titone, L.; Corsello, G. Surveillance of Multidrug-Resistant Gram-Negative Bacilli in a Neonatal Intensive Care Unit: Prominent Role of Cross Transmission. *Am. J. Infect. Control* **2007**, *35*, 222–230. [CrossRef]

31. Singh, N.P.; Choudhury, D.D.; Gupta, K.; Rai, S.; Batra, P.; Manchanda, V.; Saha, R.; Kaur, I.R. Predictors for Gut Colonization of Carbapenem-Resistant Enterobacteriaceae in Neonates in a Neonatal Intensive Care Unit. *Am. J. Infect. Control* **2018**, *46*, e31–e35. [CrossRef]

32. Han, R.; Shi, Q.; Wu, S.; Yin, D.; Peng, M.; Dong, D.; Zheng, Y.; Guo, Y.; Zhang, R.; Hu, F.; et al. Dissemination of Carbapenemases (KPC, NDM, OXA-48, IMP, and VIM) Among Carbapenem-Resistant Enterobacteriaceae Isolated From Adult and Children Patients in China. *Front. Cell. Infect. Microbiol.* **2020**, *10*, 314. [CrossRef]

33. Almeida, T.L.; Mendo, T.; Costa, R.; Novais, C.; Marçal, M.; Martins, F.; Tuna, M. Carbapenemase-Producing Enterobacteriaceae (CPE) Newborn Colonization in a Portuguese Neonatal Intensive Care Unit (NICU): Epidemiology and Infection Prevention and Control Measures. *Infect. Dis. Rep.* **2021**, *13*, 411–417. [CrossRef]

34. Magiorakos, A.P.; Srinivasan, A.; Carey, R.B.; Carmeli, Y.; Falagas, M.E.; Giske, C.G.; Harbarth, S.; Hindler, J.F.; Kahlmeter, G.; Olsson-Liljequist, B.; et al. Multidrug-Resistant, Extensively Drug-Resistant and Pandrug-Resistant Bacteria: An International Expert Proposal for Interim Standard Definitions for Acquired Resistance. *Clin. Microbiol. Infect.* **2012**, *18*, 268–281. [CrossRef]

35. Sader, H.S.; Castanheira, M.; Farrell, D.J.; Flamm, R.K.; Mendes, R.E.; Jones, R.N. Tigecycline Antimicrobial Activity Tested against Clinical Bacteria from Latin American Medical Centres: Results from SENTRY Antimicrobial Surveillance Program (2011–2014). *Int. J. Antimicrob. Agents* **2016**, *48*, 144–150. [CrossRef]

36. Zhang, R.; Dong, N.; Huang, Y.; Zhou, H.; Xie, M.; Chan, E.W.C.; Hu, Y.; Cai, J.; Chen, S. Evolution of Tigecycline- and Colistin-Resistant CRKP (Carbapenem-Resistant Klebsiella Pneumoniae) in Vivo and Its Persistence in the GI Tract Article. *Emerg. Microbes Infect.* **2018**, *7*, 1–11. [CrossRef]

37. Ojdana, D.; Gutowska, A.; Sacha, P.; Majewski, P.; Wieczorek, P.; Tryniszewska, E. Activity of Ceftazidime-Avibactam Alone and in Combination with Ertapenem, Fosfomycin, and Tigecycline against Carbapenemase-Producing Klebsiella Pneumoniae. *Microb. Drug Resist.* **2019**, *25*, 1357–1364. [CrossRef]

38. Folgori, L.; Bielicki, J. Future Challenges in Pediatric and Neonatal Sepsis: Emerging Pathogens and Antimicrobial Resistance. *J. Pediatr. Intensiv. Care* **2019**, *08*, 017–024. [CrossRef]

39. Nordmann, P.; Naas, T.; Poirel, L. Global Spread of Carbapenemase Producing Enterobacteriaceae. *Emerg. Infect. Dis.* **2011**, *17*, 1791–1798. [CrossRef]

40. ECDC. *European Centre for Disease Prevention and Control Regional Outbreak of New Delhi Metallo-Beta-Lactamase-Producing Carbapenem-Resistant Enterobacteriaceae, Italy, 2018–2019*; ECDC: Solna, Sweden, 2019; pp. 1–7.

41. Wu, W.; Feng, Y.; Tang, G.; Qiao, F.; McNally, A.; Zong, Z. NDM Metallo-β-Lactamases and Their Bacterial Producers in Health Care Settings. *Clin. Microbiol. Rev.* **2019**, *32*, e00115-18. [CrossRef]

42. Carattoli, A.; Villa, L.; Poirel, L.; Bonnin, R.A.; Nordmann, P. Evolution of IncA/C Bla CMY-2-Carrying Plasmids by Acquisition of the BlaNDM-1 Carbapenemase Gene. *Antimicrob. Agents Chemother.* **2012**, *56*, 783–786. [CrossRef]

43. Hancock, S.J.; Phan, M.D.; Peters, K.M.; Forde, B.M.; Chong, T.M.; Yin, W.F.; Chan, K.G.; Paterson, D.L.; Walsh, T.R.; Beatson, S.A.; et al. Identification of IncA/C Plasmid Replication and Maintenance Genes and Development of a Plasmid Multilocus Sequence Typing Scheme. *Antimicrob. Agents Chemother.* **2017**, *61*, e01740-16. [CrossRef]

44. Welch, T.J.; Fricke, W.F.; McDermott, P.F.; White, D.G.; Rosso, M.L.; Rasko, D.A.; Mammel, M.K.; Eppinger, M.; Rosovitz, M.J.; Wagner, D.; et al. Multiple Antimicrobial Resistance in Plague: An Emerging Public Health Risk. *PLoS ONE* **2007**, *2*, e309. [CrossRef]

45. Winokur, P.L.; Vonstein, D.L.; Hoffman, L.J.; Uhlenhopp, E.K.; Doern, G.V. Evidence for Transfer of CMY-2 AmpC β-Lactamase Plasmids between Escherichia Coli and Salmonella Isolates from Food Animals and Humans. *Antimicrob. Agents Chemother.* **2001**, *45*, 2716–2722. [CrossRef]

46. Villa, L.; García-Fernández, A.; Fortini, D.; Carattoli, A. Replicon Sequence Typing of IncF Plasmids Carrying Virulence and Resistance Determinants. *J. Antimicrob. Chemother.* **2010**, *65*, 2518–2529. [CrossRef] [PubMed]

47. Barbadoro, P.; Bencardino, D.; Carloni, E.; Omiccioli, E.; Ponzio, E.; Micheletti, R.; Acquaviva, G.; Luciani, A.; Masucci, A.; Pocognoli, A.; et al. Carriage of Carbapenem-Resistant Enterobacterales in Adult Patients Admitted to a University Hospital in Italy. *Antibiotics* **2021**, *10*, 61. [CrossRef] [PubMed]

48. Magi, G.; Tontarelli, F.; Caucci, S.; Di Sante, L.; Brenciani, A.; Morroni, G.; Giovanetti, E.; Menzo, S.; Mingoia, M. High Prevalence of Carbapenem-Resistant Klebsiella Pneumoniae ST307 Recovered from Fecal Samples in an Italian Hospital. *Fut. Microbiol.* **2021**, *16*, 703–711. [CrossRef]

49. Villa, L.; Feudi, C.; Fortini, D.; Brisse, S.; Passet, V.; Bonura, C.; Endimiani, A.; Mammina, C.; Ocampo, A.M.; Jimenez, J.N.; et al. Diversity, Virulence, and Antimicrobial Resistance of the KPCproducing Klebsiella Pneumoniae ST307 Clone. *Microb. Genom.* **2017**, *3*, e000110. [CrossRef] [PubMed]

50. Loconsole, D.; Accogli, M.; De Robertis, A.L.; Capozzi, L.; Bianco, A.; Morea, A.; Mallamaci, R.; Quarto, M.; Parisi, A.; Chironna, M. Emerging High-Risk ST101 and ST307 Carbapenem-Resistant Klebsiella Pneumoniae Clones from Bloodstream Infections in Southern Italy. *Ann. Clin. Microbiol. Antimicrob.* **2020**, *19*, 24. [CrossRef] [PubMed]

51. Mammina, C.; Palma, D.M.; Bonura, C.; Plano, M.R.A.; Monastero, R.; Sodano, C.; Calà, C.; Tetamo, R. Outbreak of Infection with Klebsiella Pneumoniae Sequence Type 258 Producing Klebsiella Pneumoniae Carbapenemase 3 in an Intensive Care Unit in Italy. *J. Clin. Microbiol.* **2010**, *48*, 1506–1507. [CrossRef]

52. Bonura, C.; Giuffrè, M.; Aleo, A.; Fasciana, T.; Di Bernardo, F.; Stampone, T.; Giammanco, A.; Palma, D.M.; Mammina, C.; Geraci, D.M.; et al. An Update of the Evolving Epidemic of BlaKPC Carrying Klebsiella Pneumoniae in Sicily, Italy, 2014: Emergence of Multiple Non-ST258 Clones. *PLoS ONE* **2015**, *10*, e0132936. [CrossRef]

53. Forde, B.M.; Roberts, L.W.; Phan, M.D.; Peters, K.M.; Fleming, B.A.; Russell, C.W.; Lenherr, S.M.; Myers, J.B.; Barker, A.P.; Fisher, M.A.; et al. Population Dynamics of an Escherichia Coli ST131 Lineage during Recurrent Urinary Tract Infection. *Nat. Commun.* **2019**, *10*, 3643. [CrossRef]

54. Arena, F.; Di Pilato, V.; Vannetti, F.; Fabbri, L.; Antonelli, A.; Coppi, M.; Pupillo, R.; Macchi, C.; Rossolini, G.M. Population Structure of Kpc Carbapenemase-Producing Klebsiella Pneumoniae in a Long-Term Acute-Care Rehabilitation Facility: Identification of a New Lineage of Clonal Group 101, Associated with Local Hyperendemicity. *Microb. Genomics* **2020**, *6*, e000308. [CrossRef]

55. Cortegiani, A.; Russotto, V.; Graziano, G.; Geraci, D.; Saporito, L.; Cocorullo, G.; Raineri, S.M.; Mammina, C.; Giarratano, A. Use of Cepheid Xpert Carba-R® for Rapid Detection of Carbapenemase-Producing Bacteria in Abdominal Septic Patients Admitted to Intensive Care Unit. *PLoS ONE* **2016**, *11*, e0160643. [CrossRef]

56. Carloni, E.; Andreoni, F.; Omiccioli, E.; Villa, L.; Magnani, M.; Carattoli, A. Comparative Analysis of the Standard PCR-Based Replicon Typing (PBRT) with the Commercial PBRT-KIT. *Plasmid* **2017**, *90*, 10–14. [CrossRef]

57. Wirth, T.; Falush, D.; Lan, R.; Colles, F.; Mensa, P.; Wieler, L.H.; Karch, H.; Reeves, P.R.; Maiden, M.C.J.; Ochman, H.; et al. Sex and Virulence in Escherichia Coli: An Evolutionary Perspective. *Mol. Microbiol.* **2006**, *60*, 1136–1151. [CrossRef]

58. Okonechnikov, K.; Golosova, O.; Fursov, M.; Varlamov, A.; Vaskin, Y.; Efremov, I.; German Grehov, O.G.; Kandrov, D.; Rasputin, K.; Syabro, M.; et al. Unipro UGENE: A Unified Bioinformatics Toolkit. *Bioinformatics* **2012**, *28*, 1166–1167. [CrossRef]

59. Kumar, S.; Stecher, G.; Li, M.; Knyaz, C.; Tamura, K. MEGA X: Molecular Evolutionary Genetics Analysis across Computing Platforms. *Mol. Biol. Evol.* **2018**, *35*, 1547–1549. [CrossRef]

60. Letunic, I.; Bork, P. Interactive Tree of Life (ITOL) v5: An Online Tool for Phylogenetic Tree Display and Annotation. *Nucleic Acids Res.* **2021**, *49*, W293–W296. [CrossRef]

61. Lindstedt, B.-A.; Finton, M.D.; Porcellato, D.; Brandal, L.T. High Frequency of Hybrid Escherichia Coli Strains with Combined Intestinal Pathogenic Escherichia Coli (IPEC) and Extraintestinal Pathogenic Escherichia Coli (ExPEC) Virulence Factors Isolated from Human Faecal Samples. *BMC Infect. Dis.* **2018**, *18*, 10–14. [CrossRef]

62. Bortolaia, V.; Kaas, R.S.; Ruppe, E.; Roberts, M.C.; Schwarz, S.; Cattoir, V.; Philippon, A.; Allesoe, R.L.; Rebelo, A.R.; Florensa, A.F.; et al. ResFinder 4.0 for Predictions of Phenotypes from Genotypes. *J. Antimicrob. Chemother.* **2020**, *75*, 3491–3500. [CrossRef]

63. Carattoli, A.; Zankari, E.; García-Fernández, A.; Larsen, M.V.; Lund, O.; Villa, L.; Aarestrup, F.M.; Hasman, H. In Silico Detection and Typing of Plasmids Using PlasmidFinder and Plasmid Multilocus Sequence Typing. *Antimicrob. Agents Chemother.* **2014**, *58*, 3895–3903. [CrossRef]

64. Joensen, K.G.; Tetzschner, A.M.M.; Iguchi, A.; Aarestrup, F.M.; Scheutz, F. Rapid and Easy in Silico Serotyping of Escherichia Coli Isolates by Use of Whole-Genome Sequencing Data. *J. Clin. Microbiol.* **2015**, *53*, 2410–2426. [CrossRef]

antibiotics

Article

Polyclonal Endemicity of Carbapenemase-Producing *Klebsiella pneumoniae* in ICUs of a Greek Tertiary Care Hospital

Efthymia Protonotariou [1], Georgios Meletis [1,*], Dimitrios Pilalas [2], Paraskevi Mantzana [1], Areti Tychala [1], Charalampos Kotzamanidis [3], Dimitra Papadopoulou [1], Theofilos Papadopoulos [4], Michalis Polemis [5], Simeon Metallidis [6] and Lemonia Skoura [1]

[1] Department of Microbiology, AHEPA University Hospital, School of Medicine, Aristotle University of Thessaloniki, S. Kiriakidi str. 1, 54636 Thessaloniki, Greece; protonotariou@auth.gr (E.P.); vimantzana@gmail.com (P.M.); aretich@gmail.com (A.T.); dimitrapapadopoulou8566@gmail.com (D.P.); mollyskoura@gmail.com (L.S.)
[2] First Propedeutic Department of Internal Medicine, AHEPA University Hospital, School of Medicine, Aristotle University of Thessaloniki, 54636 Thessaloniki, Greece; pilalas_jim@hotmail.com
[3] Hellenic Agricultural Organisation-DIMITRA, Veterinary Research Institute of Thessaloniki, Campus of Thermi, 57001 Thermi, Greece; kotzam@vri.gr
[4] Department of Microbiology and Infectious Diseases, Faculty of Veterinary Medicine, School of Health Sciences, Aristotle University of Thessaloniki, 54636 Thessaloniki, Greece; theofilos23@vet.auth.gr
[5] Central Public Health Laboratory, National Public Health Organization, 16672 Vari, Greece; m.polemis@eody.gov.gr
[6] First Department of Internal Medicine, Infectious Diseases Division, AHEPA University Hospital, School of Medicine, Aristotle University of Thessaloniki, 54636 Thessaloniki, Greece; symeonam@auth.gr
* Correspondence: meletisg@hotmail.com

Citation: Protonotariou, E.; Meletis, G.; Pilalas, D.; Mantzana, P.; Tychala, A.; Kotzamanidis, C.; Papadopoulou, D.; Papadopoulos, T.; Polemis, M.; Metallidis, S.; et al. Polyclonal Endemicity of Carbapenemase-Producing *Klebsiella pneumoniae* in ICUs of a Greek Tertiary Care Hospital. *Antibiotics* **2022**, 11, 149. https://doi.org/10.3390/antibiotics11020149

Academic Editor: Francesca Andreoni

Received: 29 December 2021
Accepted: 21 January 2022
Published: 25 January 2022

Publisher's Note: MDPI stays neutral with regard to jurisdictional claims in published maps and institutional affiliations.

Abstract: Carbapenemase-producing *Klebsiella pneumoniae* (CPKP) emerged in Greece in 2002 and became endemic thereafter. Driven by a notable variability in the phenotypic testing results for carbapenemase production in *K. pneumoniae* isolates from the intensive care units (ICUs) of our hospital, we performed a study to assess the molecular epidemiology of CPKP isolated between 2016 and 2019 using pulse-field gel electrophoresis (PFGE) including isolates recovered from 165 single patients. We investigated the molecular relatedness among strains recovered from rectal surveillance cultures and from respective subsequent infections due to CPKP in the same individual (48/165 cases). For the optimal interpretation of our findings, we carried out a systematic review regarding the clonality of CPKP isolated from clinical samples in ICUs in Europe. In our study, we identified 128 distinguishable pulsotypes and 17 clusters that indicated extended dissemination of CPKP within the hospital ICU setting throughout the study period. Among the clinical isolates, 122 harbored KPC genes (74%), 2 harbored KPC+NDM (1.2%), 38 harbored NDM (23%), 1 harbored NDM+OXA-48 (0.6%), 1 harbored NDM+VIM (0.6%) and 1 harbored the VIM (0.6%) gene. Multiple CPKP strains in our hospital have achieved sustained transmission. The polyclonal endemicity of CPKP presents a further threat for the selection of pathogens resistant to last-resort antimicrobial agents.

Keywords: *Klebsiella pneumoniae*; carbapenemases; NDM; KPC; VIM; OXA-48; molecular epidemiology; PFGE

1. Introduction

In recent years, hospital-acquired infections caused by carbapenem-resistant Gram negative bacteria, especially carbapenem-resistant *Klebsiella pneumoniae* (CRKP), have been observed worldwide causing important public health problems and posing serious infection control issues. CRKP are opportunistic pathogens that cause infections with high morbidity and mortality mainly in hospitalized patients [1,2]. In Europe, the burden of CRKP predominantly affects the south and the east. According to the annual report of the European Centre for Disease Prevention and Control on antimicrobial resistance in Europe,

66.3% of the reported invasive *K. pneumoniae* isolates in Greece during 2020 were resistant to carbapenems.

Relatively high carbapenem resistance rates were also observed in Romania (48%), Italy (29%) and Bulgaria (28%), while, in the majority of the EU countries, this proportion was below 10% [3]. High carbapenem resistance trends were also observed in other non-EU neighboring countries: Bosnia, Herzegovina, Georgia, the Russian Federation, Serbia and Turkey reported proportions between 25% and 50% whereas, Belarus, the Republic of Moldova and the Ukraine reported proportions exceeding 50% [4].

The most common mechanism of carbapenem resistance among Enterobacterales and, thus, *K. pneumoniae* is the production of carbapenemases. Carbapenemases are β-lactamases able to hydrolyze all β-lactams, including carbapenems, and are categorized into several types. The carbapenemases most commonly encountered in Greece belong in three classes: class A *K. pneumoniae* carbapenemase (KPC), class B Verona imipenemase (VIM) and New Delhi metallo-β-lactamase (NDM) and class D oxacillinase-48 (OXA-48) [5].

Carbapenemase-producing *K. pneumoniae* (CPKP) emerged in Greece in 2002; they were of VIM-type, were involved in various outbreaks and soon became endemic in many hospitals all over the country. VIM-type carbapenemase-producers often belonged to different clones with ST147 being the predominant multi-locus sequence type (MLST) [6]. In 2007, KPC-producing *K. pneumoniae* isolates were introduced in Greek hospitals and rapidly dominated [7].

Greek KPC-CPKP mostly belonged to the worldwide successful hyperepidemic clone ST258, often associated with multi-drug resistant (MDR) phenotype [8]. The emergence of NDM in CPKP strains in Greece took place in 2011; the majority of them belonged to ST11 and were involved in oligoclonal outbreaks or sporadic cases [9]. OXA-48 type carbapenemases are the most prevalent class D enzymes identified in CPKP strains. The first OXA-48 was detected in Athens, Greece in 2012 and belonged to ST11 [10].

Carbapenemase-encoding genes spread fast via horizontal gene transfer together with other resistance determinants within the *K. pneumoniae* species in hospital settings, thus, dramatically restricting the available treatment options [11]. Moreover, the local epidemiology and the limited availability for isolation of affected patients in separate rooms in Greek hospitals undermine the efforts for effective infection control strategies.

K. pneumoniae is characterized by a high variety of antimicrobial resistance genes as well as a wide ecological distribution. Thus, in addition to its significance as a nosocomial pathogen (especially the hypervirulent phenotype), *K. pneumoniae* is considered as one of the most important bacterial species contributing in the dissemination of antimicrobial resistance genes to other human pathogens [12].

K. pneumoniae has the ability to colonize various mucosal surfaces, including the upper respiratory and the gastrointestinal gut. Among hospitalized patients, colonization rates in the nasopharynx are up to 19%, while it can reach as high as 77% in the gastrointestinal tract. Gut colonization often precedes and serves as a reservoir for transmission to other body sites resulting in the development of subsequent infections [13]. The duration of gut colonization with multi-drug resistant (MDR) bacteria, such as carbapenem-resistant *K. pneumoniae* varies from 43 to 387 days [14].

Driven by a notable variability in the phenotypic testing results for carbapenemase production and the types of carbapenemases present in *K. pneumoniae* isolates in our hospital, we performed a study to assess the molecular epidemiology of CPKP isolated between 2016 and 2019. Additionally, we investigated the molecular relatedness among strains recovered by rectal surveillance cultures and by respective subsequent infections due to CPKP in the same individual. In order to put our findings in context, we also performed a systematic review regarding the clonality of CPKP isolated from clinical samples originating in intensive care units (ICUs) in Europe.

2. Results

2.1. Carbapenemase Detection

During the study period (January 2016–June 2019) 165 single-patient clinical CRKP were analyzed; the isolates were recovered from 115 patients hospitalized in ICU 1, 6 patients in ICU 2 and 44 patients in ICU 3. Clinical samples included blood (n = 36), central venous catheters (n = 27), bronchial secretions (n = 51), urine (n = 20), pus (n = 13), wound swabs (n = 12), pleural fluid (n = 1), peritoneal fluid (n = 2), nasal swab (n = 1) and cerebrospinal fluid (n = 2).

Phenotypic and molecular testing revealed that all CRKP isolates harbored at least one carbapenemase often combined with ESBL activity. Among the clinical isolates, 122 harbored the KPC gene (73.95%), 2 KPC + NDM (1.21%), 38 NDM (23.04%), 1 NDM + OXA-48 (0.60%), 1 NDM + VIM (0.60%) and 1 VIM (0.60%).

2.2. Pulse-Field Gel Electrophoresis

All the 165 clinical CRKP isolates were typeable by pulse-field gel electrophoresis (PFGE) following digestion by restriction enzyme XbaI, revealing 128 distinguishable pulsotypes (P1-P128; Figure 1).

At a similarity level of 80% or above, the majority of CPKP isolates (95.6%, 158/165) were assigned into 17 clusters (A-Q), demonstrating multiclonal dissemination. The remaining seven genomes resulted to be unrelated and were consequently classified as sporadic isolates. KPC as well as NDM genetic determinants demonstrated polyclonal dissemination being present in 14 and 11 distinct clones, respectively.

In more detail, four predominant clusters E, G, K, and M consisting of 31, 19, 16 and 41 CRKP isolates, respectively, were identified: isolates of cluster G were almost exclusively obtained from ICU 1 (18 of 19), while clusters E, K, M consisted of clinical isolates from all ICUs under study. Interestingly, within the above clusters, indistinguishable pulsotypes shared by isolates from different patients and different ICUs were identified. Furthermore, looking at indistinguishable pulsotypes, we could identify common pulsotypes among isolates obtained from different patients during different time periods of the study (P5, P27, P48 and P106).

We also used PFGE analysis for revealing the genetic association among CRKP strains from rectal and clinical samples of 48 representative patients. According to PFGE, in the majority of the cases (81.3%, 39 of 48), the clinical and rectal strains from the same patients were identical (Figure 2). Different pulsotypes were observed for pairs PAT_1422a/b, PAT_1386a/b, PAT_854a/b, PAT_476a/b, PAT_326a/b, PAT_1529a/b, PAT_735a/b, PAT_191a/b and PAT_1216a/b.

Six clinical-rectal surveillance pairs presented different PCR results (PAT_1017a/b, PAT_1436a/b, PAT_1529a/b, PAT_1569a/b, PAT_191a/b and PAT_326a/b). Among them, PAT_1017a/b, PAT_1529a/b, PAT_191a/b and PAT_326a/b showed also different PFGE profiles. Overall, different pulsotypes were observed for pairs PAT_1422a/b, PAT_1386a/b, PAT_854a/b, PAT_476a/b, PAT_1529a/b, PAT_735a/b, PAT_191a/b and PAT_1216a/b even though the PCR results were identical among the clinical and the rectal samples.

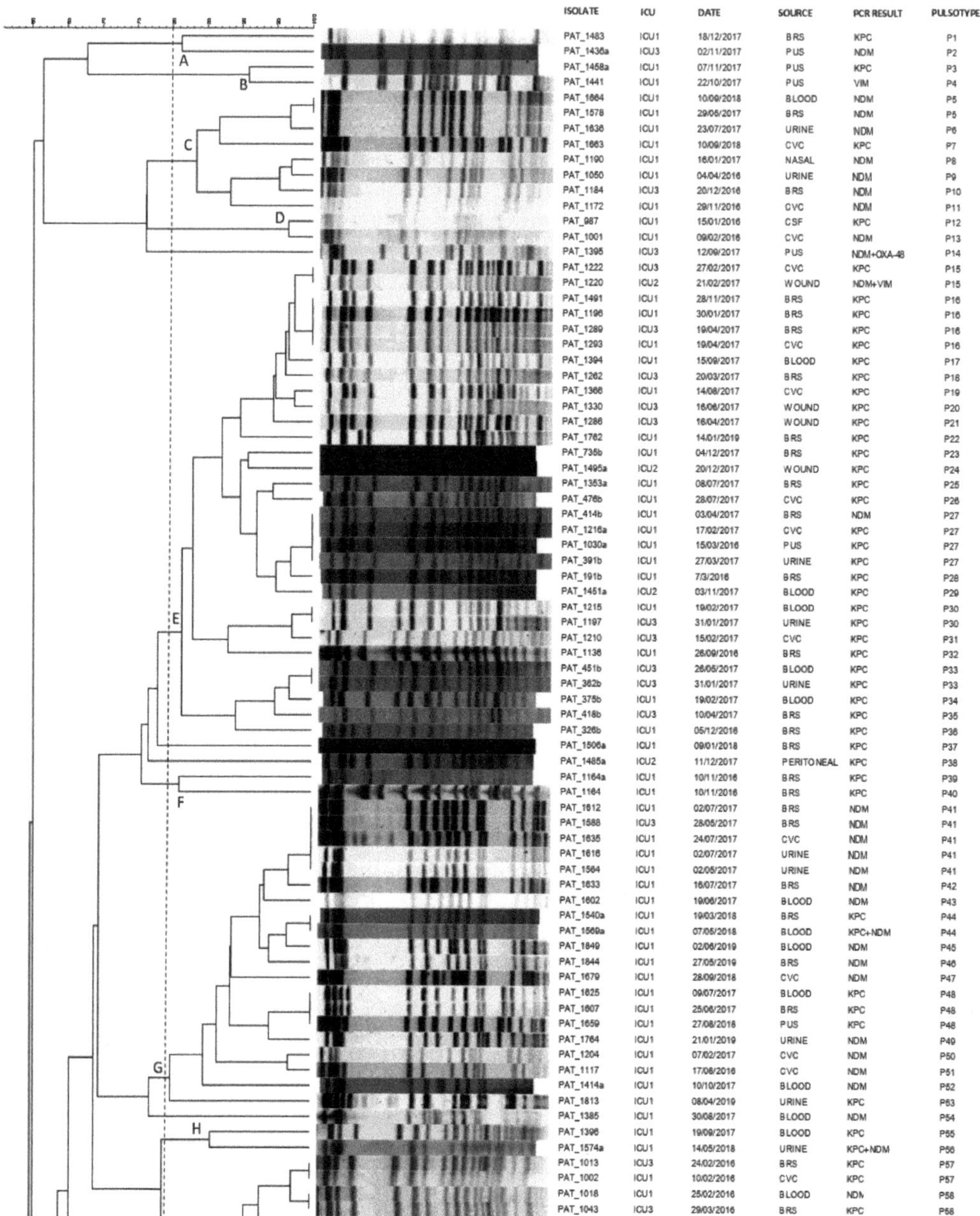

Figure 1. *Cont.*

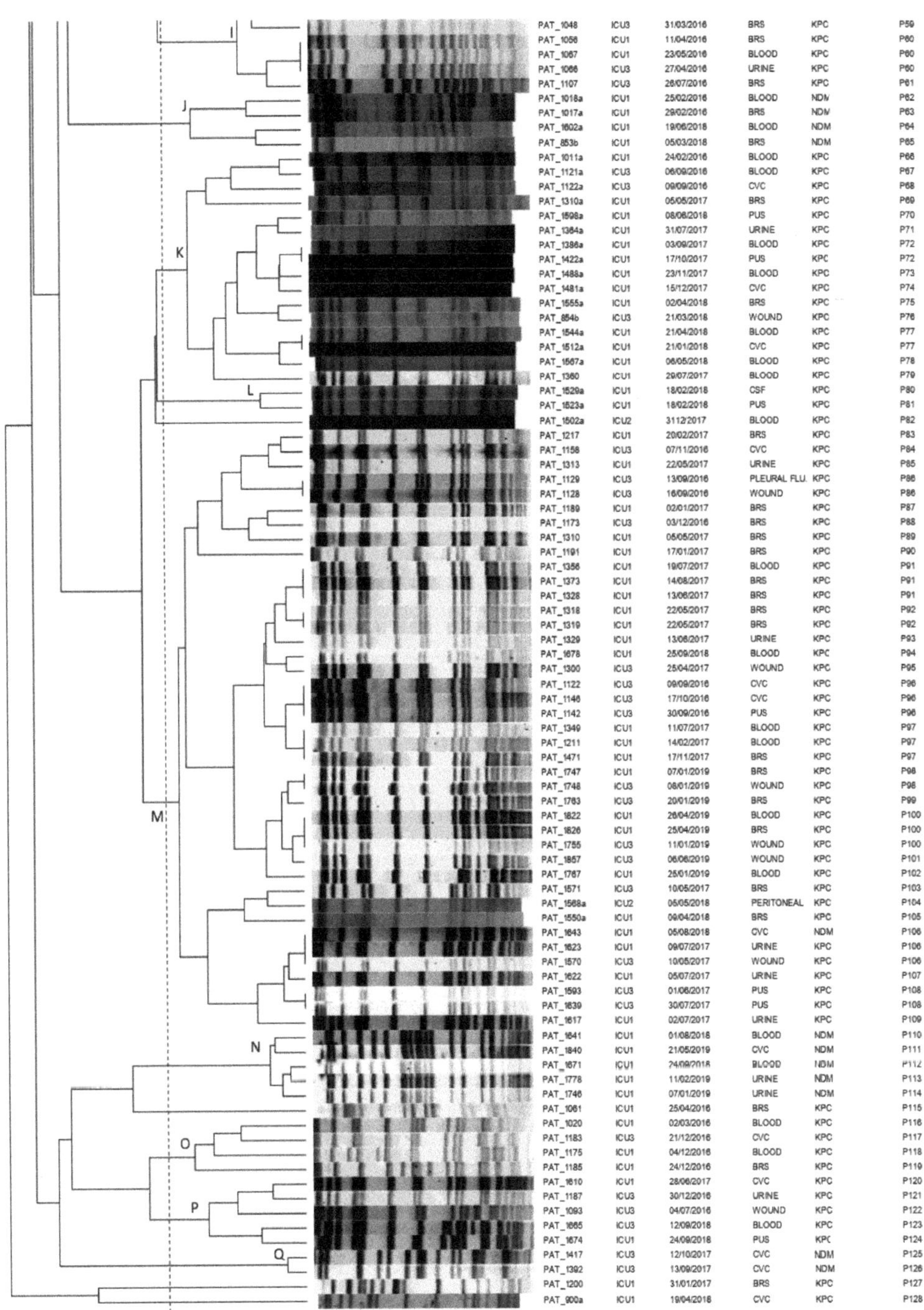

Figure 1. Pulsotypes of the 165 clinical carbapenemase-producing *K. pneumoniae* isolates. Clusters (**A–Q**) were defined at a similarity level of 80%. PAT: patient; ICU: intensive care unit; CVC: central venous catheter; BRS: bronchial secretions; CSF: cerebrospinal fluid; P: pulsotype.

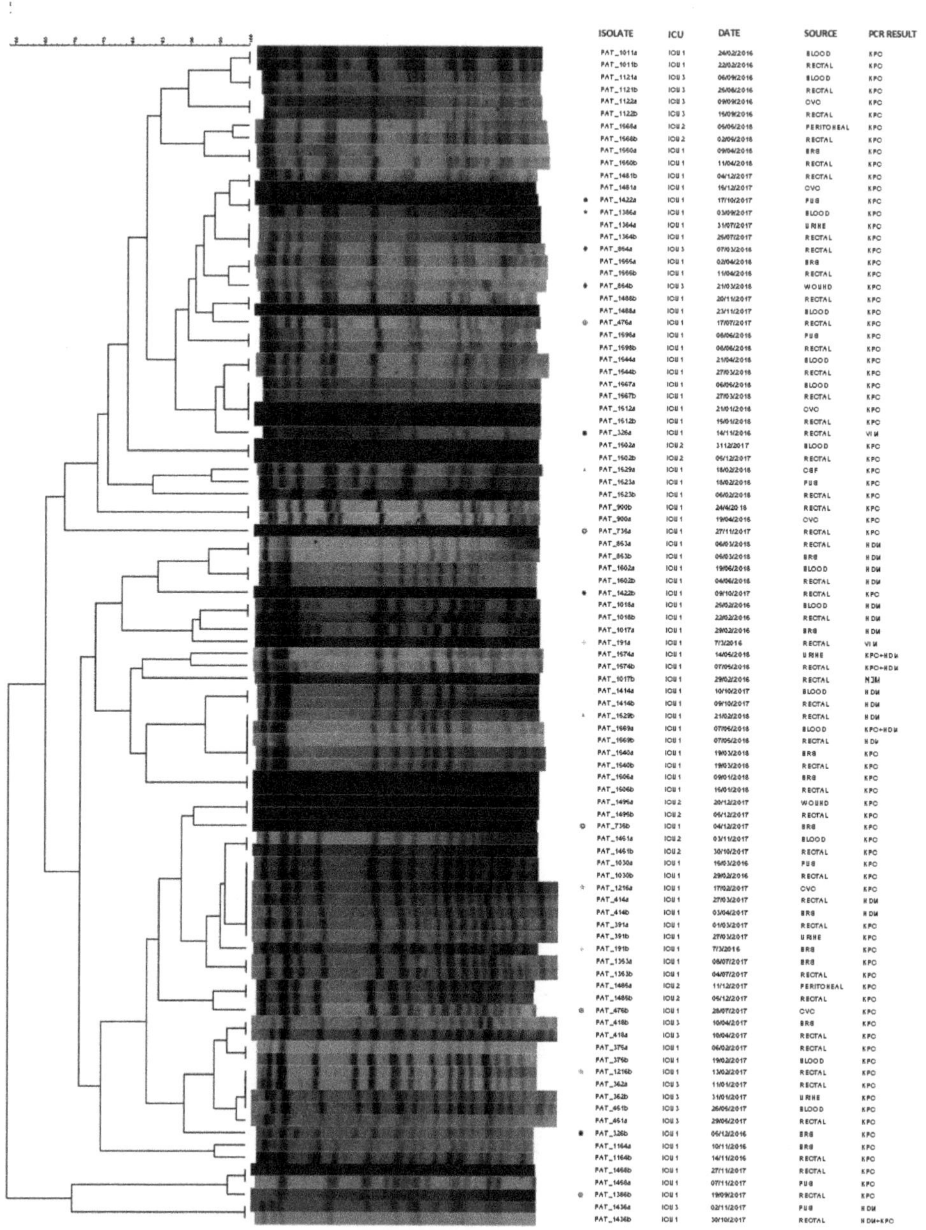

Figure 2. Pulsotypes of the 48 rectal-clinical carbapenemase-producing *K. pneumoniae* isolate pairs. Pairs with different pulsotypes are marked with the same symbol. PAT: patient; ICU: intensive care unit; CVC: central venous catheter; BRS: bronchial secretions; CSF: cerebrospinal fluid.

2.3. Non-Susceptibility Rates of CRKP Isolates in the Hospital

The non-susceptibility rates of CRKP isolates in our institution's ICUs for amikacin, aztreonam, colistin, fosfomycin, gentamicin, piperacillin/tazobactam and tigecycline in isolates during the study period are shown in Table 1 and Figure 3. CRKP isolates presented with high level of resistance to both meropenem and imipenem ($MIC_{50} \geq 16$ mg/L) throughout the study years. On the other hand, a significant increase in resistance was observed for gentamycin ranging from 15% in the first semester of 2016 to 66.7% in the first semester of 2019. In regard to tigecycline, an increase was also observed (23.5% in the 2016a semester to 61.1% in the 2019a semester). As for colistin, an increase was observed (15.8% in the 2016a semester to 27.8% in the 2019a semester).

Table 1. Imipenem resistant *K. pneumoniae* susceptibility rates per semester among single patient isolates recovered from the hospital's ICUs during the study period.

Semester	Antimicrobial	No Tested	R	I	S	R%	I%	S%
2016a	Amikacin	20	10	1	9	50%	5%	45%
	Aztreonam	19	19	0	0	100%	0%	0%
	Gentamicin	20	3	0	17	15%	0%	85%
	Piperacillin/Tazobactam	20	20	0	0	100%	0%	0%
	Colistin	19	3	0	16	15.8%	0%	84.2%
	Tigecycline	17	4	13	0	23.5%	76.5%	0%
	Fosfomycin	19	15	0	4	79.0%	0%	21.0%
2016b	Amikacin	8	5	1	2	62.5%	12.5%	25%
	Aztreonam	8	8	0	0	100%	0%	0%
	Gentamicin	8	1	0	7	12.5%	0%	87.5%
	Piperacillin/Tazobactam	8	8	0	0	100%	0%	0%
	Colistin	8	3	0	5	37.5%	0%	62.5%
	Tigecycline	8	5	2	1	62.5%	25%	12.5%
	Fosfomycin	8	2	0	6	25%	0%	75%
2017a	Amikacin	23	14	0	9	60.9%	0%	39.1%
	Aztreonam	22	22	0	0	100%	0%	0%
	Gentamicin	23	6	1	16	26.1%	4.3%	69.6%
	Piperacillin/Tazobactam	23	23	0	0	100%	0%	0%
	Colistin	22	0	0	22	0%	0%	100%
	Tigecycline	21	10	5	6	47.6%	23.8%	28.6%
	Fosfomycin	22	11	0	11	50%	0%	50%
2017b	Amikacin	32	18	3	11	56.2%	9.4%	34.4%
	Aztreonam	30	29	0	1	96.7%	0%	3.3%
	Gentamicin	32	19	3	10	59.4%	9.4%	31.2%
	Piperacillin/Tazobactam	32	32	0	0	100%	0%	0%
	Colistin	30	7	0	23	23.3%	0%	76.7%
	Tigecycline	30	13	12	5	43.3%	40%	16.7%
	Fosfomycin	30	25	0	5	83.3%	0%	16.7%
2018a	Amikacin	25	7	4	14	28%	16%	56%
	Aztreonam	24	24	0	0	100%	0%	0%
	Gentamicin	25	13	0	12	52%	0%	48%
	Piperacillin/Tazobactam	25	25	0	0	100%	0%	0%
	Colistin	24	1	0	23	4.2%	0%	95.8%
	Tigecycline	24	14	8	2	58.3%	33.3%	8.4%
	Fosfomycin	24	20	0	4	83.3%	0%	16.7%
2018b	Amikacin	30	1	3	26	3.3%	10%	86.7%
	Aztreonam	30	29	0	1	96.7%	0%	3.3%
	Gentamicin	30	21	0	9	70%	0%	30%
	Piperacillin/Tazobactam	30	30	0	0	100%	0%	0%
	Colistin	30	2	0	28	6.7%	0%	93.3%
	Tigecycline	30	14	14	2	46.7%	46.7%	6.6%
	Fosfomycin	30	24	0	6	80%	0%	20%
2019a	Amikacin	18	7	1	10	38.9%	5.5%	55.6%
	Aztreonam	18	17	0	1	94.4%	0%	5.6%
	Gentamicin	18	12	0	6	66.7%	0%	33.3%
	Piperacillin/Tazobactam	18	18	0	0	100%	0%	0%
	Colistin	18	5	0	13	27.8%	0%	72.2%
	Tigecycline	18	11	5	2	61.1%	27.8%	11.1%
	Fosfomycin	18	11	0	7	61.1%	0%	38.9%

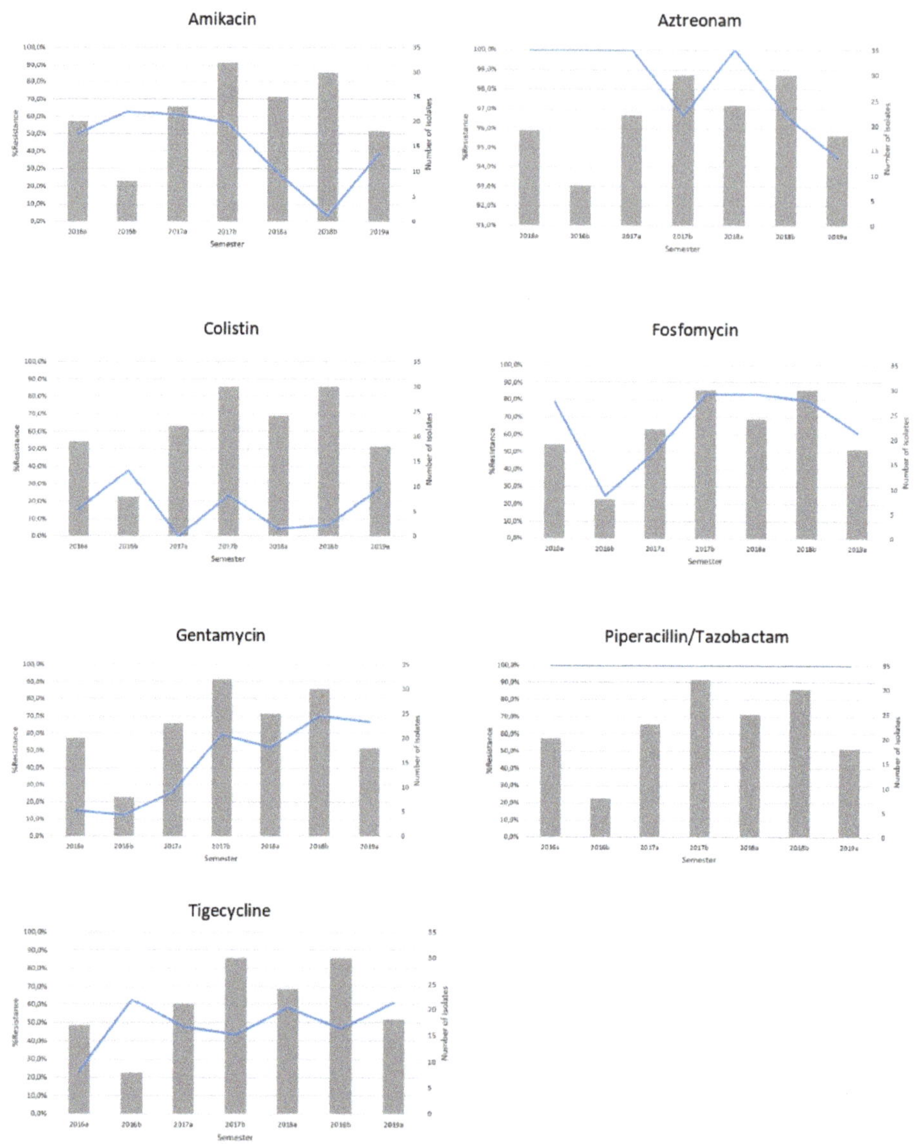

Figure 3. Non-susceptibility rates (line) and number (bars) of single patient imipenem non-susceptible *K. pneumoniae* isolates recovered from the hospital's ICUs per semester.

2.4. Systematic Review Results

Our systematic review search strategy yielded 290 results. After implementation of the exclusion criteria, 39 studies remained. Fourteen studies reported monoclonal dissemination. Data extracted from the remaining 25 studies are reported in Table 2.

Table 2. Studies reporting non-monoclonal dissemination of carbapenem-resistant *Klebsiella pneumoniae* clinical strains, including ICU populations.

Study	Setting	Time Period	Study Population (Eligible)	Sample Type (Clinical vs. Surveillance, Infection vs. Colonization)	Number Of CR-Isolates and Mechanism of Resistance	Method(s)	Number of Clusters and Isolates/Cluster
(Hernández-García et al., 2021) [15]	11 Portuguese hospitals	June 2017 to July 2018	Colistin-susceptible and -resistant MDR *Escherichia* spp. ($n = 30$) and *Klebsiella* spp. ($n = 78$) isolates	Lower respiratory, intra-abdominal and urinary tract infections of ICU patients	31 CRKP 3 CR *E. coli* KPC-3 ($n = 14$) was the most common carbapenemase followed by OXA-48 ($n = 3$) and OXA-181 ($n = 3$)	WGS	A great diversity of *Kp* high-risk clones was observed associated with the KPC-3 carbapenemase, including some lineages first reported in Portuguese Hospitals (ST13, ST34, ST405, ST1563, ST4331)
(Fontana et al., 2020) [16]	Tor Vergata University Hospital, Rome, Italy	May 2013 to Dec 2016	147 consecutive, non-replicate clinical strains of CRE from different wards	Blood cultures	bla_{KPC} was present in 121/147 (87%) strains, mainly *Kp*. The remaining strains carried bla_{VIM} or bla_{OXA-48}	WGS MLST	5 clusters with 2 to 9 strains
(Galani et al., 2020) [17]	2 ICUs of Hygeia General Hospital, Athens, Greece	Sept to Oct 2019	7 patients colonized or infected with ceftazidime-avibactam (CZA)-resistant *K. pneumoniae*	Colonization or infection	co-produced KPC-2 and the novel plasmid-borne VEB-25	WGS MLST PFGE	PFGE classified the isolates in 2 pulsotypes however, all but one, belonged to the second pulsotype
(Ferrari et al., 2019) [18]	1 cardiorespiratory ICU with 8 beds in a 900-bed Hospital in Pavia, Italy	Aug 2015 to May 2016	23 patients with 32 CRKP isolates were analyzed	12 colonized 11 infected	(9.4% carried KPC-2 and 90.6% KPC-3; All 32 analyzed isolates carried at least one ESBL gene (3.1% CTX-M-15, 3.1% SHV-1, 87.5% SHV-11, 6.3% SHV-12	WGS	Multi-clone epidemic event - 26 of the 32 isolates belong to three genome clusters and the remaining six were classified as sporadic - The first genome cluster was composed of MDR ST512 - The second infection cluster comprised four other genomes of ST512 - The third cluster ST258 colonized 12 patients

Table 2. *Cont.*

Study	Setting	Time Period	Study Population (Eligible)	Sample Type (Clinical vs. Surveillance, Infection vs. Colonization)	Number Of CR-Isolates and Mechanism of Resistance	Method(s)	Number of Clusters and Isolates/Cluster
(Mavroidi et al., 2020) [19]	Kostantinopouleio-Patission G. Hospital, Athens, Greece 280-bed general hospital (including a nine-bed ICU)	Jan 2014 to Dec 2016	248 CRKP in ICU	Bronchial secretions ($n = 105$), blood ($n = 53$), central venous catheters ($n = 39$), urine ($n = 28$)	The majority of CRKP from BSIs were OXA-48 producers ($n = 23$) and KPC producers ($n = 18$) whereas the remaining 12 isolates produced and/or MBLs (6 VIM, 3 OXA-48+VIM, and 3 NDM producers)	MLST	ST101 (OXA-48) ST258 (KPC) ST11 (NDM)
(Gona et al., 2019) [20]	1 teaching hospital in Catania, Italy	Oct 2016 to Jan 2018	Neonatal ICU. All confirmed CRKP isolates included	12 infections, 1 colonization	13 isolates all NDM+OXA-48	PFGE MLST Core genome MLST	1 pulsotype Clinical isolates included a common MLST (ST101), and 2 novel STs (ST3366 and ST3367), which differ from ST101 by a single nucleotide of rpoB gene. The cgMLST method accurately characterized transmission events of the 13 *K. pneumoniae* isolates in three clusters: A containing only ST101, B containing only ST3367, and C containing both ST3366 and ST101 due to the close relationship between ST101 and ST3366. Four isolates were included in cluster A, two isolates in cluster B, and seven isolates in cluster C.
(Karampatakis et al., 2018) [21]	Hippokration General Hospital, Thessaloniki, Greece 900 beds	Aug 2012 to Nov 2014	Conducted in a 9-bed polyvalent ICU. 143 CRKP selected randomly	Infection or colonization	44 CRKP (mostly KPC and VIM, 2 NDM, 2 OXA-48, 1 NDM+OXA-48, 1 KPC+OXA-48)	PFGE	10 pulsotypes A: 24 isolates (all KPC) A2: 1 KPC B: 11all VIM (2 VIM+KPC) C: 2 D: 1 E: 1 F: 1 G: 1 H: 1 I: 1 No relevant further information available

Table 2. *Cont.*

Study	Setting	Time Period	Study Population (Eligible)	Sample Type (Clinical vs. Surveillance, Infection vs. Colonization)	Number Of CR-Isolates and Mechanism of Resistance	Method(s)	Number of Clusters and Isolates/Cluster
(Papadimitriou-Olivgeris et al., 2018) [22]	University Hospital of Patras, Greece 800 beds.	2010–2016 Months not specified	Isolates from hospitalized patients in the ICU. It was a matched 1:2 case–control study conducted among critically ill patients in order to identify the risk factors of ColR-Kp and TigR-Kp bacteraemia	Blood infections	110 included in PFGE 91 KPC, 4 VIM, 5 KPC+VIM, 10 NDM	PFGE	3 pulsotypes A: 76 mostly KPC B: 24 mostly KPC C: 10 only NDM
(Avgoulea et al., 2018) [23]	Tzaneio Hospital, Athens, Greece 450 beds	June 2014	ICU patients; The aim of the study was to analyze the mode of spread and the characteristics of epidemic OXA-48-Kp strains responsible for bloodstream infections in ICU patients emerged in June 2014	Blood infections	19 selected OXA-48	PFGE MLST	2 pulsotypes 2 STs Pulsotype A was ST147 (the first 4 cases-PDR) Pulsotype B was ST101 (the next cases-MDR)
(Ripabelli et al., 2018) [24]	Antonio Cardarelli Hospital, Molise, Italy	2010, 2014–2016 Months not specified	30 from the ICU 10 from wards	Infection ($n = 27$) or colonization ($n = 13$)	23 WILD TYPE (2010), 17 NON-WILD TYPE (KPC) (2014–2016)	PFGE RAPD	16 clusters and 26 pulsotypes 23 clusters and 33 patterns 2010 and 2014-16 isolates were grouped in different clusters by both methods
(Bartolini et al., 2017) [25]	Padova Hospital, Italy	1/2015–9/2106	Adult patients from the ICU, surgical and medical department and patients with epidemiological link to persons with CPKP isolates	Rectal swabs and clinical samples	311 CPKP: 258 KPC, 17 OXA-48, 12 NDM	MLST	16 different CPKP strains without predominance: 35 ST-258, 85 ST-512, 32 ST-745, 54 ST-307, 22 ST-554, 5 ST-15, 11 ST-16, 3 ST-101, 3 ST-11, 1 ST-37, 1 ST-45, 1 ST-211, 1 ST-398, 1 ST-147, 1 ST-1458

Table 2. *Cont.*

Study	Setting	Time Period	Study Population (Eligible)	Sample Type (Clinical vs. Surveillance, Infection vs. Colonization)	Number Of CR-Isolates and Mechanism of Resistance	Method(s)	Number of Clusters and Isolates/Cluster
(Mavroidi et al., 2016) [26]	Kostantinopouleio-Patission G. Hospital, Athens, Greece 280-beds	July 2012 to Dec 2013	Imipenem and meropenem resistant isolates of all hospital's department	Surveillance rectal swabs and clinical samples	135 CPKP. 19 were colistin resistant and all of them harbored the bla_{KPC} gene.	MLST	The 19 COL-R CP-Kp isolates belonged to 2 STs: 18 to ST-258 and 1 ST-383 lineages.
(Bonura et al., 2015) [27]	3 acute general hospitals in Palermo, Italy	March–Aug 2014	All carbapenem resistant isolates of all hospital's department	Isolates from any sight of infection or colonisation	94 carbapenem non susceptible isolates all KPC-3 producers	PFGE and MLST	10 pulsotypes: A(4), B(1), C(subtypes:C1(15), C2(2)), D(subtypes: D1(22),D2(3),D3(1),D4(1)), E(1), F(1), G(4), H(1), I(1), O(37). 10STs. 37 ST258, 1 ST512, 27 ST307, 17 ST273, 4 ST405, 4 ST101, 1 ST15, 1 ST147, 1 ST323, 1 ST491
(Onori et al., 2015) [28]	Ospedale di Circolo e Fondazione Macchi Varese, Italy	Jan 2011 to March 2013	Infections due to carbapenem-resistant *Kp*	Clinical samples	16 CRKP isolates. 3 harbored the bla_{KPC-2} and 13 the bla_{KPC-3} variant.	WGS	2 STs. 10 isolates belonged to ST512 and 6 to ST258.
(Parisi et al., 2015) [29]	Padova Hospital, Italy	Jan 2012 to Dec 2014	Patients from the Intensive care, surgery and medical departments	Clinical and surveillance samples	496 CPKP strains out of which 436 tested with molecular methods: 432 KPC, 3 OXA-48, 1 NDM	MLST	MLST available for 238/496 isolates. In total 15 STs were identified: 90 ST258, 86 ST512, 31 ST745, 5 ST15, 2 ST101, 1 ST868, 6 ST307, 3 ST554, 1 ST392, 1 ST437, 1 ST1207, 1 ST1326, 1 ST395, 1 ST1199, 1 ST1543.
(Katsiari et al., 2015) [30]	Konstantopouleio General Hospital, Athens, Greece	2010–2012	279-bed tertiary-care hospital. Athens. 1 ICU, 9 beds, all imipenem-resistant *Kp*	clinical or surveillance	6 CRKP isolates (48 KPC-producers and 13 VIM-producers) were recovered from 58 ICU patients.	PFGE Representative isolates to MLST	Seven types (A–G) according to 85% similarity, 42 (69%) to A cluster. -MLST type ST258 Type A was further divided into 12 subtypes (A1–A12) according to 100% pattern similarity, 10/13 VIM classified in type B

Table 2. *Cont.*

Study	Setting	Time Period	Study Population (Eligible)	Sample Type (Clinical vs. Surveillance, Infection vs. Colonization)	Number Of CR-Isolates and Mechanism of Resistance	Method(s)	Number of Clusters and Isolates/Cluster
(Mezzatesta et al., 2014) [31]	1 general ICU Catania Hospital, Italy	1–31 July 2013	ICU *Kp* isolates responsible for severe infections	clinical isolates	25 *Kp* 57 patients, all harbored bla_{KPC-3}.	PFGE MLST	4 pulsotypes among all the KPC-producing *Kp* (A, B, C and D), MLST 4 distinct STs: All pulsotype A strains belonged to ST258 and pulsotype B was categorized as ST512 detected in most isolates. Pulsotypes C and D were also identified, in a few strains, as ST147 and ST395, respectively.
(Papadimitriou-Olivgeris et al., 2014) [32]	General ICU (13 beds) of the University Hospital of Patras, Greece	26 months	Hospital of Patras, Greece, a 770-bed teaching hospital.	Recovered from clinical or rectal samples from patients ($n = 273$) who stayed more than 6 days in the ICU	53 KPC-*Kp* isolates from 48 patients All 53 KPC-*Kp* isolates carried the bla_{KPC}	PFGE	Two PFGE types (A and B) were identified, with 36 (67.9%) strains belonging to PFGE type A and 17 (32.1%) to PFGE type B.
(Capone et al., 2013) [33]	9 hospitals of Rome, Italy	Dec 2010 to May 2011	1 teaching institution, 6 tertiary hospitals, 1 clinical and research institute, and 1 long-term care facility, with a total of 4000 beds, ranging from 100 to 1200 beds per centre	97 patients *Kp* strain showing reduced susceptibility to ertapenem (MIC 1 mg/L), Clinical samples urine ($n = 34$), blood ($n = 34$), lower respiratory tract ($n = 13$), surgical wound ($n = 8$), intraabdominal fluid ($n = 7$), CVC tips ($n = 12$), rectal swab ($n = 3$) and cerebrospinal fluid ($n = 1$)	Strains producing bla_{KPC-3} were identified in 89 patients, bla_{VIM} in three patients and $bla_{CTX-M-15}$ plus porin defects in the remaining five patients. 1 isolate per patient	MLST	Among strains producing KPC-3, two major clones identified by MLST: ST512 and ST258, KPC-3 was also identified in clones ST646 (new ST), ST650 (new ST), ST14 and ST101. The bla_{VIM-1} gene was identified in clones ST646, ST647 and ST648 (three new STs). Among strains producing ESBL combined with outer membrane protein (OmpK) defects, three belonged to ST37, and the other was assigned to the new ST649
(Tofteland et al., 2013) [34]	A 12-bed mixed ICU in the Arendal hospital, Norway	Nov 2007 to April 2011	KPC-producing outbreak strains	Clinical and surveillance samples/Infection or colonization	7 KPC-2 strains from 7 patients	PFGE MLST	A 6 ST258 B 1 ST461

Table 2. *Cont.*

Study	Setting	Time Period	Study Population (Eligible)	Sample Type (Clinical vs. Surveillance, Infection vs. Colonization)	Number Of CR-Isolates and Mechanism of Resistance	Method(s)	Number of Clusters and Isolates/Cluster
(Mammina et al., 2012) [35]	24 beds in two general ICUs, in 1 acute general hospital in Palermo, Italy	June to Dec 2011	All colistin-resistant *Kp* isolates during this period (possible outbreak) irrespective of their source patient and clinical sample	58 colistin-resistant *Kp* isolates were recovered from 28 patients irrespective of their source	52 isolates carried the bla_{KPC-3} and SHV-11. 6 isolates susceptible to carbapenems, resistant to fluoroquinolones and aminoglycosides	Rep-PCR	Al 52 isolates carried the bla_{KPC-3} gene belonging in sequence type ST258 Rep-PCR confirmed that the colistin-resistant isolates belonged to three different clusters, one that contained all ST258 KPC-3 producing isolates, and two clusters with unrelated patterns including the ST15 and ST273isolates
(Richter et al., 2012) [36]	2 hospitals (1580 and 300 beds) in Padua, Italy	June 2009 to Dec 2011	Phenotypic and genotypic investigation for KPC in clinical samples	Infection or colonization	189 KPC-2 or KPC-3 strains	PFGE MLST ERIC	4 PFGE profiles ST37, ST147,ST258, ST307, ST437, ST510,ST512, ST527, ST554 3 ERIC profiles
(Sánchez-Romero et al., 2012) [37]	613 bed teaching hospital, Madrid, Spain -52 ICU beds	Jan to Dec 2009	Any carbapenem non susceptible strain from ICU patients	Clinical or surveillance/ Infection or colonization	55 patients harbouring VIM-1 strains/molecular epidemiology for 99 strains	PFGE MLST	-PFGE: A 54, B 4 -MLST: 6 A isolates ST15, 3 B isolates ST340
(Souli et al., 2010) [38]	University General Hospital Attikon 635-bed teaching hospital, Athens, Greece—1 ICU (18 beds till 10/2008, 21 after)	Jan 2007 to Dec 2008	Any clinical *Kp* isolate with imipenem or meropenem MIC > 1 mg/mL producing KPC (hospital-wide)	Clinical or surveillance samples/Infection or colonization	50 KPC-2 isolates (34 ICU/16 non- ICU, 18 infections (9 ICU, 9 non-ICU)/32 colonization)	PFGE	4 PFGE types: A 41, B 6, C 1, D 2 Only A was responsible for infections
(Giakkoupi et al., 2003) [39]	3 teaching hospitals in Athens, Greece	Sep to Dec 2002	ICU patients with archived imipenem non susceptible specimens	Clinical samples/ at least 12 infections	17 bla_{VIM-1} strains from 17 patients	PFGE	4 PFGE types: the majority (5 and 10 isolates) belonged to two types

MLST: Multi-Locus Sequence Typing; WGS: Whole Genome Sequencing; ERIC: enterobacterial repetitive intergenic consensus; MDR: multi-drug resistant; PDR: pan-drug resistant; *Kp*: *Klebsiella* pneumoniae; ColR-*Kp*: Colistin resistant-*Kp*; TigR-*Kp*: Tigecycline resistant-*Kp*.

3. Discussion

The present study evaluated the type of carbapenemases and the molecular epidemiology of *K. pneumoniae* strains circulating in the ICUs of a tertiary hospital in Thessaloniki, Greece between 2016 and 2019. During the study years, KPC was the predominant carbapenemase; NDM were also present, and a few double-carbapenemase-producers were isolated. On the basis of PFGE, a total of 17 different CRKP transmission clusters were identified.

CRKP isolates have been introduced in our hospital since 2004. At that time, no phenotypic or molecular testing was performed to reveal the type of carbapenemase up to 2010. From 2010 to 2014, phenotypic testing among CRKP isolates revealed that the majority carried KPC enzymes (60%), 25% produced MBL, and 7% co-produced KPC and MBL enzymes. In the time period of 2011–2014, we also observed the first OXA-48 producers in our hospital at a rate of 1.7%. Molecular testing showed that KPC positive strains harbored the bla_{KPC-2}, while MBL-positive strains harbored bla_{VIM-1}.

During 2013–2015, an oligoclonal outbreak caused by 45 CRKP occurred in our hospital [40]. All the patients were hospitalized in the three intensive care units of the hospital, and 17 (68%) of them developed bloodstream infections; the overall mortality of the patients involved in the outbreak was 48% (12/25). Molecular testing verified that all 45 *K. pneumoniae* isolates co-harbored bla_{KPC-2} and bla_{VIM-1} and were associated with OmpK35 deficiency and OmpK36 porin loss. PFGE clustered all isolates into a single clonal type, and multi-locus sequence typing (MLST) assigned them to the emerging high-risk ST147 clonal lineage.

Starting from 2014 until 2016, while KPC producers still prevailed (approximately 75% of the CRKP), we observed a shift among MBL producers from bla_{VIM-1} towards bla_{NDM-1}. More specifically, VIM-type carbapenemases decreased from 17.6% in 2014 to 6.7% in 2016, whereas NDM-type increased from 1.2% in 2014 to 19.4% in 2016 [41]. In the years 2013–2015, the co-production of KPC and MBL enzymes corresponded to approximately 10% of all CRKP strains [35]. Finally, in 2019, we isolated a strain carrying both NDM-1 and OXA-48 genes classified as ST-11 [42].

In our study, PFGE analysis identified 128 distinguishable pulsotypes and 17 clusters indicating an extended dissemination of CRKP within the hospital setting. Moreover, the dissemination took place over a long-time frame since we included in our study isolates recovered during a 3.5-year period. The presence of identical isolates in all three ICUs highlights their successful dissemination through different hospital wards. More worryingly, the persistence of certain strains throughout the whole study period, despite the various infection control measures that were applied, reflects the difficulties that undermine the efforts for their eradication once they are well-established in a certain geographical area. Indeed, such strains may have persisted in the hospital or/and may have been re-introduced by carrier admissions.

In fact, most of the CRKP carriers that later presented a CRKP infection had identical pulsotypes between their rectal and clinical isolates. The most probable explanation for this finding is that gut colonization preceded infection. In some cases, however, the pulsotypes of rectal and clinical isolates were different indicating that the infection was caused by another nosocomial *K. pneumoniae*.

There were also two clinical-rectal pairs (PAT_1436a/b and PAT_1569a/b) that harbored different carbapenemase-encoding genes according to PCR results but had identical PFGE pulsotypes. This could be explained by the mobilization of mobile genetic elements, most likely by the loss and acquisition of plasmids [43] even though an infection by a different isolate of the same pulsotype harboring different resistance determinants could not be excluded. PFGE studies in our hospital performed in several CRKP strains during the period of 2011 to 2013 revealed that KPC strains prevailed and that the majority of them belonged to two distinct clones (unpublished data).

A similar pattern of carbapenemases was observed in Hippokration General Hospital of Thessaloniki, Greece where KPC carbapenemases have prevailed among CRKP since 2009 outnumbering the VIM-type carbapenemases that predominated previously [44].

In the same report, KPC-producers belonged to two distinct clones, the predominant of which correspond to the hyperepidemic Greek clone. In a multicenter nationwide surveillance study conducted in several Greek hospitals for CRKP from 2014 to 2016, NDM-producing isolates belonged mainly to one clone, whereas KPC, VIM, OXA-48 and double carbapenemase-producers were mainly categorized in three clones [45]. On the contrary, in our study, both KPC and NDM CRKP isolates showed extremely multiclonal profiles.

In our study, we also observed a rise in the resistance rates of tigecycline, gentamicin and colistin. This is in accordance with other studies reporting elevated resistance rates to last resort antibiotics driven, among other factors, by a vicious cycle of increased last resort antibiotic consumption and subsequent resistance [46].

Our systematic review results showed that non-monoclonal dissemination of *K. pneumoniae* strains in ICU settings has been described before in countries participating in the EARS-Net (Table 2). Of note, such observations have been reported almost exclusively by Mediterranean countries, mainly from Greece and Italy, and this is in accordance with the epidemiological situation of the region regarding carbapenem-resistance determinants. However, and despite the heterogeneity of settings and methods used, most studies reported rather oligoclonal transmission with further identification of sporadic cases.

In our study, multiple clones circulating simultaneously achieved sustainable dissemination and, according to our knowledge and our systematic review results, this is the first time that such polyclonal dissemination has been observed in Europe. This multi-clonal PFGE observation highlights a possible additional reason for their endemic persistence in our hospital even though infection control measures, including hand hygiene, surveillance for colonization among high-risk patients and contact precautions have been established.

In this context, active surveillance with rectal swab cultures is of outmost importance to control the spread of these pathogens by isolation or cohorting of the colonized patients [29]. However, the spread of CRKP in an endemic environmental niche is a dynamic and multifaceted phenomenon that involves many variables. In a similar situation, more than one CRKP clone may be simultaneously present in the hospital; whereas new admissions may be CRKP carriers most likely by previous hospitalizations in the same or other hospitals. Consequently, a multi-clonal spread is very likely to occur and, even when a previously colonized subject presents a CRKP infection, it is not certain that this infection is directly related to the strain that colonized the patient upon admission.

Our study has several limitations. A multi-centric study would be able to evaluate whether the epidemiological pattern that we observed in our single center study was an isolated phenomenon or more widespread. Including non-ICU along with ICU *K. pneumoniae* strains would yield a more complete picture for their dissemination. In our analysis, we did not include clinical patient level information, and this limits our ability to draw conclusions with regard to precipitating factors. Finally, we were not able to employ sequencing-based methods to better characterize the molecular epidemiology of the strains included in our study.

4. Materials and Methods

4.1. Study Design

This was a retrospective study that was carried out at AHEPA University Hospital, a 700-bed institution with three ICUs, a central surgical and medical ICU (8 beds, ICU 1), a surgical ICU (4 beds, ICU 2) and a cardiosurgical ICU (5 beds, ICU 3) as well as surgical and internal medicine departments. The study was approved by the institutional medical scientific board. Sample related patient data were retrieved from the laboratory database.

CRKP clinical isolates, recovered in the aforementioned ICUs between January 2016 and June 2019 from 165 single patients, were included in the study. In 48 cases, a rectal isolate (isolated upon admission in ICU for infection control purposes) and a subsequent clinical isolate (isolated by an infection that occurred during ICU stay) were considered, thus forming 48 pairs of surveillance-clinical isolates. Isolates taken from the remaining 117 patients were all recovered from clinical specimens only.

Rectal swabs taken from ICU patients upon admission were inoculated on MacConkey agar plates supplemented with meropenem and ceftazidime discs. All Gram-negative colonies that grew after 24 h of incubation near the discs were further identified and *K. pneumoniae* isolates were tested for carbapenemase production with phenotypic and molecular techniques. For clinical specimens, standard laboratory procedures were followed depending on each specimen source.

Bacterial identification and antimicrobial susceptibility testing were performed with the Vitek2 automated system (Biomerieux, Marcy-l'Étoile, France). Furthermore, the minimum inhibitory concentration of tigecycline was determined by E-test (Liofilchem, Roseto degli Abruzzi, Italy) and for colistin using the broth microdilution method (Liofilchem, Roseto degli Abruzzi, Italy). The results of all antimicrobial testing were interpreted in accordance with the CLSI criteria. For tigecycline, the breakpoints recommended by the United States Food and Drug Administration were used (susceptible: MIC $\leq$ 2 mg/L; resistant: MIC $\geq$ 8 mg/L).

4.2. Carbapenemase Detection

All isolates were phenotypically screened for carbapenemase production with the Modified Hodge Test (MHT), [47] while the type of MBL or KPC was assessed with the Combined Disk Test (CDT) [48]. Following phenotypic identification, PCR assays were performed for carbapenemase-encoding genes using specific primers for bla_{KPC}, bla_{VIM}, bla_{IMP}, bla_{NDM} and bla_{OXA-48} (Appendix A) [49].

4.3. Pulse-Field Gel Electrophoresis

The genetic relationship among the CPKP isolates was determined by PFGE according to standardized protocol [50] with the XbaI endonuclease (New England Biolabs, Beverly, MA, USA) by using a CHEF-DR III apparatus (Bio-Rad Laboratories Inc., Hercules, CA, USA) for the separation of DNA fragments. XbaI-digested DNA from *Salmonella enterica* serotype Braenderup H9812 was used as a reference size standard, while PFGE patterns were digitally analyzed using the FPQuest (Bio-Rad Laboratories Pty Ltd., Hercules, CA, USA) software package.

PFGE profiles were compared using the Dice correlation coefficient with a maximum position tolerance of 1.5% and an optimization of 1.5%. Similarity clustering analysis was performed by using the Unweighted Pair Group Method using Averages (UPGMA), and a dendrogram was generated. Two PFGE profiles were classified as indistinguishable if the DNA fragment patterns matched each other completely, while clusters were selected using a cutoff at the 80% level of genetic similarity.

4.4. Non-Susceptibility Rates of CRKP Isolates

For every semester of the study period, we determined the *K. pneumoniae* non-susceptibility rates of amikacin, aztreonam, colistin, fosfomycin, gentamicin, piperacillin/tazobactam and tigecycline in imipenem non-susceptible single patient isolates from the ICUs using the CLSI 2020 breakpoints.

4.5. Systematic Review

In order to evaluate the extent of non-monoclonal transmission of CRKP strains in the intensive care environment in countries participating in the European Antimicrobial Resistance Surveillance Network (EARS-Net), we undertook a systematic review of the recent literature. We searched MEDLINE via PubMed from 1 January 2000 to 28 April 2021, implementing the search strategy described in Appendix B. Titles and abstracts were screened for studies, which incorporated a molecular epidemiology investigation (PFGE or sequencing methods) of CRKP strains, including samples obtained from ICU patients.

We excluded reviews, case reports, studies restricted to environmental samples, studies that did not explicitly include any ICU clinical samples and studies not conducted in EARS-Net participating countries. No language and patient age restrictions were applied.

Eligibility assessment was conducted in duplicate (D.P. and G.M.) and discrepancies were adjudicated by a third author (E.P.).

Next, studies reporting monoclonal transmission were excluded, and, from the remaining studies, we extracted the following data: study setting and eligible samples, sample type (clinical versus surveillance and infection versus colonization), the number of carbapenem-resistant isolates and mechanism of resistance detected, the method used to investigate molecular epidemiology and the number and size of clusters involving ICU patients. If ICU-level data were not available, hospital-wide data were reported. Data extraction was conducted by A.T., G.M., T.P., D.P. and E.P. All steps were conducted in duplicate.

5. Conclusions

In our study, we demonstrated that CPKP in our hospital belonged to a great variety of pulsotypes and clusters, thus, indicating their extended dissemination within the hospital ICU settings. Among them, KPC carbapenemase predominated. The presence of multiple clones harboring variable resistance-determinants poses additional challenges. Further studies are required to identify suitable infection control strategies in a setting of polyclonal dissemination within a context of carbapenem resistance endemicity. Our results highlight the need to urgently reinforce infection-control measures along with antimicrobial stewardship and together with a generous increase in the nosocomial budget in order to contain the transmission of antibiotic resistant organisms.

Author Contributions: Conceptualization, E.P. and L.S.; methodology, E.P., G.M., D.P. (Dimitrios Pilalas), C.K. and T.P.; investigation, E.P., P.M., A.T., D.P. (Dimitra Papadopoulou) C.K., T.P., M.P. and S.M.; resources, D.P. (Dimitra Papadopoulou) and C.K.; software, G.M., D.P. (Dimitrios Pilalas), C.K. and M.P.; validation, E.P., G.M., C.K., formal analysis, G.M., D.P. (Dimitrios Pilala) and M.P.; data curation, E.P., G.M., D.P. (Dimitrios Pilalas), A.T., D.P. (Dimitra Papadopoulou), C.K., M.P.; writing-original draft preparation, E.P., G.M., D.P. (Dimitris Pilalas), A.T., C.K.; writing—review and editing,, E.P., G.M., D.P. (Dimitrios Pilalas), C.K., T.P., L.S.; visualization, E.P., G.M., D.P. (Dimitrios Pilalas) and L.S.; supervision, E.P. and L.S.; project administration, E.P. and L.S. All authors have read and agreed to the published version of the manuscript.

Funding: This research received no external funding.

Institutional Review Board Statement: Not applicable.

Informed Consent Statement: Not applicable.

Data Availability Statement: Not applicable.

Acknowledgments: We wish to express our gratitude to the laboratory technicians of the Microbiology Laboratory for the performance of the phenotypic and antimicrobial susceptibility tests. We are also grateful to the nurses of the Infection Control Unit of AHEPA Hospital for assistance in collecting the rectal samples.

Conflicts of Interest: The authors declare no conflict of interest.

Appendix A. Primers for Carbapenemase Detection

Gene	Primers (5′–3′)
bla_{KPC}	TGTCACTGTATCGCCGTC
	TATTTTTCCGAGATGGGTGAC
bla_{IMP}	CTACCGCAGCAGAGTCTTTG
	AACCAGTTTTGCCTTACCAT
bla_{VIM}	TCTACATGACCGCGTCTGTC
	TGTGCTTTGACAACGTTCGC
bla_{NDM-1}	GGTTTGGCGATCTGGTTTTC
	CGGAATGGCTCATCACGATC
bla_{OXA-48}	TTGGTGGCATCGATTATCGG
	GAGCACTTCTTTTGTGATGGC

Appendix B. Systematic Review Search Strategy

#1   (critical care OR ICU OR intensive care OR critical ill OR critical illness OR critically ill OR "Intensive Care Units"[Mesh] OR "Critical Care"[Mesh] OR "Critical Illness"[Mesh])

#2   carbapenem* OR meropenem OR imipenem OR ertapenem

#3   Klebsiella

#4   (epidem* OR outbreak OR clon* OR strain*)

#5   (PFGE OR puls* OR genom* OR typing OR sequenc* OR MLST OR NGS OR WGS OR cgMLST or wgMLST OR MLVA)

#6   ("1 January 2000"[Date–Publication]: "28 April 2021"[Date–Publication])

Final search(29 April 2021): #1 AND #2 AND #3 AND #4 AND #5 AND #6 → 290 results

References

1. Falcone, M.; Russo, A.; Iacovelli, A.; Restuccia, G.; Ceccarelli, G.; Giordano, A.; Farcomeni, A.; Morelli, A.; Venditti, M. Predictors of outcome in ICU patients with septic shock caused by *Klebsiella pneumoniae* carbapenemase–producing *K. pneumoniae*. *Clin. Microbiol. Infect.* **2016**, *22*, 444–450. [CrossRef] [PubMed]

2. Palacios-Baena, Z.R.; Oteo, J.; Conejo, C.; Larrosa, M.N.; Bou, G.; Fernández-Martínez, M.; Oliver, A.; Zamorano, L.; Alba Rivera, M.; Bautista, V. Comprehensive clinical and epide-miological assessment of colonisation andinfection due to carbapenemase-producing Enterobacteriaceae in Spain. *J. Infect.* **2016**, *72*, 152–160. [CrossRef] [PubMed]

3. WHO Regional Office for Europe and European Centre for Disease Prevention and Control. *Surveillance of Antimicrobial Resistance in Europe, 2020 Data. Executive Summary*; WHO Regional Office for Europe: Copenhagen, Denmark, 2021.

4. World Health Organization; Regional Office for Europe. *Central Asian and European Surveillance of Antimicrobial Resistance: Annual Report 2020*; WHO Regional Office for Europe: Copenhagen, Denmark, 2020.

5. Queenan, A.M.; Bush, K. Carbapenemases: The Versatile Beta-Lactamases. *Clin. Microbiol. Rev.* **2007**, *20*, 440–458. [CrossRef] [PubMed]

6. Karampatakis, T.; Antachopoulos, C.; Iosifidis, E.; Tsakris, A.; Roilides, E. Molecular epidemiology of carbapenem-resistant *Klebsiella pneumoniae* in Greece. *Futur. Microbiol.* **2016**, *11*, 809–823. [CrossRef]

7. Giakkoupi, P.; Papagiannitsis, C.C.; Miriagou, V.; Pappa, O.; Polemis, M.; Tryfinopoulou, K.; Tzouvelekis, L.S.; Vatopoulos, A.C. An update of the evolving epidemic of blaKPC-2-carrying *Klebsiella pneumoniae* in Greece (2009–2010). *J. Antimicrob. Chemother.* **2010**, *66*, 1510–1513. [CrossRef]

8. Pournaras, S.; Protonotariou, E.; Voulgari, E.; Kristo, I.; Dimitroulia, E.; Vitti, D.; Tsalidou, M.; Maniatis, A.N.; Tsakris, A.; Sofianou, D. Clonal spread of KPC-2 carbapenemase producing *Klebsiella pneumoniae* strains in Greece. *J. Antimicrob. Chemother.* **2009**, *64*, 348–352. [CrossRef]

9. Voulgari, E.; Gartzonika, C.; Vrioni, G.; Politi, L.; Priavali, E.; Levidiotou-Stefanou, S.; Tsakris, A. The Balkan region: NDM-1-producing *Klebsiella pneumoniae* ST11 clonal strain causing outbreaks in Greece. *J. Antimicrob. Chemother.* **2014**, *69*, 2091–2097. [CrossRef]

10. Voulgari, E.; Zarkotou, O.; Ranellou, K.; Karageorgopoulos, D.E.; Vrioni, G.; Mamali, V.; Themeli-Digalaki, K.; Tsakris, A. Outbreak of OXA-48 carbapenemase-producing *Klebsiella pneumoniae* in Greece involving an ST11 clone. *J. Antimicrob. Chemother.* **2012**, *68*. [CrossRef]

11. Meletis, G. Carbapenem resistance: Overview of the problem and future perspectives. *Ther. Adv. Infect. Dis.* **2015**, *3*, 15–21. [CrossRef]

12. Wyres, K.; Holt, K. *Klebsiella pneumoniae* as a key trafficker of drug resistance genes from environmental to clinically important bacteria. *Curr. Opin. Microbiol.* **2018**, *45*, 131–139. [CrossRef]

13. Chang, D.; Sharma, L.; Cruz, C.S.D.; Zhang, D. Clinical Epidemiology, Risk Factors, and Control Strategies of *Klebsiella pneumoniae* Infection. *Front. Microbiol.* **2021**, *12*, 750662. [CrossRef] [PubMed]

14. Mo, Y.; Hernandez-Koutoucheva, A.; Musicha, P.; Bertrand, D.; Lye, D.; Ng, O.T.; Fenlon, S.N.; Chen, S.L.; Ling, M.L.; Tang, W.Y.; et al. Duration of Carbapenemase-Producing Enterobacteriaceae Carriage in Hospital Patients. *Emerg. Infect. Dis.* **2020**, *26*, 2182–2185. [CrossRef] [PubMed]

15. Hernández-García, M.; García-Fernández, S.; García-Castillo, M.; Melo-Cristino, J.; Pinto, M.F.; Gonçalves, E.; Alves, V.; Costa, E.; Ramalheira, E.; Sancho, L.; et al. Confronting Ceftolozane-Tazobactam Susceptibility in Multidrug-Resistant Enterobacterales Isolates and Whole-Genome Sequencing Results (STEP Study). *Int. J. Antimicrob. Agents* **2020**, *57*, 106259. [CrossRef] [PubMed]

16. Fontana, C.; Angeletti, S.; Mirandola, W.; Cella, E.; Alessia, L.; Zehender, G.; Favaro, M.; Leoni, D.; Rose, D.D.; Gherardi, G.; et al. Whole genome sequencing of carbapenem-resistant *Klebsiella pneumoniae*: Evolutionary analysis for outbreak investigation. *Future Microbiol.* **2020**, *15*, 203–212. [CrossRef]

17. Galani, I.; Karaiskos, I.; Souli, M.; Papoutsaki, V.; Galani, L.; Gkoufa, A.; Antoniadou, A.; Giamarellou, H. Outbreak of KPC-2-producing *Klebsiella pneumoniae* endowed with ceftazidime-avibactam resistance mediated through a VEB-1-mutant (VEB-25), Greece, September to October 2019. *Eurosurveillance* **2020**, *25*, 2000028. [CrossRef] [PubMed]

18. Ferrari, C.; Corbella, M.; Gaiarsa, S.; Comandatore, F.; Scaltriti, E.; Bandi, C.; Cambieri, P.; Marone, P.; Sassera, D. Multiple *Klebsiella pneumoniae* KPC Clones Contribute to an Extended Hospital Outbreak. *Front. Microbiol.* **2019**, *10*, 2767. [CrossRef]
19. Mavroidi, A.; Katsiari, M.; Likousi, S.; Palla, E.; Roussou, Z.; Nikolaou, C.; Mathas, C.; Merkouri, E.; Platsouka, E.D. Changing Characteristics and In Vitro Susceptibility to Ceftazidime/Avibactam of Bloodstream Extensively Drug-Resistant *Klebsiella pneumoniae* from a Greek Intensive Care Unit. *Microb. Drug Resist.* **2020**, *26*, 28–37. [CrossRef]
20. Gona, F.; Bongiorno, D.; Aprile, A.; Corazza, E.; Pasqua, B.; Scuderi, M.G.; Chiacchiaretta, M.; Cirillo, D.M.; Stefani, S.; Mezzatesta, M.L. Emergence of two novel sequence types (3366 and 3367) NDM-1- and OXA-48-co-producing K. pneumoniae in Italy. *Eur. J. Clin. Microbiol. Infect. Dis.* **2019**, *38*, 1687–1691. [CrossRef]
21. Karampatakis, T.; Tsergouli, K.; Politi, L.; Diamantopoulou, G.; Iosifidis, E.; Antachopoulos, C.; Karyoti, A.; Mouloudi, E.; Tsakris, A.; Roilides, E. Molecular Epidemiology of Endemic Carbapenem-Resistant Gram-Negative Bacteria in an Intensive Care Unit. *Microb. Drug Resist.* **2019**, *25*, 712–716. [CrossRef]
22. Papadimitriou-Olivgeris, M.; Bartzavali, C.; Spyropoulou, A.; Lambropoulou, A.; Sioulas, N.; Vamvakopoulou, S.; Karpetas, G.; Spiliopoulou, I.; Vrettos, T.; Anastassiou, E.D.; et al. Molecular epidemiology and risk factors for colistin- or tigecycline-resistant carbapenemase-producing *Klebsiella pneumoniae* bloodstream infection in critically ill patients during a 7-year period. *Diagn. Microbiol. Infect. Dis.* **2018**, *92*, 235–240. [CrossRef]
23. Avgoulea, K.; Di Pilato, V.; Zarkotou, O.; Sennati, S.; Politi, L.; Cannatelli, A.; Themeli-Digalaki, K.; Giani, T.; Tsakris, A.; Rossolini, G.M.; et al. Characterization of Extensively Drug-Resistant or Pandrug-Resistant Sequence Type 147 and 101 OXA-48-Producing *Klebsiella pneumoniae* Causing Bloodstream Infections in Patients in an Intensive Care Unit. *Antimicrob. Agents Chemother.* **2018**, *62*, e02457-17. [CrossRef] [PubMed]
24. Ripabelli, G.; Tamburro, M.; Guerrizio, G.; Fanelli, I.; Flocco, R.; Scutellà, M.; Sammarco, M.L. Tracking Multidrug-Resistant *Klebsiella pneumoniae* from an Italian Hospital: Molecular Epidemiology and Surveillance by PFGE, RAPD and PCR-Based Resistance Genes Prevalence. *Curr. Microbiol.* **2018**, *75*, 977–987. [CrossRef] [PubMed]
25. Bartolini, A.; Basso, M.; Franchin, E.; Menegotto, N.; Ferrari, A.; De Canale, E.; Andreis, S.; Scaggiante, R.; Stefani, S.; Palù, G.; et al. Prevalence, molecular epidemiology and intra-hospital acquisition of *Klebsiella pneumoniae* strains producing carbapenemases in an Italian teaching hospital from January 2015 to September 2016. *Int. J. Infect. Dis.* **2017**, *59*, 103–109. [CrossRef] [PubMed]
26. Mavroidi, A.; Katsiari, M.; Likousi, S.; Palla, E.; Roussou, Z.; Nikolaou, C.; Maguina, A.; Platsouka, E.D. Characterization of ST258 Colistin-Resistant, blaKPC-Producing *Klebsiella pneumoniae* in a Greek Hospital. *Microb. Drug Resist.* **2016**, *22*, 392–398. [CrossRef] [PubMed]
27. Bonura, C.; Giuffrè, M.; Aleo, A.; Fasciana, T.; Di Bernardo, F.; Stampone, T.; Giammanco, A.; Palma, D.M.; Mammina, C. The MDR-GN Working Group an Update of the Evolving Epidemic of blaKPC Carrying *Klebsiella pneumoniae* in Sicily, Italy, 2014: Emergence of Multiple Non-ST258 Clones. *PLoS ONE* **2015**, *10*, e0132936. [CrossRef]
28. Onori, R.; Gaiarsa, S.; Comandatore, F.; Pongolini, S.; Brisse, S.; Colombo, A.; Cassani, G.; Marone, P.; Grossi, P.; Minoja, G.; et al. Tracking Nosocomial *Klebsiella pneumoniae* Infections and Outbreaks by Whole-Genome Analysis: Small-Scale Italian Scenario within a Single Hospital. *J. Clin. Microbiol.* **2015**, *53*, 2861–2868. [CrossRef]
29. Parisi, S.G.; Bartolini, A.; Santacatterina, E.; Castellani, E.; Ghirardo, R.; Berto, A.; Franchin, E.; Menegotto, N.; De Canale, E.; Tommasini, T.; et al. Prevalence of *Klebsiella pneumoniae* strains producing carbapenemases and increase of resistance to colistin in an Italian teaching hospital from January 2012 to December 2014. *BMC Infect. Dis.* **2015**, *15*, 244. [CrossRef] [PubMed]
30. Katsiari, M.; Panagiota, G.; Likousi, S.; Roussou, Z.; Polemis, M.; Vatopoulos, C.A.; Platsouka, D.E.; Maguina, A. Carbapenem-resistant *Klebsiella pneumoniae* infections in a Greek intensive care unit: Molecular characterisation and treatment challenges. *J. Glob. Antimicrob. Resist.* **2015**, *3*, 123–127. [CrossRef]
31. Mezzatesta, M.L.; Caio, C.; Gona, F.; Cormaci, R.; Salerno, I.; Zingali, T.; Denaro, C.; Gennaro, M.; Quattrone, C.; Stefani, S. Carbapenem and multidrug resistance in Gram-negative bacteria in a single centre in Italy: Considerations on in vitro assay of active drugs. *Int. J. Antimicrob. Agents* **2014**, *44*, 112–116. [CrossRef]
32. Papadimitriou-Olivgeris, M.; Marangos, M.; Christofidou, M.; Fligou, F.; Bartzavali, C.; Panteli, E.S.; Vamvakopoulou, S.; Filos, K.S.; Anastassiou, E.D. Risk factors for infection and predictors of mortality among patients with KPC-producing *Klebsiella pneumoniae* bloodstream infections in the intensive care unit. *Scand. J. Infect. Dis.* **2014**, *46*, 642–648. [CrossRef]
33. Capone, A.; Giannella, M.; Fortini, D.; Giordano, A.; Meledandri, M.; Ballardini, M.; Venditti, C.; Bordi, E.; Capozzi, D.; Balice, M.P.; et al. High rate of colistin resistance among patients with carbapenem-resistant *Klebsiella pneumoniae* infection accounts for an excess of mortality. *Clin. Microbiol. Infect.* **2013**, *19*, E23–E30. [CrossRef] [PubMed]
34. Tofteland, S.; Naseer, U.; Lislevand, J.H.; Sundsfjord, A.; Samuelsen, Ø. A Long-Term Low-Frequency Hospital Outbreak of KPC-Producing *Klebsiella pneumoniae* Involving Intergenus Plasmid Diffusion and a Persisting Environmental Reservoir. *PLoS ONE* **2013**, *8*, e59015. [CrossRef]
35. Mammina, C.; Bonura, C.; Di Bernardo, F.; Aleo, A.; Fasciana, T.; Sodano, C.; Saporito, M.A.; Verde, M.S.; Tetamo, R.; Palma, D.M. Ongoing spread of colistin-resistant *Klebsiella pneumoniae* in different wards of an acute general hospital, Italy, June to December 2011. *Euro Surveill.* **2012**, *17*, 20248. [CrossRef] [PubMed]
36. Richter, S.N.; Frasson, I.; Franchin, E.; Bergo, C.; Lavezzo, E.; Barzon, L.; Cavallaro, A.; Palù, G. KPC-mediated resistance in *Klebsiella pneumoniae* in two hospitals in Padua, Italy, June 2009-December 2011: Massive spreading of a KPC-3-encoding plasmid and involvement of non-intensive care units. *Gut Pathog.* **2012**, *4*, 7. [CrossRef] [PubMed]

37. Sánchez-Romero, I.; Asensio, Á.; Oteo, J.; Muñoz-Algarra, M.; Isidoro, B.; Vindel, A.; Cuevas, O.; Fernández-Romero, S.; Azañedo, L.; Campos, J.; et al. Nosocomial outbreak of VIM-1-producing *Klebsiella pneumoniae* isolates of multilocus sequence type 15: Molecular basis, clinical risk factors, and outcome. *Antimicrob. Agents Chemother.* **2012**, *56*, 420–427. [CrossRef] [PubMed]

38. Souli, M.; Galani, I.; Antoniadou, A.; Papadomichelakis, E.; Poulakou, G.; Panagea, T.; Vourli, S.; Zerva, L.; Armaganidis, A.; Kanellakopoulou, K.; et al. An Outbreak of Infection due to β-Lactamase *Klebsiella pneumoniae* Carbapenemase 2–ProducingK. pneumoniaein a Greek University Hospital: Molecular Characterization, Epidemiology, and Outcomes. *Clin. Infect. Dis.* **2010**, *50*, 364–373. [CrossRef]

39. Giakkoupi, P.; Xanthaki, A.; Kanelopoulou, M.; Vlahaki, A.; Miriagou, V.; Kontou, S.; Papafraggas, E.; Malamou-Lada, H.; Tzouvelekis, L.S.; Legakis, N.J.; et al. VIM-1 Metallo-β-Lactamase-Producing *Klebsiella pneumoniae* Strains in Greek Hospitals. *J. Clin. Microbiol.* **2003**, *41*, 3893–3896. [CrossRef]

40. Protonotariou, E.; Poulou, A.; Politi, L.; Sgouropoulos, I.; Metallidis, S.; Kachrimanidou, M.; Pournaras, S.; Tsakris, A.; Skoura, L. Hospital outbreak due to a *Klebsiella pneumoniae* ST147 clonal strain co-producing KPC-2 and VIM-1 carbapenemases in a tertiary teaching hospital in Northern Greece. *Int. J. Antimicrob. Agents* **2018**, *52*, 331–337. [CrossRef]

41. Protonotariou, E.; Poulou, A.; Vasilaki, O.; Papadopoulou, D.; Politi, L.; Kagkalou, G.; Pilalas, D.; Kachrimanidou, M.; Draganoudis, V.; Tsocha, A.; et al. Emergence of NDM-1 producing *Klebsiella pneumoniae* clinical isolates in northern Greece: A 34-month epidemiological study in a tertiary university hospital (P0270). In Proceedings of the 27th European Congress of Clinical Microbiology and Infectious Diseases, ECCMID 2017, Vienna, Austria, 22–25 April 2017.

42. Protonotariou, E.; Meletis, G.; Chatzopoulou, F.; Malousi, A.; Chatzidimitriou, D.; Skoura, L. Emergence of *Klebsiella pneumoniae* ST11 co-producing NDM-1 and OXA-48 carbapenemases in Greece. *J. Glob. Antimicrob. Resist.* **2019**, *19*, 81–82. [CrossRef]

43. Partridge, S.R.; Kwong, S.M.; Firth, N.; Jensen, S.O. Mobile Genetic Elements Associated with Antimicrobial Resistance. *Clin. Microbiol. Rev.* **2018**, *31*, e00088-17. [CrossRef] [PubMed]

44. Zagorianou, A.; Sianou, E.; Iosifidis, E.; Dimou, V.; Protonotariou, E.; Miyakis, S.; Roilides, E.; Sofianou, D. Microbiological and molecular characteristics of carbapenemase-producing *Klebsiella pneumoniae* endemic in a tertiary Greek hospital during 2004–2010. *Euro Surveill.* **2012**, *17*, 20088. [CrossRef]

45. Galani, I.; Karaiskos, I.; Karantani, I.; Papoutsaki, V.; Maraki, S.; Papaioannou, V.; Kazila, P.; Tsorlini, H.; Charalampaki, N.; Toutouza, M.; et al. Epidemiology and resistance phenotypes of carbapenemase-producing *Klebsiella pneumoniae* in Greece, 2014 to 2016. *Eurosurveillance* **2018**, *23*, 1700775. [CrossRef]

46. Ghenea, A.; Cioboată, R.; Drocaş, A.; Țieranu, E.; Vasile, C.; Moroşanu, A.; Țieranu, C.; Salan, A.-I.; Popescu, M.; Turculeanu, A.; et al. Prevalence and Antimicrobial Resistance of *Klebsiella* Strains Isolated from a County Hospital in Romania. *Antibiotics* **2021**, *10*, 868. [CrossRef]

47. Wayne, P.A. Clinical and Laboratory Standards Institute. Performance Standards for antimicrobial susceptibility testing; twenty-fifth in-formational supplement. Document M100-S25. *Inform. Suppl.* **2011**, *31*, 100–121.

48. Tsakris, A.; Poulou, A.; Pournaras, S.; Voulgari, E.; Vrioni, G.; Themeli-Digalaki, K.; Petropoulou, D.; Sofianou, D. A simple phenotypic method for the differentiation of metallo- b -lactamases and class A KPC carbapenemases in Enterobacteriaceae clinical isolates. *J. Antibicrob. Chemother.* **2010**, *65*, 1664–1671. [CrossRef] [PubMed]

49. Meletis, G.; Tzampaz, E.; Sianou, E. Phenotypic and molecular methods for the detection of antibiotic resistance mechanisms in Gram negative nosocomial pathogens. In *Trends in Infectious Diseases*; IntechOpen: London, UK, 2014; pp. 139–162.

50. Han, H.; Zhou, H.; Li, H.; Gao, Y.; Lu, Z.; Hu, K.; Xu, B. Optimization of Pulse-Field Gel Electrophoresis for Subtyping of *Klebsiella pneumoniae*. *Int. J. Environ. Res. Public Health* **2013**, *10*, 2720–2731. [CrossRef]

Article

NDM-1 Introduction in Portugal through a ST11 KL105 *Klebsiella pneumoniae* Widespread in Europe

Ângela Novais [1,2,*], Rita Veiga Ferraz [3], Mariana Viana [3], Paula Martins da Costa [3,4] and Luísa Peixe [1,2,*]

1 UCIBIO, Laboratório de Microbiologia, Departamento de Ciências Biológicas, Faculdade de Farmácia, Universidade do Porto, 4050-313 Porto, Portugal

2 Associate Laboratory i4HB—Institute for Health and Bioeconomy, Faculty of Pharmacy, University of Porto, 4050-313 Porto, Portugal

3 Centro Hospitalar do Tâmega e Sousa, 4564-007 Penafiel, Portugal; rita.ferraz@chts.min-saude.pt (R.V.F.); 60565@chts.min-saude.pt (M.V.); pcmcosta@live.com.pt (P.M.d.C.)

4 Department of Clinical Chemistry, Centro Hospitalar Universitário do Porto, 4099-001 Porto, Portugal

* Correspondence: aamorim@ff.up.pt (Â.N.); lpeixe@ff.up.pt (L.P.); Tel.: +351-220-428-580 (Â.N. & L.P.)

Citation: Novais, Â.; Ferraz, R.V.; Viana, M.; da Costa, P.M.; Peixe, L. NDM-1 Introduction in Portugal through a ST11 KL105 *Klebsiella pneumoniae* Widespread in Europe. *Antibiotics* **2022**, *11*, 92. https://doi.org/10.3390/antibiotics11010092

Academic Editor: Francesca Andreoni

Received: 19 November 2021
Accepted: 8 January 2022
Published: 12 January 2022

Publisher's Note: MDPI stays neutral with regard to jurisdictional claims in published maps and institutional affiliations.

Abstract: The changing epidemiology of carbapenem-resistant *Klebsiella pneumoniae* in Southern European countries is challenging for infection control, and it is critical to identify and track new genetic entities (genes, carbapenemases, clones) quickly and with high precision. We aimed to characterize the strain responsible for the first recognized outbreak by an NDM-1-producing *K. pneumoniae* in Portugal, and to elucidate its diffusion in an international context. NDM-1-producing multidrug-resistant *K. pneumoniae* isolates from hospitalized patients (2018–2019) were characterized using FTIR spectroscopy, molecular typing, whole-genome sequencing, and comparative genomics with available *K. pneumoniae* ST11 KL105 genomes. FT-IR spectroscopy allowed the rapid (ca. 4 h after incubation) identification of the outbreak strains as ST11 KL105, supporting outbreak control. Epidemiological information supports a community source but without linkage to endemic regions of NDM-1 producers. Whole-genome comparison with previous DHA-1-producing ST11 KL105 strains revealed the presence of different plasmid types and antibiotic resistance traits, suggesting the entry of a new strain. In fact, this ST11 KL105 clade has successfully disseminated in Europe with variable beta-lactamases, but essentially as ESBL or DHA-1 producers. We expand the distribution map of NDM-1-producing *K. pneumoniae* in Europe, at the expense of a successfully established ST11 KL105 *K. pneumoniae* clade circulating with variable plasmid backgrounds and beta-lactamases. Our work further supports the use of FT-IR as an asset to support quick infection control.

Keywords: nosocomial infections; carbapenemase-producing Enterobacterales; NDM-1; whole-genome sequencing; outbreak; Fourier transform infrared spectroscopy; infection control

1. Introduction

The dissemination of carbapenem resistant *K. pneumoniae* (CRKP) is a well-known problem, but there is a great asymmetry in the geographic distribution of carbapenemase types and *K. pneumoniae* lineages across countries [1–3]. Moreover, changing epidemiology over time is a very challenging situation that affects several countries from Southern Europe [4–7]. Knowing and to understanding these dynamics is essential in order to promptly recognize new genetic entities with enhanced antibiotic resistance, virulence and/or transmission, and to allow a prompt redesign of diagnostic tools and infection control policies to curtail further dissemination [2,8]. Whole-genome sequencing has provided the greatest resolution in establishing transmission pathways at local and global levels of CRKP [1,2], and in enlightening the phylogeny and evolutionary history of some highly frequent KP clonal groups (e.g., CG258, CG307) [8–10]. However, it is still unable to support real-time infection control decisions.

ST11 is one of the most diffused clones, and is especially prevalent in Asia and particularly in China [11]. The heterogeneity within ST11 is well recognized in studies on this geographic area, with clades reflecting capsule recombination events (KL64, KL47, KL105) [9,12]. For this reason, because Fourier transform infrared (FT-IR) spectroscopy is a suitable tool to identify *K. pneumoniae* capsular (KL) types with a fast turnaround time [13,14], it is ideal to discriminate and identify ST11 variants, which are understudied in Europe.

In Portugal, resistance to carbapenems is a problem with growing proportions, especially among *K. pneumoniae*, of which the rates among invasive isolates recorded in national surveillance networks increased ca. 80% in just three years (2014–2017) [15]. Reporting of CPE from national laboratories has been mandatory since 2013 and there have been official regulations for the management of outbreaks and for CPE infection control since 2017 [4]. The available epidemiological and population analysis data suggest the dominance all over the country of particular genetic traits (the ST147 clone, KPC-3 carbapenemase) [16–19], with an increasing diversity of species, carbapenemases types and clones [2,4,20]. KPC-3 and to a lesser extent OXA-48 or OXA-181, which are occasionally associated with GES-5 [18,20], are the most common carbapenemase types among *K. pneumoniae*. NDM-1 has been sporadically described in *Providencia stuartii*, *Morganella morganii* and *Proteus mirabilis* [21,22] but, prior to this study, no cases of colonization and/or infection with NDM-producing *K. pneumoniae* were reported. NDM-1 is especially prevalent in India, the Middle East and the Balkans, and its emergence in other countries (especially in *Escherichia coli* or *K. pneumoniae*) is considered a public health priority due to the risk of subsequent (frequently polyclonal) dissemination [3,23,24].

In this study, we used FT-IR spectroscopy to support the identification and control of the first recognized outbreak of NDM-1-producing ST11 capsular-type KL105 *K. pneumoniae* in Portugal. Furthermore, we used whole-genome sequencing to trace the variability and diffusion of this particular ST11 *K. pneumoniae* clade from an international perspective.

2. Results

2.1. Epidemiological Context and Infection Control Measures

We identified and characterized the first recognized outbreak of NDM-1-producing *K. pneumoniae* isolates in Portugal, involving a total of seven patients. The index case was from a patient (patient 1) with no history of previous hospitalization or travel abroad, whereas most other cases were linked to concomitant hospitalization in the surgical unit at the same time as the index patient (patients 2–6) (Table 1). None of the other patients had a record of foreign travel in the 3 months prior to strain isolation. Patient 3 had been previously hospitalized in a different hospital but screening at admission was negative. Patient 7 came from a household and no epidemiological link could be established with other patients, but this patient had also been previously hospitalized and tested negative at first admission.

The patient cohort, screening of putative carriers, contact precautions and the reinforcement of cleaning and disinfection practices prevented new cases. Regular screening of CPE carriers at admission and during hospitalization, as well as weekly audits to assess compliance to the protocols and implement corrective measures, were used to monitor for the appearance of new cases, and these are still in practice to control the spread of CPE.

In the three months prior to NDM-1 detection, the patients who received antibiotherapy were treated with beta-lactam/beta-lactamase inhibitors, cefuroxime, ceftazidime and ciprofloxacin.

Table 1. Epidemiological and clinical data of patients colonized/infected by NDM-1-producing *K. pneumoniae* ST11-KL105.

Patient N°	Sex/Age	Isolates	Date of Isolation	Hospital Unit	Period of Hospitalization	Patient's Origin	Sample Type	Pathology	Colonization at Admission	Antibiotherapy (Previous 3 Months)	Previous Hospitalization (Date, Unit)
1	M/71	K607 **K606 ***	27/11/18 28/11/18	Surgery	26–30 October 2018	Hospital	C (bile) C (RS)	Cholangiocarcinoma	Not tested	Piperacillin-Tazobactam	None
2	M/79	K608	28/11/18	Surgery	15 November–3 December 2018	Hospital	C (RS)	Acute pancreatitis	Not tested	None	None
3	M/55	K609	30/11/18	Surgery	24 October–5 December 2018	Hospital	C (RS)	Trauma	-	Amoxacillin-clavulanic acid	15 September–22 October 2018 (different hospital)
4	M/66	K610	30/11/18	Surgery	23–30 November 2018	Hospital	C (RS)	Adenocarcinoma	Not tested	None	None
5	M/66	K611	14/12/19	Emergency	14 December 2018–24 January 2019	Outpatient	C (RS)	Colon carcinoma	+	Ciprofloxacin	10–28 November 2018 (Surgery)
6	M/68	K612	25/12/19	Emergency	25 December 2018–11 January 2019 (Medicine)	Outpatient	C (RS)	Soft tissue infection	+	Tigecycline, Ceftazidime	31 October–9 December 2018 (Surgery)
7	M/87	**K624 *** K625	25/03/19	Emergency	25 March–18 April 2019	Household	I (urine) C (RS)	Pneumonia	-	Cefuroxime	28 January–4 March 2019

* Isolates selected for whole-genome sequencing. M = male; C = colonization; I = infection; RS = rectal swab.

2.2. Strain Relatedness

FT-IR spectroscopy was performed retrospectively on suspected isolates received at the Faculty of Pharmacy, University of Porto. After reception, our FT-IR based workflow allowed us to establish a close relationship between the first five isolates, and to communicate strain relatedness less than 4 h after bacterial culture. Comparison of the spectra with our database allowed the classification of the K-type as KL105, together with previous DHA-1 producing ST11-KL105 isolates that had been circulating in Portuguese hospitals for years (Figure S1) [25]. The four isolates identified subsequently clustered with these, confirming the K-type and strain relatedness by means of FTIR, which was further confirmed using PFGE for all studied isolates (data not shown). The genes $bla_{\mathrm{NDM-1}}$, as well as $bla_{\mathrm{CTX-M-15}}$, were detected in all isolates via PCR, and all isolates exhibited identical susceptibility profiles.

2.3. Molecular Characterization

Isolates revealed multidrug resistance profiles, being resistant to all antibiotics tested except amikacin, tigecycline and colistin. Sequenced isolates carried several antibiotic resistance genes ($bla_{\mathrm{NDM-1}}$, $bla_{\mathrm{SHV-182}}$, $bla_{\mathrm{CTX-M-15}}$, $aac(6')$-*Ib-cr*, $aac(3)$-*IId*, $aph(3'')$-*Ib*, $aph(6)$-*Id*, *oqxAB*, *qnrB1*, *catB3*, *sul2*, *tetD* and *dfrA14*), supporting the phenotypic data. They also carried chromosomal mutations in *gyrA* (S83I) and *parC* (S80I), and UhpT (E350Q), conferring reduced susceptibility to fluoroquinolones and fosfomycin, respectively. They were also enriched with virulence genes encoding enterobactin (*entB*), yersiniabactin (*irp1/2*, *ybt10* associated with ICEKp4), iron receptors (*fyuA*, *kfu*BC) and type I and type III fimbriae (*fimH*, *mrk*ABC). All strains had *wzi75*, which corresponds to a novel capsular type (KL105), and KLEBORATE confirmed 100% identity with the KL105 locus and the O2v2 O antigen locus.

2.4. Whole-Genome Data and Phylogenetic Analysis

The genomes of K606 and K624 strains had 5.5 Gb, a GC content of 57.2% and were assembled into 76–97 contigs (data not shown). The SNP-based phylogenetic tree was constructed on the basis of 4.87 Gb common positions covered by the core genome, corresponding to 88.5% of the genome of the reference strain. We observed a close relatedness between the available ST11 KL105 genomes from Europe, America and China (2013–2017) since isolates varied at 0–297 sites distributed over their 4.8 million base pair (bp) core genome (~medium 31SNP/Gb) (Figure 1, Supplementary Figure S2). All of them carried $bla_{\mathrm{SHV-182}}$ in the chromosome.

Two main branches were identified within the tree (Figure 1). One of these included highly related (<20 SNPs) KPC-2-producing strains from China and carried IncR and ColRNAI plasmids and a few additional antibiotic resistance genes. The second branch included isolates producing CTX-M-3, CTX-M-15 or DHA-1, occasionally together with NDM-1, most of them (n = 13/15; 87%) identified in Europe, and many from the Euscape project [2]. Most (92%) of these carried several antibiotic resistance genes, some of which were signatures of particular genetic platforms (e.g., *catB3*, *arr-3*, *qnrB4*). Most carried R replicons, but also FIA(HI1), FIIk and/or FIB, together with variable F replicons (FIA, FIB and/or FIIk, and colRNAI) (Figure 1). All of them consistently carried *ybt10* (ICEKp4), *ipr1/irp2* and *fyuA2* virulence genes.

The SNP-based matrix (supplementary Figure S2) showed that the two sequenced strains differ in only six SNPs, supporting their close relatedness. However, due to the absence of an epidemiological link between both patients, it is reasonable to consider an unrecognized source. Both strains showed ~100 SNPs (20 SNP/Gb) with previous DHA-1 ST11-KL105 producers circulating in Portuguese hospitals (Whole Genome Shotgun project accession N° QTTC00000000 and QTTD00000000). Three of the public genomes, corresponding to isolates from Romania and the USA, also carried $bla_{\mathrm{NDM-1}}$ and showed only 80–90 SNP differences with our outbreak strains.

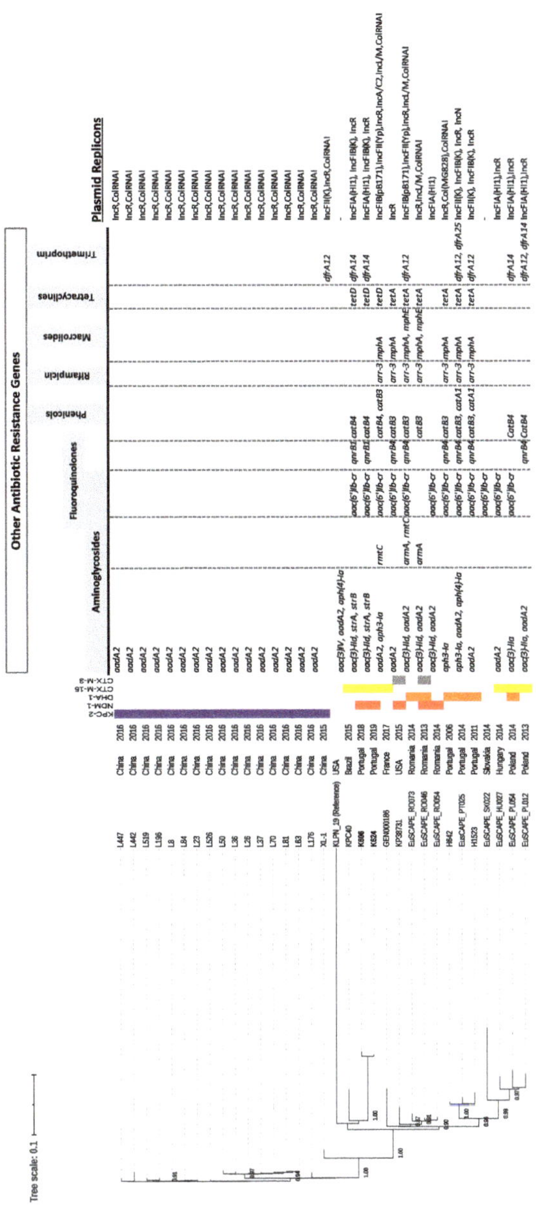

Figure 1. SNP-based phylogenetic tree of ST11-KL105 genomes and heatmap of antimicrobial resistance genotype. Genomes sequenced from this study (shown in bold) were compared with those from 28 other ST11-KL105 strains deposited in public databases (accessed on 5 November 2019). Phylogenetic tree was obtained using the SNP method of CSI phylogeny with default parameters and the reference strain KLPN_19, the closest ST11 with KL107 (accession number GCF_003861305.1). Country and year of isolation are shown, as represented in metadata, as well as their content of acquired beta-lactamases (ESBL, acquired AmpC and carbapenemases), other antibiotic resistance genes and replicons. Numbers in the left indicate bootstrap values >0.85, calculated based on the maximum-likelihood method.

2.5. Genetic Support for bla$_{NDM-1}$ Acquisition

The *bla*$_{NDM-1}$ gene was identified in pAN_K624, an 84 Kb IncR+IncFIA(HI1) multir-replicon plasmid with high identity (100%) and coverage (92%) to the pAR_0109 plasmid (Genbank acc. Number CP032212), which is of an unknown origin and which carries the typical genetic environment of *bla*$_{CTX-M-15}$. *bla*$_{NDM-1}$ is located in a Tn*125*-like multidrug resistance region (MRR) comprising deltaISAba125, *bla*$_{NDM-1}$, *ble*$_{MBL}$, *trpF*, *dsbC*, *cutA*, *groES*, *groEL*, *psp*, *qnrB1* and SDR family reductase, 99% identical to that of the reference plasmid pNDM-MAR (Genbank acc. Number JN420336). The MRR region (~8.5Kb) containing *bla*$_{NDM-1}$ was flanked by an IS*3000* element that might have mediated its insertion into the backbone of a pAR_0109-like plasmid between a Tn*5403*-like transposon and *pspF* (Figure 2).

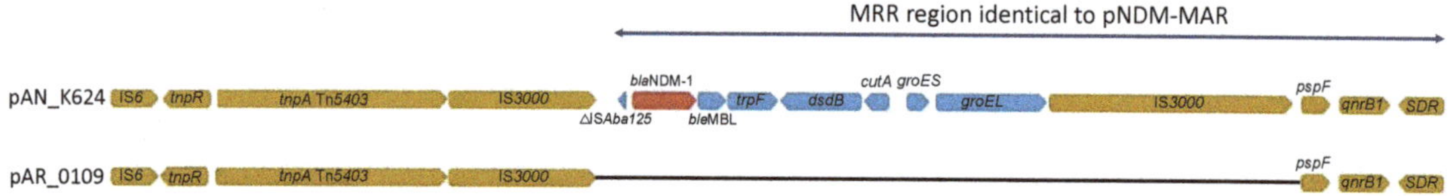

Figure 2. Genetic environment of *bla*$_{NDM-1}$ in plasmid pAN_K624, represented using Geneious version 9.1.8. The location of the insertion of the MRR region into the pAR_0109-like plasmid backbone is shown.

Additionally, these strains also carried a 149 Kb IncFIB(K) plasmid with high identity and coverage (>99%) with the ~160 Kb p002SK2_A plasmid (Genbank acc. Number CP025516), a non-antibiotic resistant plasmid identified in a K. pneumoniae ST147 isolate from wastewater in Switzerland. There were no traces of the DHA-1-producing pKPS30 IncR plasmid previously identified in ST11-KL105 isolates circulating in Portugal [25], nor of the IncC plasmid previously detected with bla$_{NDM-1}$ in species other than K. pneumoniae [21]. Likewise, none of the other NDM-1-producing ST11 KL105 public strains carried a similar plasmid replicon content, suggesting a new plasmid as the vehicle of bla$_{NDM-1}$.

3. Discussion

This study enlarges the distribution map of NDM-1 carbapenemase in Europe and raises the alarm for its emergence in our country in a multidrug-resistant ST11-KL105 *K. pneumoniae* lineage, that has previously caused the long-term dissemination of DHA-1 in Portuguese hospitals [2,25]. Simultaneously, it increases the pool of carbapenemases circulating among high-risk *K. pneumoniae* lineages in our country, which is worrisome considering the possibility of further transmission of *bla*$_{NDM-1}$ to other already circulating high-risk *K. pneumoniae* clones or even to other Enterobacterales [4,17]. In fact, other NDM-1-producing *K. pneumoniae* ST11-KL105 strains with identical PFGE patterns have been more recently identified in other hospitals in our area, suggesting further dissemination and wider expansion (data not shown).

It is of interest to highlight that none of the patients had a record of hospitalization or travel abroad in the 3 months prior to strain isolation, and the absence of an epidemiological link for the index case suggests community acquisition from an unrecognized source. A chain of nosocomial dissemination could be established, and our data further reinforce the importance of contact screening of CPE using fast and reliable methods such as FT-IR, together with quick identification of carbapenemases, in order to optimize infection control, to detect new emergences and prevent subsequent spread [4].

The quick identification of the outbreak using FT-IR spectroscopy (ca. 4 h after incubation) allowed the prompt establishment of infection control measures for outbreak control, and whole-genome sequencing supported the classification of the outbreak strain as ST11 KL105. Our comparative genomic analysis of available ST11 KL105 genomes revealed that this lineage has been found circulating in several European countries since at least 2006, in most cases carrying bla_{DHA-1}, $bla_{CTX-M-3}$ or $bla_{CTX-M-15}$ (Figure 1). This is in agreement with data from the Euscape project [2], in which ca. 62% of ST11 *K. pneumoniae* isolates were non-carbapenemase producers. Here, we additionally demonstrate that these Euscape isolates belong to the same ST11 clade expressing KL105, which differs from variants prevalent in China and eastern European countries [9,12,26].

Our comparative genomic analysis shows that NDM-1 was not acquired by ST11 KL105 strains previously circulating in Portugal [25] but through the introduction of a new ST11 KL105 strain, carrying bla_{NDM-1} in a different genetic context. In fact, comparison of NDM-1- and DHA-1-producing ST11 strains revealed variability of beta-lactamases (bla_{NDM-1}, $bla_{CTX-M-15}$ vs bla_{DHA-1}), antibiotic resistance traits (*qnrB1, catB3, dfrA14* vs *qnrB4, arr-3, catA1, dfrA12*) and plasmids (IncR+FIA(HI1) + IncFIB(K) vs IncR). The pAN_K624 plasmid (IncR+FIA(HI1)) encoding NDM-1 was also different from the IncC platform identified previously in other Enterobacterales in Portugal [21], supporting a new entry of the bla_{NDM-1} gene and the circulation of different platforms encoding NDM-1. Furthermore, our data on the plasmid genetic context suggest the introduction of bla_{NDM-1} into a multidrug resistance plasmid that is different from those carrying bla_{NDM-1} in other ST11 KL105 strains from Europe, suggesting a new platform. Also, it differs from the pKPX-1 plasmid, associated with massive NDM-1 spread in Poland [26]. In the ST11 KL105 clade, the acquisition of bla_{NDM-1} seems to be sporadic, in some cases over DHA-1 or ESBL genetic backgrounds (Figure 1, Supplementary Table S1), contributing to high genetic plasticity and plasmid shuffling in this lineage, extending previous observations [27].

Our work also highlights the need to further deepen the population analysis of *K. pneumoniae* clones to obtain maximum discrimination for the efficient tracking of outbreak strains. Indeed, there is high genetic variability within some of the most frequent clonal groups, including CG11 [2,27,28]. Despite the high capsular diversity identified among the most common clonal groups, many studies have demonstrated an extraordinary selection of lineages with specific capsular types within CG258 [28] and CG11 [9,12], suggesting that the capsule type is a good evolutionary marker of clinically relevant *K. pneumoniae* lineages [13]. Considering this evolutionary scenario, FT-IR based typing based on a representative spectral database is a reliable tool to differentiate these lineages, providing information that, together with epidemiological data, might be of great value in supporting the identification of outbreaks and real-time infection control decisions at a quick and low-cost rate [13,14,29]. Whole-genome sequencing requires a higher cost and time but it can be used downstream to provide full resolution, allowing a thorough understanding of the dynamics of the spread of antimicrobial resistance [8,10,27].

This ST11-KL105 *K. pneumoniae* lineage consistently carries $bla_{SHV-182}$, encoding a natural SHV-type beta-lactamase, according to the BLBD beta-lactamase database (http://bldb.eu accessed on 11 November 2019) [30]. In all the genomes analyzed, $bla_{SHV-182}$ was chromosomally located and might be potentially used as a marker for this specific lineage.

4. Materials and Methods

4.1. Setting

The outbreak occurred at a 500-bed hospital that serves a population of about 520,000 in the north of Portugal, together with another hospital unit at a distance of 30 km. It offers healthcare specialties within medical, surgical and emergency (adults and pediatric), as well as outpatient attendance. Control of the dissemination of CPE bacteria is implemented according to official national recommendations and includes active screening and the reinforcement of infection control procedures. At this hospital, rectal screening of CPE is performed (i) at admission when transferring from another hospital with hospitalizations >48 h or from a household or long-term care facility; (ii) when there are previous (12 months) hospitalizations, (iii) every 7 days of hospitalization. At admission, genes encoding specific carbapenemases were detected using a rapid PCR kit (Xpert CarbaR), whereas subsequent screenings on hospitalized patients were performed using a cultural method.

4.2. Index Case

The first NDM-1-producing *K. pneumoniae* was isolated in the surgical ward in the bile of a 71-year-old patient with a cholangiocarcinoma on 27 November 2018. In the absence of clinical symptoms it was considered possible drain colonization, and CPE rectal screening performed the day after revealed the carriage of a NDM-1-producing isolate. This patient had been hospitalized since 26 October 2018 but there were no criteria for screening at admission, thus this information is missing.

These isolates were resistant to third-generation cephalosporins, carbapenems, aztreonam, beta-lactam/beta-lactamase inhibitors, fluoroquinolones, trimethoprim-sulfamethoxazole, nitrofurantoin and several aminoglycosides; intermediate to fosfomycin; and susceptible only to amikacin, tigecycline and colistin.

4.3. Additional Cases

Six additional patients were identified with NDM-1-encoding *K. pneumoniae*. All patients in the same nursery were considered exposed and were screened for rectal carriage independently on the result at admission (if applicable). Three of them carried a *K. pneumoniae* isolate with the same susceptibility profile. All patients with positive screening results were transferred to cohort areas and subjected to isolation and contact precautions. Four additional isolates were subsequently identified in the rectal swabs or caused a urinary infection in three outpatients that were admitted between December 2018 and March 2019 from the emergency room. Two of them were identified as carriers of NDM-producing *K. pneumoniae* at admission, whereas the last one was negative for carriage at admission during a previous hospitalization but that missed subsequent screenings.

4.4. Bacterial Identification and Carbapenemase Detection

In total, 9 NDM-1-encoding *K. pneumoniae* isolates were identified in seven patients. These isolates were recovered from feces (n = 9), urine (n = 1) and bile (n = 1) (Table 1). Carbapenemase genes were primarily detected via molecular biology using Xpert CarbaR (Cepheid). Bacterial identification and preliminary susceptibility testing were performed using a VITEK 2 system (bioMérieux) or broth microdilution (for colistin). Available clinical and epidemiological data, as well as information on recognized risk factors for infection by CPE, were recorded and analyzed.

4.5. Antibiotic Resistance Phenotypes and Genotypes

Antibiotic susceptibility profiles to 25 antibiotics (all beta-lactams, beta-lactam/beta-lactamase combinations, fluoroquinolones, aminoglycosides, tetracyclines, fosfomycin, trimethoprim-sulfamethoxazole, chloramphenicol) were determined via disk diffusion and interpreted following EUCAST guidelines (www.eucast.org 6 January 2020). Carbapenemase production was confirmed using the Blue-Carba test [31] and the type of carbapenemase was identified via PCR directed to the most frequent carbapenemase gene families (bla_{KPC}, bla_{OXA-48}, bla_{IMP}; bla_{VIM}, bla_{NDM}) and further sequencing [17,32]. The presence of $bla_{CTX-M-15}$ was confirmed using PCR in all isolates.

4.6. Identification of Outbreak Strains

We used Fourier transform infrared (FTIR) spectroscopy to assess isolates' relationships according to our previously established workflow and models based on the comparison of bacterial discriminatory spectra. The analysis was performed retrospectively on suspected isolates received in early December 2018 and March 2019. Upon reception, isolates were re-grown in standardized culture conditions (37 °C/18 h), after which a colony was directly deposited on the ATR accessory of our FT-IR instrument (Perkin-Elmer Frontier) and air-dried. Three spectra per strain were acquired (technical replicates) from $600\ cm^{-1}$ to $4000\ cm^{-1}$, and the spectral region of interest (accumulating polysaccharide vibrations) was compared with spectra included in our in-house *K. pneumoniae* database and machine-learning model [13]. The K-type and clone of the suspected strains were then predicted with an accuracy of 100%, because of previously established correlations in internationally circulating lineages [13]. These FT-IR-based assignments were confirmed by means of PCR and sequencing of the *wzi* gene [33], as well as pulsed-field gel electrophoresis.

4.7. Whole Genome and Phylogenetic Analysis

Two NDM-1-producing isolates (the first and the last outbreak strain with no epidemiological link) were selected for whole genome sequencing. The DNA was extracted using the Wizard Genomic DNA purification kit (Promega Corporation, Madison, WI, USA) according to the manufacturer's instructions and the concentration was determined with a Qubit 3.0 Fluorometer (Invitrogen, Thermo Fisher Scientific, Waltham, MA, USA). DNA was sequenced using an Illumina Novaseq 6000 ($2\times$ 300 bp pair-ended runs, ~6 Gb genome, coverage $100\times$), reads were trimmed de novo using Trimommatic 0.39 to remove adapters (http://www.usadellab.org/cms/?page=trimmomatic accessed on 5 November 2019) and assembled using SPAdes version 3.9.0 (cab.spbu.ru/software/spades/ accessed on 5 November 2019). The quality of the reads and the assembly were assessed using FastQC v0.11.9 (http://www.bioinformatics.babraham.ac.uk/projects/fastqc/ accessed on 5 November 2019) and Quast (http://cab.cc.spbu.ru/quast/ accessed on 5 November 2019). Contigs were further annotated with RAST (https://rast.nmpdr.org accessed on 6 November 2019). This Whole Genome Shotgun project has been deposited at DDBJ/ENA/GenBank under the BioProject accession number PRJNA484888, with Biosamples SAMN16522901 and SAMN16522902.

Provided with a snapshot of the ST11 clade distribution in previous papers [9,12], we decided to focus on understanding the phylogeny and geographic distribution of the ST11 KL105 clade since these isolates were poorly represented in those series. Twenty-eight ST11-KL105 *K. pneumoniae* genomes available on GenBank database (https://www.ncbi.nlm.nih.gov/genbank accessed on 18 November 2019) were included for comparison (Supplementary Table S1). Antibiotic resistance and replicon content were extracted from Center for Genomic Epidemiology tools (ResFinder, PlasmidFinder at http://www.genomicepidemiology.org accessed on 5 November 2019) or CARD (https://card.mcmaster.ca accessed on 5 November 2019). Plasmid contigs were identified using mlplasmids [34]. Plasmid assembly and partial reconstruction was obtained, mapping ST11 contigs on reference plasmid sequences and BLAST (https://blast.ncbi.nlm.nih.gov/Blast.cgi accessed on 5 November 2019). Geneious version 9.1.8 was used to represent the $bla_{\text{NDM-1}}$ genetic environment.

The virulence gene content and *wzi*, capsular (KL) and O antigen loci were assessed using the *K. pneumoniae* Institut Pasteur database (https://bigsdb.pasteur.fr/cgi-bin/bigsdb/bigsdb.pl?db=pubmlst_klebsiella_seqdef accessed on 5 November 2019) and KLEBORATE (https://github.com/katholt/Kleborate accessed on 5 November 2019). The core genome of the two sequenced isolates (K606 and K624), the two previously described DHA-1-producing ST11-KL105 strains from Portugal [25] and those of 26 other ST11-KL105 strains deposited in the GenBank database were extracted and used to construct a SNP-based phylogenetic tree using the CSI phylogeny tool, using default parameters (https://cge.cbs.dtu.dk/services/CSIPhylogeny/ accessed on 6 November 2019) and the maximum-likelihood method to calculate bootstrap values. The closest genome corresponding to strain KLPN_19 (a ST11 strain with KL107 GCF_003861305.1) was used as a reference, and the tree was further represented with Interactive Tree of Life (iTOL, https://itol.embl.de accessed on 7 November 2019).

5. Conclusions

Data from the first recognized outbreak of NDM-1-producing *K. pneumoniae* in our country raise the alarm for the potential silent dissemination of NDM-1 through unnoticed carriage in the community, which needs to be monitored to prevent further spread. This report extends the distribution map of NDM-1 carbapenemase in Europe, which emerged through the re-introduction of ST11 KL105 *K. pneumoniae* clade into our country and a new plasmid backbone carrying bla_{NDM-1}. We provide the first comparative analysis of the ST11 KL105 *K. pneumoniae* clade and show that it is successfully established in European countries carrying a variable pool of ESBL, DHA-1 or NDM-1-encoding plasmids. Finally, we further demonstrate the usefulness of the speed and the accuracy of FT-IR spectroscopy to support outbreak identification and infection control.

Supplementary Materials: The following are available https://www.mdpi.com/article/10.3390/antibiotics11010092/s1, Table S1: List of *K. pneumoniae* ST11-KL105 genomes deposited on NCBI public databases used for phylogenetic analysis. Figure S1: Projection of FT-IR spectra from the outbreak *K. pneumoniae* isolates (gray dots) in the in-house partial-least squares discriminant model (PLSDA) [13]. Figure S2: SNP matrix obtained from core-genome DNA comparison between *K. pneumoniae* ST11 KL105 genomes, supporting phylogenetic tree.

Author Contributions: Â.N. designed the study, acquired and interpreted molecular and whole-genome sequencing data and wrote the manuscript. R.V.F., M.V. and P.M.d.C. performed the characterization of isolates and collected epidemiological information. L.P. revised the manuscript for scientific content. All authors have read and agreed to the published version of the manuscript.

Funding: This work is financed by national funds from FCT—Fundação para a Ciência e a Tecnologia, I.P., in the scope of the project UIDP/04378/2020 and UIDB/04378/2020 of the Research Unit on Applied Molecular Biosciences—UCIBIO and the project LA/P/0140/2020 of the Associate Laboratory Institute for Health and Bioeconomy—i4HB. Additionally, we would like to acknowledge the support of HEALTH-UNORTE I&D&I project (NORTE-01-0145- FEDER-000039) co-financed by the European Regional Development Fund (ERDF), through the NORTE 2020 (Programa Operacional Regional do Norte 2014/2020). Ângela Novais is supported by national funds through FCT, I.P., in the context of the transitional norm (DL57/2016/CP1346/CT0032).

Institutional Review Board Statement: Not applicable.

Informed Consent Statement: Not applicable.

Data Availability Statement: This Whole Genome Shotgun project has been deposited at DDBJ/ENA/GenBank under the BioProject accession number PRJNA484888, with Biosamples SAMN16522901 and SAMN16522902.

Acknowledgments: We thank Mariana Silva Nunes for the help in initial isolate processing and organization of information. We are also grateful to João Pedro Rueda Furlan for his support on Geneious.

Conflicts of Interest: The authors declare no conflict of interest.

References

1. Wyres, K.L.; Lam, M.M.C.; Holt, K.E. Population genomics of *Klebsiella pneumoniae*. *Nat. Rev. Microbiol.* **2020**, *18*, 344–359. [CrossRef] [PubMed]
2. David, S.; Reuter, S.; Harris, S.R.; Glasner, C.; Feltwell, T.; Argimon, S.; Abudahab, K.; Goater, R.; Giani, T.; Errico, G.; et al. Epidemic of carbapenem-resistant *Klebsiella pneumoniae* in Europe is driven by nosocomial spread. *Nat. Microbiol.* **2019**, *4*, 1919–1929. [CrossRef]
3. Lee, C.R.; Lee, J.H.; Park, K.S.; Kim, Y.B.; Jeong, B.C.; Lee, S.H. Global Dissemination of Carbapenemase-Producing *Klebsiella pneumoniae*: Epidemiology, Genetic Context, Treatment Options, and Detection Methods. *Front. Microbiol.* **2016**, *13*, 895. [CrossRef] [PubMed]
4. Guerra, A.M.; Lira, A.; Lameirão, A.; Selaru, A.; Abreu, G.; Lopes, P.; Mota, M.; Novais, Â.; Peixe, L. Multiplicity of carbapenemase-producers three years after a KPC-3-producing *Klebsiella pneumoniae* ST147-K64 hospital outbreak. *Antibiotics* **2020**, *13*, 806. [CrossRef]
5. Bonnin, R.A.; Jousset, A.B.; Chiarelli, A.; Emeraud, C.; Glaser, P.; Naas, T.; Dortet, L. Emergence of New Non-Clonal Group 258 High-Risk Clones among *Klebsiella pneumoniae* Carbapenemase-Producing K. pneumoniae Isolates, France. *Emerg. Infect. Dis.* **2020**, *26*, 1212–1220. [CrossRef] [PubMed]
6. Hernández-García, M.; Pérez-Viso, B.; Carmen Turrientes, M.; Díaz-Agero, C.; López-Fresneña, N.; Bonten, M.; Malhotra-Kumar, S.; Ruiz-Garbajosa, P.; Cantón, R. Characterization of carbapenemase-producing Enterobacteriaceae from colonized patients in a university hospital in Madrid, Spain, during the R-GNOSIS project depicts increased clonal diversity over time with maintenance of high-risk clones. *J. Antimicrob. Chemother.* **2018**, *73*, 3039–3043. [CrossRef]

7. Bonura, C.; Giuffrè, M.; Aleo, A.; Fasciana, T.; Di Bernardo, F.; Stampone, T.; Giammanco, A.; The MDR-GN Working Group; Palma, D.M.; Mammina, C. An Update of the Evolving Epidemic of bla$_{KPC}$ Carrying *Klebsiella pneumoniae* in Sicily, Italy, 2014: Emergence of Multiple Non-ST258 Clones. *PLoS ONE* **2015**, *10*, e0132936. [CrossRef]

8. Chen, L.; Mathema, B.; Pitout, J.D.; DeLeo, F.R.; Kreiswirth, B.N. Epidemic *Klebsiella pneumoniae* ST258 is a hybrid strain. *mBio* **2014**, *5*, e01355-14. [CrossRef]

9. Dong, N.; Zhang, R.; Liu, L.; Li, R.; Lin, D.; Chan, E.W.; Chen, S. Genome analysis of clinical multilocus sequence Type 11 *Klebsiella pneumoniae* from China. *Microb. Genom.* **2018**, *4*, e000149. [CrossRef] [PubMed]

10. Wyres, K.L.; Hawkey, J.; Hetland, M.A.K.; Fostervold, A.; Wick, R.R.; Judd, L.M.; Hamidian, M.; Howden, B.P.; Löhr, I.H.; Holt, K.E. Emergence and rapid global dissemination of CTX-M-15-associated *Klebsiella pneumoniae* strain ST307. *J. Antimicrob. Chemother.* **2019**, *74*, 577–581. [CrossRef]

11. Zhang, R.; Liu, L.; Zhou, H.; Chan, E.W.; Li, J.; Fang, Y.; Li, Y.; Liao, K.; Chen, S. Nationwide surveillance of clinical carbapenem-resistant Enterobacteriaceae (CRE) strains in China. *EBioMedicine* **2017**, *19*, 98–106. [CrossRef]

12. Zhou, K.; Xiao, T.; David, S.; Wang, Q.; Zhou, Y.; Guo, L.; Aanensen, D.; Holt, K.E.; Thomson, N.T.; Grundmann, H.; et al. Novel Subclone of Carbapenem-Resistant *Klebsiella pneumoniae* Sequence Type 11 with Enhanced Virulence and Transmissibility, China. *Emerg. Infect. Dis.* **2020**, *26*, 289–297. [CrossRef] [PubMed]

13. Rodrigues, C.; Sousa, C.; Lopes, J.A.; Novais, Â.; Peixe, L. A Front Line on *Klebsiella pneumoniae* Capsular Polysaccharide Knowledge: Fourier Transform Infrared Spectroscopy as an Accurate and Fast Typing Tool. *mSystems* **2020**, *5*, e00386-19. [CrossRef]

14. Silva, L.; Rodrigues, C.; Lira, A.; Leão, M.; Mota, M.; Lopes, P.; Novais, Â.; Peixe, L. Fourier transform infrared (FT-IR) spectroscopy typing: A real-time analysis of an outbreak by carbapenem-resistant *Klebsiella pneumoniae*. *Eur. J. Clin. Microbiol. Infect. Dis.* **2020**, *39*, 2471–2475. [CrossRef]

15. Direção-Geral da Saúde. Infeções e Resistências aos Antimicrobianos: Relatório Anual do Programa Prioritário 2018. Available online: https://www.dgs.pt/portal-da-estatistica-da-saude/diretorio-de-informacao/diretorio-de-informacao/por-serie-1003038-pdf.aspx?v=%3d%3dDwAAAB%2bLCAAAAAAABAArySzItzVUy81MsTU1MDAFAHzFEfkPAAAA (accessed on 5 November 2019).

16. Rodrigues, C.; Machado, E.; Ramos, H.; Peixe, L.; Novais, Â. Expansion of ESBL-producing *Klebsiella pneumoniae* in hospitalized patients: A successful story of international clones (ST15, ST147, ST336) and epidemic plasmids (IncR, IncFIIK). *Int. J. Med. Microbiol.* **2014**, *304*, 1100–11088. [CrossRef]

17. Rodrigues, C.; Bavlovič, J.; Machado, E.; Amorim, J.; Peixe, L.; Novais, Â. KPC-3-Producing *Klebsiella pneumoniae* in Portugal Linked to Previously Circulating Non-CG258 Lineages and Uncommon Genetic Platforms (Tn4401d-IncFIA and Tn4401d-IncN). *Front. Microbiol.* **2016**, *7*, 1000. [CrossRef]

18. Aires-de-Sousa, M.; Ortiz de la Rosa, J.M.; Gonçalves, M.L.; Pereira, A.L.; Nordmann, P.; Poirel, L. Epidemiology of Carbapenemase-Producing *Klebsiella pneumoniae* in a Hospital, Portugal. *Emerg. Infect. Dis.* **2019**, *25*, 1632–1638. [CrossRef] [PubMed]

19. Perdigão, J.; Modesto, A.; Pereira, A.L.; Neto, O.; Matos, V.; Godinho, A.; Phelan, J.; Charleston, J.; Sadar, A.; de Sessions, P.F.; et al. Whole-genome sequencing resolves a polyclonal outbreak by extended-spectrum beta-lactam and carbapenem-resistant *Klebsiella pneumoniae* in a Portuguese tertiary-care hospital. *Microb. Genom.* **2019**, *7*, 000349. [CrossRef]

20. Lopes, E.; Saavedra, M.J.; Costa, E.; de Lencastre, H.; Poirel, L.; Aires-de-Sousa, M. Epidemiology of carbapenemase-producing *Klebsiella pneumoniae* in northern Portugal: Predominance of KPC-2 and OXA-48. *J. Glob. Antimicrob. Resist.* **2020**, *22*, 349–353. [CrossRef] [PubMed]

21. Aires-de-Sousa, M.; Ortiz de la Rosa, J.M.; Goncalves, M.L.; Costa, A.; Nordmann, P.; Poirel, L. Occurrence of NDM-1-producing Morganella morganii and Proteus mirabilis in a single patient in Portugal: Probable in vivo transfer by conjugation. *J. Antimicrob. Chemother.* **2020**, *75*, 903–906. [CrossRef]

22. Manageiro, V.; Sampaio, D.A.; Pereira, P.; Rodrigues, P.; Veira, L.; Palos, C.; Caniça, M. Draft Genome Sequence of the First NDM-1-Producing Providencia stuartii Strain Isolated in Portugal. *Genome Announc.* **2015**, *3*, e01077-15. [CrossRef]

23. Wailan, A.M.; Paterson, D.L. The spread and acquisition of NDM-1: A multifactorial problem. *Expert. Rev. Anti. Infect. Ther.* **2014**, *12*, 91–115. [CrossRef] [PubMed]

24. Politi, L.; Gartzonika, K.; Spanakis, N.; Zarkotou, O.; Poulou, A.; Skoura, L.; Vrioni, G.; Tsakris, A. Emergence of NDM-1-producing *Klebsiella pneumoniae* in Greece: Evidence of a widespread clonal outbreak. *J. Antimicrob. Chemother.* **2019**, *74*, 2197–2202. [CrossRef] [PubMed]

25. Ribeiro, T.G.; Novais, Â.; Rodrigues, C.; Nascimento, R.; Freitas, F.; Machado, E.; Peixe, L. Dynamics of clonal and plasmid backgrounds of Enterobacteriaceae producing acquired AmpC in Portuguese clinical settings over time. *Int. J. Antimicrob. Agents* **2019**, *53*, 650–656. [CrossRef]

26. Izdebski, R.; Sitkiewicz, M.; Urbanowicz, P.; Krawczyk, M.; Gniadkowski, M. Genomic background of the *Klebsiella pneumoniae* NDM-1 outbreak in Poland, 2012-18. *J. Antimicrob. Chemother.* **2020**, *75*, 3156–3162. [CrossRef] [PubMed]

27. van Dorp, L.; Wang, Q.; Shaw, L.P.; Acman, M.; Brynildsrud, O.B.; Eldholm, V.; Wang, R.; Gao, H.; Yin, Y.; Chen, H.; et al. Rapid phenotypic evolution in multidrug-resistant *Klebsiella pneumoniae* hospital outbreak strains. *Microb. Genom.* **2019**, *5*, e000263. [CrossRef]

28. Bowers, J.R.; Kitchel, B.; Driebe, E.M.; MacCannell, D.R.; Roe, C.; Lemmer, D.; de Man, T.; Rasheed, J.K.; Engelthaler, D.M.; Keim, P.; et al. Genomic Analysis of the Emergence and Rapid Global Dissemination of the Clonal Group 258 *Klebsiella pneumoniae* Pandemic. *PLoS ONE* **2015**, *10*, e0133727. [CrossRef] [PubMed]

29. Novais, Â.; Freitas, A.R.; Rodrigues, C.; Peixe, L. Fourier transform infrared spectroscopy: Unlocking fundamentals and prospects for bacterial strain typing. *Eur. J. Clin. Microbiol. Infect. Dis.* **2019**, *38*, 427–448. [CrossRef]

30. Naas, T.; Oueslati, S.; Bonnin, R.A.; Dabos, M.L.; Zavala, A.; Dortet, L.; Retailleau, P.; Iorga, B.I. Beta-lactamase database (BLDB)—structure and function. *J. Enzyme Inhib. Med. Chem.* **2017**, *32*, 917–919. [CrossRef]

31. Pires, J.; Novais, A.; Peixe, L. Blue-carba, an easy biochemical test for detection of diverse carbapenemase producers directly from bacterial cultures. *J. Clin. Microbiol.* **2013**, *51*, 4281–4283. [CrossRef]

32. Bogaerts, P.; Rezende de Castro, R.; de Mendonça, R.; Huang, T.D.; Denis, O.; Glupczynski, Y. Validation of carbapebemase and extended-spectrum β-lactamase multiplex endpoint PCR assays according to ISO 15189. *J. Antimicrob. Chemother.* **2013**, *68*, 1576–1582. [CrossRef] [PubMed]

33. Brisse, S.; Passet, V.; Haugaard, A.B.; Babosan, A.; Kassis-Chikhani, N.; Struve, C.; Decré, D. *wzi* gene sequencing, a rapid method to determine the capsular type of *Klebsiella* strains. *J. Clin. Microbiol.* **2013**, *51*, 4073–4078. [CrossRef] [PubMed]

34. Arredondo-Alonso, S.; Rogers, M.R.C.; Braat, J.C.; Verschuuren, T.D.; Top, J.; Corander, J.; Willems, R.J.L.; Schürch, A.C. mlplasmids: A user-friendly tool to predict plasmid- and chromosome-derived sequences for single species. *Microb. Genom.* **2018**, *4*, e000224. [CrossRef] [PubMed]

Article

The Antimicrobial Resistance Characteristics of Imipenem-Non-Susceptible, Imipenemase-6-Producing *Escherichia coli*

Reo Onishi [1], Katsumi Shigemura [1,2,*], Kayo Osawa [3], Young-Min Yang [2], Koki Maeda [2], Shiuh-Bin Fang [4,5], Shian-Ying Sung [6], Kenichiro Onuma [7], Atsushi Uda [7], Takayuki Miyara [7] and Masato Fujisawa [2]

[1] Department of Public Health, Division of Infectious Diseases, Kobe University Graduate School of Health Sciences, 7-10-2 Tomogaoka Suma-ku, Kobe 654-0142, Japan; ohnishile@gmail.com

[2] Division of Urology, Kobe University Graduate School of Medicine, 7-5-2 Kusunoki-cho, Chuo-ku, Kobe 650-0017, Japan; yym1112@gmail.com (Y.-M.Y.); kokimaeda1118@gmail.com (K.M.); masato@med.kobe-u.ac.jp (M.F.)

[3] Department of Medical Technology, Kobe Tokiwa University, 2-6-2 Otani-cho, Nagata-ku, Kobe 653-0838, Japan; k-ohsawa@kobe-tokiwa.ac.jp

[4] Department of Pediatrics, Division of Pediatric Gastroenterology and Hepatology, Shuang Ho Hospital, Taipei Medical University, 291 Jhong Jheng Road, Jhong Ho District, New Taipei City 23561, Taiwan; sbfang@tmu.edu.tw

[5] Department of Pediatrics, School of Medicine, College of Medicine, Taipei Medical University, Taipei 11031, Taiwan

[6] International Ph.D. Program for Translational Science, College of Medical Science and Technology, Taipei Medical University, Taipei 11031, Taiwan; ssung@tmu.edu.tw

[7] Department of Infection Control and Prevention, Kobe University Hospital, 7-5-2 Kusunoki-cho, Chuo-ku, Kobe 650-0017, Japan; onumak@med.kobe-u.ac.jp (K.O.); a-uda@umin.ac.jp (A.U.); miyarat@med.kobe-u.ac.jp (T.M.)

* Correspondence: katsumi@med.kobe-u.ac.jp; Tel.: +81-78-382-6155

Citation: Onishi, R.; Shigemura, K.; Osawa, K.; Yang, Y.-M.; Maeda, K.; Fang, S.-B.; Sung, S.-Y.; Onuma, K.; Uda, A.; Miyara, T.; et al. The Antimicrobial Resistance Characteristics of Imipenem-Non-Susceptible, Imipenemase-6-Producing *Escherichia coli*. *Antibiotics* **2022**, *11*, 32. https://doi.org/10.3390/antibiotics11010032

Academic Editor: Francesca Andreoni

Received: 29 November 2021
Accepted: 23 December 2021
Published: 28 December 2021

Publisher's Note: MDPI stays neutral with regard to jurisdictional claims in published maps and institutional affiliations.

Abstract: Imipenemase-6 (IMP-6) type carbapenemase-producing *Enterobacteriaceae* is regarded as dangerous due to its unique lack of antimicrobial susceptibility. It is resistant to meropenem (MEPM) but susceptible to imipenem (IPM). In addition to carbapenemase, outer membrane porins and efflux pumps also play roles in carbapenem resistance by reducing the antimicrobial concentration inside cells. Extended-spectrum β-lactamase (ESBL) is transmitted with IMP-6 by the plasmid and broadens the spectrum of antimicrobial resistance. We collected 42 strains of IMP-6-producing *Escherichia coli* and conducted a molecular analysis of carbapenemase, ESBL, porin, efflux, and epidemiological characteristics using plasmid replicon typing. Among the 42 isolates, 21 strains were susceptible to IPM (50.0%) and 1 (2.4%) to MEPM. Seventeen strains (40.5%) co-produced CTX-M-2 type ESBL. We found that the relative expression of *ompC* and *ompF* significantly correlated with the MIC of IPM ($p = 0.01$ and $p = 0.03$, respectively). Sixty-eight% of CTX-M-2-non-producing strains had IncI1, which was significantly different from CTX-M-2-producing strains ($p < 0.001$). In conclusion, 50.0% of our IMP 6 producing strains were non-susceptible to IPM, which is different from the typical pattern and can be attributed to decreased porin expression. Further studies investigating other types of carbapenemase are warranted.

Keywords: IMP-6; carbapenemase; porin; efflux pump; plasmid

1. Introduction

The prevalence of imipenemase-6 (IMP-6) type carbapenemase-producing *Enterobacteriaceae* (CPE) is increasing in Japan, and this is regarded as dangerous due to its unique lack of antimicrobial susceptibility [1,2]. They are generally resistant to meropenem (MEPM), although they are susceptible to imipenem (IPM) [2,3]. Carbapenemase is the enzyme that inactivates carbapenem. CPE can produce a variety of carbapenemases, which are divided into three groups as class A, class B, and class D according to the Ambler classification [4]. In Japan, the most common carbapenemase type is IMP, one of the class B carbapenemases. It exhibits metallo-β-lactamase and is able to decompose almost all β-lactam drugs, including carbapenems, by a wide range of instrumental specificity. IMP-6 exceptionally shows characteristic inactivation as described above [5]. CPE often co-produces extended-spectrum β-lactamase (ESBL), which inactivates penicillins, cephalosporins, and monobactams [3,6] for wider spectrum antimicrobial resistance. IMP-6 generally does not inactivate penicillins and monobactams [2,7], but if an IMP-6-producing strain co-produces ESBLs, it does inactivate these antimicrobials. Thus, the co-production of carbapenemase and ESBL should be considered as an important mechanism in antimicrobial resistance.

Outer membrane porins and efflux pumps are also important mechanisms of carbapenem resistance *Enterobacteriaceae* (CRE) by reducing the antimicrobial concentration inside cells [8]. Because carbapenem is one of the most effective wide-spectrum antimicrobials, CRE infections increase treatment difficulty and worsen prognosis [9].

Porins are small pores in the outer membrane of Gram-negative bacteria (GNB). Decrease or loss of porins hinders drug entry in GNB [10]. Carbapenems enter cells through porins such as OmpC and OmpF, encoded by porin genes *ompC* and *ompF* [11]. OmpC and OmpF are the two most important outer membrane porin proteins in *Escherichia coli*. The pore size of OmpF is larger than that of OmpC, allowing more solutes, including noxious agents, to diffuse into the cell through the OmpF channel [12]. A decrease in the relative expression of *ompC* and *ompF* indicates loss or disruption of OmpC and OmpF.

Efflux pumps are responsible for membrane transport and can discharge antibacterial drugs outside the cell. Hyperfunction of efflux pump proteins is one of the most important mechanisms in carbapenem resistance. Resistance-nodulation-division (RND) efflux pumps are a major mechanism of multidrug resistance in *Enterobacteriaceae* [13]. The AcrAB-TolC RND system is the most common among the different efflux systems, which catalyze substrate efflux by an H^+ antiport mechanism [13,14]. Phenylalanine-arginine β-naphthylamide (PAβN, also called MC-207,110) is well known as a broad-spectrum efflux pump inhibitor [15]. PAβN binds to AcrB, leading to the inhibition of substrate efflux outside the cell [16,17]. In carbapenem resistance attributed to the overexpression of efflux pumps, the minimum inhibitory concentration (MIC) will decrease from the addition of PAβN because carbapenem cannot be effluxed out of the cell. Therefore, efflux pump activity in carbapenem resistance can be evaluated using PAβN [18,19].

Plasmids are extra-chromosomal circular fragments of DNA that replicate autonomously in host cells [20]. Plasmids are classified based on incompatibility (Inc) groups by Plasmid Replicon Typing [20]. Carbapenemase genes and ESBL genes are transferred easily by plasmids, contributing to the rapid spread of carbapenemase and ESBL-producing strains. IncN plasmid often carries IMP-6 with CTX-M-2 type ESBL and plays an important role in the spread of IMP-6 and CTX-M-2 co-producing strains [3,20]. Plasmid replicon typing facilitates the exploration of the epidemiological features of carbapenemase-producing strains.

Whether resistance mechanisms other than carbapenemase affect antimicrobial susceptibility in IMP-6-producing bacteria has rarely been investigated. An epidemiological analysis is also important for a comprehensive understanding of the spread of CPE. The purpose of this study was to assess antimicrobial susceptibility and the effects of the three carbapenem resistance mechanisms, ESBL production, and epidemiology in IMP-6-producing *Escherichia coli*.

2. Results

2.1. Antimicrobial Susceptibilities

Among the 42 IMP-6-producing isolates, 21 strains (50.0%) were susceptible to IPM, 1 strain (2.4%) to MEPM, no strains (0%) to ertapenem (ETP), and 11 strains (26.2%) to doripenem (DRPM). Minimum inhibitory concentrations $(MIC)_{50}$ were 1.5 μg/mL, 16 μg/mL, 16 μg/mL, and 8 μg/mL, and MIC_{90} were 4 μg/mL, 32 μg/mL, 64 μg/mL, and 32 μg/mL for IPM, MEPM, ETP, and DRPM, respectively (Table 1) (Supplemental Table S1).

Most strains were resistant to piperacillin (PIPC) (97.6%), ceftazidime (CAZ) (97.6%), cefepime (CFPM) (95.2%), ciprofloxacin (CPFX) (97.6%), and levofloxacin (LVFX) (97.6%). No strain was resistant to amikacin (AMK) (0%).

Table 1. Antimicrobial susceptibility among 42 strains of IMP-6-producing *E. coli*.

	Susceptible	Intermediate	Resistant	MIC_{50} (µg/mL)	MIC_{90} (µg/mL)
IPM	21 (50.0%)	14 (33.3%)	7 (16.7%)	1.5	4
MEPM	1 (2.4%)	2 (4.8%)	39 (92.9%)	16	32
ETP	0 (0%)	3 (7.1%)	39 (92.9%)	16	64
DRPM	11 (26.2%)	4 (9.5%)	27 (64.3%)	8	32
PIPC	1 (2.4%)	0 (0%)	41 (97.6%)	>64	>64
CAZ	0 (0%)	1 (2.4%)	41 (97.6%)	>16	>16
CFPM	1 (2.4%)	1 (2.4%)	40 (95.2%)	>16	>16
CPFX	1 (2.4%)	0 (0%)	41 (97.6%)	>2	>2
LVFX	1 (2.4%)	0 (%)	41 (97.6%)	>4	>4
AMK	41 (97.6%)	1 (2.4%)	0 (0%)	<4	16
GM	11 (26.2%)	20 (47.6%)	11 (26.2%)	8	>8

2.2. Phenotypic and Genotypic Detection of Carbapenemase and ESBL

Carbapenemase-associated genes other than bla_{IMP-6} were not detected in our strains. Twenty-nine strains (69.0%) had confirmed ESBL production and ESBL $bla_{CTX-M-2}$ genes were detected in 17 strains (40.5%), $bla_{CTX-M-14}$ in 1 strain (2.4%), and $bla_{CTX-M-15}$ in 2 strains (4.8%) (Table 2) (Supplemental Table S2).

Table 2. Characteristics of carbapenemase and ESBL among 42 strains of IMP-6-producing *E. coli*.

	Carbapenemase Producing	Carbapenemase Genes	ESBL Producing	ESBL Genes		
		bla_{IMP-6}		$bla_{CTX-M-2}$	$bla_{CTX-M-14}$	$bla_{CTX-M-15}$
Number of isolates ($n = 42$)	42	42	29	17	1	2
(%)	100.0	100.0	69.0	40.5	2.4	4.8

2.3. Relative Expression Level of ompC and ompF

Relative expression levels of *ompC* and *ompF* were measured by quantitative reverse transcription polymerase chain reaction (qRT-PCR). The relative expression levels of *ompC* significantly correlated with the MICs of IPM ($r = -0.388$, $p = 0.0112$) and DRPM ($r = -0.501$, $p = 0.000734$) (Figure 1A,D). Furthermore, the relative expression levels of *ompF* significantly correlated with the MICs of IPM ($r = -0.332$, $p = 0.0318$) and MEPM ($r = -0.529$, $p = 0.00032$) (Figure 2A,B).

2.4. Efflux Pump Activity and Carbapenem Resistance

To determine the effect of efflux on IPM resistance, the MICs of two representative carbapenems, IPM and MEPM, were measured with and without the addition of PAβN (Figure 3). If carbapenem resistance depends on the overexpression of efflux pumps, MICs of carbapenems will decrease with the addition of PAβN because carbapenem cannot be effluxed out of the cell. In this study, no isolates showed more than a four-fold decrease of MIC in the presence of PAβN compared with MICs when only IPM and MEPM were present. Therefore, no significant effect of efflux pump activity on carbapenem resistance was found in all 42 strains. It was confirmed by the control that PAβN did not inhibit bacterial growth (data not shown).

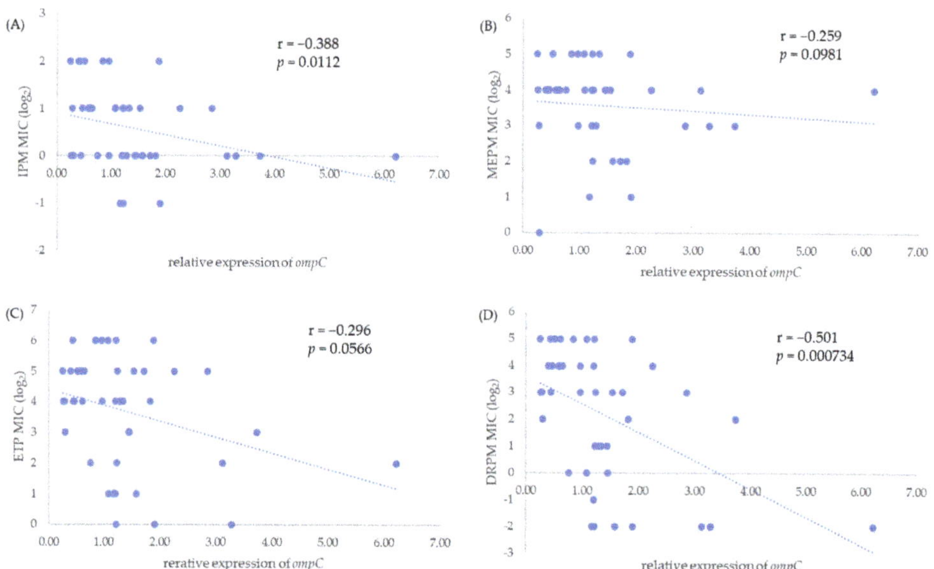

Figure 1. Correlation between relative expression levels of *ompC* and MICs of four carbapenems, IPM (**A**), MEPM (**B**), ETP (**C**), and DRPM (**D**), among IMP-6-producing *E. coli* strains.

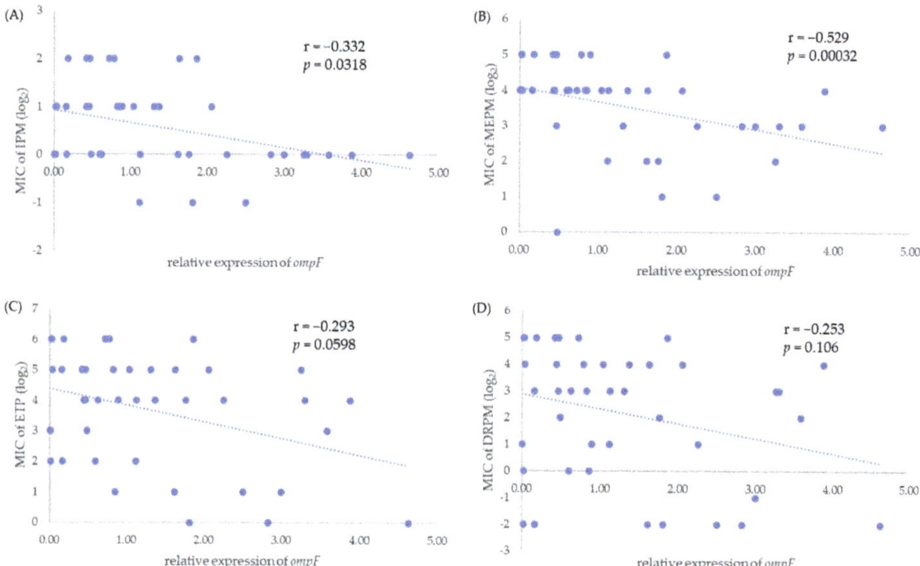

Figure 2. Correlation between relative expression levels of *ompF* and MICs of four carbapenems, IPM (**A**), MEPM (**B**), ETP (**C**), and DRPM (**D**), among IMP-6-producing *E. coli* strains.

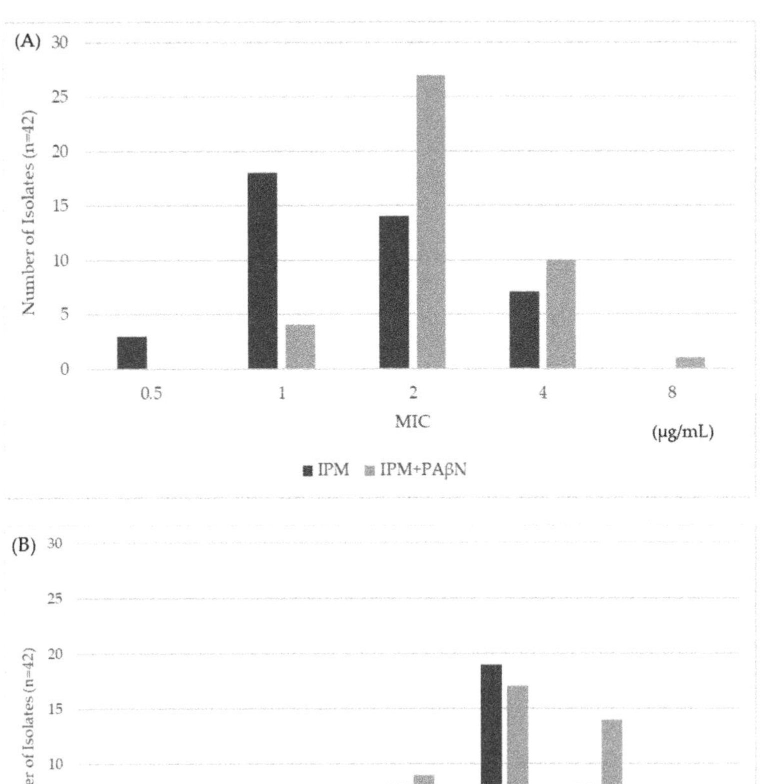

Figure 3. The MICs of IPM (**A**) and MEPM (**B**) with and without the addition of PAβN in IMP-6-producing *E. coli* strains.

2.5. Plasmid Replicon Typing

IncF was the most prevalent plasmid replicon, identified in 40 of the 42 strains (95.2%) (Table 3). A variety of plasmids were found in our IMP-6-producing *E. coli* strains, including IncFIA in 37 strains (88.1%), IncN in 36 strains (85.7%), IncFIB in 28 strains (66.7%), IncI1 in 18 strains (42.9%), IncB/O in 4 strains (9.5%), and IncA/C in 2 strains (4.8%). There was a significant difference in the rates of carrying IncI1 between CTX-M-2 co-producing strains ($n = 17$) and the others ($n = 25$) ($p < 0.001$, Table 3). We found duplications of plasmids in the isolates. The details for individual strains are presented in Supplemental Table S2.

Table 3. Plasmid replicon typing among IMP-6-producing *E. coli* strains.

	Total (*n* = 42)	IMP-6 + CTX-M-2 (*n* = 17)	IMP-6 (*n* = 25)	*p*-Value
FIA	37 (88.1%)	15 (88.2%)	22 (88.0%)	1
FIB	28 (66.7%)	9 (52.9%)	19 (76.0%)	0.184
FIC	0 (0.0%)	0 (0.0%)	0 (0.0%)	N.A.
F	40 (95.2%)	17 (100.0%)	23 (92.0%)	0.506
FII	0 (0.0%)	0 (0.0%)	0 (0.0%)	N.A.
HI-1	0 (0.0%)	0 (0.0%)	0 (0.0%)	N.A.
HI-2	0 (0.0%)	0 (0.0%)	0 (0.0%)	N.A.
I1	18 (42.9%)	1 (5.9%)	17 (68.0%)	<0.001 *
L/M	0 (0.0%)	0 (0.0%)	0 (0.0%)	N.A.
N	36 (85.7%)	16 (94.1%)	20 (80.0)	0.374
P	0 (0.0%)	0 (0.0%)	0 (0.0%)	N.A.
W	0 (0.0%)	0 (0.0%)	0 (0.0%)	N.A.
T	0 (0.0%)	0 (0.0%)	0 (0.0%)	N.A.
A/C	2 (4.8%)	0 (0.0%)	2 (8.0)	0.506
K	0 (0.0%)	0 (0.0%)	0 (0.0%)	N.A.
B/O	4 (9.5%)	2 (11.8%)	2 (8.0)	1
X	0 (0.0%)	0 (0.0%)	0 (0.0%)	N.A.
Y	0 (0.0%)	0 (0.0%)	0 (0.0%)	N.A.

N.A.: not applicable; * Statistical significance.

3. Discussion

IMP-6-producing CRE is frequently detected in Japan, especially in west Japan. A previous study from Osaka showed that 130 (97.0%) of 134 carbapenem-resistant *E. coli* produced IMP-6 [1]. Their susceptibility rates to MEPM and IPM were 0% and 100.0%, respectively, which is a typical pattern of IMP-6-producing CRE. Our IMP-6-producing *E. coli* strains were resistant to at least one of the carbapenems, and most of these strains were resistant to MEPM (92.9%). Although most previous studies indicated that IMP-6 does not inactivate IPM [2,3], 50.0% of our strains were non-susceptible to IPM. Thus, non-susceptibility to IPM may be due to a resistance mechanism other than the production of IMP-6. A few studies reported the susceptibility of IMP-6-producing *Enterobacteriaceae* to ETP and DRPM. A previous study in Korea reported that all of 10 IMP-6-producing strains were not susceptible to ETP [21], similar to our strains. Another study in Japan showed that all of 11 IMP-6-producing strains were not susceptible to DRPM [22], which is consistent with the low susceptibility (26.2%) in our strains. Altogether, these results demonstrated the non-susceptibility of IMP-6-producing strains to ETP and DRPM.

Previous studies reported that most IMP-6-producing strains co-produced CTX-M-2 [7,23], but less than half of our strains co-produced CTX-M-2. The form of transmission may have changed. Furthermore, most of our strains were resistant to PIPC with or without CTX-M-2 production, despite the fact that IMP-6-producing strain is generally susceptible to PIPC. Quite possibly, these strains co-produced another β-lactamase. For example, a previous study reported that IMP-6-producing strains co-produced TEM-1, a non-ESBL β-lactamase [2]. These strains were resistant to PIPC by co-producing TEM-1.

One of the main mechanisms for carbapenem resistance other than carbapenemase production is the decrease or loss of porin. A previous study in China showed low expression of either *ompC* or *ompF* in 46 (68.7%) of 67 carbapenemase-non-producing strains [24]. In that study, porin-related mechanisms were investigated only for carbapenemase-non-producing strains, not for carbapenemase-producing strains. Our study explored porin-related mechanisms in CPE and found that IMP-6-producing CRE had decrease or loss of porin according to the downregulation of two porin genes *ompC* and *ompF* as another resistance mechanism. Furthermore, statistical analysis showed a significant negative correlation between the expression levels of the two porin genes and MIC of some carbapenems, including *ompC* with IPM and DRPM (Figure 1A,D) as well as *ompF* with IPM and MEPM (Figure 2A,B).

Although half of the IMP-6-producing CRE strains in our study were susceptible to IPM, we still found that the MICs of IPM were correlatively higher when the porin-related genes *ompC* and *ompF* were more downregulated. Thus, decreased porin expression can also affect antimicrobial resistance in IMP-6-producing strains, with altered antimicrobial susceptibility in some cases.

Efflux activity in carbapenem resistance has been reported using several methods for confirmation of efflux activity [19,24]. A previous study showed that 7 isolates of carbapenem-resistant *Pseudomonas aeruginosa* exhibited a significant reduction in meropenem MIC with PAβN, using the same methods as our study [25]. Similar to our study, another group investigated KPC-2-producing CRE and did not find any effect of efflux pump activity in carbapenem resistance [19]. Our strains also did not show a contribution by efflux pump activity to carbapenem resistance. Taken together, these results suggest the minor effect of efflux activity on carbapenem resistance in carbapenemase-producing bacteria.

Plasmid replicon typing is a useful assay to investigate horizontal gene transfer because carbapenemase genes and ESBL genes are generally encoded on plasmids [3,20]. As mentioned above, IMP-6-producing CRE strains often possess IncN plasmid and simultaneously produce CTX-M-2 [3]. A previous study showed that 80 strains (96.4%) possessed IncN, and 73 strains (88.0%) co-produced CTX-M-2 among 83 IMP-6-producing strains [7]. In contrast, a lower portion (40.5%, 17/42) of our strains co-produced IMP-6 and CTX-M-2 despite a high carriage rate (85.7%, 36/42) of IncN. Compared with CTX-M-2 co-producing strains, IMP-6 producing but not CTX-M-2 co-producing strains significantly possessed IncI1 ($p < 0.001$). So far, IncI1 has not been reported in IMP-6-producing CRE. Transmission and carriage of IncI1 type plasmids could reduce co-production of CTX-M-2 among IMP-6-producing CRE.

Our study confirmed that IMP-6-producing strains were resistant to MEPM, ETP, and DRPM as reported so far. However, the IPM non-susceptible strains appeared to differ from the typical pattern. We demonstrated that downregulation of porin-associated genes was responsible for this IPM non-susceptibility, suggesting evolving characteristics of IMP-6-producing CPE. It is necessary to update our knowledge of the antimicrobial susceptibility of IMP-6-producing CPE to facilitate appropriate treatment and clinical practice. Downregulated porin-related genes should also be taken into account when developing antimicrobials against IMP-6-producing CPE.

There are some limitations in this study, including the small number of isolates and the unavailability of full clinical data for investigation. In addition, phenotypic studies of porin and genotypic efflux pump analysis were not performed. Despite these limitations, our study contributes to preventing the further expansion of IMP-6-producing CRE.

4. Materials and Methods

4.1. Bacterial Collection

Forty-two strains of IMP-6-producing *E. coli* were collected from Hyogo prefecture and sent to Hyogo Clinical Laboratory Corporation, Himeji, Japan, from 2012 to 2018. They were isolated from urine, sputum, and other sources. IMP-6 production was confirmed using phenotypic and genotypic methods. Carbapenemase production was detected by three phenotypic methods, the carbapenem inactivation method (CIM), double disk synergy test (DDST) with sodium mercaptoacetic acid (SMA), and the modified Hodge test. We defined a strain as carbapenemase-producing if it was positive by at least one method. CIM and modified Hodge tests were carried out according to Clinical and Laboratory Standards Institute (CLSI) recommendations [26]. SMA-DDST was performed for CAZ and IPM as previously described [27]. We detected bla_{IMP-6} by PCR and DNA sequencing as previously described [28,29].

4.2. Antimicrobial Susceptibility Tests

Antimicrobial susceptibility testing was performed by broth microdilution methods according to the CLSI guidelines [26]. MIC of *E. coli* was tested for the following antimicrobials: MEPM (Wako, Fujifilm Wako Pure Chemical Corporation, Japan), IPM (Wako, Japan), ETP (Sigma-Aldrich, St. Louis, MO, USA), and DRPM (Wako, Japan), using *E. coli* ATCC 25922 as quality control. MicroScan (Beckman Coulter Inc., Brea, CA, USA) was used to determine the MIC of the antimicrobials PIPC, CAZ, CFPM, LVFX, CPFX, AMK, and GM. We determined susceptibility according to the current CLSI clinical breakpoint [26]. We defined non-susceptible as resistant or intermediate to each antimicrobial.

4.3. Phenotypic Detections of ESBL Production in E. coli Isolates

All isolates of *E. coli* were screened for ESBL production using cefotaxime (CTX) and CAZ disc diffusion testing, according to CLSI guidelines [26]. The confirmation tests were conducted by the double-disk synergy method using CTX and CAZ disks alone and in combination with clavulanic acid [26].

4.4. Detection of Carbapenemase Genes and ESBL Genes

The presence of carbapenemase genes bla_{IMP-1}, bla_{IMP-2}, bla_{KPC}, bla_{NDM}, bla_{OXA-48} bla_{GES}, bla_{VIM-1}, and bla_{VIM-2} was determined using PCR amplification, as previously described [28,30–34]. The presence of ESBL genes bla_{CTX-M}, bla_{TEM}, and bla_{SHV} were also determined using PCR amplification, as previously described [35]. The PCR products were run on 1% agarose gel and stained with ethidium bromide (0.5 mg/mL) in a dark room. PCR purification was conducted by QIAquick PCR Purification Kit (Qiagen, Hilden, Germany) with sequencing by Eurofins Genomics, Inc. (Tokyo, Japan) [29].

4.5. q-RT-PCR for Porin Coding Genes ompC and ompF

The qRT-PCR for porin genes (*ompC* and *ompF*) was performed. RNA was extracted from bacterial pellets and DNase-treated using NucleoSpin RNA (Macherey-Nagel, Germany). cDNA was constructed using the ReverTra Ace qPCR RT Kit (Toyobo, Japan). The qRT-PCR was performed for porin genes *ompC* and *ompF* and the housekeeping gene *rpoB* using the primer previously described [11]. The reactions were run on a CFX Connect (Bio-Rad, Hercules, CA, USA) with the following cycling parameters: 95 °C for 5 min and 40 cycles of 95 °C for 15 s, 60 °C for 30 s, and 72 °C for 30 s, followed by melting with ramping from 60 °C to 95 °C in 0.2 °C increments. Melting curve analysis was performed to identify the amplicons. Each experiment was performed in triplicate, and results were presented as the mean value of three experiments. The relative expression levels were calculated by the $2^{-\Delta\Delta Ct}$ method. Expression of porin genes was normalized using the housekeeping gene *rpoB* in the same sample. Fold change in porin expression was determined by calculating the ratio of normalized porin expression in IMP-6-producing isolates to the control strain of the same species, *E. coli* ATCC 25922 [11].

4.6. Efflux Pump Inhibitory Assay

To examine the influence of efflux pumps on carbapenem resistance, efflux inhibitory assays were performed using representative carbapenem, IPM, and MEPM. The MICs of imipenem and meropenem were measured with and without 25 mg/L of the efflux pump inhibitor PAβN (Sigma-Aldrich, St Louis, MI, USA) [36]. Efflux pump activity was evaluated by fold change from the MIC without PAβN to the MIC with PAβN. If carbapenem resistance depended on the overexpression of efflux pumps, the MIC of carbapenems would be reduced by the addition of PAβN because carbapenem does not efflux. More than a four-fold decrease of the MICs in the presence of PAβN compared with the MICs in the absence of PAβN was defined as positive.

4.7. Plasmid Replicon Typing

Plasmid DNA was extracted as previously described [29]. Plasmid replicon typing was performed to see the disseminating formulation and investigate for IncF (FIA, FIB, FIC, F, FII), H, I, L/M, N, P, W, T, A/C, K, B/O, X, and Y by PCR amplification. The temperature conditions were initial denaturing at 94 °C for 5 min, followed by 30 cycles of denaturation at 94 °C for 1 min, annealing at 60 °C for 30 s, extension at 72 °C for 1 min and final extension at 72 °C for 5 min [20].

4.8. Statistical Analysis

Correlations between the relative expression of porin genes (*ompC* and *ompF*) and the MIC for each carbapenem were analyzed by Spearman's rank correlation coefficient using EZR (Saitama Medical Centre, Jichi Medical University, Saitama, Japan) [37]. The difference in plasmid carriage between CTX-M-2-producing strains and the others was analyzed by Fisher's exact test using EZR. Statistical differences among mean values were considered significant when $p < 0.05$.

5. Conclusions

Half of our IMP-6-producing strains were non-susceptible to IPM, and only 40.5% of them co-produced CTX-M-2, unlike the well-known typical pattern. Downregulation of porin-associated genes is responsible for this alteration in IPM non-susceptibility, but efflux plays a minor role. Further studies are warranted for investigation in other types of carbapenemase.

Supplementary Materials: The following supporting information can be downloaded at: https://www.mdpi.com/article/10.3390/antibiotics11010032/s1, Table S1: Minimum inhibitory concentrations against antimicrobials among 42 isolates of *Escherichia coli*. Table S2: The characteristics of carbapenemase and ESBL production among 42 isolates of Escherichia coli.

Author Contributions: Conceptualization, K.S. and K.O. (Kayo Osawa); methodology, R.O.; investigation, R.O.; resources, K.S. and K.O. (Kayo Osawa); data curation, K.S.; writing—original draft preparation, R.O.; writing—review and editing, R.O., K.S., K.O. (Kayo Osawa), Y.-M.Y., K.M., S.-B.F., S.-Y.S., K.O. (Kenichiro Onuma), A.U., T.M. and M.F.; supervision, M.F.; project administration, K.S. and M.F.; funding acquisition, K.S. All authors have read and agreed to the published version of the manuscript.

Funding: This research was funded by Japan Society for the Promotion of Science (JSPS) KAKENHI Grant Numbers 19K09670 and 21K10423.

Institutional Review Board Statement: The study was conducted according to the guidelines of the Declaration of Helsinki and approved by the ethical committee on health research of the Kobe University Graduate School of Health Sciences (No. 472-6).

Informed Consent Statement: Not applicable.

Data Availability Statement: All data generated or analyzed during this study are included in this published article.

Conflicts of Interest: The authors declare no conflict of interest. The funders had no role in the design of the study; in the collection, analyses, or interpretation of data; in the writing of the manuscript, or in the decision to publish the results.

References

1. Yamamoto, N.; Asada, R.; Kawahara, R.; Hagiya, H.; Akeda, Y.; Shanmugakani, R.K.; Yoshida, H.; Yukawa, S.; Yamamoto, K.; Takayama, Y.; et al. Prevalence of, and risk factors for, carriage of carbapenem-resistant Enterobacteriaceae among hospitalized patients in Japan. *J. Hosp. Infect.* **2017**, *97*, 212–217. [CrossRef] [PubMed]
2. Yano, H.; Kuga, A.; Okamoto, R.; Kitasato, H.; Kobayashi, T.; Inoue, M. Plasmid-Encoded Metallo-β-Lactamase (IMP-6) Conferring Resistance to Carbapenems, Especially Meropenem. *Antimicrob. Agents Chemother.* **2001**, *45*, 1343–1348. [CrossRef] [PubMed]
3. Kayama, S.; Shigemoto, N.; Kuwahara, R.; Oshima, K.; Hirakawa, H.; Hisatsune, J.; Jové, T.; Nishio, H.; Yamasaki, K.; Wada, Y.; et al. Complete Nucleotide Sequence of the IncN Plasmid Encoding IMP-6 and CTX-M-2 from Emerging Carbapenem-Resistant Enterobacteriaceae in Japan. *Antimicrob. Agents Chemother.* **2015**, *59*, 1356–1359. [CrossRef] [PubMed]
4. Ambler, R.P. The structure of beta-lactamases. *Philos. Trans. R. Soc. Lond. B Biol. Sci.* **1980**, *289*, 321–331. [CrossRef] [PubMed]
5. Doumith, M.; Ellington, M.J.; Livermore, D.M.; Woodford, N. Molecular mechanisms disrupting porin expression in ertapenem-resistant Klebsiella and Enterobacter spp. clinical isolates from the UK. *J. Antimicrob. Chemother.* **2009**, *63*, 659–667. [CrossRef]

6.	Schwaber, M.J.; Navon-Venezia, S.; Kaye, K.S.; Ben-Ami, R.; Schwartz, D.; Carmeli, Y. Clinical and economic impact of bacteremia with extended- spectrum-beta-lactamase-producing Enterobacteriaceae. *Antimicrob. Agents Chemother.* **2006**, *50*, 1257–1262. [CrossRef] [PubMed]

7.	Yonekawa, S.; Mizuno, T.; Nakano, R.; Nakano, A.; Suzuki, Y.; Asada, T.; Ishii, A.; Kakuta, N.; Tsubaki, K.; Mizuno, S.; et al. Molecular and Epidemiological Characteristics of Carbapenemase-Producing Klebsiella pneumoniae Clinical Isolates in Japan. *mSphere* **2020**, *5*, e00490-20. [CrossRef] [PubMed]

8.	Haidar, G.; Clancy, C.J.; Chen, L.; Samanta, P.; Shields, R.K.; Kreiswirth, B.N.; Nguyen, M.H. Identifying Spectra of Activity and Therapeutic Niches for Ceftazidime-Avibactam and Imipenem-Relebactam against Carbapenem-Resistant Enterobacteriaceae. *Antimicrob. Agents Chemother.* **2017**, *61*, e00642-17. [CrossRef] [PubMed]

9.	Johnson, A.P.; Woodford, N. Global spread of antibiotic resistance: The example of New Delhi metallo-β-lactamase (NDM)-mediated carbapenem resistance. *J. Med. Microbiol.* **2013**, *62*, 499–513. [CrossRef]

10.	Hamzaoui, Z.; Ocampo-Sosa, A.; Fernandez Martinez, M.; Landolsi, S.; Ferjani, S.; Maamar, E.; Saidani, M.; Slim, A.; Martinez-Martinez, L.; Boutiba-Ben Boubaker, I. Role of association of OmpK35 and OmpK36 alteration and blaESBL and/or blaAmpC genes in conferring carbapenem resistance among non-carbapenemase-producing Klebsiella pneumoniae. *Int. J. Antimicrob. Agents* **2018**, *52*, 898–905. [CrossRef]

11.	Senchyna, F.; Gaur, R.L.; Sandlund, J.; Truong, C.; Tremintin, G.; Kültz, D.; Gomez, C.A.; Tamburini, F.B.; Andermann, T.; Bhatt, A.; et al. Diversity of resistance mechanisms in carbapenem-resistant Enterobacteriaceae at a health care system in Northern California, from 2013 to 2016. *Diagn. Microbiol. Infect. Dis.* **2019**, *93*, 250–257. [CrossRef]

12.	Lin, X.; Wang, C.; Guo, C.; Tian, Y.; Li, H.; Peng, X. Differential regulation of OmpC and OmpF by AtpB in Escherichia coli exposed to nalidixic acid and chlortetracycline. *J. Proteom.* **2012**, *75*, 5898–5910. [CrossRef] [PubMed]

13.	Weston, N.; Sharma, P.; Ricci, V.; Piddock, L.J.V. Regulation of the AcrAB-TolC efflux pump in Enterobacteriaceae. *Res. Microbiol.* **2018**, *169*, 425–431. [CrossRef]

14.	Van Dyk, T.K.; Templeton, L.J.; Cantera, K.A.; Sharpe, P.L.; Sariaslani, F.S. Characterization of the Escherichia coli AaeAB efflux pump: A metabolic relief valve? *J. Bacteriol.* **2004**, *186*, 7196–7204. [CrossRef] [PubMed]

15.	Lamers, R.P.; Cavallari, J.F.; Burrows, L.L. The Efflux Inhibitor Phenylalanine-Arginine Beta-Naphthylamide (PAβN) Permeabilizes the Outer Membrane of Gram-Negative Bacteria. *PLoS ONE* **2013**, *8*, e60666. [CrossRef] [PubMed]

16.	Tikhonova, E.B.; Yamada, Y.; Zgurskaya, H.I. Sequential mechanism of assembly of multidrug efflux pump AcrAB-TolC. *Chem. Biol.* **2011**, *18*, 454–463. [CrossRef] [PubMed]

17.	Lomovskaya, O.; Warren, M.S.; Lee, A.; Galazzo, J.; Fronko, R.; Lee, M.; Blais, J.; Cho, D.; Chamberland, S.; Renau, T.; et al. Identification and characterization of inhibitors of multidrug resistance efflux pumps in Pseudomonas aeruginosa: Novel agents for combination therapy. *Antimicrob. Agents Chemother.* **2001**, *45*, 105–116. [CrossRef] [PubMed]

18.	Dupont, H.; Gaillot, O.; Goetgheluck, A.-S.; Plassart, C.; Emond, J.-P.; Lecuru, M.; Gaillard, N.; Derdouri, S.; Lemaire, B.; Girard de Courtilles, M.; et al. Molecular Characterization of Carbapenem-Nonsusceptible Enterobacterial Isolates Collected during a Prospective Interregional Survey in France and Susceptibility to the Novel Ceftazidime-Avibactam and Aztreonam-Avibactam Combinations. *Antimicrob. Agents Chemother.* **2015**, *60*, 215–221. [CrossRef]

19.	Vera-Leiva, A.; Carrasco-Anabalón, S.; Lima, C.A.; Villagra, N.; Domínguez, M.; Bello-Toledo, H.; González-Rocha, G. The efflux pump inhibitor phenylalanine-arginine β-naphthylamide (PAβN) increases resistance to carbapenems in Chilean clinical isolates of KPC-producing Klebsiella pneumoniae. *J. Glob. Antimicrob. Resist.* **2018**, *12*, 73–76. [CrossRef]

20.	Carattoli, A.; Bertini, A.; Villa, L.; Falbo, V.; Hopkins, K.L.; Threlfall, E.J. Identification of plasmids by PCR-based replicon typing. *J. Microbiol. Methods* **2005**, *63*, 219–228. [CrossRef]

21.	Lee, W.; Chung, H.-S.; Lee, Y.; Yong, D.; Jeong, S.H.; Lee, K.; Chong, Y. Comparison of matrix-assisted laser desorption ionization-time-of-flight mass spectrometry assay with conventional methods for detection of IMP-6, VIM-2, NDM-1, SIM-1, KPC-1, OXA-23, and OXA-51 carbapenemase-producing Acinetobacter spp., Pseudomonas aeruginosa, and Klebsiella pneumoniae. *Diagn. Microbiol. Infect. Dis.* **2013**, *77*, 227–230. [CrossRef]

22.	Okanda, T.; Matsumoto, T. In vitro effect of an antimicrobial combination therapy without colistin and tigecycline for CPE and non-CPE. *J. Infect. Chemother.* **2020**, *26*, 322–330. [CrossRef]

23.	Ohno, Y.; Nakamura, A.; Hashimoto, E.; Matsutani, H.; Abe, N.; Fukuda, S.; Hisashi, K.; Komatsu, M.; Nakamura, F. Molecular epidemiology of carbapenemase-producing Enterobacteriaceae in a primary care hospital in Japan, 2010–2013. *J. Infect. Chemother.* **2017**, *23*, 224–229. [CrossRef]

24.	Liu, S.; Huang, N.; Zhou, C.; Lin, Y.; Zhang, Y.; Wang, L.; Zheng, X.; Zhou, T.; Wang, Z. Molecular Mechanisms and Epidemiology of Carbapenem-Resistant Enterobacter cloacae Complex Isolated from Chinese Patients During 2004-2018. *Infect. Drug Resist.* **2021**, *14*, 3647–3658. [CrossRef] [PubMed]

25.	Ellappan, K.; Belgode Narasimha, H.; Kumar, S. Coexistence of multidrug resistance mechanisms and virulence genes in carbapenem-resistant Pseudomonas aeruginosa strains from a tertiary care hospital in South India. *J. Glob. Antimicrob. Resist.* **2018**, *12*, 37–43. [CrossRef]

26.	Melvin, P.W.; James, S.L.; April, M.B.; Shelley, C.; Sharon, K.C.; Marcelo, F.G.; Howard, G.; Romney, M.H.; Thomas, J.K.; Brandi, L.; et al. M100Ed31 I Performance Standards for Antimicrobial Susceptibility Testing, 31st ed. Available online: https://clsi.org/standards/products/microbiology/documents/m100/ (accessed on 9 October 2021).

27. Hattori, T.; Kawamura, K.; Arakawa, Y. Comparison of test methods for detecting metallo-β-lactamase-producing Gram-negative bacteria. *Jpn. J. Infect. Dis.* **2013**, *66*, 512–518. [CrossRef]

28. Tada, T.; Miyoshi-Akiyama, T.; Shimada, K.; Shimojima, M.; Kirikae, T. IMP-43 and IMP-44 metallo-β-lactamases with increased carbapenemase activities in multidrug-resistant Pseudomonas aeruginosa. *Antimicrob. Agents Chemother.* **2013**, *57*, 4427–4432. [CrossRef] [PubMed]

29. Yamasaki, S.; Shigemura, K.; Osawa, K.; Kitagawa, K.; Ishii, A.; Kuntaman, K.; Shirakawa, T.; Miyara, T.; Fujisawa, M. Genetic analysis of ESBL-producing Klebsiella pneumoniae isolated from UTI patients in Indonesia. *J. Infect. Chemother.* **2021**, *27*, 55–61. [CrossRef] [PubMed]

30. Shibata, N.; Doi, Y.; Yamane, K.; Yagi, T.; Kurokawa, H.; Shibayama, K.; Kato, H.; Kai, K.; Arakawa, Y. PCR typing of genetic determinants for metallo-beta-lactamases and integrases carried by gram-negative bacteria isolated in Japan, with focus on the class 3 integron. *J. Clin. Microbiol.* **2003**, *41*, 5407–5413. [CrossRef]

31. Mushi, M.F.; Mshana, S.E.; Imirzalioglu, C.; Bwanga, F. Carbapenemase Genes among Multidrug Resistant Gram Negative Clinical Isolates from a Tertiary Hospital in Mwanza, Tanzania. *BioMed Res. Int.* **2014**, *2014*, 303104. [CrossRef] [PubMed]

32. Hong, S.S.; Kim, K.; Huh, J.Y.; Jung, B.; Kang, M.S.; Hong, S.G. Multiplex PCR for Rapid Detection of Genes Encoding Class A Carbapenemases. *Ann. Lab. Med.* **2012**, *32*, 359–361. [CrossRef] [PubMed]

33. Farajzadeh Sheikh, A.; Rostami, S.; Jolodar, A.; Tabatabaiefar, M.A.; Khorvash, F.; Saki, A.; Shoja, S.; Sheikhi, R. Detection of Metallo-Beta Lactamases Among Carbapenem-Resistant Pseudomonas Aeruginosa. *Jundishapur J. Microbiol.* **2014**, *7*, e12289. [CrossRef] [PubMed]

34. Marchaim, D.; Navon-Venezia, S.; Schwaber, M.J.; Carmeli, Y. Isolation of imipenem-resistant Enterobacter species: Emergence of KPC-2 carbapenemase, molecular characterization, epidemiology, and outcomes. *Antimicrob. Agents Chemother.* **2008**, *52*, 1413–1418. [CrossRef] [PubMed]

35. Shahada, F.; Chuma, T.; Kosugi, G.; Kusumoto, M.; Iwata, T.; Akiba, M. Distribution of extended-spectrum cephalosporin resistance determinants in Salmonella enterica and Escherichia coli isolated from broilers in southern Japan. *Poult. Sci.* **2013**, *92*, 1641–1649. [CrossRef]

36. Peleg, A.Y.; Adams, J.; Paterson, D.L. Tigecycline Efflux as a Mechanism for Nonsusceptibility in Acinetobacter baumannii. *Antimicrob. Agents Chemother.* **2007**, *51*, 2065–2069. [CrossRef]

37. Kanda, Y. Investigation of the freely available easy-to-use software "EZR" for medical statistics. *Bone Marrow Transplant.* **2013**, *48*, 452–458. [CrossRef] [PubMed]

Article

Infection Control for a Carbapenem-Resistant Enterobacteriaceae Outbreak in an Advanced Emergency Medical Services Center

Yoshiro Sakai [1,2,3,†], Kenji Gotoh [2,3,*,†], Ryuichi Nakano [4,†], Jun Iwahashi [2], Miho Miura [3], Rie Horita [5], Naoki Miyamoto [5], Hisakazu Yano [4], Mikinori Kannae [6], Osamu Takasu [6] and Hiroshi Watanabe [2,3]

1   Department of Pharmacy, Kurume University Hospital, Kurume 831-0011, Japan; sakai_yoshirou@kurume-u.ac.jp
2   Department of Infection Control and Prevention, Kurume University School of Medicine, Kurume 831-0011, Japan; iwahashi@kurume-u.ac.jp (J.I.); hwata@med.kurume-u.ac.jp (H.W.)
3   Division of Infection Control and Prevention, Kurume University Hospital, Kurume 831-0011, Japan; miura_miho@kurume-u.ac.jp
4   Department of Microbiology and Infectious Diseases, Nara Medical University, Nara 634-8521, Japan; rnakano@naramed-u.ac.jp (R.N.); yanohisa@naramed-u.ac.jp (H.Y.)
5   Department of Clinical Laboratory Medicine, Kurume University Hospital, Kurume 831-0011, Japan; horita_rie@kurume-u.ac.jp (R.H.); miyamoto_naoki@kurume-u.ac.jp (N.M.)
6   Department of Emergency and Critical Care Medicine, Kurume University School of Medicine, Kurume 831-0011, Japan; kannae_mikinori@med.kurume-u.ac.jp (M.K.); takasu_osamu@kurume-u.ac.jp (O.T.)
*   Correspondence: gotou_kenji@kurume-u.ac.jp; Tel.: +81-942-31-7549; Fax: +81-942-31-7697
†   These authors contributed equally to this work.

Citation: Sakai, Y.; Gotoh, K.; Nakano, R.; Iwahashi, J.; Miura, M.; Horita, R.; Miyamoto, N.; Yano, H.; Kannae, M.; Takasu, O.; et al. Infection Control for a Carbapenem-Resistant Enterobacteriaceae Outbreak in an Advanced Emergency Medical Services Center. *Antibiotics* **2021**, *10*, 1537. https://doi.org/10.3390/antibiotics10121537

Academic Editor: Francesca Andreoni

Received: 17 November 2021
Accepted: 12 December 2021
Published: 15 December 2021

Publisher's Note: MDPI stays neutral with regard to jurisdictional claims in published maps and institutional affiliations.

Abstract: Background: A carbapenem-resistant Enterobacteriaceae (CRE) outbreak occurred in an advanced emergency medical service center [hereafter referred to as the intensive care unit (ICU)] between 2016 and 2017. Aim: Our objective was to evaluate the infection control measures for CRE outbreaks. Methods: CRE strains were detected in 16 inpatients located at multiple sites. Environmental cultures were performed and CRE strains were detected in 3 of 38 sites tested. Pulsed-field gel electrophoresis (PFGE), multilocus sequence typing (MLST), and detection of β-lactamase genes were performed against 25 CRE strains. Findings: Molecular typing showed the PFGE patterns of two of four *Klebsiella pneumoniae* strains were closely related and the same MLST (ST2388), and four of five *Enterobacter cloacae* strains were closely related and same MLST (ST252). Twenty-three of 25 CRE strains harbored the IMP-1 β-lactamase gene and 15 of 23 CRE strains possessed IncFIIA replicon regions. Despite interventions by the infection control team, new inpatients with the CRE strain continued to appear. Therefore, the ICU was partially closed and the inpatients with CRE were isolated, and the ICU staff was divided into two groups between inpatients with CRE and non-CRE strains to avoid cross-contamination. Although the occurrence of new cases dissipated quickly after the partial closure, a few months were required to eradicate the CRE outbreak. Conclusion: Our data suggest that the various and combined measures that were used for infection control were essential in stopping this CRE outbreak. In particular, partial closure to isolate the ICU and division of the ICU staff were effective.

Keywords: carbapenem-resistant Enterobacteriaceae (CRE); outbreak; infection control; pulsed-field gel electrophoresis (PFGE); multilocus sequence typing (MLST); carbapenemase

1. Introduction

Carbapenem-resistant Enterobacteriaceae (CRE) is a major health concern worldwide, including infections found in Japan [1–3]. Outbreaks of CRE have occurred in most parts of the world during the past decade [4–6]. When CRE is detected in hospitalized patients, contact precautions for infection control of CRE is very important. Hospitalized patients may be particularly susceptible to infections, and CRE infections are associated with increasing the risks of morbidity and mortality, prolonged hospital stay, and increasing health care costs [7,8]. As mentioned earlier, clinical infections with CRE are associated with high rates of morbidity and mortality, which is due, in part, to limited options for therapy [7]. When CRE is detected in hospitalized patients, contact precautions for infection control of CRE are very important [8]. The treatment options for CRE infections remain very limited, and colistin and tigecycline are considered the drugs of choice to treat infections caused by CRE [9]. However, the emergence of bacteria that are resistant to these antibiotics has also been recognized worldwide [10–12]. In addition, outbreaks of colistin-resistant CRE have occurred [11,13].

The management of CRE in hospital settings is not only costly but presents a significant challenge. While reliably detecting CRE in the laboratory is an important first step, it can be hampered by the fact that resistance occurs through a variety of different mechanisms. By accurately understanding the homology and resistance mechanisms of CRE, it is possible to know whether the infection is nosocomial or spreading in the community. This information is useful in deciding whether infection control measures should be implemented on a ward basis, on a hospital basis, or including the community [14]. On the other hand, knowing the mechanism of resistance can contribute to the appropriate selection of therapeutic agents. KPC-producing CRE can be treated with antimicrobial agents containing avibactam and vaborbactam, so colistin and tigecycline can be preserved [15,16]. This is also important from the viewpoint of antimicrobial stewardship.

The advanced emergency medical service center in Kurume University Hospital has experienced several outbreaks due to resistant bacteria such as methicillin-resistant *Staphylococcus aureus* and vancomycin-intermediate *S. aureus*, and an infection control team (ICT) is usually implemented [17,18]. In this report, we describe a CRE outbreak in our advanced emergency medical service center (hereafter referred to as the ICU) and discuss the stepwise infection control measures that were implemented, along with our evaluation of the effectiveness of these measures.

2. Results

2.1. Bacterial Strains and Patient Characteristics

Sixteen CRE strains were isolated from the stools of nine inpatients; from the nasal cavity of four inpatients; and from the pus, sputum, and skin of each of these inpatients between August and December 2016 (Figure 1). The mean age of the 16 inpatients (11 males and 5 females) was 65.7 years, and their actual ages ranged from 24 to 86 years. The mean detection period of the CRE strain after admission was 23.9 days, which represented a range of from 1 to 170 days. CRE had been detected at the point of hospitalization in 4 of the 16 inpatients. During the CRE outbreak, 13 CRE strains were recognized as colonization. However, three CRE strains were isolated from inpatients with pneumonia or bacteremia, and one inpatient died from bacteremia due to CRE.

An environmental culture was performed in November 2016, and CRE strains were detected in 3 of 38 sites (3 different sinks).

2.2. MIC

In judging the effectiveness of CRE treatments, eight isolates showed meropenem MICs ≥ 2 mg/L, another eight isolates showed imipenem MICs ≥ 2 mg/L, and cefmetazole recorded MICs ≥ 64 mg/L according to the reporting criteria of the Infectious Disease Act of Japan (Table 1).

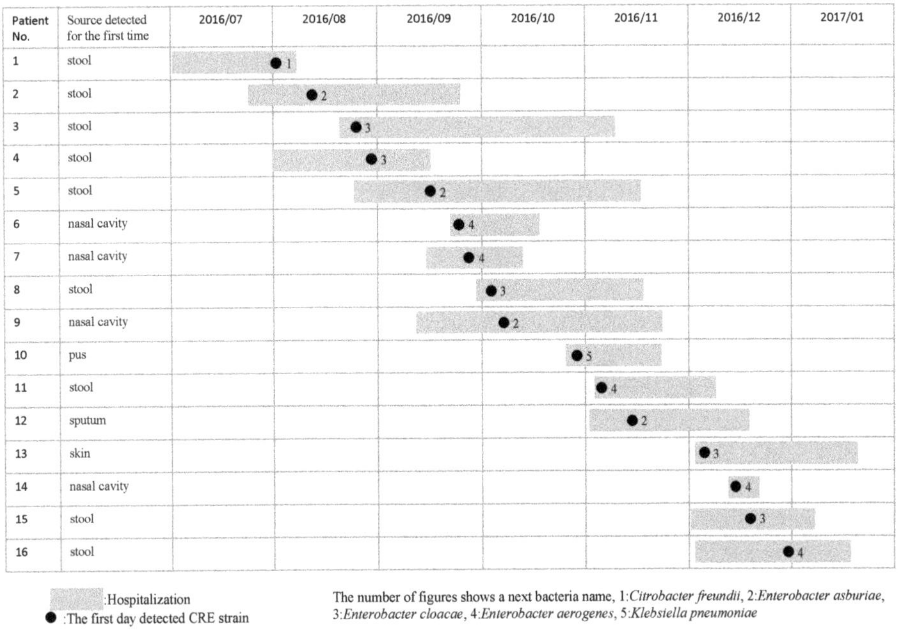

Figure 1. The time course for a CRE outbreak in the ICU. Gray shadow: period of hospitalization. Black circle: the first day of CRE strain detection.

Table 1. Antibiotics susceptibolity profiles (minimum inhibitory concentrations, mg/L).

Patient No.	ABPC	PIPC	CTX	CAZ	CFPM	CMZ	IPM	MEPM	AZT	ABPC/SBT	PIPC/TAZ	GM	AMK	MINO	LVFX	ST
1	>16	<8	>2	>8	4	>32	2	>2	<4	>16	<16	4	<4	4	2	<2
2	>16	>64	>2	>8	>16	>32	>2	>2	<4	>16	64	>8	<4	4	1	<2
3	>16	>64	>2	>8	>16	>32	>2	>2	<4	>16	>64	>8	<4	4	1	<2
4	>16	<8	<1	<4	<2	>32	2	<1	<4	16	<16	<2	<4	4	<0.5	<2
5	>16	>64	>2	>8	>16	>32	>2	>2	>8	>16	>64	8	<4	<2	<0.5	<2
6	<8	<8	<1	<4	<2	>32	2	<1	<4	<8	<16	<2	<4	<2	<0.5	<2
7	>16	<8	<1	<4	<2	>32	2	<1	<4	16	<16	<2	<4	<2	<0.5	<2
8	>16	>64	>2	>8	>16	>32	>2	>2	>8	>16	>64	>8	<4	>8	4	<2
9	>16	>64	>2	>8	>16	>32	>2	>2	<4	>16	64	>8	<4	>8	4	>2
10	>16	16	>2	>8	>16	>32	2	>2	<4	>16	<16	8	<4	4	4	>2
11	>16	<8	<1	<4	<2	>32	2	<1	<4	16	<16	<2	<4	<2	<0.5	<2
12	>16	<8	>2	>8	4	>32	>2	>2	<4	>16	<16	>8	<4	<2	1	<2
13	>16	<8	2	<4	<2	>32	2	<1	<4	>16	<16	<2	<4	4	<0.5	<2
14	>16	<8	<1	<4	<2	>32	2	<1	<4	16	<16	<2	<4	<2	<0.5	<2
15	>16	<8	>2	<4	<2	>32	2	<1	<4	>16	<16	<2	<4	<2	<0.5	<2
16	>16	<8	<1	<4	<2	>32	2	<1	<4	16	<16	<2	<4	<2	<0.5	<2

ABPC: ampicillin, PIPC: piperacillin, CTX: cefotaxime, CAZ: ceftazidime, CFPM: cefepime, CMZ: cefmetazole, IPM: imipenem, MEPM: meropenem, AZT: aztreonam, ABPC/SBT: ampicillin/sulbactam, PIPC/TAZ: piperacillin/tazobactam, GM: gentamicin, AMK: amikacin, MINO: minocycline, LVFX: levofloxacin, ST: sulfamethoxazole-trimethoprim.

2.3. Interpretation of Molecular Typing by PFGE and MLST Analysis

Molecular typing by the PFGE patterns of 25 CRE strains was divided into eight patterns (A–I). The PFGE patterns of *K. pneumoniae* in strains three and four, *Enterobacter asburiae* from strains 5 to 16, and *E. cloacae* from strains 19 to 22 all were closely related (Figure 2). MLST analysis was performed for all four of the *K. pneumoniae* strains and for 5 of the *E. cloacae* strains. Of *K. pneumoniae* the isolates identified ST286 (2 strains) and ST2388 (2 strains). Of *E. cloacae*, the isolates identified ST252 (4 strains) and ST384 (1 strain).

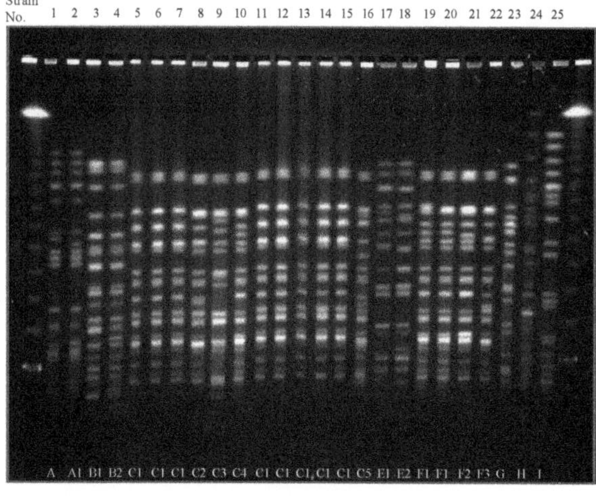

Strain No.	Date (mo/day/yr)	Patient No.	Source	Bacteria	MLST
1	08/31/2016	3	dennis tube	*K. pneumoniae*	ST286
2	11/07/2016	3	nasal cavity	*K. pneumoniae*	ST286
3	10/28/2016	8	stool	*K. pneumoniae*	ST2388
4	10/28/2016	10	pus	*K. pneumoniae*	ST2388
5	08/31/2016	3	dennis tube	*E. asburiae*	ND
6	09/12/2016	3	sputum	*E. asburiae*	ND
7	09/26/2016	3	stool	*E. asburiae*	ND
8	11/04/2016	5	stool	*E. asburiae*	ND
9	11/07/2016	5	sputum	*E. asburiae*	ND
10	10/13/2016	9	nasal cavity	*E. asburiae*	ND
11	11/14/2016	9	sputum	*E. asburiae*	ND
12	11/08/2016	9	blood	*E. asburiae*	ND
13	11/11/2016	9	blood	*E. asburiae*	ND
14	11/18/2016	9	blood	*E. asburiae*	ND
15	11/14/2016	12	sputum	*E. asburiae*	ND
16	11/15/2016	-	sink	*E. asburiae*	ND
17	11/15/2016	-	sink	*E. asburiae*	ND
18	11/15/2016	-	sink	*E. asburiae*	ND
19	08/26/2016	3	stool	*E. cloacae*	ST252
20	09/12/2016	3	nasal cavity	*E. cloacae*	ST252
21	10/17/2016	8	stool	*E. cloacae*	ST252
22	10/31/2016	5	stool	*E. cloacae*	ST252
23	12/07/2016	13	skin	*E. cloacae*	ST384
24	10/21/2016	9	stool	*K. oxytoca*	ND
25	11/10/2016	11	stool	*E. aerogenes*	ND

ND: not determined

Figure 2. PFGE patterns of *Xba*I-digested DNA from 25 CRE isolates (22 from inpatients and 3 from environments). Molecular typing showed that the PFGE patterns of 25 CRE isolates were divided into eight patterns (A–I). Those of *K. pneumoniae* in strains three and four, *E. asburiae* from strains 5 to 16, and *E. cloacae* strains from strains No.19 to 22 were closely related. Similarly, the MLST patterns of 2 *K. pneumoniae* in strains three and four were identical as ST2388 and those of four *E. cloacae* strains from strains 19 to 22 were identical as ST252.

2.4. Distribution of β-Lactamase Genes

The distribution of β-lactamase genes is shown in Table 2. Twenty-three isolates were positive for CIM, and all of the isolates harbored the IMP-1 β-lactamase gene. Two other CIM-negative isolates harbored no carbapenemase gene and were categorized as non-CPE (carbapenemase-producing Enterobacteriaceae). These two strains were resistant to carbapenems probably due to overexpression of AmpC β-lactamase combined with a disrupted outer membrane (porin) permeability or other mechanisms.

Table 2. Distribution of β-lactamase genes for 25 CRE isolates (22 from inpatients and 3 from environments). Twenty-three isolates were positive for CIM and all of the isolates harbored the IMP-1 β-lactamase gene. Two other CIM-negative isolates harbored no carbapenemase gene. Incompatibility group typing revealed two types of plasmids in the CPEs. Fifteen of the 23 CPE isolates possessed IncFIIA replicon regions, including *K. pneumoniae* (*n* = 2), *E. asburiae* (*n* = 11), and *E. cloacae* (*n* = 2). Two other isolates belonged to IncN in *K. pneumoniae*, and the Inc type could not be determined for the six remaining isolates.

Strain No.	Carbapenemase	CTX-M	ESBL	CIM	Inc
1	IMP-1	ND	SHV	+	FIIA
2	IMP-1	ND	SHV	+	FIIA
3	IMP-1	ND	SHV	+	N
4	IMP-1	ND	SHV	+	N
5	IMP-1	ND	ND	+	FIIA
6	IMP-1	ND	TEM, SHV	+	FIIA
7	IMP-1	ND	TEM, SHV	+	FIIA
8	IMP-1	ND	ND	+	FIIA
9	IMP-1	ND	SHV	+	ND
10	IMP-1	ND	TEM, SHV	+	FIIA
11	IMP-1	ND	ND	+	FIIA
12	IMP-1	ND	ND	+	FIIA
13	IMP-1	ND	ND	+	FIIA
14	IMP-1	ND	ND	+	FIIA
15	IMP-1	ND	ND	+	FIIA
16	IMP-1	ND	ND	+	FIIA
17	IMP-1	ND	ND	+	ND
18	IMP-1	ND	ND	+	ND
19	IMP-1	ND	TEM, SHV	+	ND
20	IMP-1	ND	TEM	+	ND
21	IMP-1	ND	TEM, SHV	+	FIIA
22	IMP-1	ND	TEM, SHV	+	FIIA
23	ND	ND	ND	-	ND
24	IMP-1	ND	TEM	+	ND
25	ND	ND	ND	-	FIIA

EBSL: extended-spectrum β-lactamase, CIM: Carbapenem Inactivation Method, ND: not deceted.

Incompatibility group typing revealed two types of plasmids in the CPEs. Fifteen of the 23 CPE isolates possessed IncFIIA replicon regions, including *K. pneumoniae* (*n* = 2), *E. asburiae* (*n* = 11), and *E. cloacae* (*n* = 2). Two other isolates belonged to IncN in *K. pneumoniae*, and the remaining six isolates were not determined according to Inc type. We assumed that the IMP-1 β-lactamase gene encoding the IncFIIA plasmid was disseminated among the species.

2.5. Intervention by the ICT

Initiatives for cohort isolation, active surveillance, environmental culture, monitoring, and education for the ICU staff were performed by the ICT. However, new inpatients with the CRE strain continued to appear despite such interventions. We notified the government of the outbreak and received guidance, but the outbreak was not contained. We invited several additional infection control experts from other facilities to take measures, but the outbreak was still not contained. Therefore, the ICU was partially closed after discussions with the government in November 2016, and the inpatients with the CRE strain were isolated in order to prevent further horizontal transmission. The ICU staff was divided into two groups between inpatients with CRE and non-CRE strains to avoid cross-contamination. Although the occurrence of new cases dissipated quickly after the partial closure, it took several months to eradicate the CRE outbreak, and the hospital suffered economically.

3. Discussion

In this study, we characterized the epidemiological, microbiological, and molecular analysis of CRE outbreaks in an advanced emergency medical service center in Japan. CRE has recently been detected in the world and its outbreaks have increased [19]. Because CRE had been detected in 4 of the 16 inpatients before hospitalization, we have routinely performed surveillance cultures for all patients upon admission. Thus, it is important to monitor patients with resistant bacteria before hospitalization [8,20,21]. If there is an increasing trend in the frequency of isolates of resistant organisms such as CRE, vancomycin resistance *Enterococcus faecium* and vancomycin resistant *Staphylococcus aureus* in the tests at the time of admission, it is also necessary to exchange information and collaborate on the isolation status of resistant organisms at medical facilities in the surrounding areas.

The routes of infection for resistant organisms such as CRE are mainly the result of direct or indirect contact that can be spread in a ward via the transiently colonized hands of healthcare workers [8,20], and CRE is known to exist in hospital water environments such as sinks [6]. In addition, our results showed that the CRE organisms detected in blood culture and in the sink were identified as the same bacteria by MLST analysis. Furthermore, closely related strains have been detected in several different sinks. Considering the results, daily cleaning of hospital water environments [19–21] and hand hygiene [21,22] are important for infection control. In response to this outbreak, we took these factors into consideration and conducted daily rounds and infection control, focusing on cleaning the water environment. However, it was not enough to contain the CRE outbreak. The water-free ICU is now being proposed as a management method for the water-borne outbreak, and Some reports have shown that removing sinks in intensive care units has reduced the prevalence of multidrug-resistant Gram-negative bacteria [23,24]. Gram-negative bacteria can survive for a long time in sinks, where they acquire resistant genes through contact with bacteria that have resistant genes. By washing hands in the contaminated sink, water droplets are dispersed into the environment and adhere to the clothes of healthcare workers, which is thought to spread the drug-resistant bacteria in the ICU. If we had stopped using sinks, the CRE outbreak might have been contained earlier.

During the CRE outbreak, molecular analysis by PFGE was performed repeatedly to evaluate horizontal transmission, and members of the ICU staff were immediately informed of the results. PFGE seems useful for evaluating the presence of horizontal transmission in hospital-acquired infection [17,18]. Furthermore, we used MLST for analysis in addition to PFGE for a portion of the CRE strains. The two methods detected the same sequence in most strains.

In this study, 23 of 25 CRE strains produced IMP-1, but the remaining two isolates had no carbapenemase. Despite the fact that KPC, OXA-48, and NDM are found globally, these are rarely found in Japan, where IMP-1 and IMP-6 are exclusively the predominant forms of carbapenemases [25,26]. Since the resistance gene is known to spread across strains producing carbapenemase [27], infection control against inpatients with the CRE strain is important in preventing outbreaks. Our results also suggest that 17/23 strains of CPE had transmissible plasmid. Further, because the IMP-1 β-lactamase gene encoding the IncFIIA plasmid was disseminated among the species, infection control against CPE is particularly important regardless of the bacterial species. Regarding the choice of therapeutic agents, it is important to investigate the type of carbapenemase in CRE outbreaks. These CREs that produced IMP type carbapenemase cannot be treated with antibacterial agents including beta-lactamase inhibitors such as avibactam and vaborbactam. Because The CRE strains in this study were susceptible to aztreonam, quinolones, tetracycline, and aminoglycosides, we were able to intervene appropriately regarding the choice of treatment. The appropriate use of antimicrobial agents is essential to inhibit the emergence of resistant strains, including CRE.

During the CRE outbreak, active surveillance, environmental culture, monitoring, and education for the ICU staff were performed by the ICT, but we were unable to stop the expansion of CRE. As a result, we partially closed the ICU, which allowed us to strictly segregate staff caring for CRE-affected and unaffected inpatients. Nevertheless, several months were required to finally terminate the CRE outbreak, and the hospital suffered economically as a result.

This study is limited by the fact that it is a single-center experience of a CRE outbreak, and the number of cases is small. In order to prevent the spread of CRE after even one case is isolated, infection control measures and laboratory testing systems similar to those for outbreaks are necessary. Our analysis of the organisms and estimation of the route of infection will be useful for other institutions.

In conclusion, despite the employment of various infection control measures, partial closure for isolation plus division of the ICU staff was essential in terminating this CRE outbreak.

4. Methods

4.1. Ethical Approval

All studies described herein were approved by the Human Ethics Review Boards of Kurume University (17161). At the time of admission to the ICU, we have obtained consent from the patient or family for checking resistant organisms' carriage and for active surveillance in all cases.

4.2. Setting and Outbreak Description

In the Kurume University Hospital, there are 25 diagnosis and treatment departments that serve 24 wards with 1018 beds, which includes an ICU with 43 beds. The ICU accepts many severe patients from ambulance and helicopter emergency medical services. A CRE strain was first detected from the stool of an inpatient in the ICU in August 2016. Isolation in a private room and the reinforcement of direct or indirect contact infection measures were performed for this inpatient with the CRE strain. However, additional inpatients with the CRE strain eventually emerged. Three new inpatients with the CRE strain were simultaneously identified at the beginning of September 2016, and the infection control team (ICT) classified the intervention with the status of an outbreak.

4.3. Bacterial Strains and Patients

Twenty-five CRE isolates from 16 inpatients and three environments in the ICU between August and December 2016 were enrolled in this study.

4.4. Identification Test and Minimum Inhibitory Concentration

An identification test was conducted using MicroScan WalkAway96 plus NBP 6.23J (Siemens Healthcare Diagnostics Inc., Tokyo, Japan). The minimum inhibitory concentrations (MIC) of ampicillin, piperacillin, cefotaxime, ceftazidime, cefepime, cefmetazole, imipenem, meropenem, aztreonam, ampicillin/sulbactam, piperacillin/tazobactam, gentamicin, amikacin, minocycline, levofloxacin, and sulfamethoxazole-trimethoprim were determined in reference to MicroScan Neg NENC1J (Siemens Healthcare Diagnostics Inc., Tokyo, Japan) via the broth-dilution method, in accordance with the guidelines of the Clinical and Laboratory Standards Institute [28]. The criteria for CRE were based on laboratory findings of Japanese criteria as follows: the MIC for meropenem was $\geq$2 mg/L, or the MIC for imipenem was $\geq$2 mg/L and the MIC for cefmetazole was $\geq$64 mg/L.

4.5. Pulsed-Field Gel Electrophoresis

Pulsed-field gel electrophoresis (PFGE) against 25 CRE strains (22 from inpatients and three from environments) was performed, as described previously [17]. The DNA was digested with *Xba*I (Takara Shuzo Co., Shiga, Japan). CHEF Mapper pulsed-field electrophoresis systems (Bio-Rad Life Science Group, Hercules, CA, USA) were used with a potential of 6 V/cm, with switch times of 2.16 and 44.69 s, and run-times of 20 h. After staining with ethidium bromide, the PFGE patterns were interpreted based on the criteria described by Tenover et al. [29,30].

4.6. Multilocus Sequence Typing

Multilocus sequence typing (MLST) was performed for the isolates of *Klebsiella pneumoniae* and *Enterobacter cloacae*. All strains of *K. pneumoniae* and *E. cloacae* were assessed by MLST in accordance with the protocol on the MLST website. The primers of seven housekeeping genes were based on information from the following website: https://pubmlst.org/ecloacae/, https://bigsdb.pasteur.fr/klebsiella/klebsiella.html (accessed on 10 July 2019). The sequence types were assigned using the MLST website.

4.7. Detection of β-Lactamase Genes

Carbapenemase production was confirmed using the carbapenem inactivation method (CIM) [31]. The presence of β-lactamase genes including carbapenemases (bla_{IMP}, bla_{VIM}, bla_{KPC}, $bla_{OXA-48-like}$, and bla_{NDM}) and ESBL (bla_{TEM}, bla_{SHV}, and bla_{CTX-M}) was assessed using PCR and DNA sequencing as previously described [32,33].

4.8. Plasmid Incompatibility Typing

Plasmids incompatibility (Inc) groups were determined using the PCR replicon-typing scheme, as previously described [34].

5. Conclusions

Despite the employment of various infection control measures, partial closure for isolation plus division of the ICU staff was essential in terminating this CRE outbreak.

Author Contributions: Y.S., K.G. and R.N. designed the experiments, conducted the main experiments, and prepared the original draft; H.Y., H.W., M.K. and O.T. supervised and revised the manuscript; J.I., N.M., Y.S., M.M. and R.H. analyzed the data. All authors have read and agreed to the published version of the manuscript.

Funding: This research received no external funding.

Institutional Review Board Statement: All studies described herein were approved by the Human Ethics Review Boards of Kurume University (17161).

Informed Consent Statement: Informed consent was obtained from all subjects involved in the study.

Acknowledgments: We thank Ayako Namitome for her help in completing this study.

Conflicts of Interest: The authors declare no conflict of interest.

References

1. Nordmann, P.; Naas, T.; Poirel, L. Global spread of carbapenemase producing Enterobacteriaceae. *Emerg. Infect. Dis.* **2011**, *17*, 1791–1798. [CrossRef] [PubMed]
2. Logan, L.K.; Weinstein, R.A. The epidemiology of carbapenem-resistant Enterobacteriaceae: The impact and evolution of a global menace. *J. Infect. Dis.* **2017**, *215* (Suppl. S1), S28–S36. [CrossRef]
3. Ishii, Y.; Aoki, K.; Tateda, K.; Kiyota, H. Multicenter collaboration study on the ß-lactam resistant Enterobacteriaceae in Japan— The 65th anniversary public interest purpose project of the Japanese Society of Chemotherapy. *J. Infect. Chemother.* **2017**, *23*, 583–586. [CrossRef]
4. Guh, A.Y.; Bulens, S.N.; Mu, Y.; Jacob, J.T.; Reno, J.; Scott, J.; Wilson, L.E.; Vaeth, E.; Lynfield, R.; Shaw, K.M.; et al. Epidemiology of carbapenem-resistant Enterobacteriaceae in 7 US communities, 2012–2013. *JAMA* **2015**, *314*, 1479–1487. [CrossRef]
5. Zhang, Y.; Wang, Q.; Yin, Y.; Chen, H.; Jin, L.; Gu, B.; Xie, L.; Yang, C.; Ma, X.; Li, H.; et al. Epidemiology of carbapenem-resistant Enterobacteriaceae infections: Report from the China CRE network. *Antimicrob. Agents Chemother.* **2018**, *62*, e01882-17. [CrossRef] [PubMed]
6. Kizny Gordon, A.E.; Mathers, A.J.; Cheong, E.Y.; Gottlieb, T.; Kotay, S.; Walker, A.S.; Peto, T.E.; Crook, D.W.; Stoesser, N. The hospital water environment as a reservoir for carbapenem-resistant organisms causing hospital-acquired infections-a systematic review of the literature. *Clin. Infect. Dis.* **2017**, *64*, 1435–1444. [CrossRef] [PubMed]
7. Jacob, J.T.; Klein, E.; Laxminarayan, R.; Beldavs, Z.; Lynfield, R.; Kallen, A.J.; Ricks, P.; Edwards, J.; Srinivasan, A.; Fridkin, S.; et al. Centers for Disease Control and Prevention (CDC). Vital signs: Carbapenem-resistant Enterobacteriaceae. *Morb. Mortal. Wkly. Rep.* **2013**, *62*, 165–170.
8. Friedman, N.D.; Carmeli, Y.; Walton, A.L.; Schwaber, M.J. Carbapenem-resistant Enterobacteriaceae: A strategic roadmap for infection control. *Infect. Control Hosp. Epidemiol.* **2017**, *38*, 580–594. [CrossRef]

9.   Sheu, C.C.; Chang, Y.T.; Lin, S.Y.; Chen, Y.H.; Hsueh, P.R. Infections caused by carbapenem-resistant Enterobacteriaceae: An update on therapeutic options. *Front. Microbiol.* **2019**, *10*, 80. [CrossRef]

10.   Malchione, M.D.; Torres, L.M.; Hartley, D.M.; Koch, M.; Goodman, J.L. Carbapenem and colistin resistance in Enterobacteriaceae in southeast Asia: Review and mapping of emerging and overlapping challenges. *Int. J. Antimicrob. Agents* **2019**, *54*, 381–399. [CrossRef] [PubMed]

11.   Jayol, A.; Poirel, L.; Dortet, L.; Nordmann, P. National survey of colistin resistance among carbapenemase-producing Enterobacteriaceae and outbreak caused by colistin-resistant OXA-48-producing *Klebsiella pneumoniae*, France, 2014. *Eurosurveillance* **2016**, *21*, 30339. [CrossRef] [PubMed]

12.   Peck, K.R.; Kim, M.J.; Choi, J.Y.; Kim, H.S.; Kang, C.-I.; Cho, Y.K.; Park, D.W.; Lee, H.J.; Lee, M.S.; Ko, K.S. In vitro time-kill studies of antimicrobial agents against blood isolates of imipenem-resistant *Acinetobacter baumannii*, including colistin-or tigecycline-resistant isolates. *J. Med. Microbiol.* **2012**, *61*, 353–360. [CrossRef]

13.   Gonçalves, I.R.; Ferreira, M.; Araujo, B.; Campos, P.; Royer, S.; Batistão, D.; Souza, L.; Brito, C.; Urzedo, J.; Gontijo-Filho, P.; et al. Outbreaks of colistin-resistant and colistin-susceptible KPC-producing *Klebsiella pneumoniae* in a Brazilian intensive care unit. *J. Hosp. Infect.* **2016**, *94*, 322–329. [CrossRef] [PubMed]

14.   Robert, F.P.; Alaric, W.D.; Gautam, D. The rapid spread of carbapenem-resistant Enterobacteriaceae. *Drug Resist. Updat.* **2016**, *29*, 30–46.

15.   Shields, R.K.; Nguyen, M.H.; Chen, L.; Press, E.G.; Kreiswirth, B.N.; Clancy, C.J. Pneumonia and Renal Replacement Therapy Are Risk Factors for Ceftazidime-Avibactam Treatment Failures and Resistance among Patients with Carbapenem-Resistant Enterobacteriaceae Infections. *Antimicrob. Agents Chemother.* **2018**, *62*, e02497-17. [CrossRef]

16.   Ackley, R.; Roshdy, D.; Meredith, J.; Minor, S.; Anderson, W.E.; Capraro, G.A.; Polk, C. Meropenem-Vaborbactam versus Ceftazidime-Avibactam for Treatment of Carbapenem-Resistant Enterobacteriaceae Infections. *Antimicrob. Agents Chemother.* **2020**, *64*, e02313-19. [CrossRef]

17.   Hidaka, H.; Miura, M.; Masunaga, K.; Qin, L.; Uemura, Y.; Sakai, Y.; Watanabe, H.; Hashimoto, K.; Kawano, S.; Yamashita, N.; et al. Infection control for a methicillin-resistant *Staphylococcus aureus* outbreak in an advanced emergency medical service center, as monitored by molecular analysis. *J. Infect. Chemother.* **2013**, *19*, 884–890. [CrossRef]

18.   Sakai, Y.; Qin, L.; Miura, M.; Masunaga, K.; Tanamachi, C.; Iwahashi, J.; Kida, Y.; Takasu, O.; Sakamoto, T.; Watanabe, H. Successful infection control for a vancomycin-intermediate *Staphylococcus aureus* outbreak in an advanced emergency medical service centre. *J. Hosp. Infect.* **2016**, *92*, 385–391. [CrossRef]

19.   French, C.E.; Coope, C.; Conway, L.; Higgins, J.P.T.; McCulloch, J.; Okoli, G.; Patel, B.C.; Oliver, I. Control of carbapenemase-producing Enterobacteriaceae outbreaks in acute settings: An evidence review. *J. Hosp. Infect.* **2017**, *95*, 3–45. [CrossRef]

20.   Munoz-Price, L.S.; Hayden, M.K.; Lolans, K.; Won, S.; Calvert, K.; Lin, M.; Sterner, A.; Weinstein, R.A. Successful control of an outbreak of *Klebsiella pneumoniae* carbapenemase-producing *K. pneumoniae* at a long-term acute care hospital. *Infect. Control Hosp. Epidemiol.* **2010**, *31*, 341–347. [CrossRef]

21.   Hussein, K.; Rabino, G.; Eluk, O.; Warman, S.; Reisner, S.; Geffen, Y.; Halif, L.; Paul, M. The association between infection control interventions and carbapenem-resistant Enterobacteriaceae incidence in an endemic hospital. *J. Hosp. Infect.* **2017**, *97*, 218–225. [CrossRef]

22.   Friedrich, A.W. Control of hospital acquired infections and antimicrobial resistance in Europe: The way to go. *Wien. Med. Wochenschr* **2019**, *169* (Suppl. S1), 25–30. [CrossRef]

23.   Hopman, J.; Tostmann, A.; Wertheim, H.; Bos, M.; Kolwijck, E.; Akkermans, R.; Sturm, P.; Voss, A.; Pickkers, P. Reduced rate of intensive care unit acquired gram-negative bacilli after removal of sinks and introduction of 'water-free' patient care. *Antimicrob. Resist. Infect. Control.* **2017**, *6*, 59. [CrossRef]

24.   Shaw, E.; Gavaldà, L.; Càmara, J.; Gasull, R.; Gallego, S.; Tubau, F.; Granada, R.; Ciercoles, P.; Domínguez, M.A.; Manez, R.; et al. Control of endemic multidrug-resistant Gram-negative bacteria after removal of sinks and implementing a new water-safe policy in an intensive care unit. *J. Hosp. Infect.* **2018**, *98*, 275–281. [CrossRef] [PubMed]

25.   Yamamoto, N.; Asada, R.; Kawahara, R.; Hagiya, H.; Akeda, Y.; Shanmugakani, R.K.; Yoshida, H.; Yukawa, S.; Yamamoto, K.; Takayama, Y.; et al. Prevalence of, and risk factors for, carriage of carbapenem-resistant Enterobacteriaceae among hospitalized patients in Japan. *J. Hosp. Infect.* **2017**, *97*, 212–217. [CrossRef] [PubMed]

26.   Mori, N.; Kagawa, N.; Aoki, K.; Ishi, Y.; Tateda, K.; Aoki, Y. Clinical and molecular analyses of bloodstream infections caused by IMP metallo-β-lactamase-producing Enterobacteriaceae in a tertiary hospital in Japan. *J. Infect. Chemother.* **2020**, *26*, 144–147. [CrossRef]

27.   Queenan, A.M.; Bush, K. Carbapenemases: The versatile beta-lactamases. *Clin. Microbiol. Rev.* **2007**, *20*, 440–458. [CrossRef]

28.   Clinical and Laboratory Standards Institute. *Performance Standards for Antimicrobial Susceptibility Testing*; Twenty-Second Informational Supplement; CLSI document M100-S22; Clinical and Laboratory Standards Institute: Wayne, PA, USA, 2012.

29.   Ribot, E.M.; Fair, M.A.; Gautom, R.; Cameron, D.N.; Hunter, S.B.; Swaminathan, B.; Barrett, T.J. Standardization of pulsed-field gel electrophoresis protocols for the subtyping of *Escherichia coli* O157:H7, *Salmonella*, and *Shigella* for PulseNet. *Foodborne Pathog. Dis.* **2006**, *3*, 59–67. [CrossRef] [PubMed]

30.   Tenover, F.C.; Arbeit, R.D.; Goering, R.V.; Mickelsen, P.A.; Murray, B.; Persing, D.H.E.; Swaminathan, B. Interpreting chromosomal DNA restriction patterns produced by pulsed-field gel electrophoresis: Criteria for bacterial strain typing. *J. Clin. Microbiol.* **1995**, *33*, 2233–2239. [CrossRef]

31. van der Zwaluw, K.; de Haan, A.; Pluister, G.N.; Bootsma, H.J.; de Neeling, A.J.; Schouls, L.M. The carbapenem inactivation method (CIM), a simple and low-cost alternative for the Carba NP test to assess phenotypic carbapenemase activity in gram-negative rods. *PLoS ONE* **2015**, *10*, e0123690. [CrossRef] [PubMed]

32. Poirel, L.; Walsh, T.R.; Cuvillier, V.; Nordmann, P. Multiplex PCR for detection of acquired carbapenemase genes. *Diagn. Microbiol. Infect. Dis.* **2011**, *70*, 119–123. [CrossRef] [PubMed]

33. Dallenne, C.; Da Costa, A.; Decré, D.; Favier, C.; Arlet, G. Development of a set of multiplex PCR assays for the detection of genes encoding important β-lactamases in Enterobacteriaceae. *J. Antimicrob. Chemother.* **2010**, *65*, 490–495. [CrossRef] [PubMed]

34. Carattoli, A.; Bertini, A.; Villa, L.; Falbo, V.; Hopkins, K.L.; Threlfall, E.J. Identification of plasmids by PCR-based replicon typing. *J. Microbiol. Methods* **2005**, *63*, 219–228. [CrossRef] [PubMed]

Article

Complete Genome Sequences of *Klebsiella michiganensis* and *Citrobacter farmeri*, KPC-2-Producers Serially Isolated from a Single Patient

Jehane Y. Abed [1,2], Maxime Déraspe [1,2], Ève Bérubé [1], Matthew D'Iorio [3,†], Ken Dewar [4,5], Maurice Boissinot [1], Jacques Corbeil [1,6], Michel G. Bergeron [1,2] and Paul H. Roy [1,7,*]

1. Centre de Recherche en Infectiologie, Centre de Recherche du CHU de Québec, Université Laval, 2705 boul. Laurier, Suite R-0709, Québec, QC G1V 4G2, Canada; Jehane.Abed@crchudequebec.ulaval.ca (J.Y.A.); maxime@deraspe.net (M.D.); eve.berube@crchudequebec.ulaval.ca (È.B.); maurice.boissinot@crchudequebec.ulaval.ca (M.B.); jacques.corbeil@crchudequebec.ulaval.ca (J.C.); Michel.G.Bergeron@crchudequebec.ulaval.ca (M.G.B.)
2. Département de Microbiologie et Immunologie, Pavillon Vandry, Université Laval, Québec, QC G1V 0A6, Canada
3. McGill Genome Centre, 740 Avenue Docteur-Penfield, Montréal, QC H3A 0G1, Canada; matt.diorio@mail.mcgill.ca
4. Department of Human Genetics, McGill University, 3640 rue University, Rm 2/38F, Montréal, QC H3A 0C7, Canada; ken.dewar@mcgill.ca
5. McGill Centre for Microbiome Research, 3605 de la Montagne, Montréal, QC H3G 2M1, Canada
6. Département de Médecine Moléculaire, Pavillon Vandry, Université Laval, Québec, QC G1V 0A6, Canada
7. Département de Biochimie, de Microbiologie et de Bio-Informatique, Pavillon Vachon, Université Laval, Québec, QC G1V 0A6, Canada
* Correspondence: paul.roy@crchudequebec.ulaval.ca; Tel.: +1-418-843-7134
† McGill University PhD Program in Quantitative Life Sciences.

Citation: Abed, J.Y.; Déraspe, M.; Bérubé, È.; D'Iorio, M.; Dewar, K.; Boissinot, M.; Corbeil, J.; Bergeron, M.G.; Roy, P.H. Complete Genome Sequences of *Klebsiella michiganensis* and *Citrobacter farmeri*, KPC-2-Producers Serially Isolated from a Single Patient. *Antibiotics* **2021**, *10*, 1408. https://doi.org/10.3390/antibiotics10111408

Academic Editor: Francesca Andreoni

Received: 4 September 2021
Accepted: 14 November 2021
Published: 18 November 2021

Publisher's Note: MDPI stays neutral with regard to jurisdictional claims in published maps and institutional affiliations.

Abstract: Carbapenemase-producing *Enterobacterales*, including KPC-2 producers, have become a major clinical problem. During an outbreak in Quebec City, Canada, KPC-2-producing *Klebsiella michiganensis* and *Citrobacter farmeri* were isolated from a patient six weeks apart. We determined their complete genome sequences. Both isolates carried nearly identical IncN2 plasmids with *bla*KPC-2 on a Tn*4401b* element. Both strains also carried IncP1 plasmids, but that of *C. farmeri* did not carry a Beta-lactamase gene, whereas that of *K. michiganensis* carried a second copy of bla_{KPC-2} on Tn*4401b*. These results suggest recent plasmid transfer between the two species and a recent transposition event.

Keywords: *Klebsiella michiganensis*; *Citrobacter farmeri*; KPC-2; carbapenemase; plasmid; transposon

1. Introduction

Carbapenemase-producing enterobacteria (CRE) have become a major problem throughout the world. The most frequently found carbapenemases are the class A KPC, class D OXA-48 and its variants, and metallo-Beta-lactamases like IMP, VIM, and NDM [1]. Each carbapenemase has a distinct epidemiology. KPC carbapenemases are plasmid-mediated and often found on a Tn*4401b* transposon [2,3]. KPC has spread throughout the United States and into Canada, with outbreaks in Toronto, Montreal, and Quebec City [4,5]. We obtained two strains identified as KPC producers from the Hôtel-Dieu hospital in Quebec City, where five KPC-producing strains were isolated in 2017 from four patients on four different floors. These two strains were anal swab isolates that colonized but did not infect the patient, a 69-year-old male. The first isolate was identified as *Klebsiella oxytoca* (later reclassified as *K. michiganensis*), and the second, isolated six weeks later, was identified as *Citrobacter farmeri*. We report here the sequences of the chromosomes, the four plasmids of *C. farmeri*, and the two plasmids of *K. michiganensis*, which are very similar to two of the plasmids of *C. farmeri*, except that both *K. michiganensis* plasmids carry Tn*4401b* and encode bla_{KPC-2}, whereas only one of the *C. farmeri* plasmids does.

2. Results

2.1. Genomic Sequencing

The complete genome sequences of *Klebsiella michiganensis* CCRI-24235 and *Citrobacter farmeri* CCRI-24236 were determined by PacBio and polished with Illumina to resolve homopolymer undercounts (see Materials and Methods). The two chromosomes, two plasmids from *K. michiganensis* and four plasmids from *C. farmeri,* were confirmed as being circular by trimming the terminal duplication of the linear assemblies.

2.2. Chromosomes

The chromosome of *K. michiganensis* CCRI-24235 (Figure 1A) was 5,977,739 nt in length. The genome of greatest similarity was that of *K. michiganensis* E718 (CP003683) [6], with an average nucleotide identity (ANI) of 99.49%. The chromosome of *C. farmeri* (Figure 1B) was 5,022,624 nt in length, and among the complete genomes, it was most similar to the *C. farmeri* strain AUSDM00008141 (CP022695) [7] with an ANI of 98.86%. It was also very similar to whole genome shotgun (wgs) genomes of *C. farmeri* 1001216B_150713_F2 and CB00091 (WGS Projects JADMON and JADVHI, respectively) with ANIs of 99.05%. No acquired resistance genes were found in the two chromosomes.

2.3. Plasmids pCCRI24235-1 and pCCRI24236-2

Plasmid pCCRI24235-1 from *K. michiganensis* was 88,159 nucleotides in length. Plasmid pCCRI24236-2 from *C. farmeri* was 82,438 nt in length. The two plasmids were identical except for a sequence duplication of the insertion sequence IS*CR1* and an adjacent region, 5721 nt in length. They belong to IncN2 and are very similar to pEC448_OXA-163 from *Escherichia coli* (CP015078; brown arc in Figure 2A). The plasmids contained a Tn*4401b* transposon 10,006 nt in length, encoding a *bla*$_{KPC-2}$ gene. Figure 2A is a map of pCCRI24235-1 and shows the transposon, the *bla*$_{KPC-2}$ gene, and the duplication absent from pCCRI24236-2. In addition to the *bla*$_{KPC-2}$ gene, these plasmids had resistance genes *bla*$_{TEM-1}$, *mphA*, *mefA*, *sul1*, *qnrB2*, *sapA*, *qacEdelta1*, *dfrA25*, and a mercury resistance operon. The integron region is very similar to pE51_003 from *E.* coli (CP042537; dark yellow arc in Figure 2A), while the region containing the *bla*$_{KPC-2}$ and *bla*$_{KTEM-1}$ genes is very similar to pKPC_CAV1042-44 from *K. pneumoniae* (CP018668; light yellow arc in Figure 2A).

2.4. Plasmids pCCRI24235-2 and pCCRI24236-3

Plasmid pCCRI24235-2 from *K. michiganensis* was 62,417 nucleotides in length. Plasmid pCCRI24236-3 from *C. farmeri* was 52,406 nt in length. The two plasmids were identical except that the former contained a Tn*4401b* transposon identical to that of pCCRI24235-1 and pCCRI24236-2, and encoding the *bla*$_{KPC-2}$ gene, while the latter lacked the transposon. They belong to a new clade of IncP1 (see Discussion) and are very similar to the *E. coli* plasmid pHS102707 (KF701335; brown arc in Figure 2B) [8]. Figure 2B is a map of pCCRI24235-2 showing the transposon. Plasmids pCCRI24235-2 and pCCRI24236-3 had no other resistance genes except for the tellurium resistance gene *telA*.

2.5. Plasmids pCCRI24236-1 and pCCRI24236-4

Plasmid pCCRI24236-1 from *C. farmeri* was 198,299 nucleotides in length. It was identified by PlasmidFinder (see Materials and Methods) as belonging to a novel unknown incompatibility group on the basis of similarity of its *repB* gene to that of pKPC-CAV1321-244 (CP011611), and the whole sequence is closest to *C. freundii* plasmids pRHBSTW-00153-2 (CP055565) and pRHBSTW-00370_2 (CP056574). Figure 3A shows a map of pCCRI24236-1, which had a variety of heavy metal resistance genes, including a copper resistance operon *pcoABCDRSE*. However, tellurium resistance gene *telA* (interrupted by IS*Ecl1*), copper/silver resistance gene *silE* (interrupted by IS*1*), and an arsenical pump-driving ATPase-encoding gene (N-terminal truncated) were pseudogenes.

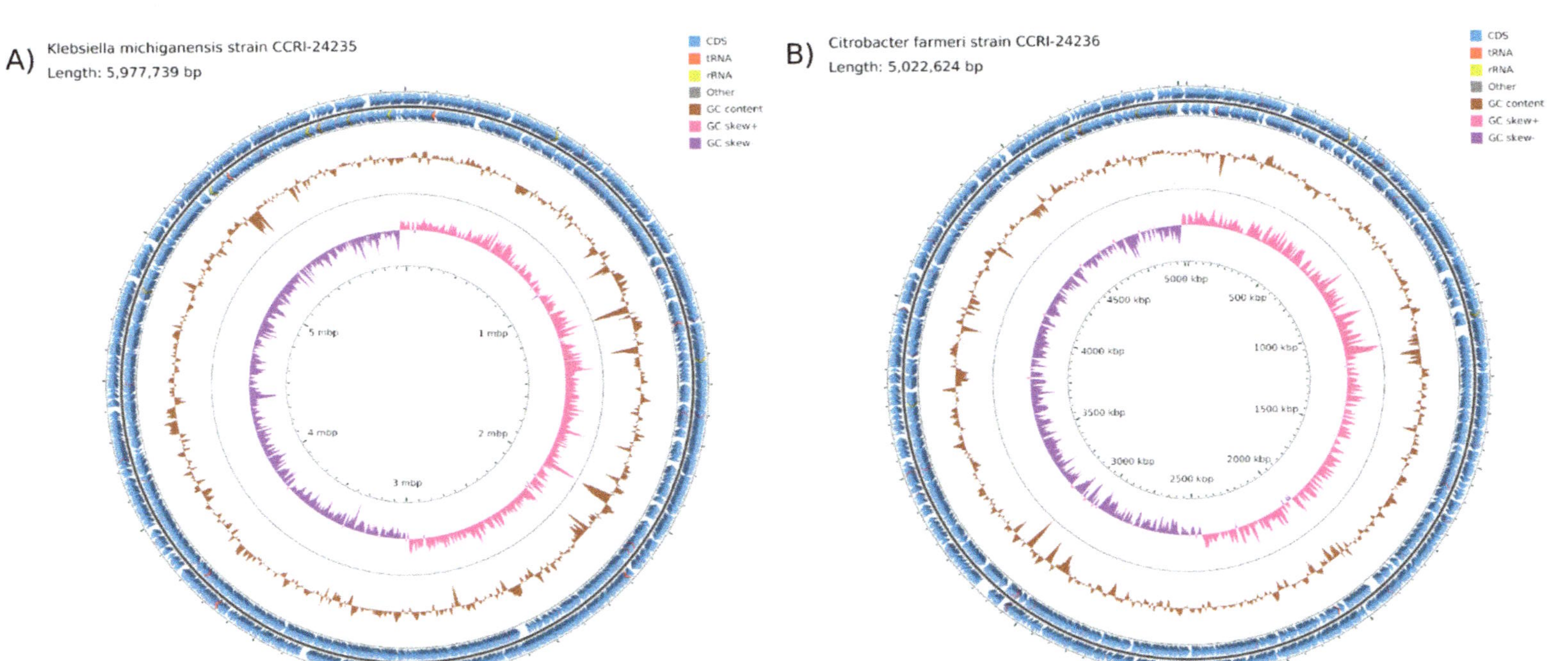

Figure 1. Map of chromosomes of *K. pneumoniae* CCRI-24235 (**A**) and *C. farmeri* 24236 (**B**). The scales are indicated on the innermost circles. The second circles are G+C skew in pink (+) and purple (−), and circles 3 show G+C content (deviation from the average) in brown (+, outward and −, inward). The next two circles illustrate positions of CDSs in minus (circle 4) and plus (circle 5) strands in dark blue.

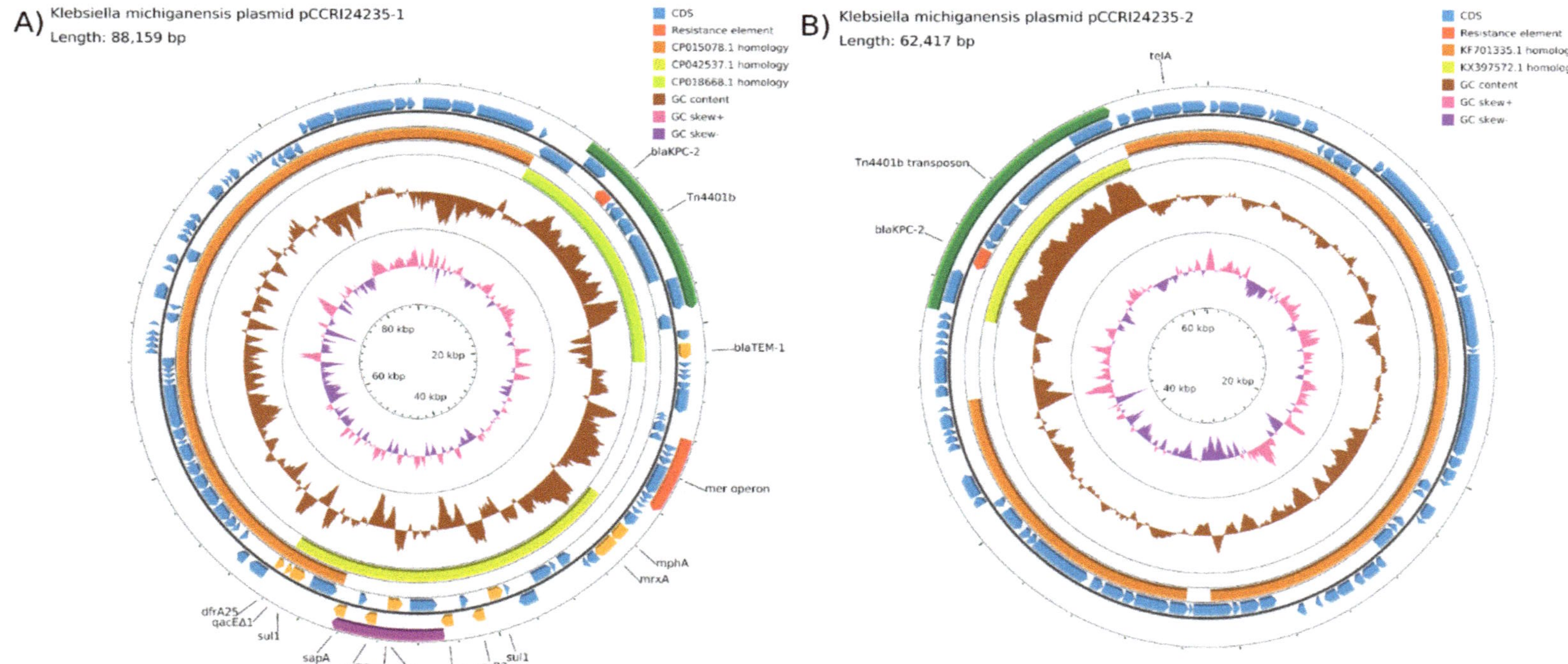

Figure 2. (**A**) Map of plasmid pCCRI24235-1. The scales are indicated on the innermost circles. The second circles are G+C skew in pink (+) and purple (−), and circles 3 show G+C content (deviation from the average) in brown (+, outward and −, inward). The next two circles illustrate regions homologous to related plasmids. The following two circles show positions of CDSs in minus (circle 4) and plus (circle 5) strands in dark blue. The *bla*KPC-2 gene is indicated in red. Transposon Tn*4401b* is indicated by a green arc. The segment absent in pCCRI24236-2 (DUP) is indicated by a purple arc. (**B**) Map of plasmid pCCRI24235-2. Transposon Tn*4401b*, indicated in green, is absent in pCCRI24236-3.

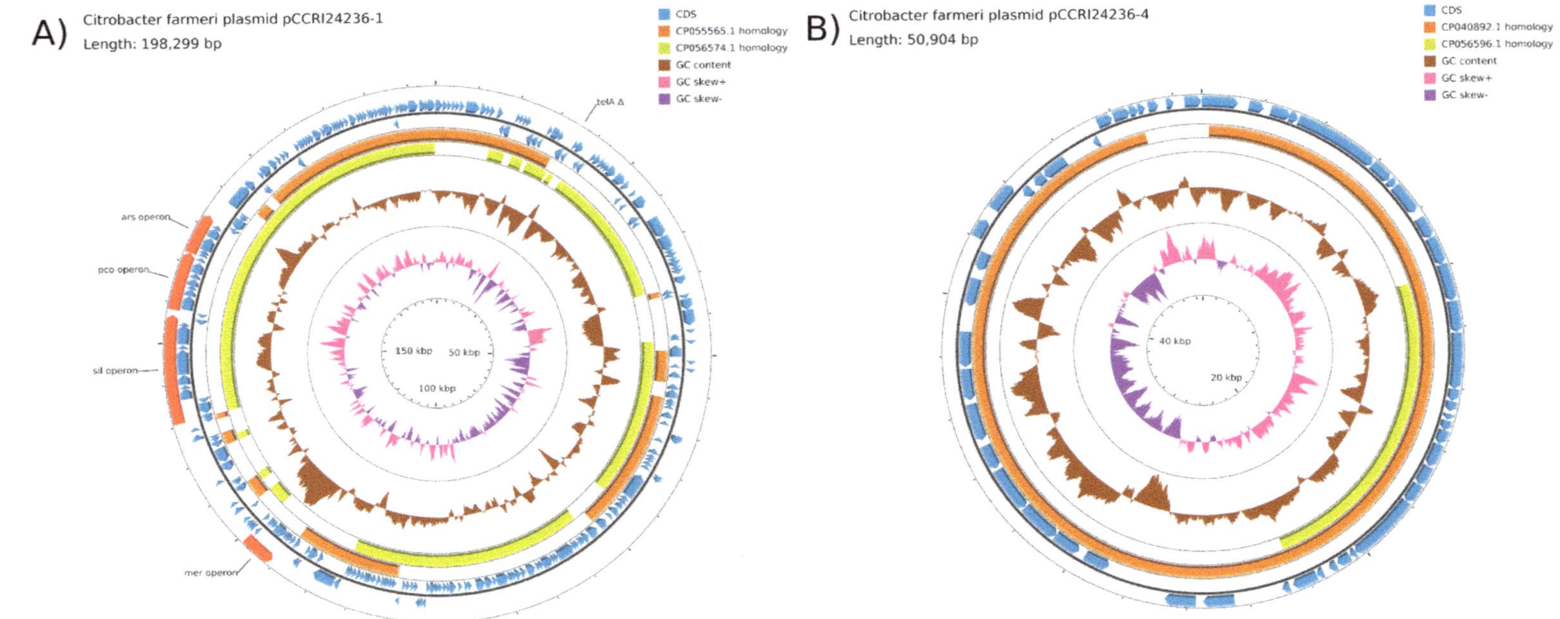

Figure 3. Map of plasmids pCCRI24236-1 (**A**) and pCCRI24236-4 (**B**). The scales are indicated on the innermost circles. The second circles are G+C skew in pink (+) and purple (−), and circles 3 show G+C content (deviation from the average) in brown (+, outward and −, inward). The next two circles illustrate regions homologous to related plasmids. The following two circles show positions of CDSs in minus (circle 6) and plus (circle 7) strands in dark blue.

Plasmid pCCRI24236-4 from *C. farmeri* (Figure 3B) was 50,904 nucleotides in length, belonged to the IncX5 incompatibility group, and was very similar to *Escherichia coli* plasmid pEc1677 (MG516910) [9]. No resistance genes were found.

3. Discussion

The two strains, *K. michiganensis* CCRI-24235 and *C. farmeri* CCRI-24236, share a pair of very similar plasmids, with two additional plasmids in the *C. farmeri* strain. The plasmids pCCRI24235-1 and pCCRI24236-2 are IncN plasmids with *bla*KPC-2 on Tn*4401b*, a context first found in KPC-producing isolates in the 1990s [10] that is still very common. Tn*4401* contains *bla*$_{\mathrm{KPC-2}}$, a transposase and resolvase, and insertion sequences IS*Kpn6* and IS*Kpn7* [11]. A variety of other elements, collectively called NTEKPC, encode *bla*$_{\mathrm{KPC}}$ [12]. Plasmids pCCRI24235-1 and pCCRI24236-2 are identical except for a 5.7-kb duplication of IS*CR1* and adjacent genes in the former. Plasmid pCCRI24235-1 may have evolved from pCCRI24236-2 by a one-ended transposition event mediated by the IS*CR1* transposase [13]. IS*CR1* is found in several integrons downstream of the *sul1* sulfonamide resistance genes [14], and a promoter downstream of the transposase is involved in the expression of downstream resistance genes [15]. In our plasmids, the downstream *sapA* gene is in the opposite orientation. IncP1 plasmids pCCRI24236-3 and pCCRI24235-2 are identical except for the presence of Tn*4401b* in the latter. The location of Tn*4401b* in pCCRI24235-2 represents a novel and unique target. The transposition of Tn*4401b* is likely to have taken place in *K. michiganensis* from pCCRI23235-1 into a plasmid otherwise identical to the *C. farmeri* plasmid pCCRI24236-3, resulting in a second copy of *bla*$_{\mathrm{KPC-2}}$ in the *K. michiganensis* strain. The quasi-identity of the two plasmid pairs suggests the possibility of their transfer by conjugation between the two species, although indirect transfer via a third species cannot be ruled out. Plasmid transfer may be between-patient or within-patient events [16], but there were too few isolates from the hospital in 2017 to elucidate the series of events.

Although pCCRI24235-2 and pCCRI24236-3 belong to IncP1, they are in a novel clade that includes pHS102707 (KF701335) [8], pHNFP671 (KP324830), and pMCR1511 (KX377410) [17]. While plasmids of most clades of IncP-1 can be found in *Pseudomonas aeruginosa*, plasmids of this novel clade have not. Plasmid pCCRI24236-1 from *C. farmeri* belongs to a novel unknown incompatibility group; similar plasmids are found in Enterobacteriaceae but not in *Pseudomonas*.

Our results add to the small number of *C. farmeri* complete genomes and show that this species is a factor in KPC dissemination. KPC-producing CRE are still clinically important, although, unlike MBL, they are usually sensitive to certain Beta-lactam/Beta-lactamase inhibitors such as ceftazidime/avibactam. However, various new mechanisms of KPC-mediated ceftazidime/avibactam resistance have been reported [1].

4. Materials and Methods

Strains *K. michiganensis* CCRI-24235 and *C. farmeri* CCRI-24236 were obtained from the microbiology laboratory at Hôtel-Dieu de Québec hospital, where they had been isolated from anal swabs of a patient, six weeks apart in 2017, during an outbreak of KPC-2 producers in Québec City. DNA was prepared according to the PacBio Template Preparation and Sequencing Guide (Pacific Biosciences, Menlo Park, CA, USA) and sequenced by the single-molecule real-time.

The (SMRT) technique was completed using an RS II instrument (Pacific Biosciences) at the McGill University and Genome Quebec Innovation Centre. DNA was also extracted using a KingFisher/Qiagen blood kit and prepared for Illumina MiSeq sequencing using a Nextera XT kit. The genome was first assembled de novo using the Hierarchical Genome Assembly process (HGAP) [18], and the Illumina data were used to correct and validate the entire sequence; the only errors encountered in the PacBio data were homopolymer undercounts. Chromosomes were automatically annotated using an in-house method based on Prodigal [19]; plasmids were manually annotated using Artemis [20] and Blastp on NCBI (https://blast.ncbi.nih.gov/Blast.cgi accessed on 1 March 2021) ANI's were calculated using OrthoAnIu from EZBioCloud [21]: (https://www.ezbiocloud.net/tools/ani, accessed on 1 July 2021). Plasmid incompatibility groups were determined using PlasmidFinder [22] at the site of the Center for Genomic Epidemiology: (https://cge.dtu.dk/services/PlasmidFinder, accessed on 1 June 2021).

Author Contributions: Conceptualization, J.Y.A., M.G.B., and P.H.R.; data curation, J.Y.A., M.D. (Maxime Déraspe), M.D. (Matthew D'Iorio), K.D., and P.H.R.; investigation, J.Y.A., M.B., J.C., and P.H.R.; methodology, J.Y.A., È.B., and P.H.R.; project administration, M.B., and P.H.R.; software, M.D. (Maxime Déraspe), M.D. (Matthew D'Iorio), and K.D.; supervision, M.B., J.C., M.G.B., and P.H.R.; visualization, P.H.R.; writing—original draft, P.H.R.; writing—review and editing, J.Y.A., and M.B. All authors have read and agreed to the published version of the manuscript.

Funding: This research received no external funding.

Institutional Review Board Statement: Not applicable.

Informed Consent Statement: Not applicable.

Data Availability Statement: The chromosomal and plasmid sequences are available in GenBank, accession numbers: *K. michiganensis* CP081351-CP081353; *C. farmeri* CP081314-CP081318.

Acknowledgments: We thank Anne Desjardins (Hôtel-Dieu de Québec) for the strains, Annie Ruest (Hôtel-Dieu de Québec) for information on KPC-producing isolates at HDQ in 2017, and Gary Leveque (Canadian Centre for Computational Genomics) for help with PacBio assemblies. This research was enabled in part by support provided by Calcul Québec (www.calculquebec.ca, accessed on 1 July 2021) and Compute Canada (www.computecanada.ca, accessed on 1 July 2021). Jacques Corbeil thanks the Canada Research Chair program for support. Jehane Y. Abed was funded by a doctoral scholarship from the Sentinel North program at Université Laval (Canada First Research Excellence Fund).

Conflicts of Interest: The authors declare no conflict of interest.

References

1. Hansen, G.T. Continuous Evolution: Perspective on the Epidemiology of Carbapenem Resistance Among Enterobacterales and Other Gram-Negative Bacteria. *Infect. Dis. Ther.* **2021**, *10*, 75–92. [CrossRef] [PubMed]
2. Drawz, S.M.; Bonomo, R.A. Three decades of β-lactamase inhibitors. *Clin. Microbiol. Rev.* **2010**, *23*, 160–201. [CrossRef]
3. Chavda, K.D.; Chen, L.; Fouts, D.E.; Sutton, G.; Brinkac, L.; Jenkins, S.G.; Bonomo, R.A.; Adams, M.D.; Kreiswirth, B.N. Comprehensive genome analysis of carbapenemase-producing *Enterobacter* spp.: New insights into phylogeny, population structure, and resistance mechanisms. *MBio* **2016**, *7*, e02093-16. [CrossRef] [PubMed]
4. Mataseje, L.F.; Abdesselam, K.; Vachon, J.; Mitchel, R.; Bryce, E.; Roscoe, D.; Boyd, D.A.; Embree, J.; Katz, K.; Kibsey, P.; et al. Results from the Canadian Nosocomial infection surveillance program on Carbapenemase-producing Enterobacteriaceae, 2010 to 2014. *Antimicrob. Agents Chemother.* **2016**, *60*, 6787–6794. [CrossRef]
5. Lefebvre, B.; Lévesque, S.; Bourgault, A.M.; Mulvey, M.R.; Mataseje, L.; Boyd, D.; Doualla-Bell, F.; Tremblay, C. Carbapenem non-susceptible Enterobacteriaceae in Quebec, Canada: Results of a laboratory surveillance program (2010–2012). *PLoS ONE* **2015**, *10*, e0125076. [CrossRef] [PubMed]
6. Liao, T.L.; Lin, A.C.; Chen, E.; Huang, T.W.; Liu, Y.M.; Chang, Y.H.; Lai, J.F.; Lauderdale, T.L.; Wang, J.T.; Chang, S.C.; et al. Complete genome sequence of Klebsiella oxytoca E718, a New Delhi metallo-β-lactamase-1-producing nosocomial strain. *J. Bacteriol.* **2012**, *194*, 19. [CrossRef] [PubMed]
7. Kwong, J.C.; Lane, C.R.; Romanes, F.; da Silva, A.G.; Easton, M.; Cronin, K.; Waters, M.J.; Tomita, T.; Stevens, K.; Schultz, M.B.; et al. Translating genomics into practice for real-time surveillance and response to carbapenemase-producing Enterobacteriaceae: Evidence from a complex multi-institutional KPC outbreak. *PeerJ* **2018**, *6*, e4210. [CrossRef]
8. Li, G.; Zhang, Y.; Bi, D.; Shen, P.; Ai, F.; Liu, H.; Tian, Y.; Ma, Y.; Wang, B.; Rajakumar, K.; et al. First report of a clinical, multidrug-resistant Enterobacteriaceae isolate coharboring fosfomycin resistance gene fosA3 and carbapenemase gene *bla*KPC-2 on the same transposon, Tn1721. *Antimicrob. Agents Chemother.* **2015**, *59*, 338–343. [CrossRef]
9. Dolejska, M.; Papagiannitsis, C.C.; Medvecky, M.; Davidova-Gerzova, L.; Valcek, A. Characterization of the Complete Nucleotide Sequences of IMP-4-Encoding Plasmids, Belonging to Diverse Inc Families, Recovered from Enterobacteriaceae Isolates of Wildlife Origin. *Antimicrob. Agents Chemother.* **2018**, *62*, 1–7. [CrossRef]
10. Eilertson, B.; Chen, L.; Chavda, K.D.; Kreiswirth, B.N. Genomic characterization of two KPC-producing Klebsiella isolates collected in 1997 in New York City. *Antimicrob. Agents Chemother.* **2017**, *61*, 1–6. [CrossRef]

11. Naas, T.; Cuzon, G.; Villegas, M.-V.; Lartigue, M.-F.; Quinn, J.P.; Nordmann, P. Genetic Structures at the Origin of Acquisition of the b-lactamase *bla*KPC Gene. *Antimicrob. Agents Chemother.* **2008**, *52*, 1257–1263. [CrossRef] [PubMed]

12. Chen, L.; Mathema, B.; Chavda, K.D.; DeLeo, F.R.; Bonomo, R.A.; Kreiswirth, B.N. Carbapenemase-producing Klebsiella pneumoniae: Molecular and genetic decoding. *Trends Microbiol.* **2014**, *22*, 686–696. [CrossRef]

13. Toleman, M.A.; Bennett, P.M.; Walsh, T.R. IS CR Elements: Novel Gene-Capturing Systems of the 21st Century? *Microbiol. Mol. Biol. Rev.* **2006**, *70*, 296–316. [CrossRef] [PubMed]

14. Arduino, S.M.; Roy, P.H.; Jacoby, G.A.; Orman, B.E.; Pineiro, S.A.; Centron, D. *bla*CTX-M-2 Is Located in an Unusual Class 1 Integron (In35) Which Includes Orf513. *Antimicrob. Agents Chemother.* **2002**, *46*, 2303–2306. [CrossRef] [PubMed]

15. Mammeri, H.; Van De Loo, M.; Poirel, L.; Martinez-Martinez, L.; Nordmann, P. Emergence of Plasmid-Mediated Quinolone Resistance in Escherichia coli in Europe. *Antimicrob. Agents Chemother.* **2005**, *49*, 71–76. [CrossRef] [PubMed]

16. León-Sampedro, R.; DelaFuente, J.; Díaz-Agero, C.; Crellen, T.; Musicha, P.; Rodríguez-Beltrán, J.; de la Vega, C.; Hernández-García, M.; López-Fresneña, N.; Ruiz-Garbajosa, P.; et al. Pervasive transmission of a carbapenem resistance plasmid in the gut microbiota of hospitalized patients. *Nat. Microbiol.* **2021**, *6*, 606–616. [CrossRef]

17. Zhao, F.; Feng, Y.; Lü, X.; McNally, A.; Zong, Z. IncP Plasmid Carrying Colistin Resistance Gene mcr-1 in Klebsiella pneumoniae from Hospital Sewage. *Antimicrob. Agents Chemother.* **2017**, *61*, e02229-16. [CrossRef]

18. Chin, C.S.; Alexander, D.H.; Marks, P.; Klammer, A.A.; Drake, J.; Heiner, C.; Clum, A.; Copeland, A.; Huddleston, J.; Eichler, E.E.; et al. Nonhybrid, finished microbial genome assemblies from long-read SMRT sequencing data. *Nat. Methods* **2013**, *10*, 563–569. [CrossRef]

19. Hyatt, D.; Chen, G.-L.; LoCascio, P.F.; Land, M.L.; Larimer, F.W.; Hauser, L.J. Integrated nr Database in Protein Annotation System and Its Localization. *Nat. Commun.* **2010**, *6*, 1–8. [CrossRef]

20. Carver, T.; Berriman, M.; Tivey, A.; Patel, C.; Böhme, U.; Barrell, B.G.; Parkhill, J.; Rajandream, M.A. Artemis and ACT: Viewing, annotating and comparing sequences stored in a relational database. *Bioinformatics* **2008**, *24*, 2672–2676. [CrossRef]

21. Yoon, S.-H.; Ha, S.-M.; Lim, J.; Kwon, S.; Chun, J. A large-scale evaluation of algorithms to calculate average nucleotide identity. *Antonie Van Leeuwenhoek* **2017**, *110*, 1281–1286. [CrossRef] [PubMed]

22. Carattoli, A.; Zankari, E.; Garciá-Fernández, A.; Larsen, M.V.; Lund, O.; Villa, L.; Aarestrup, F.M.; Hasman, H. In Silico detection and typing of plasmids using plasmidfinder and plasmid multilocus sequence typing. *Antimicrob. Agents Chemother.* **2014**, *58*, 3895–3903. [CrossRef] [PubMed]

 *antibiotics*

Article

Antimicrobial Activity of a Lipidated Temporin L Analogue against Carbapenemase-Producing *Klebsiella pneumoniae* Clinical Isolates

Emanuela Roscetto [1,†], Rosa Bellavita [2,†], Rossella Paolillo [1], Francesco Merlino [2], Nicola Molfetta [2], Paolo Grieco [2], Elisabetta Buommino [2,‡] and Maria Rosaria Catania [1,*,‡]

1. Department of Molecular Medicine and Medical Biotechnology, University of Naples Federico II, Via S. Pansini 5, 80131 Naples, Italy; emanuela.roscetto@unina.it (E.R.); rossella.paolillo@unina.it (R.P.)
2. Department of Pharmacy, University of Naples Federico II, Via Montesano 49, 80131 Naples, Italy; rosa.bellavita@unina.it (R.B.); francesco.merlino@unina.it (F.M.); nicolamolfetta31@gmail.com (N.M.); paolo.grieco@unina.it (P.G.); elisabetta.buommino@unina.it (E.B.)
* Correspondence: mariarosaria.catania@unina.it; Tel.: +39-081-7464577
† These authors contributed equally to this work.
‡ These authors contributed equally to this work.

Abstract: Over the years, the increasing acquisition of antibiotic resistance genes has led to the emergence of highly resistant bacterial strains and the loss of standard antibiotics' efficacy, including β-lactam/β-lactamase inhibitor combinations and the last line carbapenems. *Klebsiella pneumoniae* is considered one of the major exponents of a group of multidrug-resistant ESKAPE pathogens responsible for serious healthcare-associated infections. In this study, we proved the antimicrobial activity of two analogues of Temporin L against twenty carbapenemase-producing *K. pneumoniae* clinical isolates. According to the antibiotic susceptibility assay, all the *K. pneumoniae* strains were resistant to at least one other class of antibiotics, in addition to beta-lactams. Peptides **1B** and **C** showed activity on all test strains, but the lipidated analogue **C** expressed the greater antimicrobial properties, with MIC values ranging from 6.25 to 25 µM. Furthermore, the peptide **C** showed bactericidal activity at MIC values. The results clearly highlight the great potential of antimicrobial peptides both as a new treatment option for difficult-to-treat infections and as a new strategy of drug-resistance control.

Keywords: *Klebsiella pneumoniae*; ESKAPE; multidrug resistance; carbapenemases; healthcare-associated infections; antimicrobial peptides; Temporin L

Citation: Roscetto, E.; Bellavita, R.; Paolillo, R.; Merlino, F.; Molfetta, N.; Grieco, P.; Buommino, E.; Catania, M.R. Antimicrobial Activity of a Lipidated Temporin L Analogue against Carbapenemase-Producing *Klebsiella pneumoniae* Clinical Isolates. *Antibiotics* **2021**, *10*, 1312. https://doi.org/10.3390/antibiotics10111312

Academic Editor: Francesca Andreoni

Received: 11 October 2021
Accepted: 26 October 2021
Published: 28 October 2021

Publisher's Note: MDPI stays neutral with regard to jurisdictional claims in published maps and institutional affiliations.

1. Introduction

In the order *Enterobacteriales*, *Klebsiella pneumoniae* is one of the most important causes of bloodstream, urinary and respiratory tract infections in vulnerable hosts [1]. An empirical antibiotic treatment is often required due to the severity of the infections and/or the patient's critical conditions and the use of broad-spectrum antibiotics is necessary because of the possibility of a multidrug-resistant bacteria aetiology.

The worldwide spread of difficult-to-treat extended-spectrum beta-lactamase-producing enterobacteria has led to the use of carbapenems in empirical therapy [2], but the treatment with carbapenems has led to the rapid selection of carbapenem-resistant *Enterobacteriales* (CRE) [3].

Enterobacteriales may have different mechanisms of resistance to carbapenems. The most widespread one is the production of beta-lactamases with high affinity for carbapenems (carbapenemases). Another common mechanism of resistance is the hyperproduction of β-lactamases with limited affinity and/or hydrolytic activity toward carbapenems combined with structural alterations such as porin loss [4].

Carbapenemase production is also the most epidemiologically relevant resistance mechanism, as the genes for carbapenemases are carried by plasmids and therefore horizontally transmissible [5].

Klebsiella pneumoniae is the most common species harbouring transmissible carbapenemase [6]. In the Ambler classification system, carbapenemases are distributed in three classes depending on their chemical structure: classes A and D include serine-carbapenemases, whereas class B includes metallo-beta-lactamases [7]. *Klebsiella pneumoniae* carbapenemases (KPCs) are the most common transmissible genes among *Enterobacteriales* [8].

Treatment options for CRE are aminoglycosides, polymyxins or tigecycline, but some of these drugs have non-negligible adverse effects. Furthermore, many enterobacteria have additional plasmid-borne resistance genes, consequently resulting in resistance to several other antimicrobial groups. This issue induces clinicians to administer a combination therapy of two or more drugs [9]. The threat posed by CRE to human health is evidenced by their placement by WHO in the most critical group of multidrug-resistant bacteria for which the development of new antibiotics is urgently needed [10]. The new therapeutic options against CRE are drugs belonging to already known classes of antibiotics or new beta-lactam/beta-lactamase inhibitor combinations [11].

Antimicrobial peptides (AMPs) could provide a valid chance to overcome and control the antibiotic resistance [12]. Among these compounds, the temporins, isolated from the skin of *Rana temporaria* [13], represent one of the largest AMPs families. Temporin L (TL) is the most studied isoform for its potent activity both against Gram-positive and Gram-negative bacteria and yeasts. Due to its high cytotoxicity, TL has been the subject of different structure–activity relationship (SAR) studies to obtain novel analogues with an improved therapeutic index [14–16]. In this context, a previous SAR study consisting of the application of lipidation strategy on a potent Temporin L analogue, named peptide 1B [17,18], has led to the discovery of the lipidated peptide C featured by an alkyl chain of five carbons in para position of Phe^1 in its N-terminus [19]. The addition of fatty acid conferred to peptide C self-assembling properties improved the effectiveness in inhibiting the growth of both *Staphylococcus aureus* (ATCC 25923) and *Klebsiella pneumoniae* (ATCC BAA-1705) cells, with a minimum inhibitory concentration (MIC) of 6.25 μM. Interestingly, it did not show a significant cytotoxic effect even at the high concentration of 25 μM [19]. In this study, we evaluated the activity of peptides 1B and C towards clinical carbapenem-resistant *K. pneumoniae* isolates harboring kpc or metallo-beta-lactamase genes.

The conventional antibiotic susceptibility was tested for the following antibiotics: amoxicillin/clavulanic acid, cefotaxime, ceftazidime, piperacillin/tazobactam, gentamicin, amikacin, trimethoprim-sulfamethoxazole, ciprofloxacin, meropenem, ertapenem and ceftazidime/avibactam. The clinical strains were classified in different category S/I/R (susceptible/intermediate/resistant) according to EUCAST 2021 breakpoints (https://www.eucast.org/clinical_breakpoints, accessed on 28 April 2021). Table 2 shows the antimicrobial susceptibility profile of clinical strains: except for KN*Kp*, all the isolates were resistant to amoxicillin/clavulanic acid, ceftazidime, cefotaxime, piperacillin/tazobactam and to the carbapenems ertapenem and meropenem (except for KPC*Kp*6 and KPC*Kp*10 that showed an intermediate resistance to meropenem). They were also resistant to ciprofloxacin, excluding KPC*Kp*5 and NDM*Kp*2. Particularly, KPC*Kp*8 and KPC*Kp*15 showed a resistant profile towards all tested antibiotics. We observed that the aminoglycosides amikacin and gentamicin and the combination ceftazidime/avibactam were the most effective compounds, acting against 52.2% (12/23), 56.5% (13/23), 8.7 % (2/23) of clinical strains, respectively. The 22 carbapenem-resistant strains were subjected to molecular tests (Xpert Carba-R-test), which allow the identification of the resistance determinants involved (KPC, VIM, IPM-1, NDM, OXA-48). We found that 81.8% of the strains (named KPC*Kp*1–KPC*Kp*18) produced KPC-type carbapenemases, whereas 18.2% produced VIM (VIM*Kp*) or NDM (NDM*Kp*1, NDM*Kp*2, NDM*Kp*3) metallo-beta-lactamases (MBL). Overall,

99% of the *K. pneumoniae* clinical isolates were multidrug-resistant (MDR, resistant to at least three antibiotics belonging to different antibiotics categories).

2. Results

2.1. Antibiotic Susceptibility and RAPD Profiles

Twenty-three *K. pneumoniae* clinical strains were tested in this study and their sources are listed in Table 1. The KPC*Kp*1–KPC*Kp*18 and KN*Kp* strains all come from the Intensive Care Unit (ICU), whereas the strains VIM*Kp*, NDM*Kp*1, NDM*Kp*2 and NDM*Kp*3 come from different wards.

Table 1. *K. pneumonaie* clinical strains used in antimicrobial assays.

Strain Name	Source	Strain Name	Source
KN*Kp*	AF	KPC*Kp*12	TR
KPC*Kp*1	TR	KPC*Kp*13	E
KPC*Kp*2	TFr	KPC*Kp*14	B
KPC*Kp*3	UC	KPC*Kp*15	TF
KPC*Kp*4	UC	KPC*Kp*16	TR
KPC*Kp*5	AF	KPC*Kp*17	U
KPC*Kp*6	U	KPC*Kp*18	U
KPC*Kp*7	E		
KPC*Kp*8	TF	VIM*Kp*	AF
KPC*Kp*9	C	NDM*Kp*1	Tfr
KPC*Kp*10	E	NDM*Kp*2	U
KPC*Kp*11	P	NDM*Kp*3	E

Abbreviations: AF, pharyngeal aspirate; B, bronchus aspirate; C, catheter; CF, cystic fibrosis; E, blood colture; P, prothesis; TF, pharyngeal swab; TFr, wound swab; TR, rectal swab; UC, catheter urine; U, urine.

Table 2. Antibiotic-susceptibility profile of *K. pneumoniae* clinical strains.

Strains	AMC	CTX	CAZ	TZP	GM	AK	SXT	CIP	MEM	ERT	CAZ/AVI
KN*Kp*	4 (S)	≤1 (S)	≤0.5 (S)	≤4 (S)	≤1 (S)	≤4 (S)	≤1 (S)	≤0.25 (S)	≤0.125 (S)	≤0.25 (S)	1 (S)
KPC*Kp*1	>16 (R)	>32 (R)	>32 (R)	>64 (R)	≤1 (S)	≤1 (S)	4 (I)	>2 (R)	>8 (R)	>4 (R)	8 (S)
KPC*Kp*2	>16 (R)	>32 (R)	>32 (R)	>64 (R)	≤1 (S)	≤1 (S)	4 (I)	>2 (R)	>8 (R)	>4 (R)	1 (S)
KPC*Kp*3	>32 (R)	>4 (R)	>8 (R)	>16 (R)	≤1 (S)	4 (S)	>4 (R)	>1 (R)	>8 (R)	>1 (R)	4 (S)
KPC*Kp*4	>32 (R)	>4 (R)	>8 (R)	>16 (R)	>4 (R)	4 (S)	>4 (R)	>1 (R)	>8 (R)	>1 (R)	1 (S)
KPC*Kp*5	>16 (R)	>32 (R)	>32 (R)	>64 (R)	>8 (R)	4 (S)	>8 (R)	0.25 (S)	>8 (R)	>4 (R)	4 (S)
KPC*Kp*6	>32 (R)	>4 (R)	>32 (R)	>64 (R)	2 (S)	8 (S)	>4 (R)	>1 (R)	8 (I)	>1 (R)	1 (S)
KPC*Kp*7	>32 (R)	>4 (R)	>8 (R)	>16 (R)	4 (I)	≤4 (S)	≤1 (S)	>1 (R)	>8 (R)	>1 (R)	1 (S)
KPC*Kp*8	>16 (R)	>32 (R)	>32 (R)	>64 (R)	>8 (R)	32 (R)	>8 (R)	>2 (R)	>8 (R)	>4 (R)	>8 (R)
KPC*Kp*9	>16 (R)	>32 (R)	>32 (R)	>64 (R)	≤1 (S)	4 (S)	>8 (R)	>2 (R)	>8 (R)	>4 (R)	4 (S)
KPC*Kp*10	>16 (R)	>32 (R)	>32 (R)	>64 (R)	≤1 (S)	≤1 (S)	>8 (R)	>2 (R)	8 (I)	>4 (R)	4 (S)
KPC*Kp*11	>32 (R)	>4 (R)	>8 (R)	>16 (R)	>4 (R)	>16 (R)	>8 (R)	>1 (R)	>8 (R)	>1 (R)	1 (S)
KPC*Kp*12	>16 (R)	>32 (R)	>32 (R)	>64 (R)	2 (S)	>32 (R)	≤1 (S)	>2 (R)	>8 (R)	>4 (R)	4 (S)
KPC*Kp*13	>16 (R)	>32 (R)	>32 (R)	>64 (R)	>8 (R)	>16 (R)	>8 (R)	>2 (R)	>8 (R)	>4 (R)	4 (S)
KPC*Kp*14	>16 (R)	>32 (R)	>16 (R)	>64 (R)	>8 (R)	32 (R)	>8 (R)	>2 (R)	>8 (R)	>4 (R)	1 (S)
KPC*Kp*15	>16 (R)	>32 (R)	>32 (R)	>64 (R)	>8 (R)	32 (R)	>8 (R)	>2 (R)	>8 (R)	>1 (R)	>8 (R)
KPC*Kp*16	>16 (R)	>32 (R)	>32 (R)	>64 (R)	2 (S)	4 (S)	>8 (R)	>2 (R)	>8 (R)	>4 (R)	4 (S)
KPC*Kp*17	>16 (R)	>32 (R)	>32 (R)	>64 (R)	≤1 (S)	≤1 (S)	≤1 (S)	>2 (R)	>8 (R)	>4 (R)	4 (S)
KPC*Kp*18	>32 (R)	>4 (R)	>8 (R)	>16 (R)	2 (S)	4 (S)	≤1 (S)	>1 (R)	>8 (R)	>1 (R)	4 (S)

Table 2. *Cont.*

Strains	AMC	CTX	CAZ	TZP	GM	AK	SXT	CIP	MEM	ERT	CAZ/AVI
VIM*Kp*	>32 (R)	>4 (R)	>8 (R)	>16 (R)	2 (S)	≤4 (I)	>4 (R)	1 (R)	8 (R)	>1 (R)	>8 (R)
NDM*Kp*1	>16 (R)	>32 (R)	>32 (R)	>64 (R)	>8 (R)	16 (I)	≤1 (S)	>2 (R)	8 (R)	>1 (R)	>8 (R)
NDM*Kp*2	>16 (R)	32 (R)	>32 (R)	>64 (R)	>8 (R)	>16 (R)	>8 (R)	≤0.06 (S)	>8 (R)	>4 (R)	>8 (R)
NDM*Kp*3	>16 (R)	>32 (R)	>32 (R)	>64 (R)	>8 (R)	32 (R)	≤1 (S)	>2 (R)	>8 (R)	>1 (R)	>8 (R)

Abbreviations: AMC, Amoxicillin/clavulanic acid; AK, Amikacin; CAZ, Ceftazidime; CAZ/AVI, Ceftazidime/avibactam; CIP, Ciprofloxacin; CTX, Cefotaxime; ERT, Ertapenem; GM, Gentamicin; I, Intermediate; MEM, Meropenem; R, Resistant; S, Susceptible; SXT, Trimethoprim-Sulfamethoxazole; TZP, Piperacillin/tazobactam.

All the *K. pneumoniae* clinical strains coming from the same ward (ICU) (KPC*Kp*1–KPC*Kp*18) were genotyped through the Random Amplified Polymorphic DNA (RAPD) analysis based on the Polymerase Chain Reaction technique (PCR), and the genomic profiles were compared with the antibiotic susceptibility patterns. As shown in Figure 1, the strains showed intra-specific variations between the genomic profiles, except for KPC*Kp*1 and KPC*Kp*2, sharing the same antibiotic susceptibility profile (Table 2), as well as KPC*Kp*13 and KPC*Kp*14. KPC*Kp*5 showed a genetic profile comparable to that of KPC*Kp*1 and KPC*Kp*2, but a different susceptibility to GM, SXT and CIP. Similarly, KPC*Kp*11 and KPC*Kp*12 showed a comparable genetic profile, but differences in sensitivity to GM and SXT. Regarding the strains KPC*Kp*15 and KPC*Kp*16, RAPD analysis showed some similarity, but by comparing their antibiotic susceptibility profile we found that these were different. KPC*Kp*15 showed resistance to GM, AK and CAZ/AVI, whereas KPC*Kp*16 did not. On the contrary, KPC*Kp*17 and KPC*Kp*18 showed a similar profile for antibiotic susceptibility, but a different RAPD profile. Thus, from the results of RAPD analysis, we chose to work only on 21 KPC*Kp* strains that showed both different genetic and antibiotic susceptibility profiles, excluding the strains KPC*Kp*2 and KPC*Kp*14. The strains selected were further investigated.

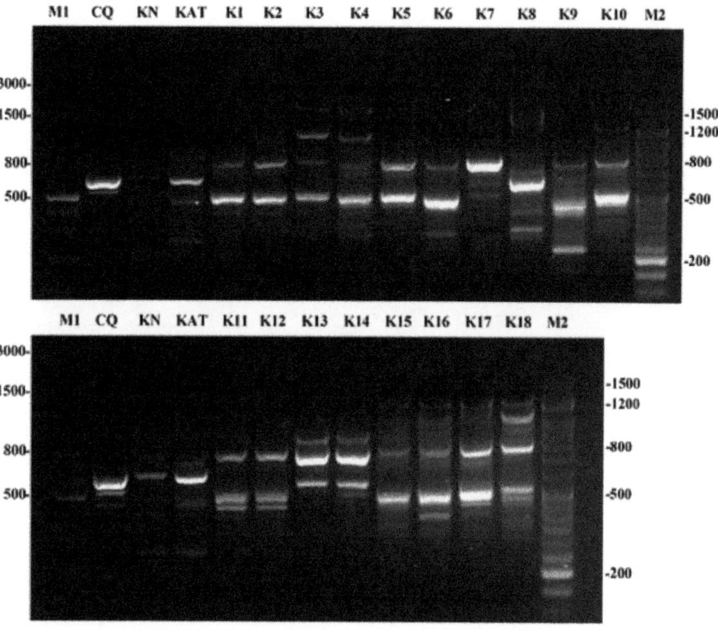

Figure 1. RAPD analysis of both ATCC and clinical strains from ICU. M1: 100 bp ladder; M2: 50 bp.

2.2. Antimicrobial Activity of Peptides

The antimicrobial activity of peptides **1B** and **C** was tested against ATCC 13883 (KCQ) and ATCC BAA-1705 (KAT) as *K. pneumoniae* reference strains and *K. pneumoniae* clinical isolates. The peptides resulted to be active against all the tested strains and MIC values are reported in Table 3. Both peptides inhibited the growth of carbapenem-sensitive strains (KCQ and KN*Kp*) at MIC values of 6.25 μM; on the other hand, both peptides were able to effectively inhibit the growth of carbapenemase-producing strains with MIC values ranging from 12.5 μM to 100 μM for peptide **1B**, and MIC values ranging from 6.25 μM to 25 μM for peptide **C**. Furthermore, peptide **C** showed bactericidal activity at MIC values, whereas **1B** was bacteriostatic at MIC values, and bactericidal at 2 × MIC values.

Table 3. Minimum inhibitory concentrations (MIC) of peptides **1B** and **C** against *K. pneumoniae* test strains.

Strains	MIC (μM)		MIC (μg/mL)
	1B	**C**	**Polymyxin E**
KCQ	6.25	6.25	<2 (S)
KAT	12.5	6.25	<2 (S)
KN*Kp*	6.25	6.25	<2 (S)
KPC*Kp*1	50	6.25	8 (R)
KPC*Kp*3	100	25	<2 (S)
KPC*Kp*4	25	25	<2 (S)
KPC*Kp*5	50	12.5	<2 (S)
KPC*Kp*6	12.5	12.5	<2 (S)
KPC*Kp*7	12.5	12.5	<2 (S)
KPC*Kp*8	50	12.5	<2 (S)
KPC*Kp*9	12.5	12.5	<2 (S)
KPC*Kp*10	50	12.5	<2 (S)
KPC*Kp*11	25	12.5	<2 (S)
KPC*Kp*12	25	25	<2 (S)
KPC*Kp*13	25	12.5	<2 (S)
KPC*Kp*15	50	25	8 (R)
KPC*Kp*16	25	25	<2 (S)
KPC*Kp*17	25	12.5	<2 (S)
KPC*Kp*18	25	12.5	<2 (S)
VIM*Kp*	50	25	<2 (S)
NDM*Kp*1	25	25	<2 (S)
NDM*Kp*2	50	25	<2 (S)
NDM*Kp*3	100	25	<2 (S)

Polymyxin E used as control conventional antimicrobial.

3. Discussion

Klebsiella pneumoniae is a frequent colonizer of the human gut and a major cause of healthcare-related infections whose treatment is complicated by the constant increase in antibiotic resistance. *K. pneumoniae* was included in the "ESKAPE" group (*Enterococcus faecium, Staphylococcus aureus, K. pneumoniae, Acinetobacter baumannii, Pseudomonas aeruginosa* and *Enterobacter species*) [18]. These pathogens acquired resistances through time, becoming one of the major health concerns of the modern day.

Carbapenems represent the last-resort beta-lactams, and carbapenem-resistant *K. pneumoniae* strains are currently spread all over the world [8]. The most relevant mechanism of carbapenem resistance is the production of carbapenemases. *Klebsiella pneumoniae* carbapenemases (KPCs) are the most common enzymes reported worldwide and capable of deactivating all of the beta-lactams [20]. Among the metallo-beta-lactamases (MBL), New Delhi MBL (NDM), Verona integron-encoded MBL (VIM) and imipenemase MBL (IMP) are the most common enzymes identified worldwide [8]. MBL-producers are continuously isolated in new regions, notably *K. pneumoniae* strains harboring the *ndm* gene [21].

The worrying spread of carbapenemase-producing enterobacteria is due to the prevalent localization of these genes on mobile genetic elements. *K. pneumoniae* can both acquire and carry a great number of genetic mobile elements, thus accumulating resistance genes and expanding its accessory genome, with the evolution of multi drug- and extensively drug-resistant strains [22].

Most of the recently approved drugs for the treatment of CRE are new combinations of an old beta-lactam with a second-generation beta-lactam inhibitor (BL/BLI). These combinations are ineffective on MBL-producing strains, whereas they generally show activity on strains harboring KPC carbapenemases [23]. Among these new combinations, ceftazidime/avibactam was approved by the FDA in 2015. Although it was recently introduced, KPC-producing *K. pneumoniae* isolates resistant to ceftazidime/avibactam have already been reported [24,25].

For that reason, the attention shifted on the identification of natural-derived peptides, whose mechanisms of action strongly differ from the classic antibiotics.

Among the novel generations of antimicrobial compounds, the antimicrobial peptides (AMPs) play a significant role in this context [26,27]. AMPs are widely produced by different kinds of living forms, and their structure typically consist of a variable-length amino acids chain (10 to 60 a.a.) [28,29]. The positive charge due to the presence of basic residues (lys and arg), the hydrophobic residues (about 50%) and the amphipathic nature are commonly shared features that characterize those molecules [30]. Considering that the bacterial membrane has been identified as a physical target of AMPs [31], the development of resistance mechanisms is greatly hampered. On the other hand, the cellular toxicity and the pharmacokinetic issues represent the main drawbacks of these compounds [32]. To augment their activity against bacteria and decrease the cytotoxicity, lipidation strategy was employed [19].

In our previous study, two derivatives of the Temporin L from *Rana temporaria*, named **1B** and **C**, were tested on *Pseudomonas aeruginosa*, *Klebsiella pneumoniae* and *Staphylococcus aureus*, showing good activity on these pathogens. Based on these preliminary results, in this study the peptides **1B** and **C** were evaluated on 20 clinical strains of *K. pneumoniae*, all carbapenemase-producers. Of these, 16 strains carried a KPC carbapenemase, whereas four isolates harbored an MBL. Initially, 18 KPC-producing clinical strains were included in the study, all from the intensive care unit. They were genotyped by RAPD: most of the tested strains showed different RAPD profiles, confirming the high heterogeneity of *K. pneumoniae* [33]. For those strains showing comparable RAPD profiles, antibiotic-susceptibility patterns were considered. On this basis, strains 2 and 14 were excluded as both RAPD and antibiotic-susceptibility profiles were comparable to strains 1 and 13, respectively. *K. pneumoniae* strains with comparable RAPD profile, but with different antibiotic-susceptibility pattern for at least one interpretative category were instead included in the study.

Peptides **1B** and **C** showed activity on all the tested strains. The lipidated analogue **C** was more active than peptide **1B**, probably due to the modification applied on its molecular structure, with MIC values ranging from 6.25 to 25 μM against the KPC-producing strains, and MIC values of 25 μM against the MBL-producing strains. Interestingly, at the highest concentration of 25 μM used in this study, it has previously been shown that peptide **C** was not cytotoxic both on human keratinocytes and erythrocytes [19]. MBL producers are very difficult to treat as therapeutic options are even more limited, but isolation of KPC-producing strains resistant to the new BL/BLI combinations complicates antibiotic treatment. In our study, 2 (KPC*Kp*8 and KPC*Kp*15) of the 16 KPC-producers were resistant to ceftazidime/avibactam. Moreover, these results seem particularly interesting to us as all the carbapenemase-producing strains tested were also resistant to classes of antibiotics other than beta-lactams. Notably, KPC*Kp*1 strain was also resistant to polymyxin E, drug shelved for its side effects and then reintroduced into human therapy as a salvage treatment against multidrug-resistant Gram-negative bacteria [34]; the KPC*Kp*15 strain was resistant to all drugs used.

4. Materials and Methods

4.1. Synthesis

The peptides **1B** [H-Phe-Val-Pro-Trp-Phe-Ser-Lys-Phe-DLeu-DLys-Arg-Ile-Leu-NH$_2$] and **C** [H-Phe(4-NHCO(CH$_2$)$_3$CH$_3$)-Val-Pro-Trp-Phe-Ser-Lys-Phe-DLeu-DLys-Arg-Ile-Leu-NH$_2$] were synthesized using Fmoc-based ultrasonic-assisted solid phase peptide synthesis (US–SPPS) methodology [35]. The elongation of the peptide sequence consisted in repeated cycles of Fmoc-deprotection and coupling reactions. Specifically, the Fmoc group was removed treating the resin with a solution of 20% piperidine in DMF (0.5 × 1 min) by ultrasonic irradiations, whereas each coupling reaction was performed using N$^\alpha$-Fmoc-amino acid (3 equiv), HBTU (3 equiv), HOBt (3 equiv) and DIEA (6 equiv) in DMF for 5 min by ultrasound waves. After the peptide assembly, the conjugation of valeric acid in para position of Phe^1 of peptide C was performed as previously reported [19]. In particular, the nitro group in para position of Phe^1 was reduced treating the resin with a 1M solution of SnCl$_2$ in DMF for 12 h and then, the valeric acid (3 equiv) was added using HBTU (3 equiv), HOBt (3 equiv) and DIEA (6 equiv) in DMF for 2 h on automated shaker. Finally, peptides were treated with a cleavage cocktail (TFA:TIS:H$_2$O, 95:2.5:2.5) to be released from the resin and cleaved from their protecting groups, and then they were purified and characterized by RP-HPLC using linear gradients of MeCN (0.1% TFA) in water (0.1% TFA), from 10 to 90% over 20 min.

4.2. Bacterial Strains and Culture Conditions

Strains of *Klebsiella pneumoniae* evaluated in this study included reference strains such as carbapeneme-susceptible ATCC 13883 (KCQ) and carbapeneme-resistant ATCC BAA-1705 (KAT), and 23 clinical strains (Table 1) belonging to a collection of anonymous isolates, previously established at the Department of Molecular Medicine and Medical Biotechnology (University of Naples Federico II) during two-year period (March 2020–March 2021). Among the clinical strains, the first 19 listed strains all come from the Intensive Care Unit and all but one (KN*Kp*) were resistant to carbapenems. The last 4 listed strains were from different wards (Cystic Fibrosis Center, Oncology, Geriatrics, Cardiac Surgery) and were metallo-β-lactamases (MBLs) producers. Identification was performed by biochemical characterization using the Vitek II system (BioMérieux, Marcy-l'Étoile, France) and was confirmed by MS MALDI-TOF (Bruker Daltonics, Bremen, Germany). Antibiotic susceptibility profiles were assessed using automated systems (Vitek 2; Phoenix—Becton Dickinson, Sparks, NV, USA) and broth microdilution method. Carbapenemase gene detection was performed using the Xpert Carba-R-test (Cepheid, Sunnyvale, CA, USA), real-time PCR diagnostic tests that allow to detect and differentiate the most prevalent carbapenemase gene families. All strains were stored as 15% (*v/v*) glycerol stocks at −80 °C. Before each experiment, cells were sub-cultured from stocks on Tryptic Soy Agar (TSA) (Becton Dickinson) plates to 37 °C for 24 h.

4.3. Antimicrobial Assays: Determination of the Minimum Inhibitory Concentration (MIC) and the Minimum Bactericidal Concentration (MBC)

The antibacterial activity of peptides **1B** and **C** was determined using a standard method of microdilution in broth, following the procedure already described [36]. For each strain, the bacterial suspension was prepared at 0.5 McFarland standard (corresponding to about 10^8 CFU/mL) in Mueller Hinton broth (MHB—Becton Dickinson) and subsequently adjusted to about 1.5 × 10^6 CFU/mL. One hundred microliter aliquots of this suspension were dispensed into 96-well microtiter plates. A 2x stock solution of temporin was serially diluted (twofold dilutions) with MHB and added to the wells to a final concentration between 3.125 μM and 100 μM. The plates were incubated at 37 °C for 19 h with stirring (300 rpm). The turbidity of the medium was measured with a spectrophotometer at 595 nm (Bio-Rad Laboratories S.r.l., Hercules, CA, USA). Wells with only MHB were used as a negative control and wells without peptide as a growth control. Polymyxin E (Sigma-Aldrich, Milan, Italy) was selected as a control from the conventional antimicrobials and tested at concentrations ranging from 2 μg/mL to 8 μg/mL. The MIC was defined as the lowest concentration of the compound that resulted in 100% growth inhibition after 19 h of incubation. The MBC was determined by transferring 50 μL aliquots of each well with concentrations equal to or higher than the MIC, onto TSA plates and incubating the plates at 37 °C for 24 h. The lowest compound concentration that yielded no bacterial growth on agar plates was defined as the MBC. Each compound was tested alone in triplicate; each experiment was performed twice.

4.4. RAPD Analysis

Random Amplification of Polymorphic DNA was performed on both ATCC strains and clinical isolates. Among all the tested primer, the HI-RP (5′-AACTCGGCGACCAGC TACAA-3′) primer was selected and used for the amplification [15]. The final RAPD conditions for HI-RP were: 0.5 μL of genomic DNA, 20 μL H_2O, 2.5 μL buffer, 1 μL dNTP, 1 μL primers and 0.1 μL of Taq DNA polymerase (Biotech Rabbit) in a final volume of 25 μL. The amplification program included an initial step at 94 °C for 5 min, followed by 40 cycles of 30 s at 95 °C, 1 min at 36 °C, and1 min at 72 °C, with a final extension cycle at 72 °C for 7 min. Reactions were performed using a thermo cycler (T100 THERMAL CYCLER-BioRad, Hercules, CA, USA). The PCR products were analysed by electrophoresis on 2% agarose gel in TBE and stained with ethidium bromide. The gels were photographed under UV light to record the results.

5. Conclusions

In conclusion, the obtained data show that the selected lipidated peptide C seems very promising for the development of a new drug with extensive antimicrobial activity and confirm and underline the potential role of temporins, and AMPs globally and efficiently counter the outbreaks of new multidrug-resistant pathogens. These compounds represent a valid mean to support both the management of serious infections and contrast the further expansion of antibiotic resistances.

Author Contributions: R.B. and F.M. performed peptide synthesis. E.R., N.M. and E.B. performed the experiments. R.P. isolated and identified the clinical strains. M.R.C., P.G. and E.B. designed the study, drafted and wrote the manuscript. M.R.C. and E.B. edited and revised the manuscript. All authors have read and agreed to the published version of the manuscript.

Funding: This research did not receive any specific grant from funding agencies in the public, commercial, or not-for-profit sectors.

Data Availability Statement: Data are contained within the manuscript.

Conflicts of Interest: The authors declare no conflict of interest.

References

1. Vading, M.; Nauclér, P.; Kalin, M.; Giske, C.G. Invasive infection caused by *Klebsiella pneumoniae* is a disease affecting patients with high comorbidity and associated with high long-term mortality. *PLoS ONE* **2018**, *13*, e0195258. [CrossRef]
2. Vardakas, K.Z.; Tansarli, G.S.; Rafailidis, P.I.; Falagas, M.E. Carbapenems versus alternative antibiotics for the treatment of bacteraemia due to *Enterobacteriaceae* producing extended-spectrum β-lactamases: A systematic review and meta-analysis. *J. Antimicrob. Chemother.* **2012**, *67*, 2793–2803. [CrossRef]
3. Walsh, T.R. Emerging carbapenemases: A global perspective. *Int. J. Antimicrob. Agents* **2010**, *36*, 8–14. [CrossRef]
4. Pitout, J.D.D.; Nordmann, P.; Poirel, L. Carbapenemase-Producing *Klebsiella pneumoniae*, a Key Pathogen Set for Global Nosocomial Dominance. *Antimicrob. Agents Chemother.* **2015**, *59*, 5873–5884. [CrossRef] [PubMed]
5. Cui, X.; Zhang, H.; Du, H. Carbapenemases in *Enterobacteriaceae*: Detection and Antimicrobial Therapy. *Front. Microbiol.* **2019**, *10*, 1823. [CrossRef]
6. Logan, L.K.; Weinstein, R.A. The Epidemiology of Carbapenem-Resistant *Enterobacteriaceae*: The Impact and Evolution of a Global Menace. *J. Infect. Dis.* **2017**, *215* (Suppl. 1), S28–S36. [CrossRef]
7. Ambler, R.P. The structure of beta-lactamases. *Philos. Trans. R. Soc. Lond. B Biol. Sci.* **1980**, *289*, 321–331.
8. Bonomo, R.A.; Burd, E.M.; Conly, J.; Limbago, B.M.; Poirel, L.; Segre, J.L.; Westblade, L.F. *Carbapenemase*-Producing Organisms: A Global Scourge. *Clin. Infect. Dis.* **2018**, *66*, 1290–1297. [CrossRef]

9. Trecarichi, E.M.; Tumbarello, M. Therapeutic options for carbapenem-resistant *Enterobacteriaceae* infections. *Virulence* **2017**, *8*, 470–484. [CrossRef]

10. WHO. Global Priority List of Antibiotic-Resistant Bacteria to Guide Research, Discovery, and Development of New Antibiotics. World Health Organization, 2017. Available online: https://www.who.int/medicines/publications/WHO-PPL-Short_Summary_25Feb-ET_NM_WHO.pdf (accessed on 3 August 2021).

11. Bassetti, M.; Vena, A.; Sepulcri, C.; Giacobbe, D.R.; Peghin, M. Treatment of Bloodstream Infections Due to Gram-Negative Bacteria with Difficult-to-Treat Resistance. *Antibiotics* **2020**, *9*, 632. [CrossRef] [PubMed]

12. Mangoni, M.L.; Grazia, A.D.; Cappiello, F.; Casciaro, B.; Luca, V. Naturally occurring peptides from *Rana temporaria*: Antimicrobial properties and more. *Curr. Top. Med. Chem.* **2016**, *16*, 54–64. [CrossRef]

13. Rinaldi, A.C.; Mangoni, M.L.; Rufo, A.; Luzi, C.; Barra, D.; Zhao, H.; Kinnunen, P.K.; Bozzi, A.; Di Giulio, A.; Simmaco, M. Temporin L: Antimicrobial, haemolytic and cytotoxic activities, and effects on membrane permeabilization in lipid vesicles. *Biochem. J.* **2002**, *368*, 91–100. [CrossRef]

14. Merlino, F.; Carotenuto, A.; Casciaro, B.; Martora, F.; Loffredo, M.R.; Di Grazia, A.; Yousif, A.M.; Brancaccio, D.; Palomba, L.; Novellino, E.; et al. Glycine-replaced derivatives of [Pro(3),DLeu(9)]TL, a temporin L analogue: Evaluation of antimicrobial, cytotoxic and hemolytic activities. *Eur. J. Med. Chem.* **2017**, *139*, 750–761. [CrossRef] [PubMed]

15. Bellavita, R.; Casciaro, B.; Di Maro, S.; Brancaccio, D.; Carotenuto, A.; Falanga, A.; Cappiello, F.; Buommino, E.; Galdiero, S.; Novellino, E.; et al. First-in-class cyclic Temporin L analogue: Design, synthesis, and antimicrobial assessment. *J. Med. Chem.* **2021**, *64*, 11675–11694. [CrossRef] [PubMed]

16. Buommino, E.; Carotenuto, A.; Antignano, I.; Bellavita, R.; Casciaro, B.; Loffredo, M.R.; Merlino, F.; Novellino, E.; Mangoni, M.L.; Nocera, F.P.; et al. The outcomes of decorated prolines in the discovery of antimicrobial peptides from Temporin-L. *Chem. Med. Chem.* **2019**, *14*, 1283–1290. [CrossRef] [PubMed]

17. Bellavita, R.; Vollaro, A.; Catania, M.R.; Merlino, F.; De Martino, L.; Nocera, F.P.; DellaGreca, M.; Lembo, F.; Grieco, P.; Buommino, E. Novel antimicrobial peptide from Temporin L in the treatment of *Staphylococcus pseudintermedius* and *Malassezia pachydermatis* in polymicrobial inter-kingdom infection. *Antibiotics* **2020**, *9*, 530. [CrossRef]

18. Bellavita, R.; Raucci, F.; Merlino, F.; Piccolo, M.; Ferraro, M.G.; Irace, C.; Santamaria, R.; Iqbal, A.J.; Novellino, E.; Grieco, P.; et al. Temporin L-derived peptide as a regulator of the acute inflammatory response in zymosan-induced peritonitis. *Biomed. Pharmacother.* **2020**, *123*, 109788. [CrossRef]

19. Bellavita, R.; Falanga, A.; Buommino, E.; Merlino, F.; Casciaro, B.; Cappiello, F.; Mangoni, M.L.; Novellino, E.; Catania, M.R.; Paolillo, R.; et al. Novel temporin L antimicrobial peptides: Promoting self-assembling by lipidic tags to tackle superbugs. *J. Enzym Inhib. Med. Chem.* **2020**, *35*, 1751–1764. [CrossRef]

20. Plazak, M.E.; Tamma, P.D.; Heil, E.L. The antibiotic arms race: Current and emerging therapy for *Klebsiella pneumoniae* carbapenemase (KPC)—Producing bacteria. *Expert. Opin. Pharmacother.* **2018**, *19*, 2019–2031. [CrossRef]

21. Lee, C.R.; Lee, J.H.; Park, K.S.; Kim, Y.B.; Jeong, B.C.; Lee, S.H. Global Dissemination of Carbapenemase-Producing *Klebsiella pneumoniae*: Epidemiology, Genetic Context, Treatment Options, and Detection Methods. *Front. Microbiol.* **2016**, *7*, 895. [CrossRef]

22. Navon-Venezia, S.; Kondratyeva, K.; Carattoli, A. *Klebsiella pneumoniae*: A major worldwide source and shuttle for antibiotic resistance. *FEMS Microbiol. Rev.* **2017**, *41*, 252–275. [CrossRef]

23. Yusuf, E.; Bax, H.I.; Verkaik, N.J.; van Westreenen, M. An Update on Eight "New" Antibiotics against Multidrug-Resistant Gram-Negative Bacteria. *J. Clin. Med.* **2021**, *10*, 1068. [CrossRef]

24. Göttig, S.; Frank, D.; Mungo, E.; Nolte, A.; Hogardt, M.; Besier, S.; Wichelhaus, T.A. Emergence of ceftazidime/avibactam resistance in KPC-3-producing *Klebsiella pneumoniae* in vivo. *J. Antimicrob. Chemother.* **2019**, *74*, 3211–3216. [CrossRef]

25. Gaibani, P.; Re, M.C.; Campoli, C.; Viale, P.L.; Ambretti, S. Bloodstream infection caused by KPC-producing *Klebsiella pneumoniae* resistant to ceftazidime/avibactam: Epidemiology and genomic characterization. *Clin. Microbiol. Infect.* **2020**, *26*, 516.e1–516.e4. [CrossRef]

26. Wang, J.; Dou, X.; Song, J.; Lyu, Y.; Zhu, X.; Xu, L.; Li, W.; Shan, A. Antimicrobial peptides: Promising alternatives in the post feeding antibiotic era. *Med. Res. Rev.* **2019**, *39*, 831–859. [CrossRef] [PubMed]

27. Falanga, A.; Galdiero, M.; Galdiero, S. Membranotropic cell penetrating peptides: The outstanding journey. *Int. J. Mol. Sci.* **2015**, *16*, 25323–25337. [CrossRef]

28. Kumar, P.; Kizhakkedathu, J.N.; Straus, S.K. Antimicrobial peptides: Diversity, mechanism of action and strategies to improve the activity and biocompatibility in vivo. *Biomolecules* **2018**, *8*, 4. [CrossRef]

29. Falanga, A.; Lombardi, L.; Franci, G.; Vitiello, M.; Iovene, M.R.; Morelli, G.; Galdiero, M.; Galdiero, S. Marine antimicrobial peptides: Nature provides templates for the design of novel compounds against pathogenic bacteria. *Int. J. Mol. Sci.* **2016**, *17*, 785. [CrossRef] [PubMed]

30. Hollmann, A.; Martinez, M.; Maturana, P.; Semorile, L.C.; Maffia, P.C. Antimicrobial peptides: Interaction with model and biological membranes and synergism with chemical antibiotics. *Front. Chem.* **2018**, *6*, 204. [CrossRef] [PubMed]

31. Bechinger, B.; Gorr, S.U. Antimicrobial Peptides: Mechanisms of Action and Resistance. *J. Dent. Res.* **2017**, *96*, 254–260. [CrossRef]

32. Zharkova, M.S.; Orlov, D.S.; Golubeva, O.Y.; Chakchir, O.B.; Eliseev, I.E.; Grinchuk, T.M.; Shamova, O.V. Application of antimicrobial peptides of the innate immune system in combination with conventional antibiotics-a novel way to combat antibiotic resistance? *Front. Cell. Infect. Microbiol.* **2019**, *9*, 128. [CrossRef] [PubMed]

33. Wasfi, R.; Elkhatib, W.F.; Ashour, H.M. Molecular typing and virulence analysis of multidrug resistant *Klebsiella pneumoniae* clinical isolates recovered from Egyptian hospitals. *Sci. Rep.* **2016**, *6*, 38929. [CrossRef]
34. Boisson, M.; Gregoire, N.; Couet, W.; Mimoz, O. Colistin in critically ill patients. *Minerva Anestesiol.* **2013**, *79*, 200–208. [PubMed]
35. Merlino, F.; Tomassi, S.; Yousif, A.M.; Messere, A.; Marinelli, L.; Grieco, P.; Novellino, E.; Cosconati, S.; Di Maro, S. Boosting Fmoc solid-phase peptide synthesis by ultrasonication. *Org. Lett.* **2019**, *21*, 6378–6382. [CrossRef] [PubMed]
36. Roscetto, E.; Masi, M.; Esposito, M.; Di Lecce, R.; Delicato, A.; Maddau, L.; Calabrò, V.; Evidente, A.; Catania, M.R. Anti-Biofilm Activity of the Fungal Phytotoxin Sphaeropsidin A Against Clinical Isolates of Antibiotic-Resistant Bacteria. *Toxins* **2020**, *12*, 444. [CrossRef] [PubMed]

 *antibiotics*

Article

A Study in a Regional Hospital of a Mid-Sized Spanish City Indicates a Major Increase in Infection/Colonization by Carbapenem-Resistant Bacteria, Coinciding with the COVID-19 Pandemic

Estefanía Cano-Martín [1], Inés Portillo-Calderón [2], Patricia Pérez-Palacios [2], José María Navarro-Marí [3], María Amelia Fernández-Sierra [1] and José Gutiérrez-Fernández [3,4,*]

[1] Department of Preventive Medicine and Public Health, Virgen de las Nieves University Hospital & ibs, Granada-Instituto de Investigación Biosanitaria de Granada, Avda. de las Fuerzas Armadas, 2, 18014 Granada, Spain; estefania.cano.sspa@juntadeandalucia.es (E.C.-M.); mamelia.fernandez.sspa@juntadeandalucia.es (M.A.F.-S.)

[2] Unidad de Gestión Clínica de Enfermedades Infecciosas y Microbiología Clínica, Hospital Universitario Virgen Macarena & Instituto de Biomedicina de Sevilla (IBIs), Hospital Universitario Virgen Macarena/CSIC/Universidad de Sevilla, 41020 Sevilla, Spain; inesm.portillo.sspa@juntadeandalucia.es (I.P.-C.); pperez5@us.es (P.P.-P.)

[3] Laboratory of Microbiology, Virgen de las Nieves University Hospital. & ibs, Granada-Instituto de Investigación Biosanitaria de Granada, Avda. de las Fuerzas Armadas, 2, 18014 Granada, Spain; josem.navarro.sspa@juntadeandalucia.es

[4] Department of Microbiology, School of Medicine, University of Granada & ibs, Granada-Instituto de Investigación Biosanitaria de Granada, Avda. de la Investigación, 11, 18016 Granada, Spain

* Correspondence: josegf@ugr.es

Citation: Cano-Martín, E.; Portillo-Calderón, I.; Pérez-Palacios, P.; Navarro-Marí, J.M.; Fernández-Sierra, M.A.; Gutiérrez-Fernández, J. A Study in a Regional Hospital of a Mid-Sized Spanish City Indicates a Major Increase in Infection/Colonization by Carbapenem-Resistant Bacteria, Coinciding with the COVID-19 Pandemic. *Antibiotics* **2021**, *10*, 1127. https://doi.org/10.3390/antibiotics10091127

Academic Editor: Francesca Andreoni

Received: 20 July 2021
Accepted: 14 September 2021
Published: 18 September 2021

Publisher's Note: MDPI stays neutral with regard to jurisdictional claims in published maps and institutional affiliations.

Abstract: Bacterial resistance to antibiotics has proven difficult to control over the past few decades. The large group of multidrug-resistant bacteria includes carbapenemase-producing bacteria (CPB), for which limited therapeutic options and infection control measures are available. Furthermore, carbapenemases associate with high-risk clones that are defined by the sequence type (ST) to which each bacterium belongs. The objectives of this cross-sectional and retrospective study were to describe the CPB population isolated in a third-level hospital in Southern Spain between 2015 and 2020 and to establish the relationship between the ST and the epidemiological situation defined by the hospital. CPB were microbiologically studied in all rectal and pharyngeal swabs and clinical samples received between January 2015 and December 2020, characterizing isolates using MicroScan and mass spectrometry. Carbapenemases were detected by PCR and Sanger sequencing, and STs were assigned by multilocus sequence typing (MLST). Isolates were genetically related by pulsed-field gel electrophoresis using Xbal, Spel, or Apal enzymes. The episodes in which each CPB was isolated were recorded and classified as involved or non-involved in an outbreak. There were 320 episodes with CPB during the study period: 18 with *K. pneumoniae*, 14 with *Klebisella oxytoca*, 9 with *Citrobacter freundii*, 11 with *Escherichia coli*, 46 with *Enterobacter cloacae*, 70 with *Acinetobacter baumannii*, and 52 with *Pseudomonas aeruginosa*. The carbapenemase groups detected were OXA, VIM, KPC, and NDM with various subgroups. Synchronous relationships were notified between episodes of *K. pneumoniae* and outbreaks for ST15, ST258, ST307, and ST45, but not for the other CPB. There was a major increase in infections with CPB over the years, most notably during 2020, coinciding with the COVID-19 pandemic. This study highlights the usefulness of gene sequencing techniques to control the spread of these microorganisms, especially in healthcare centers. These techniques offer faster results, and a reduction in their cost may make their real-time application more feasible. The combination of epidemiological data with real-time molecular sequencing techniques can provide a major advance in the transmission control of these CPB and in the management of infected patients. Real-time sequencing is essential to increase precision and thereby control outbreaks and target infection prevention measures in a more effective manner.

Keywords: carbapenems resistance; carbapenemases; Gram-negative bacteria; infection; colonization; COVID-19

1. Introduction

There has been an alarming rise in bacterial resistance to antibiotics over the past two decades, representing a "silent pandemic" that appears unstoppable. This resistance has become especially frequent in healthcare-related infections (HCRIs) worldwide [1–3]. However, there has also been an increase in the emergence and transmission of multidrug-resistant bacteria in the community setting over the past few years, including residential facilities and care homes. The more frequent hospitalization of these residents has favored exchanges between hospital and community, increasing the capacity to spread resistant bacteria in each setting [3]. The World Health Organization (WHO) has declared infection by these microorganisms as an emerging disease that poses a major public health threat worldwide [1–3].

The large group of carbapenem-resistant bacteria includes carbapenemase-producing bacteria (CPB), for which there are limited therapeutic options, and infection control measures have proven ineffective to prevent the dissemination of these bacteria to date [4]. The WHO list of global priority pathogens includes carbapenem-resistant *Acinetobacter baumannii* (CRAb) and *Pseudomonas aeruginosa* (CRPa) and carbapenem-producing *Enterobacteriaceae* (CPE) [2]. The most frequently observed carbapenem-resistance mechanism is the production of carbapenemases [5]. Control of these infections is further hampered by the horizontal transmission of these enzymes via plasmids among different *Enterobacteriaceae* [4].

In 2018, one-third of *Acinetobacter* spp. isolates obtained in the European Union (EU) were resistant [6], and numerous outbreaks increased morbidity and mortality rates, mainly in intensive care units (ICUs). A worldwide increase was observed in the frequency of patients colonized or infected by CPE, which was above the EU mean in Spain [5]. The European Center for Disease Control (ECDC) reported an increase in combined resistances in *Escherichia coli* and *Klebsiella pneumoniae* over the past few years [3]. *K. pneumoniae* is the most prevalent CPE [2] and responsible for the majority of outbreaks in healthcare centers [7]. CPE infections were first detected in Spain in 2005, and their number had multiplied 16-fold by 2019 [8].

In addition, carbapenemases are enzymes that associate with certain high-risk clones [8] that are defined by the sequence type (ST) to which each bacterium belongs. The STs most frequently associated with carbapenemases in *K. pneumoniae* are ST258, which is predominant worldwide [8], and ST307, which is also associated with a higher mortality [9]. In Germany, it was reported that *E. coli* ST131 is a high-risk clone that should be monitored very closely [10], alongside ST38, located in various European countries [10]. In Spain, the STs most frequently associated with carbapenemases are ST11, 13, 15, 16, 101, 147, 340, 384, 388, 405, 437, 464, 512, 846 and 1235 [8].

Carbapenemases can be divided into metallocarbapenemases (zinc-dependent class B) and non-metallocarbapenemases (zinc-independent classes A, C, and D) [11]. The first carbapenemase detected was in class A *K. pneumoniae* [11], and some carbapenemases appear more frequently than others. The first carbapenemase-producing isolates detected in Spain belonged to the KPC group [8]. OXA-48 group carbapenemases are currently the most abundant, mainly among *Klebsiella* spp. and *Enterobacter* spp. [5], and are the most common resistance mechanism among CPEs in Spain [5]. Other carbapenemases frequently reported in Spain are VIM-1, KPC-2, IMP, and NDM-1 [8].

Since 2020, a rise in CPB resistance in our setting coincided with the COVID-19 pandemic, which has been accompanied by populations at risk of severe complications and long-term sequelae and by longer hospital stays and the prescription of an elevated amount of antibiotics. These factors have led to an elevated risk of nosocomial CBP infections, as recently reported [12–14]. In addition, COVID-19 was especially prevalent in our hospital catchment area from 2020, and the patients frequently required hospitalization, as recorded by the Andalusian Health Service (https://www.juntadeandalucia.es/institutodeestadisticaycartografia/salud/datosSanitarios.html (accessed on 1 July 2021); and https://www.sspa.juntadeandalucia.es/servicioandaluzdesalud/todas-noticia/informacion-sobre-el-numero-de-casos-de-coronavirus-511?utm_source=servicioandaluzdesalud&utm_campaign=Boletin%20Novedades&utm_medium=mail&utm_content=20210901&utm_term=Informacion%20sobre%20el%20numero%20de%20casos%20de%20coronavirus (accessed on 1 July 2021).

The objectives of this study were to describe the CPB population isolated in a third-level hospital between 2015 and 2020 and to establish the relationship between STs and the epidemiological situation as defined by the hospital.

2. Material and Methods

This retrospective cross-sectional study included adult patients admitted to the Departments of Internal Medicine and its Specialties, ICUs (general and cardiac), and the Department of General Surgery and its Specialties of the Virgen de las Nieves University Hospital in Granada (Spain). This hospital provides specialized care to a population of around 331,220 inhabitants. No exclusion criteria were applied, except for repeat microbiological studies of the same episode. For the colonization study, the presence of CPB was studied in rectal (RS) and pharyngeal (PS) swabs received by the Clinical Microbiology Laboratory between 1 January 2015 and 31 December 2020 (5415 RS and 1034 PS for 3107 episodes studied before 2020 and 2308 during 2020). For the study of "possible infection episodes", clinical samples from different localizations were studied by applying standard clinical microbiology procedures, detecting CPB as previously described [15]. In brief, samples were seeded on selective culture medium CHROMID® ESBL (BioMérieux, Marcy-l'Étoile, France) and incubated at 37 °C in aerobiosis for 48 h. Isolates were identified by using the MicroScan system (Beckman Coulter, Brea, CA, USA) and mass spectrometry (Maldi-Tof®, Brucker Daltonik GmbH, Bremen, Germany). Resistance was characterized with the MicroScan system, using currently available Neg Combo panels, and interpreted according to the clinical cutoff points defined by the European Committee on Antimicrobial Susceptibility Testing (EUCAST) [16]. Carbapenemase was detected using the colorimetric Neo-Rapid CARB Kit® (Rosco Diagnostica A/S, Taastrup, Denmark) and immunochromatography (NG5-Test Carba, NG Biotech, Guipry-France to detect KPC, NDM, VIM, IMP, and OXA-48-like enzymes, and K-Set, Coris BioConcept, Gembloux, Belgium to detect OXA-23) in isolates meeting EUCAST cutoff point criteria for CPB screening. In parallel, isolates identified in swab and clinical samples were sent to the reference laboratory for molecular typing of nosocomial pathogens and genotypic detection of antimicrobial resistance mechanisms of interest under the regional Integrated Program for the Prevention and Control of Healthcare-related Infections and Appropriate Utilization of Antibiotics (acronym in Spanish, PIRASOA) led by the Microbiology Department of Virgen Macarena Hospital in Seville.

2.1. Microbiological Study of Carbapenemase-Producing Bacteria (CPB) under the PIRASOA Program

The susceptibility to ertapenem, imipenem, and meropenem was investigated by disk diffusion in Mueller Hinton agar, using EUCAST clinical cutoff points to interpret the results [16]. Carbapenemase activity inhibition [17] was studied by disk diffusion using meropenem, meropenem/boronic acid, meropenem/dipicolinic acid, and meropenem/cloxacillin disks as well as a temocillin disk (Rosco Diagnostica, Taastrup, Denmark). Carbapenemase and MLST genes were studied by PCR using specific primers and Sanger sequencing until 2018 and subsequently massive sequencing (Illumina Inc., San Diego, CA, USA). Sequences were analyzed with CLC Genomics Workbench, v10 software (Qiagen Iberia, Las Rozas de Madrid, Madrid, Spain). Determinants of resistance were detected using ResFinder (https://cge.cbs.dtu.dk/services/ResFinder) (accessed on 1 July 2021). and CARD (https://card.mcmaster.ca/) (accessed on 1 July 2021). databases, and MLST was identified using the MLST finder 2.0 database (https://cge.cbs.dtu.dk/services/MLST) (accessed on 1 July 2021). Clonal relationships among isolates were evaluated by pulsed-field gel electrophoresis (PFGE). Complete chromosomal DNA digestion in agarose gel was performed with XbaI (*Enterobacterales*), SpeI (*Pseudomonas* spp. and *Stenotrophomonas*), and ApaI (*Acinetobacter* spp.) according to the species. The resulting restriction fragments were separated in the CHIEF DR-II system (Bio-Rad Laboratories, Alcobendas, Madrid, Spain) with 1% agarose gel. The gels were subsequently stained with ethidium bromide, illuminated with ultraviolet light, and photographed in an automatic Gel Logic 200 Imaging System (Kodak, Rochester, NY, USA).

The conversion, normalization, and analysis of band patterns were performed using Bionumerics 7.6 software (AppliedMaths, Jollyville Rd., Austin, TX, USA), analyzing the patterns as previously described. Band position tolerance and optimization were set at 1%. An unweighted pair group method with arithmetic mean (UPGMA) was employed to generate a dendrogram and the Dice coefficient was used to measure genetic similarity among isolates. PFGE patterns with $\geq 90.0\%$ similarity were considered in the same group as closely related isolates.

2.2. Epidemiological Study

Episodes in which each carbapenemase-producing microorganism was isolated were recoded to avoid repetition with the same patient. An episode was defined as each hospital stay in which one or several different carbapenemase-producing microorganisms were isolated, only considering the first isolate of each microorganism during the hospital stay for infection or colonization study. When there were multiple isolated of the same microorganism during the same episode, one was selected according to the following criteria and order:

1. Isolate described in a PIRASOA report;
2. Isolate corresponding to infection study;
3. Isolate not reported in infection study, corresponding to colonization study.

The PIRASOA program describes the ST of the microorganism and its genetic similarity to other microorganisms of the same species from any hospital in Andalusia. It also reports whether the relationship with other microorganisms was recent or derived from a common transmission focus. It was determined whether a CPB had a synchronous (SR) or asynchronous relationship (AR) with others of the same species when the following criteria were met: (1) they belong to the same clone of an ST, (2) patients coincided in their hospital stay, and/or (3) the genetic study indicates direct transmission among patients and/or very recent exposure to a common reservoir. An SR was considered when the first criterion and one other criterion were met and an AR when only one criterion was met.

It was also recorded whether the CPB was involved or not in an outbreak of HCRI, defined by two or more cases of HCRI due to the same microorganism and associated in space and time with suspicion of an epidemiological link. The emergence of a single case of HCRI by a new, or unknown, or re-emergent infectious agent of mandatory declaration was considered an outbreak of nosocomial infection, henceforth "nosocomial outbreak" [18].

2.3. Influence of COVID-19 Infection during 2020

Data were compared between 2019 and 2020 on tested patients with a positive result for COVID-19 ($n = 178$ episodes), including their age, sex, length of hospital stay, and in-hospital consumption of imipenem (IP), meropenem (MP), and/or piperacillin–tazobactam (PTZ) as daily dose per 1000 stays (DDD/1000 stays).

2.4. Statistical Study

The involvement of episodes in which an SR was reported in a declared nosocomial outbreak was examined by constructing contingency tables and applying Fisher's exact test. The comparison between sexes was analyzed in the same way. The comparison by age was studied with the Student's *t*-test after establishing the normality of data distribution with the Kolmogorov–Smirnov test. IBM SPSS Statistics v. 19 was used for data analyses. $p < 0.05$ was considered significant.

3. Results

Between 1 January 2015 and 31 December 2020, 320 episodes of CPB were identified: 118 episodes of *K. pneumoniae*, 14 of *Klebisella oxytoca*, 9 of *Citrobacter freundii*, 11 of *E. coli*, 46 of *Enterobacter cloacae*, 70 of *A. baumannii*, and 52 of *P. aeruginosa*. Among these, 39 episodes were excluded because there was no definitive PIRASOA report and four because the clinical history of the patient was missing from the Andalusian Health Service database. Among the 43 episodes (13.4%) lost to the study, 34.9% involved *K. pneumoniae*, 6.9% *K. oxytoca*, 4.6% *C. freundii*, 6.9% *E. coli*, 2.3% *E. cloacae*, 13.9% *A. baumannii*, and 30.2% *P. aeruginosa*. Out of the final sample of 277 episodes, 103 involved *K. pneumoniae*, 11 *K. oxytoca*, 7 *C. freundii*, 8 *E. coli*, 45 *E. cloacae*, 64 *A. baumannii*, and 39 *P. aeruginosa*.

3.1. Description of Episodes

The most frequently isolated CPB in the studied episodes was *K. pneumoniae* (37.19% of total episodes), followed by *A. baumannii* (23.10%) and *E. cloacae* (16.25%) (Table 1). *K. pneumoniae* was detected in more numerous episodes every year except for 2018, when more episodes with *P. aeruginosa* were detected, and 2019, when there were more episodes with *A. baumannii*. Over the six-year study period, episodes with CPB were most frequent in 2020 (37.54%). There has been an increase in *A. baumannii* in episodes over the past two years and an increase in *E. cloacae* over the past year. The number of episodes each year rose from 6 in 2015 (2.17% of total episodes) to 123 in 2020 (37.54%). Globally, there were 156 episodes of infection and 121 of colonization. However, as depicted in Table 1, infection episodes were much more frequent than colonization episodes in 2016 and 2017, there was a similar frequency of infection and colonization episodes in 2018 and 2019, and colonization episodes (57) were much more frequent in 2020 (47).

The carbapenemase groups detected in this study were OXA, VIM, KPC, and NDM with their subgroups (Table 2). Table 3 exhibits the relationships between ST and carbapenemase for each species. There were four peak episodes with *K. pneumoniae* of ST258, ST307, ST15, and ST45, mainly associated with KPC-3, OXA-48, NDM-5, and OXA-48, respectively. There were fewer episodes with *K. oxytoca* and *C. freundii*, and only four STs were found in *K. oxytoca*, highlighting the association of ST36 with VIM-1, and six STs in *C. freundii*, highlighting the production of OXA-48 in three of the STs. Table 3 highlights the production of OXA-48 in *E. coli* with ST58, ST69, ST405, and ST648. However, the most frequent carbapenemase in *E. cloacae* was VIM-1, closely related to ST78, with OXA-48 being the second most frequent. Notably, the production of OXA-23 was associated with ST2 in *A. baumannii*. A wide variety of carbapenemases was detected in *P. aeruginosa*, highlighting the relationship of IMP-8 with ST348, the most frequent ST. ST175 and ST253 were less frequently detected, mainly associated with OXA-50 and IMP-16, respectively.

Table 1. Colonization/infection episodes of each carbapenemase-producing bacteria in each year.

Microorganisms	Years/Episodes												Total	
	2015		2016		2017		2018		2019		2020			
	Infection	Colonization	Infection	Colonization	Infection	Colonization	Infection	Colonization	Infection	Colonization	Infection	Colonization	Infection	Colonization
K. pneumoniae	1	3	16	5	10	2	6	2	12	5	21	20	66	37
K. oxytoca	0	0	3	0	2	0	2	2	0	1	0	0	7	4
C. freundii	0	0	0	0	0	0	0	0	0	2	2	3	2	5
E. coli	0	0	0	0	1	0	1	1	0	1	2	2	4	4
E. cloacae	0	2	4	0	3	1	6	2	4	2	12	9	29	16
A. baumannii	0	0	0	0	0	0	2	2	17	17	8	18	27	37
P. aeruginosa	0	0	0	0	7	0	6	6	6	7	2	5	21	18
Total	1 (16.7%)	5 (83.3%)	23 (82.1%)	5 (17.9%)	23 (85.2%)	4 (14.8%)	23 (60.5%)	15 (39.5%)	39 (52.7%)	35 (47.3%)	47 (45.2%)	57 (54.8%)	156 (56.3%)	121 (43.7%)

Table 2. Subgroups of carbapenemases detected in each carbapenemase-producing bacteria (CPB).

Microorganisms	Carbapenemase Types																Total
	OXA							VIM			IMP			KPC		NDM	
	OXA-1	OXA-23	OXA-48	OXA-50	OXA-58	OXA-244	OXA-245	VIM-1	VIM-2	VIM-63	IMP-8	IMP-16	IMP-23	KPC-2	KPC-3	NDM-5	
K. pneumoniae	1		56				1	4			1			1	32	8	102
K. oxytoca			1					8						2			11
C. freundii *			4					2		1				1	2		10
E. coli			5			1		2									8
E. cloacae			16					28				1					45
A. baumannii		47			17												64
P. aeruginosa *				1				2	3	1	14	10	9				40
Total (%)	1 (0.36%)	47 (16.79%)	82 (29.99%)	1 (0.36%)	17 (6.07%)	1 (0.36%)	1 (0.36%)	46 (16.43%)	3 (1.07%)	2 (0.71%)	16 (5.71%)	10 (3.57%)	9 (3.21%)	4 (1.43%)	32 (11.43%)	8 (2.86%)	280 (100%)

* CPB in which more than one carbapenemase is detected in an episode.

Table 3. Description in the microorganisms of carbapenemase and sequence types.

SEQUENCIO-TYPES	CARBAPENEMASE TYPES																
	KPC-2	KPC-3	0XA-1	OXA-23	0XA-48	0XA-50	0XA-58	0XA-244	0XA-245	VIM-1	VIM-2	VIM-63	IMP-8	IMP-16	IMP-23	NDM-5	Not Available
ST1									1								
ST1				1													
ST2				46			17										
ST10								1									
ST11			1							1							

Table 3. *Cont.*

SEQUENCIO-TYPES	CARBAPENEMASE TYPES																
	KPC-2	KPC-3	OXA-1	OXA-23	OXA-48	OXA-50	OXA-58	OXA-244	OXA-245	VIM-1	VIM-2	VIM-63	IMP-8	IMP-16	IMP-23	NDM-5	Not Available
ST15	1				3											8	
ST18		2															
ST22					2												
ST24														1			
ST25										1							
ST36										9							
ST37																	1
ST45					11												
ST50					1					1							
ST58					1												
ST69					1												
ST78										22							
ST90										4							
ST108					1												
ST111										1							
ST114					13												
ST128										1							
ST145										1							
ST147					1												
ST170	2																
ST175	1											1					
ST175											2				9		
ST214					1												
ST238					1							1					
ST253						1								10			
ST258		29															
ST277											1						

Table 3. *Cont.*

SEQUENCIO-TYPES	CARBAPENEMASE TYPES																
	KPC-2	KPC-3	0XA-1	OXA-23	0XA-48	0XA-50	0XA-58	0XA-244	0XA-245	VIM-1	VIM-2	VIM-63	IMP-8	IMP-16	IMP-23	NDM-5	Not Available
ST307					29												
ST307/cgST3303					3												
ST307/cgST5556					1												
ST321													1				
ST348													14				
ST392					1												
ST405/cgST5158					3												
ST405					2												
ST512		1															
ST513					1												
ST525										1							
ST648					1												
ST845										2							
ST896					3												
ST1262					1												
ST1774					1												
ST2242												1					
ST8327										1							
Not available										1							

K. oxytoca
E. coli
E. cloacae
A. baumannii
K. pneumoniae
C. freundii
P. aeruginosa

3.2. Description of the Association among ST, Type of Relationship, and Involvement in a Nosocomial Outbreak of Infection with CPB

This association was more frequent in *K. pneumoniae*, *E. cloacae*, *A. baumannii*, and *P. aeruginosa* (Table 4). In *K. pneumoniae*, SRs were only reported for ST15, ST258, ST307, and ST45, the STs with episodes included in outbreaks. ARs alone were reported for the other STs detected, none of which had more than three episodes. An association was found between the presence of an SR in episodes and involvement in an outbreak ($p = 0.000007299$). In *E. cloacae*, PIRASOA reported an SR in only two of the eight STs detected (ST114 and ST78), but only ST114 was involved in an outbreak, and the association between the presence of an SR and involvement in an outbreak was just short of statistical significance ($p = 0.06829$). In *A. baumannii*, only two STs (ST1 and ST2) were found, being ST1 in all episodes except for one; *A. baumannii* had an SR with others in half of the episodes and an AR in the remaining half. No significant association was found ($p = 1$) between the presence of an SR and involvement in an outbreak. Finally, in *P. aeruginosa*, six STS were detected and four of these were SRs: ST175, ST253, ST348, and ST845, but only ST253 was involved in an outbreak. No significant association was found ($p = 1$) between the presence of an SR and involvement in an outbreak.

Table 4. Relationship between sequence types of each CPB, type of relationship, and involvement in a nosocomial outbreak.

	Sequence Type	AR	SR	Total	OB NO	OB YES	Total
K. pneumoniae	ST1	1	0	1	1	0	1
	ST11	2	0	2	2	0	2
	ST128	1	0	1	1	0	1
	ST147	1	0	1	1	0	1
	ST15	4	8	12	6	6	12
	ST1774	1	0	1	1	0	1
	ST25	1	0	1	1	0	1
	ST258	8	21	29	26	3	29
	ST307	8	21	29	20	9	29
	ST307/cgST3303	3	0	3	3	0	3
	ST307/cgST5556	1	0	1	1	0	1
	ST321	1	0	1	1	0	1
	ST37	1	0	1	1	0	1
	ST392	1	0	1	1	0	1
	ST405/cgST5158	3	0	3	3	0	3
	ST45	4	7	11	7	4	11
	ST512	1	0	1	1	0	1
	ST525	1	0	1	1	0	1
	ST896	3	0	3	3	0	3
	Total	46	57	103	81	22	103
E. cloacae	ST108	1		1	1		1
	ST111	1		1	1		1
	ST114	3	10	13	10	3	13
	ST1262	1		1	1		1
	ST24	1		1	1		1
	ST50	2		2	2		2
	ST78	13	9	22	22		22
	ST90	4		4	4		4
	Total	26	19	45	42	3	45
A. baumannii	ST1	1	0	1	1	0	1
	ST2	31	32	63	45	18	63
	Total	32	32	64	46	18	64

Table 4. *Cont.*

P. aeruginosa	ST175	3	8	11	11		11
	ST2242	1		1	1		1
	ST253	4	6	10	4	6	10
	ST277	1		1	1		1
	ST348	1	13	14	14		14
	ST845	1	1	2	2		2
	Total	11	28	39	33	6	39

Type of relationship: synchronous (SR) or asynchronous relationship (AR). Outbreak: OB.

As shown in Table 5, episodes with SR did not correspond to episodes involved in an outbreak, and episodes with AR were even involved in outbreaks in some cases.

Table 5. Comparison between identified SRs and those involved in a nosocomial outbreak of CPB.

CPBs	SR	OB YES	AR	OB YES
K. pneumoniae	57	21	46	1
K. oxytoca	4	0	7	0
C. freundii	2	0	5	0
E. cloacae	19	3	26	0
A. baumannii	32	12	32	6
P. aeruginosa	28	4	11	2

Type of relationship: synchronous (SR) or asynchronous relationship (AR). Outbreak: OB.

3.3. Influence of Infection with COVID-19 during 2020

The absolute number of CPB infections was higher during 2020 than during 2019, attributable to a marked rise in the number of samples studied rather than a higher percentage of positive results (4.5% in 2020 vs. 5.6% in 2019). Table 6 compares clinical data of patients testing during 2019 (74) and 2020 (104), positive for COVID-19 infection, including age ($p = 0.014$), sex (not significant), hospital stay, hospital occupation, and prescriptions of PTZ, IMP, and PTZ (daily dose per 1000 stays).

Table 6. Clinical data for episodes of carbapenemase-producing bacteria.

Parameters	2019	2020
Age, years	63.36	57.48
Standard deviation	16.601	17.254
Males, %	65	60
Females, %	35	40
Hospital stay, days	7.75	8.79
Hospital occupation, %	77.56	77.24
Piperacillin–tazobactam *	65.13	66.88
Imipenem *	6.71	5.13
Meropenem *	47.45	61.21

* daily dose per 1000 stays (DDD/1000).

4. Discussion

This study describes carbapenemase-producing microorganisms isolated over the past six years in the microbiological laboratory of a tertiary hospital in Southern Spain. The main findings were an increasingly elevated transmission of these bacteria in both hospital and community settings and a major rise in the detection of CPB.

The capacity to detect CPB and their resistances has significantly improved through technological advances and the rigorous implementation of screening and control protocols. However, the upsurge of CPB cases in 2020 can be attributed at least in part to the COVID-19 pandemic, which led to a major rise in the number of studied episodes due to an increase in hospital stays, among other reasons. The resulting overload of hospital departments hampered the application of infection prevention measures, and there was an increased administration of antibiotics (often inappropriate) to prevent and treat bacterial over-infection in patients with pneumonia [19]. The inappropriate use of antibiotics is one cause of the increasingly frequent emergence of resistance to carbapenem, among other antibiotics. However, improvements in the diagnostic capabilities of microbiology laboratories over time should also be taken into account, which would also influence an increase in the detection of infectious episodes.

Four groups of carbapenemases (OXA, VIM, KPC, and NDM) and corresponding subgroups were detected, all previously reported in Spain. Some were found in more than one microorganism species (OXA-48, IMP-8, VIM-1, VIM-63, and KPC-3) and the others were detected in a single species. The most abundant subgroup was OXA-48 (in 29.79% of episodes), which is also the most prevalent subgroup worldwide [5], followed by VIM-1 (17.12%), OXA-23 (16.44%), and KPC-3 (12.33%). The most frequently isolated bacteria in these episodes was *K. pneumoniae*, one of the most widely disseminated CPB worldwide and largely responsible for the alarm caused by CPB, given the limited therapeutic options and high mortality rates [7].

The relationships between STs and carbapenemases were explored in the detected CPB, followed by a search for ST–carbapenemase combinations that have been reported in Spain for each CPB, highlighting isolates of: *K. pneumoniae* ST15-OXA-48 [20,21], ST258-KPC-3 [8,22], ST307-OXA-48 [23,24], and ST512-KPC-3 [23,25–27]; *K. oxytoca* ST36-VIM 1 [28]; *E. coli* ST58-OXA-48 [29]; *E. cloacae* ST114-OXA-48 [30]; *A. baumannii* ST1-OXA-23 [31]; and *P. aeruginosa* ST175-VIM-2 [32,33].

No other combinations were found, either because they were not previously detected in a Spanish hospital or were detected but not published. CPB associated with STs or carbapenemases were found, but no relationship was observed between STs and carbapenemase.

Although not as notorious as *K. pneumoniae*, the other CPB described in this study form part of a group of multidrug-resistant microorganisms that pose a major public health threat [34] and show increased diversity over time, as reported in a Portuguese study [4]. The frequency of their detection is rising among colonized and infected hospital patients, despite the implementation of infection prevention and control measures and routine colonization screenings [4]. There is also a trend towards more frequent nosocomial outbreaks, especially of *K. pneumoniae* [7], and towards their longer persistence as colonizers of hospitalized patients and in the hospital setting. This creates "silent" reservoirs and carriers of CPB over prolonged time periods, hampering control of their dissemination [7,11]. The comprehensive and exhaustive implementation of available control measures is essential. In our setting, these include the surveillance and follow-up of HCRIs and nosocomial outbreaks and active communication from the microbiology departments on microorganisms of concern.

We highlight the importance of genetic sequencing techniques for controlling the spread of these microorganisms, given the speed with which results are obtained. A reduction in their cost may allow their real-time performance to be more widely available at health centers. Data obtained on the STs and carbapenemase subgroup of a CPB indicate its pathogenicity, dissemination capacity, local prevalence, therapeutic options, and responses to treatment and can also yield information on changes in the resistances of these bacteria [4,7]. These techniques can be highly effective for the detection and control of nosocomial outbreaks. When only the species and antibiogram of microorganisms are identified but not their ST, it is not known whether multiple unrelated STs might be involved, and this question would be resolved by real-time molecular analysis. Without this information, an outbreak is declared when there appear to be two or more cases of hospital-acquired infection. Accordingly, an outbreak is declared in the hospital when there are two or more patients with the infection who shared hospital space and/or staff/equipment at some time since their admission (epidemiological link). If cases continue to emerge among patients successively admitted at different times to the same unit, the possibility of a common reservoir is investigated to establish the epidemiological link, i.e., an SR. The association observed between the presence of an SR and involvement in outbreaks suggests that the epidemiological criteria for declaring an outbreak were correct because relationships were observed among microorganisms that were subsequently found to have the same lineage. However, it cannot be assumed that CPB with identical STs are involved in the same chain of transmission without considering the epidemiological evidence [35].

In conclusion, there has been a major increase in infections with CPB over the years, especially during 2020, coinciding with the COVID-19 pandemic. The combination of epidemiological data with real-time molecular sequencing data represents a major advance in the control of CPB transmission and the management of infected patients. The increased precision offered by molecular sequencing techniques and the possibility of their real-time performance can contribute to a greater control of nosocomial outbreaks and a more effective targeting of infection prevention measures.

Author Contributions: Conceptualization, J.G.-F.; methodology, E.C.-M., I.P.-C., P.P.-P., J.M.N.-M., M.A.F.-S., J.G.-F.; writing—original draft preparation, E.C.-M., J.G.-F.; writing—review and editing, E.C.-M., I.P.-C., P.P.-P., J.M.N.-M., M.A.F.-S., J.G.-F. All authors have read and agreed to the published version of the manuscript.

Funding: This research received no external funding.

Institutional Review Board Statement: The study protocol was conducted in agreement with the Helsinki Declaration and ethical norms for epidemiological investigations. Ethical review and approval were waived for this study, due to the non-interventionist nature of the study, in which the biological material was only used for the standard diagnosis of infections as prescribed by the attending physicians, no investigation was performed in addition to routine procedures, and the laboratory performed no additional sampling or any modification of routine diagnostic protocols. The Clinical Management Unit of the Department of Clinical Microbiology of the hospital granted permission to access and utilize the data.

Informed Consent Statement: The informed consent of patients for integrated result analysis was not required, in accordance with WHO ethical guidelines for research in humans. Analyzed data were drawn from a completely anonymous database in which individuals were only identified by their unique health record number in the Andalusian health system, considering the first episode for each patient in the analyses.

Data Availability Statement: Not applicable.

Acknowledgments: We would like to thank Dª Manuela Expósito-Ruíz for the statistical analysis carried out, as well as Dª Pilar Aznarte-Padial for the information provided on the use of antibiotics.

Conflicts of Interest: The authors declare no conflict of interest.

References

1. Londoño, R.J.; Macias, O.I.C.; Ochoa, J.F.L. Factores de riesgo asociados a infecciones por bacterias multirresistentes derivadas de la atención en salud en una institución hospitalaria de la ciudad de Medellín 2011–2014. *Infectio* **2016**, *20*, 77–83. [CrossRef]
2. Sleiman, A.; Fayad, A.G.A.; Banna, H.; Matar, G.M. Prevalence and molecular epidemiology of carbapenem-resistant *Gram-negative bacilli* and their resistance determinants in the Eastern Mediterranean Region over the last decade. *J. Glob. Antimicrob. Resist.* **2021**, *25*, 209–221. [CrossRef] [PubMed]
3. Rodríguez-Cundín, P.; Antolín, F.; Wallamnn, R.; Fabo, M.; Portal, T.; Rebollo-Rodrigo, H. Epidemiología de los gérmenes multirresistentes. *Rev. Med. Vladecilla* **2016**, *1*, 26–30.
4. Guerra, A.M.; Lira, A.; Lameirão, A.; Selaru, A.; Abreu, G.; Lopes, P.; Mota, M.; Novais, Â.; Peixe, L. Multiplicity of carbapenemase-producers three years after a KPC-3-producing *K. pneumoniae* ST147-K64 hospital outbreak. *Antibiotics* **2020**, *9*, 806. [CrossRef] [PubMed]
5. Rivera-Izquierdo, M.; Láinez-Ramos-Bossini, A.J.; Rivera-Izquierdo, C.; López-Gómez, J.; Fernández-Martínez, N.F.; Redruello-Guerrero, P.; Martín-delosReyes, L.M.; Martínez-Ruiz, V.; Moreno-Roldán, E.; Jiménez-Mejías, E. OXA-48 carbapenemase-producing enterobacterales in Spanish hospitals: An updated comprehensive review on a rising antimicrobial resistance. *Antibiotics* **2021**, *10*, 89. [CrossRef]
6. Lötsch, F.; Albiger, B.; Monnet, D.L.; Struelens, M.J.; Seifert, H.; Kohlenberg, A. Epidemiological situation, laboratory capacity and preparedness for carbapenem-resistant *Acinetobacter baumannii* in Europe, 2019. *Eurosurveillance* **2020**, *25*, 2001735. [CrossRef] [PubMed]

7. Snitkin, E.S.; Zelazny, A.M.; Thomas, P.J.; Stock, F.; Henderson, D.K.; Palmore, T.N.; Segre, J.A.; NISC Comparative Sequencing Program Group. Tracking a hospital outbreak of carbapenem-resistant *Klebsiella pneumoniae* with whole-genome sequencing. *Sci. Transl. Med.* **2012**, *4*, 148ra116. [CrossRef]

8. Soria-Segarra, C.; González-Bustos, P.; López-Cerero, L.; Fernández-Cuenca, F.; Rojo-Martín, M.D.; Fernández-Sierra, M.A.; Gutiérrez-Fernández, J. Tracking KPC-3-producing ST-258 *Klebsiella pneumoniae* outbreak in a third-level hospital in Granada (Andalusia, Spain) by risk factors and molecular characteristics. *Mol. Biol. Rep.* **2020**, *47*, 1089–1097. [CrossRef] [PubMed]

9. Bocanegra-Ibarias, P.; Garza-González, E.; Padilla-Orozco, M.; Mendoza-Olazarán, S.; Pérez-Alba, E.; Flores-Treviño, S.; Camacho-Ortiz, A.; Silva-Sanchez, J.; Garza-Ramons, U. The successful containment of a hospital outbreak caused by NDM-1-producing Klebsiella pneumoniae ST307 using active surveillance. *PLoS ONE* **2019**, *14*, e0209609. [CrossRef]

10. Welker, S.; Boutin, S.; Miethke, T.; Heeg, K.; Nurjadi, D. Emergence of carbapenem-resistant ST131 *Escherichia coli* carrying blaOXA-244 in Germany, 2019 to 2020. *Eurosurveillance* **2020**, *25*, 2001815. [CrossRef]

11. Li, J.; Huang, Z.Y.; Yu, T.; Tao, X.Y.; Hu, Y.M.; Wang, H.C.; Zou, M.X. Isolation and characterization of a sequence type 25 carbapenem-resistant hypervirulent *Klebsiella pneumoniae* from the mid-south region of China. *BMC Microbiol.* **2019**, *19*, 219. [CrossRef] [PubMed]

12. Mazzariol, A.; Benini, A.; Unali, I.; Nocini, R.; Smania, M.; Bertoncelli, A.; De Sanctis, F.; Ugel, S.; Donadello, K.; Polati, E.; et al. Dynamics of SARS-CoV2 infection and multi-drug resistant bacteria superinfection in patients with assisted mechanical ventilation. *Front. Cell. Infect. Microbiol.* **2021**, *11*, 730. [CrossRef] [PubMed]

13. Weiner-Lastinger, L.M.; Pattabiraman, V.; Konnor, R.Y.; Patel, P.R.; Wong, E.; Xu, S.Y.; Smith, B.; Edwards, J.R.; Dudeck, M.A. The impact of coronavirus disease 2019 (COVID-19) on healthcare-associated infections in 2020: A summary of data reported to the National Healthcare Safety Network. *Infect. Control. Hosp. Epidemiol.* **2021**, 1–14. [CrossRef]

14. Livermore, D.M. Antibiotic resistance during and beyond COVID-19. *JAC Antimicrob. Resist.* **2021**, *3* (Suppl. 1), i5–i16. [CrossRef]

15. Soria-Segarra, C.; Delgado-Valverde, M.; Serrano-García, M.L.; López-Hernández, I.; Navarro-Marí, J.M.; Gutiérrez-Fernández, J. Infections in patients colonized with carbapenem-resistant Gram-negative bacteria in a medium Spanish city. *Rev. Esp. Quim.* **2021**, *34*, 269–401. [CrossRef]

16. The European Committee on Antimicrobial Susceptibility Testing. Clinical Breakpoints—Bacteria (v 10.0). 2020. Available online: https://www.eucast.org/fileadmin/src/media/PDFs/EUCAST_files/Breakpoint_tables/v_10.0_Breakpoint_Tables.pdf (accessed on 10 November 2020).

17. Tenover, F.C.; Arbeit, R.D.; Goering, R.V.; Mickelsen, P.A.; Murray, B.E.; Persing, D.H.; Swaminathan, B. Interpreting chromosomal DNA restriction patterns produced by pulsed-field gel electrophoresis: Criteria for bacterial strain typing. *J. Clin. Microbiol.* **1995**, *33*, 2233–2239. [CrossRef]

18. Red Nacional de Vigilancia Epidemiológica (RENAVE). Protocolo de Vigilancia de Brotes de IRAS (Protocolo-BROTES). Madrid, 2016. Available online: https://www.isciii.es/QueHacemos/Servicios/VigilanciaSaludPublicaRENAVE/EnfermedadesTransmisibles/Documents/PROTOCOLOS/PROTOCOLOS%20EN%20BLOQUE/PROTOCOLOS%20IRAS%20Y%20RESISTENCIAS/PROTOCOLOS%20NUEVOS%202019%20IRAS/Protocolo-BROTES_Nov2017_rev_Abril2019.pdf (accessed on 1 July 2021).

19. Calderón-Parra, J.; Muiño-Miguez, A.; Bendala-Estrada, A.D.; Ramos-Martínez, A.; Muñez-Rubio, E.; Carracedo, E.F.; Montes, J.T.; Rubio-Rivas, M.; Arnalich-Fernandez, F.; Pérez, J.L.B.; et al. Inappropriate antibiotic use in the COVID-19 era: Factors associated with inappropriate prescribing and secondary complications. Analysis of the registry SEMI-COVID. *PLoS ONE* **2021**, *16*, e0251340. [CrossRef]

20. Ruiz-Garbajosa, P.; Hernández-García, M.; Beatobe, L.; Tato, M.; Méndez, M.I.; Grandal, M.; Aranzabal, L.; Alonso, S.; Lópaz, M.; Astray, J.; et al. A single-day point-prevalence study of faecal carriers in long-term care hospitals in Madrid (Spain) depicts a complex clonal and polyclonal dissemination of carbapenemase-producing Enterobacteriaceae. *J. Antimicrob. Chemother.* **2015**, *71*, 348–352. [CrossRef]

21. Machuca, J.; López-Cerero, L.; Fernández-Cuenca, F.; Mora-Navas, L.; Mediavilla-Gradolph, C.; López-Rodríguez, I.; Pascual, A. OXA-48-like-producing *Klebsiella pneumoniae* in Southern Spain in 2014–2015. *Antimicrob. Agents Chemother.* **2019**, *63*, e01396-18. [CrossRef]

22. Sorlózano-Puerto, A.; Esteva-Fernández, D.; Oteo-Iglesias, J.; Navarro-Marí, J.M.; Gutierrez-Fernandez, J. A new case report of urinary tract infection due to KPC-3-producing *Klebsiella pneumoniae* (ST258) in Spain. *Arch. Esp. Urol.* **2016**, *69*, 437–440. [PubMed]

23. Vázquez-Ucha, J.C.; Seoane-Estévez, A.; Rodiño-Janeiro, B.K.; González-Bardanca, M.; Conde-Pérez, K.; Martínez-Guitián, M.; Bou, G. Activity of imipenem/relebactam against a Spanish nationwide collection of carbapenemase-producing Enterobacterales. *J. Antimicrob. Chemother.* **2021**, *76*, 1498–1510. [CrossRef]

24. Fuster, B.; Salvador, C.; Tormo, N.; García-González, N.; Gimeno, C.; González-Candelas, F. Molecular epidemiology and drug-resistance mechanisms in carbapenem-resistant *Klebsiella pneumoniae* isolated in patients from a tertiary hospital in Valencia, Spain. *J. Glob. Antimicrob. Resist.* **2020**, *22*, 718–725. [CrossRef]

25. López-Cerero, L.; Egea, P.; Gracia-Ahufinger, I.; González-Padilla, M.; Rodríguez-López, F.; Rodríguez-Baño, J.; Pascual, A. Characterisation of the first ongoing outbreak due to KPC-3-producing *Klebsiella pneumoniae* (ST512) in Spain. *Int. J. Antimicrob. Agents* **2014**, *44*, 538–540. [CrossRef]

26. Oteo, J.; Pérez-Vázquez, M.; Bautista, V.; Ortega, A.; Zamarrón, P.; Saez, D.; Spanish Antibiotic Resistance Surveillance Program Collaborating Group. The spread of KPC-producing Enterobacteriaceae in Spain: WGS analysis of the emerging high-risk clones of Klebsiella pneumoniae ST11/KPC-2, ST101/KPC-2 and ST512/KPC-3. *J. Antimicrob. Chemother.* **2016**, *71*, 3392–3399. [CrossRef]

27. Gonzalez-Padilla, M.; Torre-Cisneros, J.; Rivera-Espinar, F.; Pontes-Moreno, A.; López-Cerero, L.; Pascual, A.; Natera, C.; Rodríguez, M.; Salcedo, I.; Rodríguez-López, F.; et al. Gentamicin therapy for sepsis due to carbapenem-resistant and colistin-resistant Klebsiella pneumoniae. *J. Antimicrob. Chemother.* **2014**, *70*, 905–913. [CrossRef]

28. Pérez-Vazquez, M.; Oteo-Iglesias, J.; Sola-Campoy, P.J.; Carrizo-Manzoni, H.; Bautista, V.; Lara, N.; Aracil, B.; Alhambra, A.; Martínez-Martínez, L.; Campos, J.; et al. Characterization of carbapenemase-producing *Klebsiella oxytoca* in Spain, 2016–2017. *Antimicrob. Agents Chemother.* **2019**, *63*, e02529-18. [CrossRef] [PubMed]

29. De Toro, M.; Fernández, J.; García, V.; Mora, A.; Blanco, J.; De La Cruz, F.; Rodicio, M.R. Whole genome sequencing, molecular typing and in vivo virulence of OXA-48-producing *Escherichia coli* isolates including ST131 H 30-Rx, H 22 and H 41 subclones. *Sci. Rep.* **2017**, *7*, 12103. [CrossRef]

30. Delgado-Valverde, M.; Conejo, M.D.C.; Serrano, L.; Fernández-Cuenca, F.; Pascual, A. Activity of cefiderocol against high-risk clones of multidrug-resistant Enterobacterales, *Acinetobacter baumannii*, *Pseudomonas aeruginosa* and *Stenotrophomonas maltophilia*. *J. Antimicrob. Chemother.* **2020**, *75*, 1840–1849. [CrossRef] [PubMed]

31. Guerrero-Lozano, I.; Fernández-Cuenca, F.; Galán-Sánchez, F.; Egea, P.; Rodríguez-Iglesias, M.; Pascual, Á. Description of the OXA-23 β-lactamase gene located within Tn2007 in a clinical isolate of *Acinetobacter baumannii* from Spain. *Microb. Drug Resist.* **2015**, *21*, 215–217. [CrossRef] [PubMed]

32. Viedma, E.; Juan, C.; Villa, J.; Barrado, L.; Orellana, M.Á.; Sanz, F.; Joaquin, R.O.; Oliver, A.; Chaves, F. VIM-2–producing multidrug-resistant *Pseudomonas aeruginosa* ST175 clone, Spain. *Emerg. Infect. Dis.* **2012**, *18*, 1235. [CrossRef]

33. Perez-Vazquez, M.; Sola-Campoy, P.J.; Zurita, Á.M.; Avila, A.; Gomez-Bertomeu, F.; Solis, S.; Jesús, O.I. Carbapenemase-producing *Pseudomonas aeruginosa* in Spain: Interregional dissemination of the high-risk clones ST175 and ST244 carrying blaVIM-2, blaVIM-1, blaIMP-8, blaVIM-20 and blaKPC-2. *Int. J. Antimicrob. Agents* **2020**, *56*, 106026. [CrossRef] [PubMed]

34. Shaidullina, E.; Shelenkov, A.; Yanushevich, Y.; Mikhaylova, Y.; Shagin, D.; Alexandrova, I.; Ershova, O.; Akimkin, V.; Kozlov, R.; Edelstein, M. Antimicrobial resistance and genomic characterization of OXA-48- and CTX-M-15-co-producing hypervirulent *Klebsiella pneumoniae* ST23 recovered from nosocomial outbreak. *Antibiotics* **2020**, *9*, 862. [CrossRef] [PubMed]

35. Woodford, N.; Turton, J.; Livermore, D.M. Multiresistant Gram-negative bacteria: The role of high-risk clones in the dissemination of antibiotic resistance. *FEMS Microbiol. Rev.* **2011**, *35*, 736–755. [CrossRef] [PubMed]

 *antibiotics*

Article

High-Level Carbapenem Resistance among OXA-48-Producing *Klebsiella pneumoniae* with Functional OmpK36 Alterations: Maintenance of Ceftazidime/Avibactam Susceptibility

Pilar Lumbreras-Iglesias [1,2], María Rosario Rodicio [2,3], Pablo Valledor [4], Tomás Suárez-Zarracina [5] and Javier Fernández [1,2,4,*]

1 Department of Clinical Microbiology, Hospital Universitario Central de Asturias, 33011 Oviedo, Spain; pilar.lumbreras08@gmail.com
2 Traslational Microbiology Group, Instituto de Investigación Sanitaria del Principado de Asturias (ISPA), 33011 Oviedo, Spain; rrodicio@uniovi.es
3 Department of Functional Biology, Microbiology Area, University of Oviedo, 33006 Oviedo, Spain
4 Research & Innovation, Artificial Intelligence and Statistical Department, Pragmatech AI Solutions, 33003 Oviedo, Spain; pablo.valledor@pragmatech.ai
5 Department of Internal Medicine, Hospital Universitario Central de Asturias, 33011 Oviedo, Spain; tomassecades@hotmail.com
* Correspondence: javifdom@gmail.com

Citation: Lumbreras-Iglesias, P.; Rodicio, M.R.; Valledor, P.; Suárez-Zarracina, T.; Fernández, J. High-Level Carbapenem Resistance among OXA-48-Producing *Klebsiella pneumoniae* with Functional OmpK36 Alterations: Maintenance of Ceftazidime/Avibactam Susceptibility. *Antibiotics* 2021, 10, 1174. https://doi.org/10.3390/antibiotics10101174

Academic Editor: Francesca Andreoni

Received: 17 August 2021
Accepted: 22 September 2021
Published: 27 September 2021

Publisher's Note: MDPI stays neutral with regard to jurisdictional claims in published maps and institutional affiliations.

Abstract: The aim of this work was to analyze outer membrane porin-encoding genes (*ompK35* and *ompK36*) in a collection of OXA-48 producing *Klebsiella pneumoniae*, to assess the effect of porin alterations on the susceptibility to ceftazidime/avibactam, and to describe a screening methodology for phenotypic detection of OXA-48-producing *K. pneumoniae* with disrupted porins. Antimicrobial susceptibility was tested by Microscan and Etest. The genomes of 81 OXA-48-producing *K. pneumoniae* were sequenced. MLST, detection of antimicrobial resistance genes, and analysis of *ompK35* and *ompK36* were performed *in silico*. Tridimensional structures of the OmpK36 variants were assessed. Receiver operating characteristics curves were built to visualize the performance ability of a disk diffusion assay using carbapenems and cefoxitin to detect OmpK36 functional alterations. A wide variety of OmpK36 alterations were detected in 17 OXA-48-producing *K. pneumoniae* isolates. All displayed a high-level meropenem resistance (MIC $\geq$ 8 mg/L), and some belonged to high-risk clones, such as ST15 and ST147. Alterations in *ompK35* were also observed, but they did not correlate with high-level meropenem resistance. All isolates were susceptible to ceftazidime/avibactam and porin alterations did not affect the MICs of the latter combination. Cefoxitin together with ertapenem/meropenem low inhibition zone diameters (equal or lower than 16 mm) could strongly suggest alterations affecting OmpK36 in OXA-48-producing *K. pneumoniae*. OXA-48-producing *K. pneumoniae* with porin disruptions are a cause of concern; ceftazidime/avibactam showed good *in vitro* activity against them, so this combination could be positioned as the choice therapy to combat the infections caused by this difficult-to-treat isolates.

Keywords: *K. pneumoniae*; carbapenem resistance; OXA-48; porins; ceftazidime/avibactam

1. Introduction

The use of carbapenems has been increasing in hospitals in the past years as a consequence of the rise of extended spectrum β-lactamase (ESBL) and/or AmpC-producing *Enterobacterales* [1]. The extensive clinical use of these drugs has led in turn to the emergence of carbapenem-resistant *Enterobacterales* (CRE), boosted by the spread of carbapenemases, such as serine β-lactamases (KPC and OXA-48) or metallo-β-lactamases (VIM, IMP and NDM), by means of mobile genetic elements [2]. Carbapenem resistance can also be mediated by permeability defects in the outer membrane of ESBL- or AmpC-producing strains [3], which can besides increase the minimal inhibitory concentrations (MICs) of carbapenems in carbapenemase-producing *Enterobacterales* (CPE). These defects can be associated with alterations in porin-encoding genes or their promoter regions, such as point mutations, deletions or insertions which can hinder the synthesis of a functional protein, hence obstructing the entry of molecules into the periplasm [4]. *Klebsiella pneumoniae* is one of the main reservoirs of carbapenemases in the hospital environment [5]. OmpK35 and OmpK36 are the major porins in this species, and they constitute the main pathway through which carbapenems enter the cell [6].

The increasing carbapenem resistance among *Enterobacterales* poses a major threat to public health because of its difficult treatment. Several clinical studies have demonstrated that carbapenems, alone or combined with other drugs, are options for the treatment of infections caused by CRE when MICs of these compounds are low [7,8]. This is the case of most infections caused by OXA-48-producing *Enterobacterales*, which weakly hydrolyze carbapenems [9,10]. However, when MICs of meropenem are higher than 8 mg/L, the probability of reaching therapeutic success is low according to stochastic modeling data, and thus carbapenems are contraindicated [7]. In response to the emergence and spread of CRE, new antimicrobials or combinations of a β-lactam plus a β-lactamase inhibitor are either under development or have been approved in the last years. This is the case of ceftazidime/avibactam, a combination of a third-generation cephalosporin with a synthetic β-lactamase inhibitor that prevents the activities of Ambler class A and C β-lactamases and some Ambler class D enzymes including OXA-48 [11].

In view of the worrisome medical situation raised by CRE, and the recognized contribution of porin alterations to carbapenem resistance, the aims of the present study were (i) to molecularly analyze outer membrane porin-encoding genes in a collection of OXA-48 producing *K. pneumoniae* isolates; (ii) to assess the effect of the observed alterations on the susceptibility to ceftazidime/avibactam; (iii) to develop a screening methodology for phenotypic detection of OXA-48 producing *Klebsiella* spp. with porin disruptions.

2. Results and Discussion

2.1. Carbapenems and Ceftazidime/Avibactam Susceptibility, β-Lactamases and Molecular Epidemiology

Bacterial identification by MALDI-TOF recognized the 81 isolates as *K. pneumoniae*, results which were confirmed *in silico* by analysis of the sequenced genomes. Strains of *K. pneumoniae* were classified in 15 sequence types (ST), being ST326 (22/81, 27.2%) and ST147 (20/81, 24.7%) the most prevalent (Table 1 and Table S1). The presence of *bla*$_{OXA-48}$ was confirmed in all of them. Antimicrobial susceptibility testing, performed with the Microscan system, showed that 100%, 24.69% and 20.99% isolates were resistant to ertapenem, imipenem and meropenem, respectively. In addition to the five isolates selected from other samples, twelve recovered from blood cultures displayed high level of meropenem resistance (MIC ≥ 8 mg/L), which could not be explained only by the OXA-48 production. This suggested the existence of additional mechanisms. Apart from β-lactamase production, mechanisms which can lead to a high-level carbapenem resistance comprise efflux pumps, PBP (penicillin binding protein) alterations, and mutations that modify the expression and/or function of outer membrane porins, the latter being a mechanism frequently found in *Enterobacterales* [6]. All isolates of the study were susceptible to ceftazidime/avibactam, including those highly resistant to meropenem, with MICs ranging from 0.19 mg/L to 0.75 mg/L (three of the latter did not coproduced ESBLs and were also susceptible to ceftazidime alone). Relevant features of the 17 isolates displaying high-level carbapenem resistance are shown in Table 1, while the same features relative to the remaining isolates are compiled in Supplementary Table S2.

Table 1. Microbiological features of OXA-48-producing *Klebsiella pneumoniae* isolates displaying high carbapenem resistance due to porin alterations.

Strain	ST	Sample	Previous Carbapenem Exposure (days) [1]	ESBL Enzyme	FOX MIC/IZD [2]	MER MIC/IZD [2]	ERT MIC/IZD [2]	IPM vMIC/IZD [2]	CAZ/AVI MIC	*ompK35*	*ompK36*
Kp_HUCA_Bac_5	326	Blood culture	Yes (10)	CTX-M-15	>16/6	>32/6	>4/6	>8/6	0.5	WT	::IS5
Kp_HUCA_Bac_7	326	Blood culture	No	CTX-M-15	>16/6	16/6	>4/6	>8/6	0.75	WT	::IS5
Kp_HUCA_Bac_10	326	Blood culture	Yes (21)	CTX-M-15	>16/6	>32/8	>4/6	>8/9	0.5	WT	ΔC201
Kp_HUCA_Bac_13	326	Blood culture	Yes (11)	CTX-M-15	>16/6	>32/9	>4/7	>8/11	0.75	WT	G259T
Kp_HUCA_Bac_18	326	Blood culture	No	CTX-M-15	>16/7	32/8	>4/6	>8/9	0.5	WT	ΔC201
Kp_HUCA_Bac_24	326	Blood culture	Yes (5)	CTX-M-15	>16/11	>32/6	>1/6	>8/9	0.5	::623T	::IS5
Kp_HUCA_Bac_64	16	Blood culture	Yes (5)	CTX-M-15	16/11	8/16	>4/16	8/18	0.5	WT	ΔA74
Kp_HUCA_Bac_65	147	Blood culture	No	SHV-12	≤8/15	8/16	>1/16	4/19	0.38	WT	::IS5
Kp_HUCA_Bac_76	405	Blood culture	Yes (7)	CTX-M-15	>16/12	>32/6	>1/6	>8/6	0.75	::623T	T612A
Kp_HUCA_Bac_85	273	Blood culture	Yes (19)	Non-ESBL	>16/6	>32/6	>1/6	>8/6	0.5	::623T	::29TG
Kp_HUCA_Bac_88	405	Blood culture	Yes (6)	Non-ESBL	16/18	8/16	>1/16	8/17	0.19	WT	::818ACAAAGCGCAGAAC TTCGAACC TGGGCTTTGCAA
Kp_HUCA_Bac_90	101	Blood culture	Yes (2)	CTX-M-15	>16/6	24/6	>1/6	4/17	0.5	ΔG185	::403GACGGC
Kp_HUCA_2	326	Rectal swab	No	CTX-M-15	>16/6	>32/6	>1/6	>8/8	0.5	WT	G844T
Kp_HUCA_3	15	Sputum	No	CTX-M-15	>16/6	>32/6	>1/6	>8/14	0.75	::623T	::403GACGGC
Kp_HUCA_4	15	Urine	Yes (15)	CTX-M-15	>16/7	>32/6	>1/6	>8/6	0.75	WT	Δ(886)-184
Kp_HUCA_5	193	Urine	No	Non-ESBL	>16/8	16/12	>1/10	>8/12	0.5	WT	::IS1
Kp_HUCA_8	326	Rectal swab	Yes (4)	CTX-M-15	>16/6	32/6	>1/6	>8/9	0.5	WT	ΔA4

ST, sequence type; ESBL, extended-spectrum beta-lactamase; FOX, cefoxitin; MER, meropenem; ERT, ertapenem; IPM, imipenem; CAZ/AVI, ceftazidime/avibactam; MIC, minimal inhibitory concentration; IZD, inhibition zone diameter; WT, wild type.[1] This column indicates if the patient received a carbapenem during hospital admission before the isolate was recovered, with the number of days receiving this drug shown in brackets.[2] MIC are expressed in mg/L and IZD in mm.

Moreover, 82% of the isolates displaying high-level carbapenem resistance (14/17) coproduced ESBLs, with roughly the same rate (52/64, 81%) found among the isolates with low carbapenem MICs (MICs $\leq$ 4 mg/L). All of the isolates which did not coproduce ESBLs were susceptible to ceftazidime alone, including those three that displayed high-level carbapenem resistance. bla_{SHV-12} and $bla_{CTX-M-15}$ were the most frequent genes detected among the isolates that showed an ESBL phenotype (Table 1 and Table S1). Based on the antimicrobial susceptibility results, and considering that three of the highly-carbapenem resistant isolates did not co-produce ESBLs, there does not appear to be a necessary correlation between production of these enzymes and high MICs of meropenem. For this reason, a detailed analysis of the genes encoding the main porins of *K. pneumoniae*, OmpK35 and OmpK36, and their upstream regions was performed, in order to infer possible permeability alterations in the outer cell membrane.

2.2. Molecular Analysis of OmpK35 and OmpK36 in OXA-48 Producers

Bioinformatic analysis of *ompK35* from the 17 isolates with high-level carbapenem resistance revealed a single nucleotide insertion (T) at 623 position in four of them, and a single nucleotide deletion (G) at 185 position in another one (Table 1). Both alterations led to a change in the reading frame. On the other hand, analysis of the *ompK35* sequences of the isolates exhibiting low meropenem MICs showed that three of them carried a IS1-like element inserted upstream of the gene, one had a G890A transition in the coding region, which resulted in a premature stop codon, and another one had a single nucleotide deletion (G) at position 575 (Table S1). Accordingly, it seems that at least not all the changes detected in *ompK35* are enough to substantially increase carbapenem MICs (Table S1). These results are in line with previous observations by other authors, reporting that alterations affecting *ompK35* are not capable to generate high resistance to carbapenems [12–14].

Regarding OmpK36, functional alterations have been broadly reported among KPC-producing *K. pneumoniae* [15–17]; however, reports and analysis of porin disruptions among OXA-48 producers are still scarce [18]. In the present study, all isolates displaying high meropenem MICs, including the 12 strains recovered from blood cultures and the five strains selected from other samples, had alterations affecting the OmpK36 coding region or the upstream DNA, which could lead to non-functional proteins or hinder the expression of the gene (Table 1 and Figure 1). The remaining isolates carried no mutations or had nucleotide changes considered as polymorphisms.

Among high-level carbapenem resistant isolates, four contained an IS5-like element located within the coding region (three isolates) or immediately upstream of the RSB (ribosome binding site; one isolate) of OmpK36; the latter alteration was similar to one previously reported in KPC-2-producing *K. pneumoniae* ST258, where it was shown to be associated with significantly lower OmpK36 expression levels and increased MICs of carbapenems [15]. A single isolate carried an IS1-like insertion within OmpK36. Additional alterations included: (i) nonsense mutations that generate premature stop codons yielding truncated and probably non-functional proteins (3 isolates); (ii) single nucleotide deletions that led to frameshift mutations (4 isolates, with two of them, both ST326, having the same deletion at the same position); (iii) a large deletion of 886 nucleotides including the first 702 nucleotides of the ORF and 184 of the upstream DNA (1 isolate; not shown in Figure 1); (iv) insertions of 2 and 34 nucleotides at positions 29 and 818, respectively, which resulted on a frame shift (each displayed by a single isolate); (v) insertion of 6 nucleotides at position 403, resulting in the incorporation of two additional amino acids (DG) on a highly conserved region (L3) of the porin (two isolates). The same insertion was associated with carbapenem resistance in KPC-2-producing *K. pneumoniae* ST423 and ST11 isolates [16]. In addition, Lunha et al. reported a very similar amino acid insertion (GD) at the same position within the L3 region of OmpK36 in ST37 and ST11 OXA-48-producing *K. pneumoniae*, and linked this alteration to an increase in the carbapenem resistance level. Such resistance increase was also observed in KPC-2-producing *K. pneumoniae* ST258 strains with the same

GD insertion in OmpK36, and attributed to a diminishment in the pore diameter caused by the two additional amino acids [17].

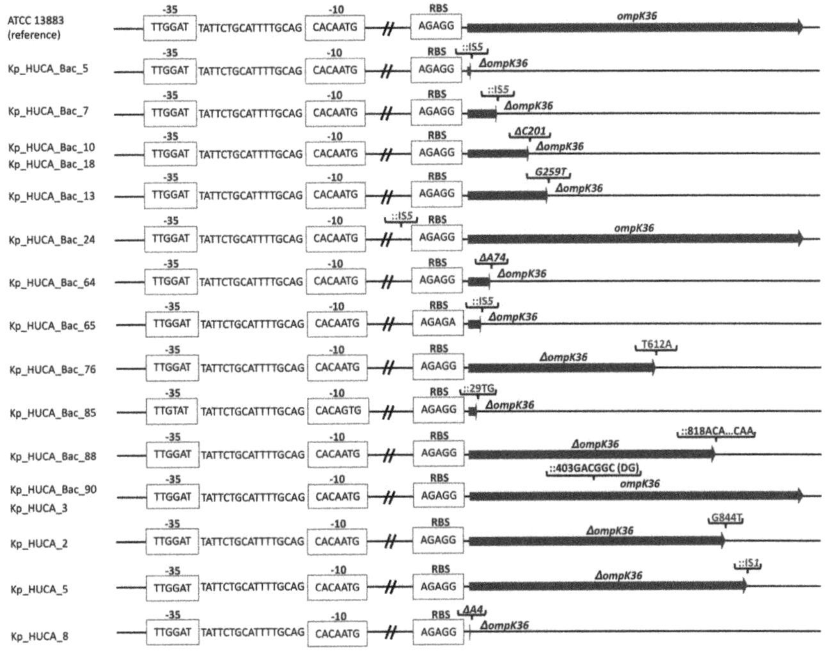

Figure 1. Genetic alterations in *ompK36* and upstream DNA found in OXA-48-producing *Klebsiella pneumoniae* isolates displaying high-level resistance to meropenem. Kp_HUCA_4 is not represented as it carries a large deletion that includes the promoter region and the first 702 nucleotides of the ORF. RBS, ribosome binding site; IS, insertion sequence;, insertion; Δ, deletion.

The tridimensional structures of OmpK36 variant proteins predicted by the Phyre2 web portal are shown in Figure S1. In most of them, it can be visually verified how the conformation of the porin is affected by the observed alterations.

Our series represent a wide variety of OmpK36 alterations detected among different clones of OXA-48-producing *K. pneumoniae*. Some of them, such as ST15 or ST147, are considered of high-risk since they are clones which have a special ability for successful expansion, and are typically involved in outbreaks [19,20], which poses an additional concern. OmpK36 disruptions described in our work are probably responsible for the high MICs of meropenem of the strains in which they were detected, in comparison with MICs of strains with functional OmpK36 proteins. The increase of meropenem MICs in OXA-48 producers is especially marked since, as previously indicated, this enzyme by itself only causes weak carbapenem hydrolysis and a consequent low level of resistance to these drugs [9,10]. Carbapenems must first penetrate the outer membrane in order to reach the PBPs and, because these drugs are relatively hydrophilic, their entry occurs through the water-filled porin channels [21]. Obviously, their inactivation via periplasmic carbapenemases will be more effective in increasing resistance if the influx is decreased through the loss of porins [22,23]. Apart from allowing carbapenems to cross the outer membrane, OmpK36 constitutes the entry way into the cell of some nutrients and other physiological important substances; thus, the loss of this porin is not free of charge. In fact, Wong et al. recently demonstrated that OmpK36-mediated carbapenem resistance attenuates the virulence of *K. pneumoniae* ST258 [17].

Selection of isolates displaying high-level carbapenem resistance due to outer membrane alterations may be conditioned by several factors, including the selective pressure exerted by previous exposure to carbapenems of individual patients or by the hospital environment [17]. In this sense, 11 (64.7%) of the isolates with high-level carbapenem resistance in our study were recovered from patients who had been previously treated with carbapenems during the same hospital stay (Table 1). The high-level of meropenem resistance displayed by these isolates (meropenem MICs $\geq$ 8 mg/L), exclude these drugs as therapeutic alternatives to treat the infections they cause. In fact, as already mentioned in the introduction, PK/PD modeling has shown that the probability of reaching the target pharmacodynamic parameter is low, even if high dose and extended infusion are administered [7]. Fortunately, all isolates in the present study were susceptible to ceftazidime/avibactam and displayed low MICs of these drugs, regardless of the OmpK36 variant. This might be due to the fact that, unlike other β-lactamase inhibitors, the outer membrane porins are not the major route by which avibactam enters the periplasm [24], and it has been previously reported that ceftazidime MICs are not affected by OmpK36 alterations [14]. Wong et al., also, did not find a decrease in the susceptibility to ceftazidime/avibactam among *K. pneumoniae* ST258 with alterations affecting OmpK35 or OmpK36 [17].

2.3. Disk Diffusion Assay Using Carbapenems and Cefoxitin for Detection of Functional Alteration in OmpK36:

Detection of strains with functional alterations in porins among OXA-48-producing *K. pneumoniae* may be of particular interest in clinical microbiology laboratories, especially when affecting OmpK36. However, next-generation sequencing techniques are not available in the routine of these laboratories, so the establishment of a phenotypic screening for the detection of such strains would be very useful to predict genotypes from phenotypes and, also, for epidemiological purposes. Most of the isolates with alterations in OmpK36 displayed reduced inhibition zone diameters to cefoxitin and carbapenems. Thus, our data suggest that resistance to both drugs could be considered a surrogate marker of functional alterations in OmpK36 in OXA-48-producing *K. pneumoniae*. In addition to carbapenems, it is well known that this porin constitutes the main pathway through which cefoxitin penetrates the outer membrane [4], and alterations affecting OmpK36 have been associated with the development of resistance to this drug [25]. In order to set a threshold to detect isolates carrying functional alterations in OmpK36, a disk diffusion assay testing ertapenem, imipenem, meropenem and cefoxitin was performed, and ROC curves were built to evaluate sensitivity and specificity of their detection at different thresholds (Figure 2). All isolates displaying inhibition zone diameter lower or equal to 16 mm to both ertapenem and meropenem, carried alterations in OmpK36 with or without changes in OmpK35. However, isolates displaying higher diameters did not have alterations in the former porin. Accordingly, the sensitivity and specificity for detection of OmpK36 functional alterations was of 100% when applying this threshold. In contrast, when using an imipenem and cefoxitin disk diffusion assay, the strains with alterations in OmpK36 could not be clearly separated from those with the wild type porin (i.e., 100% sensitivity and specificity were not obtained at any threshold). Nevertheless, as mentioned above, a reduction on the inhibition zone to cefoxitin, together with an inhibition zone diameter to meropenem and ertapenem lower or equal to 16, strongly suggest alterations in this porin.

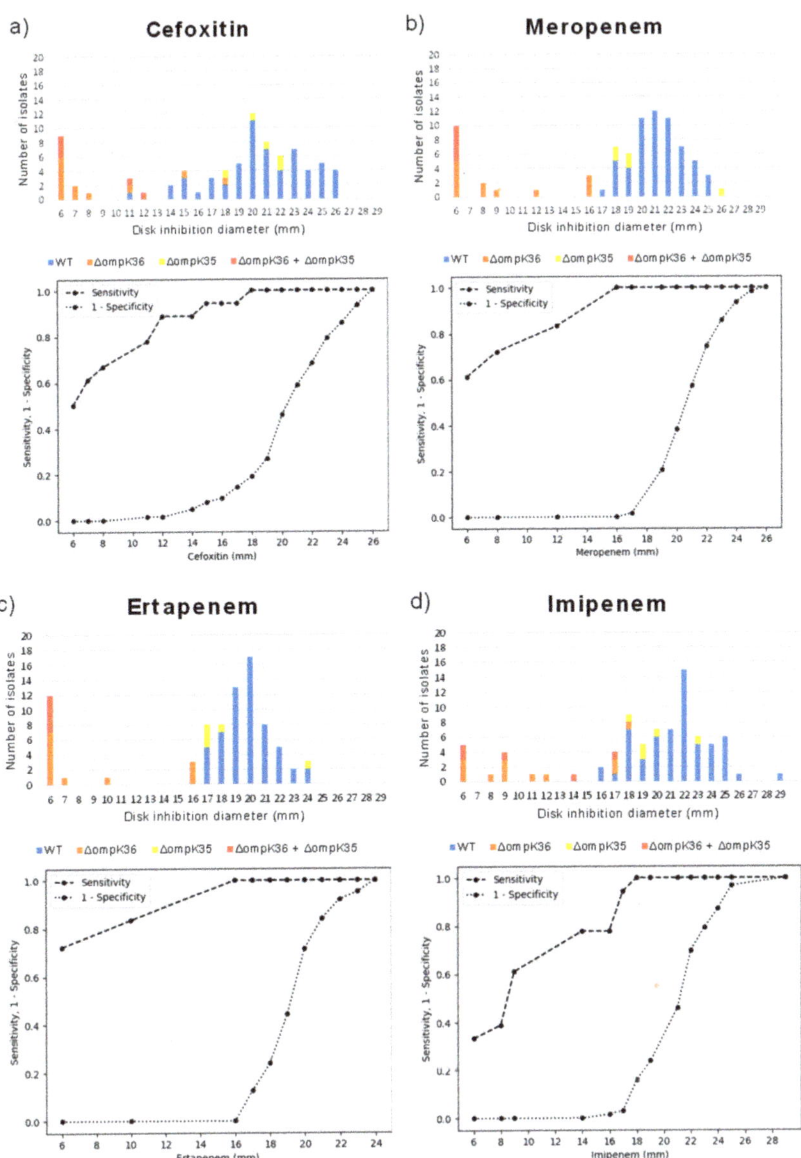

Figure 2. Distribution of cefoxitin 30 µg (**a**), meropenem 10 µg (**b**), ertapenem 10 µg (**c**) and imipenem 10 µg (**d**) inhibition zones (mm) of OXA-48 producing *Klebsiella pneumoniae* with and without porin alterations. Below of each graphic, a receiver operating curve (ROC), showing the evolution of sensitivity and 1-specificity to identify OmpK36 functional alterations in relation to the different thresholds applied for each antibiotic, is represented.

3. Materials and Methods

All carbapenem-resistant *K. pneumoniae* isolates recovered from blood cultures of different patients admitted to a tertiary hospital, Hospital Universitario Central de Asturias (HUCA), in northern Spain over a five-year period (2014–2019) were collected (*n* = 76). Additionally, five *K. pneumoniae* isolates with high meropenem MICs, recovered from different samples of patients admitted to the HUCA during the same period were studied. Bacterial identification was performed by MALDI-TOF/MS (Bruker Daltonics, Billerica, MA) using α-Cyano-4-hydroxycinnamic acid as a matrix and following the manufacturer instructions available on https://www.bruker.com/en/services/training, (accessed on 14 December 2020). *In silico* identification was performed by using KmerFinder, available at the Center for Genomic Epidemiology site (https://www.genomicepidemiology.org/; CGE, 2020; accessed on 25 January 2021). Antimicrobial susceptibility was determined by the Microscan system (Beckman Coulter, Brea, CA, USA) and, also, by Etest® (bioMérieux, Marcy l'Etoile, France) for meropenem and ceftazidime/avibactam. Results were interpreted according to the EUCAST guidelines [26]. Screening of carbapenemases was performed by means of a previously described algorithm [27].

Genomic DNA from the 81 isolates (76 recovered from blood cultures and five from other samples) was extracted with the NZY Microbial gDNA Isolation kit (NZYTech, Lisbon, Portugal), and then sequenced by Illumina technology to generate 125 bp paired-end reads in a HiSeq 1500. Quality control of the reads was performed using FastQC software (Babraham Bioinformatics, Cambridgeshire, UK) and de novo assembly was carried out with VelvetOptimizer [28].

Multi-locus sequence types and the presence of resistance genes were determined *in silico* by the use of the MLST 2.0 and ResFinder 3.2 tools, respectively [29]. The sequences of the two major porin-encoding genes of *Klebsiella* (*ompK35* and *ompK36*), including their upstream regions, were analyzed by bioinformatic tools, such as Clone Manager Professional v9.2 (CloneSuit9), Clustal Omega and Jalview [30,31]. PCR amplification of *ompK36* followed by Sanger sequencing was performed using previously described primers [15], when required for confirmation of whole genome sequencing results. The tridimensional structures of OmpK36 variants were assessed on the Phyre2 web portal for protein modelling, prediction and analysis, except for sequences of less than 30 amino acid residues, which are excluded by Phyre2 specifications.

Receiver operating characteristic (ROC) curves were built to visualize the performance ability of a disk diffusion assay using ertapenem (10 μg), meropenem (10 μg), imipenem (10 μg) and cefoxitin (30 μg) (Bio-Rad, Hercules, CA, USA), to detect OmpK36 functional alterations. In order to identify the optimal cut-off values to discriminate strains carrying functional alterations in this porin, a figure showing the evolution of sensitivity and 1-specificity in relation to the threshold changes was plotted for each of these antibiotics. This analysis was performed using Python programming language (Python Software Foundation. Python Language Reference, version 3.7., available at http://www.python.org, accessed on 8 March 2021, scikit-learn library) and visualized using matplotlib [32].

Genomes of all isolates studied in this work have been deposited in GenBank and their accession numbers are shown in Supplementary Table S1.

The present study was approved by the ethics committee of the Principality of Asturias.

4. Conclusions

In summary, we found that OmpK36 alterations could play an important role on the resistance to carbapenems of OXA-48-producing *K. pneumoniae*, consistent with the fact that this porin constitutes the main pathway of entrance of these drugs into the cell. Although our study only includes isolates recovered from a single hospital, they represent a great variety of *K. pneumoniae* clones, and different mutations in *ompK36* were found. As far as we know, our study represents the most extensive and complete molecular analysis of OXA-48-producing *K. pneumoniae* with alterations in the outer membrane porins. We have also demonstrated that these alterations do not affect the susceptibility to ceftazidime/avibactam, and thus this drug combination could be positioned as the choice therapy to combat the infections caused by this difficult to treat isolates. Finally, phenotypic detection of isolates carrying functional alterations on OmpK36 in clinical microbiology laboratories is important, and cefoxitin together with ertapenem/meropenem low inhibition zone diameters would strongly suggest alterations in this porin.

Supplementary Materials: The following are available online at https://www.mdpi.com/article/10.3390/antibiotics10101174/s1, Figure S1: Tridimensional analysis, assessed by the Phyre2 web, of OmpK36 protein structures of OXA-48-producing *Klebsiella pneumoniae* isolates displaying high-level resistance to meropenem. This web was not able to create structures for Kp_HUCA_Bac_5, Kp_HUCA_Bac_24, Kp_HUCA_Bac_65, Kp_HUCA_Bac_85, Kp_HUCA_4 and Kp_HUCA_8 since the proteins predicted had less than 30 amino acids., Table S1: GenBank accession numbers of the genomes of OXA-48-producing Klebsiella pneumoniae isolates, Table S2: Microbiological features of OXA-48-producing Klebsiella pneumoniae isolates displaying meropenem MICs lower than 8 mg/L.

Author Contributions: Conceptualization, J.F. and M.R.R.; methodology, J.F., P.L.-I., P.V. and T.S.-Z.; software, P.V.; resources, J.F.; writing—original draft preparation, P.L.-I. and J.F.; writing—review and editing, J.F. and M.R.R.; funding acquisition, J.F. All authors have read and agreed to the published version of the manuscript.

Funding: This research was funded by project FIS PI17-00728 (Fondo de Investigación Sanitaria, Instituto de Salud Carlos III, Ministerio de Economía y Competitividad, Spain), co-funded by the European Regional Development Fund of the European Union: A Way to Making Europe; and by Pfizer as an Independent Scientific Research Grant.

Institutional Review Board Statement: The study was conducted according to the guidelines of the Declaration of Helsinki, and approved by the Ethics Committee of the Principality of Asturias.

Informed Consent Statement: Not applicable.

Data Availability Statement: Data is contained within the article and/or supplemental materials. Genomes have been deposited in GenBank. Additional data are freely available on request from the corresponding author.

Conflicts of Interest: The authors declare no conflict of interest. The funders had no role in the design of the study; in the collection, analyses, or interpretation of data; in the writing of the manuscript, or in the decision to publish the results.

References

1. Vardakas, K.; Tansarli, G.; Rafailidis, P.; Falagas, M. Carbapenems versus alternative antibiotics for the treatment of bacteraemia due to *Enterobacteriaceae* producing extended-spectrum β-lactamases: A systematic review and meta-analysis. *J. Antimicrob. Chemother.* **2012**, *67*, 2793–2803. [CrossRef] [PubMed]
2. Tamma, P.; Han, J.; Rock, C.; Harris, A.; Lautenbach, E.; Hsu, A.; Avdic, E.; Cosgrove, S. Carbapenem therapy is associated with improved survival compared with piperacillin-tazobactam for patients with extended-spectrum β-lactamase bacteremia. *Clin. Infect. Dis.* **2015**, *60*, 1319–13125. [CrossRef] [PubMed]
3. Haidar, G.; Clancy, C.; Chen, L.; Samanta, P.; Shields, R.; Kreiswirth, B.; Nguyen, M. Identifying spectra of activity and therapeutic niches for ceftazidime-avibactam and imipenem-relebactam against carbapenem-resistant *Enterobacteriaceae*. *Antimicrob. Agents Chemother.* **2017**, *61*, e00642–17. [CrossRef]
4. Hamzaoui, Z.; Ocampo-Sosa, A.; Fernandez Martinez, M.; Landolsi, S.; Ferjani, S.; Maamar, E.; Saidani, M.; Slim, A.; Martinez-Martinez, L.; Boutiba-Ben Boubaker, I. Role of association of OmpK35 and OmpK36 alteration and blaESBL and/or blaAmpC genes in conferring carbapenem resistance among non-carbapenemase producing *Klebsiella pneumoniae*. *Int. J. Antimicrob. Agents* **2018**, *52*, 898–905. [CrossRef]
5. Van Duin, D.; Doi, Y. The global epidemiology of carbapenemase-producing *Enterobacteriaceae*. *Virulence* **2016**, *8*, 460–469. [CrossRef]
6. Papp-Wallace, K.; Endimiani, A.; Taracila, M.; Bonomo, R. Carbapenems: Past, Present, and Future. *Antimicrob. Agents Chemother.* **2011**, *55*, 4943–4960. [CrossRef]
7. Rodríguez-Baño, J.; Gutiérrez-Gutiérrez, B.; Machuca, I.; Pascual, A. Treatment of infections caused by extended-spectrum-beta-lactamase-, AmpC-, and carbapenemase-producing *Enterobacteriaceae*. *Clin. Microb. Rev.* **2018**, *31*, e00079–17. [CrossRef] [PubMed]
8. Gutiérrez-Gutiérrez, B.; Salamanca, E.; de Cueto, M.; Hsueh, P.; Viale, P.; Paño-Pardo, J.; Venditti, M.; Tumbarello, M.; Daikos, G.; Cantón, R.; et al. Effect of appropriate combination therapy on mortality of patients with bloodstream infections due to carbapenemase-producing *Enterobacteriaceae* (INCREMENT): A retrospective cohort study. *Lancet Infect. Dis.* **2017**, *17*, 726–734. [CrossRef]
9. Poirel, L.; Potron, A.; Nordmann, P. OXA-48-like carbapenemases: The phantom menace. *J. Antimicrob. Chemother.* **2012**, *67*, 1597–1606. [CrossRef]
10. Pitout, J.; Peirano, G.; Kock, M.; Strydom, K.; Matsumura, Y. The Global Ascendency of OXA-48-Type Carbapenemases. *Clin. Microb. Rev.* **2019**, *33*, 00102–19. [CrossRef] [PubMed]

11. Zhanel, G.; Lawson, C.; Adam, H.; Schweizer, F.; Zelenitsky, S.; Lagacé-Wiens, P.; Denisuik, A.; Rubinstein, E.; Gin, A.; Hoban, D.; et al. Ceftazidime-Avibactam: A Novel Cephalosporin/β-lactamase Inhibitor Combination. *Drugs* **2013**, *73*, 159–177. [CrossRef]

12. García-Fernaández, A.; Miriagou, V.; Papagiannitsis, C.; Giordano, A.; Venditti, M.; Mancini, C.; Carattoli, A. An ertapenem-resistant extended-spectrum-β-lactamase-producing *Klebsiella pneumoniae* clone carries a novel OmpK36 porin variant. *Antimicrob. Agents Chemother.* **2010**, *54*, 4178–4184. [CrossRef]

13. Netikul, T.; Kiratisin, P. Genetic characterization of carbapenem-resistant *Enterobacteriaceae* and the spread of carbapenem-resistant *Klebsiella pneumonia* ST340 at a university hospital in Thailand. *PLoS ONE* **2015**, *10*, e0139116. [CrossRef]

14. Tsai, Y.; Fung, C.; Lin, J.; Chen, J.; Chang, F.; Chen, T.; Siu, L. Klebsiella pneumoniae outer membrane porins OmpK35 and OmpK36 play roles in both antimicrobial resistance and virulence. *Antimicrob. Agents Chemother.* **2011**, *55*, 1485–1493. [CrossRef] [PubMed]

15. Clancy, C.; Chen, L.; Hong, J.; Cheng, S.; Hao, B.; Shields, R.; Farrell, A.; Doi, Y.; Zhao, Y.; Perlin, D.; et al. Mutations of the *ompK36* porin gene and promoter impact responses of sequence type 258, KPC-2-producing *Klebsiella pneumoniae* strains to doripenem and doripenem-colistin. *Antimicrob. Agents Chemother.* **2013**, *57*, 5258–5265. [CrossRef] [PubMed]

16. Zhang, Y.; Jiang, X.; Wang, Y.; Li, G.; Tian, Y.; Liu, H.; Ai, F.; Ma, Y.; Wang, B.; Ruan, F.; et al. Contribution of β-lactamases and porin proteins OmpK35 and OmpK36 to carbapenem resistance in clinical isolates of KPC-2-producing *Klebsiella pneumoniae*. *Antimicrob. Agents Chemother.* **2013**, *58*, 1214–1217. [CrossRef] [PubMed]

17. Wong, J.; Romano, F.; Kelly, L.; Kwong, H.; Low, W.; Brett, S.; Clements, A.; Beis, K.; Frankel, G. OmpK36-mediated Carbapenem resistance attenuates ST258 *Klebsiella pneumoniae* in vivo. *Nat. Com.* **2019**, *10*, 3957. [CrossRef]

18. Lunha, K.; Chanawong, A.; Lulitanond, A.; Wilailuckana, C.; Charoensri, N.; Wonglakorn, L.; Saenjamla, P.; Chaimanee, P.; Angkititrakul, S.; Chetchotisakd, P. High-level carbapenem-resistant OXA-48-producing *Klebsiella pneumoniae* with a novel OmpK36 variant and low-level, carbapenem-resistant, non-porin-deficient, OXA-181-producing *Escherichia coli* from Thailand. *Diag. Microb. Infect. Dis.* **2016**, *85*, 221–226. [CrossRef]

19. Navon-Venezia, S.; Kondratyeva, K.; Carattoli, A. *Klebsiella pneumoniae*: A major worldwide source and shuttle for antibiotic resistance. *FEMS Microb. Rev.* **2017**, *41*, 252–275. [CrossRef]

20. Peirano, G.; Chen, L.; Kreiswirth, B.; Pitout, J. Emerging antimicrobial-resistant high-risk *Klebsiella pneumoniae* Clones ST307 and ST147. *Antimicrob. Agents Chemother.* **2020**, *64*, e01148–20. [CrossRef]

21. Cornaglia, G.; Mazzariol, A.; Fontana, R.; Satta, G. Diffusion of carbapenems through the outer membrane of *Enterobacteriaceae* and correlation of their activities with their periplasmic concentrations. *Microb. Drug. Res.* **1996**, *2*, 273–276. [CrossRef]

22. Sugawara, E.; Kojima, S.; Nikaido, H. *Klebsiella pneumoniae* major porins OmpK35 and OmpK36 allow more efficient diffusion of β-lactams than their Escherichia coli homologs OmpF and OmpC. *J. Bacter.* **2016**, *198*, 3200–3208. [CrossRef] [PubMed]

23. Nikaido, H. Molecular basis of bacterial outer membrane permeability revisited. *Microb. Mol. Biol. Rev.* **2003**, *67*, 593–656. [CrossRef] [PubMed]

24. Pagès, J.; Peslier, S.; Keating, T.; Lavigne, J.; Nichols, W.; Nichols, W. Role of the outer membrane and porins in susceptibility of β-lactamase-producing *Enterobacteriaceae* to ceftazidime-avibactam. *Antimicrob. Agents Chemother.* **2015**, *60*, 1349–1359. [CrossRef]

25. Martínez-Martínez, L.; Hernández-Allés, S.; Albertí, S.; Tomás, S.; Benedi, V.; Jacoby, G. In vivo selection of porin-deficient mutants of *Klebsiella pneumoniae* with increased resistance to cefoxitin and expanded-spectrum-cephalosporins. *Antimicrob. Agents Chemother.* **1996**, *40*, 342–348. [CrossRef]

26. European Committee on Antimicrobial Susceptibility Testing (EUCAST). Breakpoint Tables for Interpretation of Mics and Zone Diameters. Version 11.0. 2021. Available online: https://www.eucast.org/clinical_breakpoints/ (accessed on 2 March 2021).

27. Rodríguez-Lucas, C.; Rodicio, M.; Rosete, Y.; Fernández, J. Prospective evaluation of an easy and reliable work flow for the screening of OXA-48-producing *Klebsiella pneumoniae* in endemic settings. *J. Hosp. Infect.* **2020**, *105*, 659–662. [CrossRef]

28. Zerbino, D.; Birney, E. Velvet: Algorithms for *de novo* short read assembly using de Bruijn graphs. *Genome Res.* **2008**, *18*, 821–829. [CrossRef]

29. CGE. Center for Genomic Epidemiology. 2020. Available online: http://www.genomicepidemiology.org (accessed on 25 January 2021).

30. Madeira, F.; Park, Y.; Lee, J.; Buso, N.; Gur, T.; Madhusoodanan, N.; Basutkar, P.; Tivey, A.; Potter, S.; Finn, R.; et al. The EMBL-EBI search and sequence analysis tools APIs in 2019. *Nucl. Acids Res* **2019**, *47*, W636–W641. [CrossRef]

31. Waterhouse, A.; Procter, J.; Martin, D.; Clamp, M.; Barton, G. Jalview Version 2—A multiple sequence alignment editor and analysis workbench. *Bioinformatics* **2009**, *25*, 1189–1191. [CrossRef]

32. Hunter, J. Matplotlib: A 2D graphics environment. *Comput. Sci. Eng.* **2017**, *9*, 90–95. [CrossRef]

antibiotics

MDPI

Article

Whole-Genome Sequencing for Molecular Characterization of Carbapenem-Resistant Enterobacteriaceae Causing Lower Urinary Tract Infection among Pediatric Patients

Hassan Al Mana [1], Sathyavathi Sundararaju [2], Clement K. M. Tsui [2,3,4], Andres Perez-Lopez [2,3], Hadi Yassine [1,5], Asmaa Al Thani [1], Khalid Al-Ansari [3,6] and Nahla O. Eltai [1,*]

[1] Biomedical Research Center, Qatar University, Doha 2713, Qatar; h.almana@qu.edu.qa (H.A.M.); hyassine@qu.edu.qa (H.Y.); aaja@qu.edu.qa (A.A.T.)

[2] Division of Microbiology, Department of Pathology Sidra Medicine, Doha 2713, Qatar; ssundararaju@sidra.org (S.S.); ktsui@sidra.org (C.K.M.T.); aperezlopez@sidra.org (A.P.-L.)

[3] Department of Pathology and Laboratory Medicine, Weill Cornell Medical College in Qatar, Doha 2713, Qatar; kalansari@sidra.org

[4] Division of Infectious Diseases, Faculty of Medicine, University of British Columbia, Vancouver, BC V6T 1Z4, Canada

[5] Department of Biomedical Science, College of Health Sciences, QU Health, Qatar University, Doha 2713, Qatar

[6] Department of Emergency, Sidra Medicine, Doha 2713, Qatar

* Correspondence: nahla.eltai@qu.edu.qa

Citation: Al Mana, H.; Sundararaju, S.; Tsui, C.K.M.; Perez-Lopez, A.; Yassine, H.; Al Thani, A.; Al-Ansari, K.; Eltai, N.O. Whole-Genome Sequencing for Molecular Characterization of Carbapenem-Resistant Enterobacteriaceae Causing Lower Urinary Tract Infection among Pediatric Patients. *Antibiotics* **2021**, *10*, 972. https://doi.org/10.3390/antibiotics10080972

Academic Editors: Francesca Andreoni and Gabriella Orlando

Received: 20 June 2021
Accepted: 5 August 2021
Published: 12 August 2021

Publisher's Note: MDPI stays neutral with regard to jurisdictional claims in published maps and institutional affiliations.

Abstract: Antibiotic resistance is a growing public health problem globally, incurring health and cost burdens. The occurrence of antibiotic-resistant bacterial infections has increased significantly over the years. Gram-negative bacteria display the broadest resistance range, with bacterial species expressing extended-spectrum β-lactamases (ESBLs), AmpC, and carbapenemases. All carbapenem-resistant *Enterobacteriaceae* (CRE) isolates from pediatric urinary tract infections (UTIs) between October 2015 and November 2019 ($n = 30$). All isolates underwent antimicrobial resistance phenotypic testing using the Phoenix NMIC/ID-5 panel, and carbapenemase production was confirmed using the NG-Test CARBA 5 assay. Whole-genome sequencing was performed on the CREs. The sequence type was identified using the Achtman multi-locus sequence typing scheme, and antimicrobial resistance markers were identified using ResFinder and the CARD database. The most common pathogens causing CRE UTIs were *E. coli* (63.3%) and *K. pneumoniae* (30%). The most common carbapenemases produced were OXA-48-like enzymes (46.6%) and NDM enzymes (40%). Additionally, one *E. coli* harbored IMP-26, and two *K. pneumoniae* possessed mutations in *ompK37* and/or *ompK36*. Lastly, one *E. coli* had a mutation in the *marA* porin and efflux pump regulator. The findings highlight the difference in CRE epidemiology in the pediatric population compared to Qatar's adult population, where NDM carbapenemases are more common.

Keywords: carbapenem-resistance; *Enterobacteriaceae*; Qatar; CRE; OXA-48

1. Introduction

Antibiotic resistance is a growing public health problem globally, incurring health and cost burdens. The occurrence of antibiotic-resistant bacterial infections has increased significantly over the years. In 2013, the Center for Disease Control and Prevention (CDC) issued an antibiotic resistance threats report estimating approximately two million infections annually in the United States [1]. By 2017, the number increased to approximately 2.8 million, and deaths increased from 23,000 to 35,900 [2]. Beta-lactams are the most used antibiotics worldwide and include the penicillins, cephalosporins, monobactams, and carbapenems; they all share a typical beta-lactam ring. Gram-negative bacteria display the broadest range of resistance, with bacterial species expressing extended-spectrum β-lactamases (ESBLs), AmpC, and carbapenemases [3]. Of these, carbapenem-resistant

Enterobacteriaceae (CRE) are of the most concern. Both the CDC and the World Health Organization (WHO) assign CREs the highest urgency levels [2,4].

Carbapenem antibiotics are the choice for treating infections caused by ESBL or AmpC-producing bacteria [5]. Carbapenem resistance is caused mainly by carbapenemase enzymes. Carbapenemase production can be intrinsic, such as Metallo-β-lactamases (MBLs) expressed by *Stenotrophomonas maltophilia*, *Bacillus cereus*, and *Aeromonas* species [6]. In *Enterobacteriaceae*, the most common carbapenemases are categorized into three groups, Ambler classes A, B, and D. The most common group is class A, which includes the *Klebsiella pneumoniae* (*K. pneumoniae*) carbapenemases (KPC) and Imipenem-hydrolyzing β-lactamase (IMI), with KPC being the most prevalent overall [3,7]. Class B contains the Metallo-β-lactamases (MBL), New Delhi Metallo-lactamases (NDM), Imipenem-resistant *Pseudomonas* enzyme (IMP), and the Verona integron-mediated Metallo-lactamase (VIM). Finally, class D contains Oxacillin-hydrolyzing carbapenemases (OXA), of which OXA-48 is the most commonly isolated [3,6]. Carbapenemase genes can be either intrinsic or acquired [8,9]. These carbapenemases are typically plasmid-mediated. In addition to carbapenemases, some members of the *Enterobacteriaceae* family may possess intrinsic carbapenem resistance, arising from mutations in porins or efflux pumps [6].

Several factors increase the risk of CRE infections, including immune suppression, advanced age, intensive care unit (ICU) admission, and previous exposure to antimicrobials [10,11]. *Escherichia coli* (*E. coli*) and *K. pneumoniae* are the most common human pathogens, causing various infections [12,13]. Urinary tract infections (UTIs) are among the most commonly diagnosed infections in children [14]. Empirical treatment for UTIs initially used amoxicillin; however, the increase in resistance of the pathogens, namely *E. coli*, prompted the change to other antibiotics [14]. Recently, with the increased rate of ESBL UTIs, the prescription of carbapenems has increased. However, reports from various regions indicate an increase in the rate of carbapenem resistance [15].

Furthermore, a recently published systematic review, taking into account human, animal, and environmental samples, reported the possibility of the Middle East being an endemic region for CREs [16]. Nevertheless, reports on the prevalence of CREs among the pediatric population, while increasing, are still sporadic [17]. There are limited data describing carbapenem-resistant *Enterobacteriaceae* (CRE) from the Arabian Gulf region among the pediatric population. In Qatar, only one study conducted on different clinical samples from different age groups, presented at the infectious disease forum in 2019, reported that NDM and OXA-48 are the predominant carbapenemases in Qatar and are associated with high overall mortality [18]. The epidemiology of CREs in the adult population is well characterized. As for the pediatric population, while the number of reports is increasing, they remain sporadic, and there is a shortage of data on epidemiology [17]. To that end, we aimed to investigate the genotypic profile of CREs among the pediatric population with UTIs in Qatar.

2. Results

2.1. Demographics and Etiology

The study of the population's demographics is summarized in Table 1. The male to female ratio is approximately 1:2.63, with males constituting 27.6% (*n* = 8) and females constituting 72.4% (*n* = 21). CRE infections were more prevalent in non-Qataris (86.2%) compared to Qataris (13.8%). Additionally, 93% of the isolated CREs come from patients aged from 2 months to 13 years. The majority of the CRE UTIs were caused by *E. coli* (*n* = 19, 63.3%), followed by *K. pneumoniae* (*n* = 9, 30%), and the remaining were caused by *Enterobacter hormaechei* (*E. hormaechei*) (*n* = 2, 6.7%). One patient had a mixed infection by *E. coli* and *K. pneumoniae*.

Table 1. Demographic profile of the study population (*n = 29*).

Gender	Total Number (%)	Nationality	
		Qatari	Non-Qatari
Male	8 (27.6%)	0 (0%)	8 (27.6%)
Female	21 (72.4%)	4 (13.8%)	17 (58.6%)
Total	29 (100%)	4 (13.8%)	25 (86.2%)
Age			
<2	5 (17.2%)	0 (0%)	5 (17.2%)
2–5	13 (44.8%)	1 (3.4%)	12 (41.4%)
6–15	11 (37.9%)	3 (10.3%)	8 (27.6%)
Total	29 (100%)	4 (13.8%)	25 (86.2%)

2.2. Phenotypic Resistance Profiles of the CRE Isolates

The antibiotic resistance profile of all 30 CRE-producing *Enterobacteriaceae* is depicted in Figure 1. All CRE-producing isolates showed 100% resistance to ampicillin, amoxicillin/clavulanate, cefazolin, cephalothin, and ciprofloxacin. High resistance, more than 80%, was detected against cephalosporines, including cefuroxime, ceftriaxone, ceftazidime, and cefepime. High resistance was also noticed against the β-lactam/β-lactamase inhibitor combinations; 87.5% were resistant to piperacillin/tazobactam, whereas 100% were resistant to amoxicillin/clavulanate. All isolates demonstrated resistance to at least one carbapenem, with the highest resistance against ertapenem (96.9%). All isolates were susceptible to colistin and tigecycline. The resistance to other antibiotics, namely, nitrofurantoin, amikacin, cefoxitin, gentamicin, levofloxacin, trimethoprim/sulfamethoxazole, was 31.3%, 31.3%, 32.5%, 40.6%, 62.5%, and 56.3%, respectively.

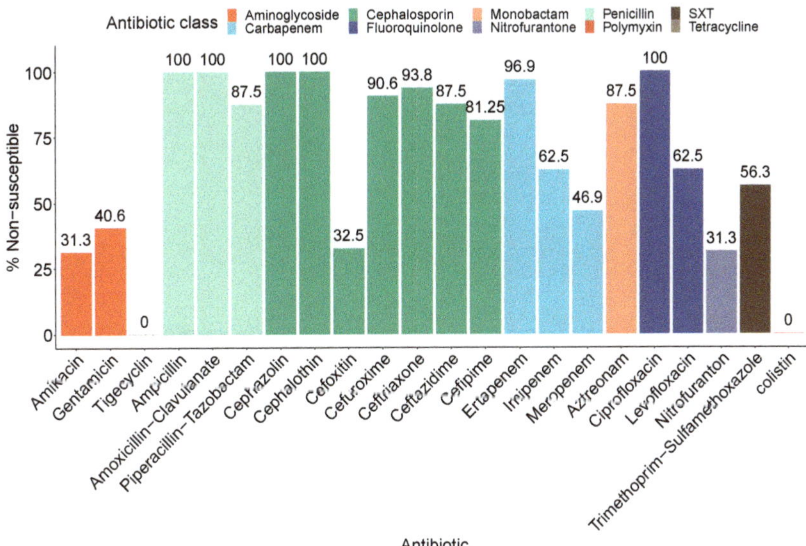

Figure 1. Phenotypic profile of the carbapenem-resistant *Enterobacteriaceae* isolated from children (0–15 years old) with urinary tract infections. The isolates were tested for antibiotic resistance against 22 clinically relevant antibiotics using the Phoenix NMIC/ID-5 panel (BD Biosciences, Heidelberg, Germany). The figure depicts the percentage of isolates that are non-susceptible to each antibiotic. The bars are colored according to the antibiotic class.

2.3. Molecular Genotyping Profile of CRE Isolates

The genome assemblies included in the study are available at the NCBI website (https://www.ncbi.nlm.nih.gov/bioproject/, accessed on 24 January 2021) under BioProject: PRJNA690895. The accessions for the isolates are detailed in Table S2. The clonal diversity among the isolates was determined; the *E. coli* isolates had 11 different strain types (STs) (ST11021, ST38, ST162, ST448, ST131, ST2083, ST95, ST227, ST410, ST2346, ST10); the *K. pneumoniae* isolates had nine different strain types (ST196, ST45, ST987, ST218, ST101, ST147, ST35, ST870, ST3712); the *E. hormaechei* had the ST 269 and 171 (Table 2).

Table 2. Antibiotic resistance genes present in the isolates.

Isolate ID	Collection Date (Month-Year)	Sequence Type (ST)	Carbapenem Resistance Genes (Mlplasmid Posterior Probability for Carbapenemases)
		Escherechia coli	
EC-QU-7	June-2015	162	bla_{NDM-1} *(0.126)*
EC-QU-5	November-2015	38	bla_{OXA-48} *(0.971)*
EC-QU-14	November-2015	448	bla_{NDM-5} *(0.63)*
EC-QU-1	April-2016	11,021	bla_{NDM-4} *(0.937)*
EC-QU-10	December-2016	11,021	*marA* mutation
EC-QU-16	January-2017	131	bla_{IMP-26} *(0.126)*
EC-QU-6	March-2017	11,021	bla_{NDM-4} *(0.764)*
EC-QU-19	September-2018	2083	bla_{NDM-5} *(0.65)*
EC-QU-21	October-2018	162	bla_{NDM-5} *(0.79)*
EC-QU-23	January-2019	38	$bla_{OXA-244}$ *(0.852)*
EC-QU-25	January-2019	38	$bla_{OXA-244}$ *(0.972)*
EC-QU-26	May-2019	38	$bla_{OXA-244}$ *(0.80)*
EC-QU-27	July-2019	95	$bla_{OXA-181}$ *(0.84)*
EC-QU-31	August-2019	410	$bla_{OXA-484}$ *(0.78)*
EC-QU-30	September-2019	227	bla_{OXA-48} *(0.94)*
EC-QU-29	September-2019	38	$bla_{OXA-244}$ *(0.991)*
EC-QU-28	September-2018	131	$bla_{OXA-244}$ *(0.94)*
EC-QU-33	September-2019	2346	$bla_{OXA-244}$ *(0.983)*
EC-QU-35	October-2019	10	$bla_{OXA-244}$ *(0.731)*
		Klebsiella pneumonaie	
KPN-QU-9	October-2015	45	*ompK37* mutation
KPN-QU-15	January-2017	218	*ompK37 and ompK36* mutations
KPN-QU-17	March-2017	101	bla_{NDM-1} *(0.884)*
KPN-QU-3	March-2017	196	bla_{NDM-1} *(0.992)*
KPN-QU-11	April-2017	987	*ompK37* mutation bla_{OXA-48} *(0.947)*
KPN-QU-20	April-2018	147	bla_{NDM-5} *(0.942)*
KPN-QU-22	January-2019	35	*ompK37, ompK36* mutations $bla_{OXA-181}$ *(0.913)*
KPN-QU-37	August-2019	3712	*ompK37* mutation *bla* $_{OXA-181}$ *(0.97)*
KPN-QU-36	October-2019	870	*ompK37 and ompK36* mutations bla_{NDM-1} *(0.982)*
		Enterobacter hormaechi	
EBH-QU-2	October-2016	269	bla_{NDM-7} [a]
		Enterobacter cloacae	
EBH-QU-4	December-2016	171	bla_{NDM-1} [a]

[a] Posterior probability was not determined for these isolates.

The antimicrobial markers carried in the isolates are summarized in Table 2 and Table S1. Seven *E. coli* isolates harbored $bla_{OXA-244}$, three *E. coli* isolates carried bla_{NDM-5}, two harbored bla_{OXA-48}, two fostered bla_{NDM-4} while bla_{NDM-1}, bla_{IMP-26}, $bla_{OXA-181}$, $bla_{OXA-484}$, and *marA* mutations were each housed by one *E. coli*. Carbapenem resistance in four *K. pneumoniae* isolates was ascribed to the combination of $bla_{OXA-48/OXA-181/NDM-1}$ β-Lactamase production, and porin *ompK36/37* insertional inactivation. Two *K. pneumoniae* harbored bla_{NDM-1}, one isolate carried bla_{NDM-5}. Carbapenem resistance in one *K. pneumoniae* was assigned to the *ompK37* mutation and in another isolate is due to the *ompK36* and *ompK37* mutations. *E. hormaechei* harbored *bla*$_{NDM-1}$ and *bla*$_{NDM-7}$ (Table 2).

The locations of the carbapenemases were determined for *E.coli* and *K. pneumonia* using a support vector machine (SVM) algorithm to determine whether the contig of the gene lies on the chromosome or a plasmid. The posterior probabilities of the carbapenemase lying on the plasmid are presented in Table 2. Using a probability cutoff of 0.7 (minimum validated for the algorithm) 21 out of 27 (77%) isolates harbored plasmid type. The algorithm gave no probability for the location of the gene in the two *Enterobacter* isolates as the algorithm was trained on *Enterococcus species*. Only 15.6% (5/32) have other genetic mutations that have been previously associated with carbapenem resistance (Table 2).

3. Discussion

CRE UTIs in the pediatric population in Qatar are caused by a diverse set of *E. coli* and *K. pneumoniae* sequence types (STs). The Pediatric Emergency Center sees patients from across the country. The isolates are representative of the general pediatric population. However, no spatial association could be drawn as most patients were not hospitalized, and no information was available on their geographic origin. Moreover, for most isolates, there was no temporal relationship between members of the same species and/or ST except for two isolates (EC-QU-23 and EC-QU-25, Table 2).

CRE infections are associated with mortality rates as high as 65% in the adult population; the rates vary between reports in the pediatric population, with higher rates among neonates [17]. In this study, two patients died within one year of developing CRE UTI (one in 3 months and the other in 11 months). The leading cause of death is attributed to their complicated underlying co-morbidities, namely, congenital heart disease and pulmonary disorders for the first patient and Wiskott–Aldritch for the second.

Carbapenemases are typically carried on plasmids, which enable horizontal gene transfer and their dissemination between bacteria. However, there have been reports of carbapenemases integrated into the chromosome [19]. Multiple isolates showed a probability of the carbapenemase being harbored on the plasmid below the cutoff (0.7). Two isolates, EC-QU-7 and EC-QU-16, had very low probabilities, indicating that the genes are likely on the chromosome. The presence of a carbapenemase on the chromosome can increase dissemination further by allowing vertical transmission while also retaining the ability to transfer horizontally due to the presence of integrons [19].

The most common carbapenemase genes observed in this study are those that encode OXA-48-like carbapenemase lactamases. A previous report on CRE infections in Qatar in the general population (median age 57) and using various sample sources reported NDM-1 and NDM-7 as the most common carbapenemases [18]. However, investigating a sub-population, pediatric UTIs in the study case show a shift in the distribution of carbapenemases. The worldwide dissemination of carbapenemases, particularly in the Middle East, has been suggested due to the importation of CREs through international travel [20]. The contrast in the carbapenemase distribution in this study with the previous report may suggest varying importation sources. The higher prevalence of NDM carbapenemases in that report may be attributed to the higher proportion of migrant workers from the Indian subcontinent, where NDM carbapenemases are endemic [21,22]. However, none of the patients in the study had a recent travel history except for one patient that traveled to India two months before specimen collection (January 2017). The patient had a mixed infection by *E. coli* and *K. pneumoniae* carrying NDM-4 and NDM-1, respectively. Additional information on the carriage rates in the pediatric population and their family members, and their travel history, may provide more concrete data to build a conclusion on.

Nevertheless, there is an apparent relationship between the type of carbapenemase and the nationality of the patient. Of the 12 NDM producers, 66.67% were isolated from patients from the Indian subcontinent (including the patient with travel history to India) compared to 14.3% of OXA-48-like carbapenemase producers (Figure 2). The majority of OXA-48-

like enzyme producers were isolated from patients from the Arabian Peninsula, Levant, or North Africa, where OXA-48-like outbreaks or transmission have been reported [23,24].

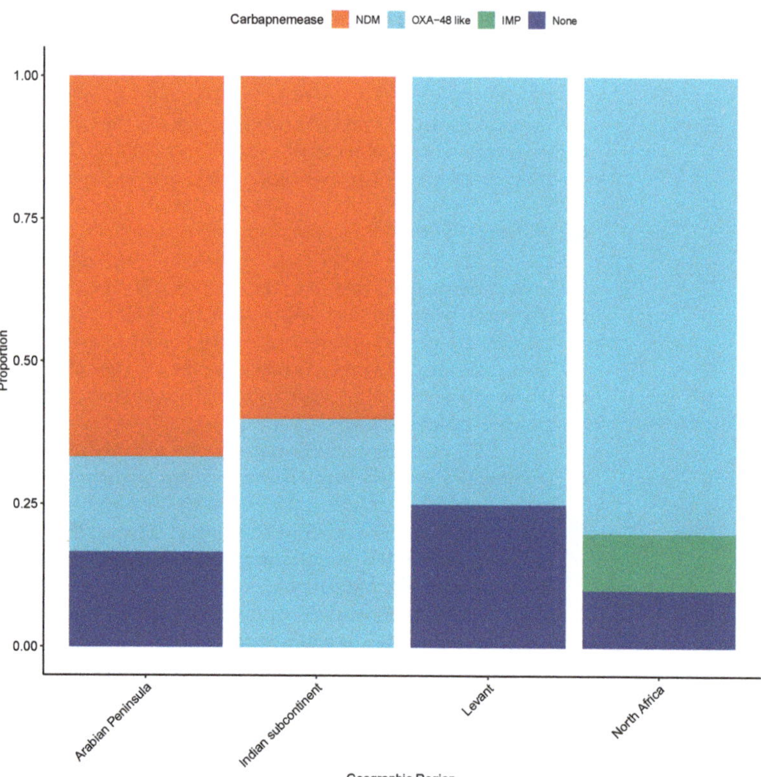

Figure 2. Distribution of carbapenemases by patient ethnicity. Patients from the Arabian Peninsula include the patients from Qatar and the United Arab Emirates. The Levant includes patients from Jordan, Syria, Lebanon, North Africa includes patients from Egypt and Sudan, and the Indian subcontinent includes patients from India and Pakistan.

The most common OXA-48-like carbapenemase observed is OXA-244 (50% of OXA48 like enzymes), followed by OXA-181 and OXA-48. OXA-244 is a single amino acid substitution variant of OXA-48 with a weaker carbapenemase activity that has been frequently reported across Europe [25–27]. The lower carbapenemase activity results in lower MICs and more challenging detection, contributing to its silent dissemination [28]. All the isolates carrying OXA-244, all of which were *E. coli*, were non-susceptible to ertapenem and susceptible to imipenem and meropenem, except for one isolate non-susceptible to all three. The *E. coli* isolate that is non-susceptible to all three carbapenems may possess other AMR markers not detected in this study. Notably, all the isolates co-carried a CTX-M type ESBL and/or an AmpC β-lactamase, leading to resistance to third generation cephalosporins. The isolates carrying either carbapenemase were not susceptible to ertapenem or imipenem. Two of the three OXA-181-harboring isolates were from patients whose origin is the Indian subcontinent, which is the main reservoir for this carbapenemase [23].

Three of the isolates did not carry any carbapenemases. One *K. pneumoniae* isolate had mutations in the outer membrane porin OmpK37, and another had mutations in OmpK37 and OmpK36. Modifications of OmpK36/37 contribute to resistance to third generation cephalosporins by reducing antibiotic influx [29–31]. However, porin mutations combined with ESBL and/or AmpC β-lactamase have been reported to lead to carbapenem non-susceptibility [32]. The two *K. pneumoniae* isolates carried both an ESBL and AmpC β-lactamase and were non-susceptible to ertapenem and meropenem. The third isolate without a carbapenemase was an *E. coli* with a mutation in the multiple antibiotic resistance A (*marA*) gene. *marA* encodes a transcriptional factor that regulates the expression of multi-drug efflux pumps and porins [33–35]. Overexpression of *marA* can result in reduced susceptibility to carbapenems [36]. In addition to *marA*, the isolate also carried several efflux pumps and the DHA-1 AmpC β-lactamase. The combination may potentially explain the observed carbapenem non-susceptibility in a similar mechanism to the porin mutation and ESBL/AmpC combination seen in *K. pneumoniae*.

4. Materials and Methods

4.1. Clinical Isolates and Control Strains

Institutional Biosafety Committee approval for this study was obtained from the Medical Research Centre (MRC), Hamad Medical Corporation (HMC), Doha, Qatar, protocol no. 16434/16. In total, 30 carbapenem-resistant bacterial isolates were collected between October 2015 and November 2019 from children presented to the Pediatric Emergency Center at Al-Saad, HMC, Qatar, with lower UTIs. All urine analyses were performed on patients who presented with symptoms, mainly fever and dysuria. The urinary catheter was applied for all patients less than or equal to 2 years of age, cerebral palsy (CP) patients, and patients under intermittent catheterization. Otherwise, urine was obtained from the mid-stream catch. Samples that did not yield significant bacterial growth, those with multiple organisms, and samples with suspected contamination as per lab report were excluded from the study, and no duplicate samples were collected. All of the reported cases had UTI as their primary diagnosis. In addition, for each patient, demographic data such as age, nationality, and gender were collected.

Enterobacteriaceae species were then isolated using readymade Cystine Lactose Electrolyte-Deficient media (IMES, Doha, Qatar) and identified by MALDI-TOF (Bruker Daltonik GmbH, Leipzig, Germany). Initial antimicrobial susceptibility testing was performed by Phoenix using the NMIC/ID-5 panel (BD Biosciences, Heidelberg, Germany) according to the manufacturer's recommendations. Briefly, panels were inoculated with 0.5 McFarland pure culture, placed into the instrument, and incubated at 35 °C. The instrument tests the panel every 20 min up to 16 h if necessary. MIC values of each antimicrobial agent are automatically read as susceptible, intermediate, or resistant (SIR). All intermediates were considered susceptible, and susceptibility testing was performed for twenty-two clinically relevant antibiotics. Both automated tests were performed at Hamad General Hospital Microbiology laboratory. Thirty-two out of 36 detected carbapenem-resistant isolates were subjected to further genotypic analysis (strain characteristics of these isolates are described in Table 1), and non-*Enterobacteriaceae* were excluded.

Standard strains, *E. coli* ATCC® 25922™ and *E. coli* ATCC® 35218™ were used as controls for antimicrobial drug susceptibility testing. *Enterobacter cloacae* complex.

NCTC® 13925™, *K. pneumoniae* NCTC® 13440™, *K. pneumoniae* NCTC® 13443™, *E. coli* NCTC® 13476 ™ IMP, and *K. pneumoniae* NCTC® 13442™ were used as positive controls for class A carbapenemases (IMI), class B Carbapenemases (Metallo-β-lactamases: VIM-1, NDM-1, IMP) and class D Carbapenemases (OXA-48), respectively. All intermediate-resistant isolates were considered susceptible. These clinical isolates were preserved at −80 °C for further analysis.

4.2. Carbapenemase Phenotypic Confirmation

The isolates resistant to meropenem, ertapenem, or imipenem were further tested with the NG-Test CARBA 5 assay (NG Biotech, Guipry, France), following the manufacture's protocol. The assay detects the presence of NDM, VIM, IMP, KPC, and OXA-48-like carbapenemases. In the event that an isolate was negative for the five carbapenemases, it was tested with both ertapenem and meropenem E-tests (Liofilchem, Roseto degli

Abruzzi TE, Italy) and interpreted following the Clinical Laboratory Standards Institute's (CLSI) guidelines.

4.3. Molecular Characterization

DNA extraction, whole-genome sequencing, and bioinformatic analysis.

Genomic DNA was extracted from all *E. coli*, *K. pneumoniae*, and *Enterobacter cloacae* (*E. cloacae*) isolates that were positive for one of the five carbapenemase genes. Extraction was performed using the QIAamp® UCP Pathogen mini kit (Qiagen, Düsseldorf, Germany) following the manufacturer's protocol. Briefly, genomic DNA was purified from each isolate and later quantified using a Qubit dsDNA high sensitivity assay (Thermo Fisher, Waltham, MA, USA). Whole-genome sequencing was performed on the Illumina MiSeq platform, using Nextera XT (Illumina, San Diego, CA, USA), for paired-end (PE) library construction; the DNA was tagmented, amplified with index primers, and purified with AMPure XP beads (Beckman Coulter, Brea, CA, USA). Finally, the DNA library was normalized, pooled, and sequenced using the MiSeq platform with 300bp PE reads (MiSeq Reagent Kit V3). The raw sequences were subjected to a quality check and analyzed using CLC genomics workbench v20.0.4 (https://digitalinsights.qiagen.com; accessed on 20 August 2020). Briefly, the reads were quality assessed, trimmed, and followed by de novo assembly. Resistance genes were identified using ResFinder v 4.1 [37] and CARD's comprehensive antibiotic resistance database [38]. Only resistance genes that showed a perfect match, with 100% identity and coverage for a given gene in the database, were reported in this study. Sequence type was identified by an exact match against the chosen locus scheme Achtman via pubMLST database for molecular typing (www.pubMLST.org accessed on 20 August 2020). Identification of mobile genetic elements, and their relation to antibiotic resistance were analyzed through Mobile Element Finder [39], and the location of the gene was determined using mlplasmids v1.0.0 [40]. In brief, mlplasmid uses SVM algorithm pentamer frequencies to classify the contigs harboring the genes of interest as plasmid- or chromosome-derived using maximum likelihood.

5. Conclusions

The most common carbapenemases in pediatric UTIs in Qatar are OXA-48-like carbapenemases. CREs expressing these carbapenemases may pose a threat of silent transmission as they typically confer low-level resistance to carbapenems, particularly in the case of OXA-224. Additionally, the presence of CREs that gain resistance through mechanisms other than carbapenemases highlights the limitation of the common screening methods that rely only on the presence of carbapenemases. While the sample is comprehensive, including all CRE UTIs over the period between October 2015 and November 2019, the small sample size and lack of all-inclusive clinical information preclude in-depth analysis of the clinical manifestations and implications on the treatment of such infections or measuring meaningful associations. Nevertheless, the results warrant further investigation on the epidemiology of CRE infections in the pediatric population, particularly in terms of carriage, to elucidate the distribution and transmission dynamics of CREs.

Supplementary Materials: The following are available online at https://www.mdpi.com/article/10.3390/antibiotics10080972/s1, Table S1: Non-β-lactamase antimicrobial resistance genes present in the isolates, Table S2: Sample assembly accessions.

Author Contributions: Conceptualization, N.O.E.; methodology, N.O.E., H.A.M., S.S., C.K.M.T.; formal analysis, N.O.E., H.A.M.; resources, N.O.E., K.A.-A., A.P.-L.; data curation, N.O.E., H.A.M. writing—original draft preparation, N.O.E., H.A.M.; writing—review and editing, N.O.E., H.A.M., H.Y., A.A.T., A.P.-L., K.A.-A., C.K.M.T., S.S.; project administration, N.O.E.; funding acquisition, N.O.E. All authors have read and agreed to the published version of the manuscript.

Funding: This research was funded by Qatar University, grant number "QUUG-BRC-2017" to Nahla O. Eltai.

Institutional Review Board Statement: Not applicable.

Informed Consent Statement: Not applicable.

Data Availability Statement: The genome assemblies included in the study are available at the NCBI website (https://www.ncbi.nlm.nih.gov/bioproject/, accessed on 24 January 2021) under BioProject: PRJNA690895. The accessions for the isolates are detailed in Table S2.

Acknowledgments: Authors would like to thank the Microbiology laboratory at Hamad General Hospital for their technical support.

Conflicts of Interest: The authors declare no conflict of interest. The funders had no role in the study's design; the collection, analyses, interpretation of data; the writing of the manuscript, or in the decision to publish the results.

References

1. Centers for Disease Control and Prevention. *Antibiotic Resistance Threats in the United States, 2013*; US Department of Health and Human Services, Centers for Disease Control and Prevention: Atlanta, GA, USA, 2013.
2. Centers for Disease Control and Prevention. *Antibiotic Resistance Threats in the United States, 2019*; US Department of Health and Human Services, Centers for Disease Control and Prevention: Atlanta, GA, USA, 2019.
3. Smith, H.Z.; Kendall, B. *Carbapenem-Resistant Enterobacteriaceae (CRE)*; StatPearls: Treasure Islanfe, FL, USA, 2019.
4. Tacconelli, E.; Magrini, N.; Kahlmeter, G.; Singh, N. Global priority list of antibiotic-resistant bacteria to guide research, discovery, and development of new antibiotics. *Lancet Infect. Dis.* **2017**, *27*, 318–327.
5. Vardakas, K.Z.; Tansarli, G.S.; Rafailidis, P.I.; Falagas, M.E. Carbapenems versus alternative antibiotics for the treatment of bacteraemia due to Enterobacteriaceae producing extended-spectrum β-lactamases: A systematic review and meta-analysis. *J. Antimicrob. Chemother.* **2012**, *67*, 2793–2803. [CrossRef] [PubMed]
6. Codjoe, F.S.; Donkor, E.S. Carbapenem Resistance: A Review. *Med. Sci.* **2017**, *6*, 1. [CrossRef] [PubMed]
7. Marsik, F.J.; Nambiar, S. Review of carbapenemases and AmpC-beta lactamases. *Pediatr. Infect. Dis. J.* **2011**, *30*, 1094–1095. [CrossRef]
8. Patel, G.; Bonomo, R. "Stormy waters ahead": Global emergence of carbapenemases. *Front. Microbiol.* **2013**, *4*, 48. [CrossRef] [PubMed]
9. Walsh, F. The multiple roles of antibiotics and antibiotic resistance in nature. *Front. Microbiol.* **2013**, *4*, 255. [CrossRef] [PubMed]
10. Gasink, L.B.; Edelstein, P.H.; Lautenbach, E.; Synnestvedt, M.; Fishman, N.O. Risk factors and clinical impact of Klebsiella pneumoniae carbapenemase–producing K. pneumoniae. *Infect. Control Hosp. Epidemiol.* **2009**, *30*, 1180. [CrossRef]
11. Arnold, R.S.; Thom, K.A.; Sharma, S.; Phillips, M.; Johnson, J.K.; Morgan, D.J. Emergence of Kl1–3ebsiella pneumoniae carbapenemase (KPC)-producing bacteria. *South Med. J.* **2011**, *104*, 40–45. [CrossRef]
12. Tzouvelekis, L.; Markogiannakis, A.; Psichogiou, M.; Tassios, P.; Daikos, G. Carbapenemases in Klebsiella pneumoniae and other Enterobacteriaceae: An evolving crisis of global dimensions. *Clin. Microbiol. Rev.* **2012**, *25*, 682–707. [CrossRef] [PubMed]
13. Nordmann, P.; Dortet, L.; Poirel, L. Carbapenem resistance in Enterobacteriaceae: Here is the storm! *Trends Mol. Med.* **2012**, *18*, 263–272. [CrossRef]
14. Zorc, J.J.; Kiddoo, D.A.; Shaw, K.N. Diagnosis and management of pediatric urinary tract infections. *Clin. Microbiol. Rev.* **2005**, *18*, 417–422. [CrossRef] [PubMed]
15. Logan, L.K.; Weinstein, R.A. The epidemiology of carbapenem-resistant Enterobacteriaceae: The impact and evolution of a global menace. *J. Infect. Dis.* **2017**, *215* (Suppl. 1), S28–S36. [CrossRef] [PubMed]
16. Touati, A.; Mairi, A. Epidemiology of Carbapenemase-producing Enterobacterales in the Middle East: A systematic review. *Expert Rev. Anti-Infect. Ther.* **2020**, *18*, 241–250. [CrossRef]
17. Chiotos, K.; Han, J.H.; Tamma, P.D. Carbapenem-resistant Enterobacteriaceae infections in children. *Curr. Infect. Dis. Rep.* **2016**, *18*, 2. [CrossRef]
18. Abid, F.B.; Tsui, C.K.M.; Doi, Y.; Deshmukh, A.; McElheny, C.L.; Bachman, W.C.; Fowler, E.L.; Albishawi, A.; Mushtaq, K.; Ibrahim, E.B.; et al. Molecular characterization of clinical carbapenem-resistant Enterobacterales from Qatar. *Eur. J. Clin. Microbiol. Infect. Dis. Off. Publ. Eur. Soc. Clin. Microbiol.* **2021**, *40*, 1779–1785. [CrossRef] [PubMed]
19. Queenan, A.M.; Bush, K. Carbapenemases: The Versatile β-Lactamases. *Clin. Microbiol. Rev.* **2007**, *20*, 440–458. [CrossRef] [PubMed]
20. Zahedi-Bialvaei, A.; Samadi-Kafil, H.; Ebrahimzadeh-Leylabadlo, H.; Asgharzadeh, M.; Aghazadeh, M. Dissemination of carbapenemases producing Gram negative bacteria in the Middle East. *Iran. J. Microbiol.* **2015**, *7*, 226–246. [PubMed]
21. Eltai, N.O.; Yassine, H.M.; al Thani, A.A.; Madi, M.A.A.; Ismail, A.; Ibrahim, E.; Alali, W.Q. Prevalence of antibiotic resistant Escherichia coli isolates from fecal samples of food handlers in Qatar. *Antimicrob. Resist. Infect. Control* **2018**, *7*, 78. [CrossRef]
22. Walsh, T.R. New Delhi metallo-β-lactamase-1: Detection and prevention. *CMAJ Can. Med. Assoc. J.* **2011**, *183*, 1240–1241. [CrossRef]
23. Poirel, L.; Potron, A.; Nordmann, P. OXA-48-like carbapenemases: The phantom menace. *J. Antimicrob. Chemother.* **2012**, *67*, 1597–1606. [CrossRef]

24. Sonnevend, Á.; Ghazawi, A.A.; Hashmey, R.; Jamal, W.; Rotimi, V.O.; Shibl, A.M.; Al-Jardani, A.; Al-Abri, S.S.; Tariq, W.U.Z.; Weber, S.; et al. Characterization of Carbapenem-Resistant Enterobacteriaceae with High Rate of Autochthonous Transmission in the Arabian Peninsula. *PLoS ONE* **2015**, *10*, e0131372. [CrossRef]

25. Oteo, J.; Hernández, J.M.; Espasa, M.; Fleites, A.; Sáez, D.; Bautista, V.; Pérez-Vázquez, M.; Fernández-García, M.D.; Delgado-Iribarren, A.; Sánchez-Romero, I. Emergence of OXA-48-producing Klebsiella pneumoniae and the novel carbapenemases OXA-244 and OXA-245 in Spain. *J. Antimicrob. Chemother.* **2013**, *68*, 317–321. [CrossRef]

26. Potron, A.; Poirel, L.; Dortet, L.; Nordmann, P. Characterisation of OXA-244, a chromosomally-encoded OXA-48-like β-lactamase from Escherichia coli. *Int. J. Antimicrob. Agents* **2016**, *47*, 102–103. [CrossRef]

27. Masseron, A.; Poirel, L.; Falgenhauer, L.; Imirzalioglu, C.; Kessler, J.; Chakraborty, T.; Nordmann, P. Ongoing dissemination of OXA-244 carbapenemase-producing Escherichia coli in Switzerland and their detection. *Diagn. Microbiol. Infect. Dis.* **2020**, *97*, 115059. [CrossRef] [PubMed]

28. Hoyos-Mallecot, Y.; Naas, T.; Bonnin, R.A.; Patino, R.; Glaser, P.; Fortineau, N.; Dortet, L. OXA-244-Producing Escherichia coli Isolates, a Challenge for Clinical Microbiology Laboratories. *Antimicrob. Agents Chemother.* **2017**, *61*, e00818-17. [CrossRef]

29. Doménech-Sánchez, A.; Hernández-Allés, S.; Martínez-Martínez, L.; Benedí, V.J.; Albertí, S. Identification and characterization of a new porin gene of Klebsiella pneumoniae: Its role in beta-lactam antibiotic resistance. *J. Bacteriol.* **1999**, *181*, 2726–2732. [CrossRef] [PubMed]

30. Sugawara, E.; Kojima, S.; Nikaido, H. Klebsiella pneumoniae Major Porins OmpK35 and OmpK36 Allow More Efficient Diffusion of β-Lactams than Their Escherichia coli Homologs OmpF and OmpC. *J. Bacteriol.* **2016**, *198*, 3200–3208. [CrossRef]

31. Wong, J.L.C.; Romano, M.; Kerry, L.E.; Kwong, H.O.; Low, W.E.; Brett, S.J.; Clements, A.; Beis, K.; Frankel, G. OmpK36-mediated Carbapenem resistance attenuates ST258 Klebsiella pneumoniae in vivo. *Nat. Commun.* **2019**, *10*, 3957. [CrossRef]

32. David, S.; Romano, M.; Kerry, L.E.; Kwong, H.O.; Low, W.E.; Brett, S.J.; Clements, A.; Beis, K.; Frankel, G. Epidemic of carbapenem-resistant Klebsiella pneumoniae in Europe is driven by nosocomial spread. *Nat. Microbiol.* **2019**, *4*, 1919–1929. [CrossRef]

33. McMurry, L.M.; George, A.M.; Levy, S.B. Active efflux of chloramphenicol in susceptible Escherichia coli strains and in multiple-antibiotic-resistant (Mar) mutants. *Antimicrob. Agents Chemother.* **1994**, *38*, 542–546. [CrossRef]

34. White, D.G.; Goldman, J.D.; Demple, B.; Levy, S.B. Role of the acrAB locus in organic solvent tolerance mediated by expression of marA, soxS, or robA in Escherichia coli. *J. Bacteriol.* **1997**, *179*, 6122–6126. [CrossRef] [PubMed]

35. Aono, R.; Tsukagoshi, N.; Yamamoto, M. Involvement of outer membrane protein TolC, a possible member of the mar-sox regulon, in maintenance and improvement of organic solvent tolerance of Escherichia coli K-12. *J. Bacteriol.* **1998**, *180*, 938–944. [CrossRef]

36. Chetri, S.; Choudhury, A.; Zhang, S.; Garst, A.D.; Song, X.; Liu, X.; Chen, T.; Gill, R.T.; Wang, Z. Transcriptional response of mar, sox and rob regulon against concentration gradient carbapenem stress within Escherichia coli isolated from hospital acquired infection. *BMC Res.* **2020**, *13*, 168. [CrossRef]

37. Bortolaia, V.; Kaas, R.S.; Ruppe, E.; Roberts, M.C.; Schwarz, S.; Cattoir, V.; Philippon, A.; Allesoe, R.L.; Rebelo, A.R.; Florensa, A.F.; et al. ResFinder 4.0 for predictions of phenotypes from genotypes. *J. Antimicrob. Chemother.* **2020**, *75*, 3491–3500. [CrossRef] [PubMed]

38. Alcock, B.P.; Raphenya, A.R.; Lau, T.T.; Tsang, K.K.; Bouchard, M.; Edalatmand, A.; Huynh, W.; Nguyen, A.V.; Cheng, A.A.; Liu, S.; et al. CARD 2020: Antibiotic resistome surveillance with the comprehensive antibiotic resistance database. *Nucleic Acids Res.* **2020**, *48*, D517–D525. [CrossRef] [PubMed]

39. Johansson, M.H.; Bortolaia, V.; Tansirichaiya, S.; Aarestrup, F.M.; Roberts, A.P.; Petersen, T.N. Detection of mobile genetic elements associated with antibiotic resistance in Salmonella enterica using a newly developed web tool: MobileElementFinder. *J. Antimicrob. Chemother.* **2020**, *76*, 101–109. [CrossRef] [PubMed]

40. Arredondo-Alonso, S.; Rogers, M.R.; Braat, J.C.; Verschuuren, T.D.; Top, J.; Corander, J.; Willems, R.J.L.; Schürch, A.C. Mlplasmids: A user-friendly tool to predict plasmid-and chromosome-derived sequences for single species. *Microb. Genom.* **2018**, *4*, e000224. [CrossRef]

Article

Carriage of Carbapenem-Resistant Enterobacterales in Adult Patients Admitted to a University Hospital in Italy

Pamela Barbadoro [1,2], Daniela Bencardino [3], Elisa Carloni [4], Enrica Omiccioli [4], Elisa Ponzio [1], Rebecca Micheletti [1], Giorgia Acquaviva [1], Aurora Luciani [1], Annamaria Masucci [5], Antonella Pocognoli [5], Francesca Orecchioni [5], Marcello Mario D'Errico [1,2], Mauro Magnani [3] and Francesca Andreoni [3,*]

1. Department of Biomedical Science and Public Health, Università Politecnica delle Marche, 60122 Ancona, Italy; p.barbadoro@staff.univpm.it (P.B.); e.ponzio@staff.univpm.it (E.P.); rebeccamicheletti@libero.it (R.M.); g.acquaviva@pm.univpm.it (G.A.); a.luciani@pm.univpm.it (A.L.); m.m.derrico@staff.univpm.it (M.M.D.)
2. SOD Igiene Ospedaliera-AOU Ancona Associated Hospitals, 60126 Ancona, Italy
3. Department of Biomolecular Sciences, University of Urbino, 61029 Fano, Italy; daniela.bencardino@uniurb.it (D.B.); mauro.magnani@uniurb.it (M.M.)
4. Diatheva srl, 61030 Cartoceto, Italy; e.carloni@diatheva.com (E.C.); e.omiccioli@diatheva.com (E.O.)
5. SOS Microbiologia Laboratorio Analisi, AOU Ancona Associated Hospitals, 60126 Ancona, Italy; Annamaria.Masucci@ospedaliriuniti.marche.it (A.M.); Antonella.Pocognoli@ospedaliriuniti.marche.it (A.P.); Francesca.Orecchioni@ospedaliriuniti.marche.it (F.O.)
* Correspondence: francesca.andreoni@uniurb.it; Tel.: +39-0722-3049-78

Citation: Barbadoro, P.; Bencardino, D.; Carloni, E.; Omiccioli, E.; Ponzio, E.; Micheletti, R.; Acquaviva, G.; Luciani, A.; Masucci, A.; Pocognoli, A.; et al. Carriage of Carbapenem-Resistant Enterobacterales in Adult Patients Admitted to a University Hospital in Italy. *Antibiotics* **2021**, *10*, 61. https://doi.org/10.3390/antibiotics10010061

Received: 11 November 2020
Accepted: 8 January 2021
Published: 10 January 2021

Abstract: The emerging spread of carbapenemase-producing Enterobacterales (CPE) strains, in particular, *Klebsiella pneumoniae* and *Escherichia coli*, has become a significant threat to hospitalized patients. Carbapenemase genes are frequently located on plasmids than can be exchanged among clonal strains, increasing the antibiotic resistance rate. The aim of this study was to determine the prevalence of CPE in patients upon their admission and to analyze selected associated factors. An investigation of the antibiotic resistance and genetic features of circulating CPE was carried out. Phenotypic tests and molecular typing were performed on 48 carbapenemase-producing strains of *K. pneumoniae* and *E. coli* collected from rectal swabs of adult patients. Carbapenem-resistance was confirmed by PCR detection of resistance genes. All strains were analyzed by PCR-based replicon typing (PBRT) and multilocus sequence typing (MLST) was performed on a representative isolate of each PBRT profile. More than 50% of the strains were found to be multidrug-resistant, and the bla_{KPC} gene was detected in all the isolates with the exception of an *E. coli* strain. A multireplicon status was observed, and the most prevalent profile was FIIK, FIB KQ (33%). MLST analysis revealed the prevalence of sequence type 512 (ST512). This study highlights the importance of screening patients upon their admission to limit the spread of CRE in hospitals.

Keywords: Enterobacteriaceae; PCR-based replicon typing; antibiotic-resistance; sequence types; multilocus sequence typing; plasmids

1. Introduction

The rapid spread of carbapenem-resistant Enterobacterales (CRE) mediated by carbapenemase enzymes represents a serious problem in hospitals worldwide [1]. Carbapenemases are beta-lactamases that have the ability to hydrolyze penicillins, cephalosporins, monobactams and carbapenems rendering them ineffective as antibiotics [2]. The use of broad-spectrum antimicrobials is a risk factor for the colonization of CRE in healthcare settings, and the lack of alternative therapies increases the mortality and morbidity rates as well as the costs of prolonged hospitalizations. Currently, the European epidemiology of CRE is variable. It is endemic in some countries, such as Italy, Greece and Romania, whereas its spread is still limited in most other European countries, notwithstanding a growing incidence in Spain, Portugal and Bulgaria [3].

The frequent exchange of plasmids carrying carbapenemase genes occurring among the strains increases the risk of CRE infections [4]. The European Antibiotic Surveillance Network (EARS-Net) data for 2018 reported frequent cases of carbapenem-resistance, in particular, related to the spread of carbapenemases-producing *Klebsiella pneumoniae* (KPC-Kp) and *Escherichia coli* (CP-Ec), with higher levels associated with the former. An increasing incidence of carbapenem-resistance in the EU/EEA population was reported between 2015 and 2018, with a growing number of deaths caused by *K. pneumoniae* infections. Despite its limited incidence, the distribution of *E. coli* resistant to carbapenems also needs to be monitored, considering the global impact of antimicrobial resistance [5]. In Italy, CRE have been spreading since 2010, and recent national surveillances data show that 95% of carbapenem resistance is attributable to *K. pneumoniae* and *E. coli* isolated from bacteraemia.

The main clinical characteristics and risk factors associated with CRE colonization of vulnerable patients include comorbidities, recurrent hospitalization, lengthy hospitalization and complex therapeutic management [6]. However, the seriousness of CRE carriage may vary according to each patient's particular clinical situation. For example, it is considered serious in patients with complicated intra-abdominal and urinary tract infections that require long hospital stays [3]. As suggested by the European Centre for Disease Prevention and Control (ECDC 2018) [5], a range of hygiene control measures must be implemented in healthcare settings. Moreover, in light of the importance of the role of patient transfers in CPE spread in the healthcare network, such measures must be supported by the systematic screening of patients upon their admission and during their hospitalization [7].

The identification of CPE carriage through the active rectal surveillance of patients is an effective way to limit and control CPE spread in healthcare settings. When setting up a CPE surveillance program, relevant considerations include the program's level of automation, its costs, the time it requires to execute, as well as how easy it is to use. Rapid molecular methods may be advisable to support screening performed in clinical microbiology laboratories [3].

The investigation described herein aims to determine the prevalence of CRE in patients upon their admission to a teaching hospital in Central Italy and to analyze selected associated factors. Moreover, a characterization of the strains was performed using PCR-based replicon typing (PBRT), PCR resistance genes and multilocus sequence typing.

2. Results

2.1. Antimicrobial Resistance

In the study period, 2478 patients were screened on admission by rectal swab. Overall 48 tested positive upon their first admission; hence, the prevalence of CRE was 1.93%. The isolated strains were predominantly *K. pneumoniae* (94%, $n = 45$), while the detection of *E. coli* was infrequent ($n = 3$). All strains were carbapenem-resistant and a phenotypic test revealed a resistance mediated by KPC carbapenemase enzymes in 47 strains, classifying them as carbapenemase-producing Enterobacterales (CPE). Genotypic characterization of the main carbapenemase-resistance genes showed that all strains harbored the bla_{KPC} resistance gene with the exception of an *E. coli* strain; bla_{KPC} was combined with bla_{VIM} in a single strain of *K. pneumoniae*. Twenty-one isolates (44%) were classified as multidrug-resistant (MDR) as indicated by Magiorakos et al. [8]. Only seven isolates, identified as *K. pneumoniae*, were susceptible to all of the tested antibiotics. Moreover, the highest resistance rate was towards cefuroxime (54%) followed by ciprofloxacin (44%), levofloxacin (42%) and ampicillin sulbactam (42%). The genotypic and phenotypic patterns of antibiotic-resistance and PBRT profiles of the isolates are summarized in Table 1.

Table 1. Genotypic and phenotypic profiles of strains isolated from rectal swabs.

[a] Isolate	[b] PBRT Profile	[c] CR Genes	[d] Antibiotic Resistance										[e] ST
			CN	AK	TOB	TZP	FOS	AMS	CIP	LEV	CXM	SXT	
Kp_6		bla_{KPC}	R	R	R	S	S	S	R	S	R	S	512
Kp_14		bla_{KPC}	S	S	S	S	S	S	S	S	R	S	
Kp_48		bla_{KPC}	S	S	S	S	S	S	S	R	R	S	
Kp_56		bla_{KPC}	S	S	S	S	S	S	S	R	R	S	
Kp_58		bla_{KPC}	S	S	S	S	R	S	R	R	R	S	
Kp_194		bla_{KPC}	S	S	S	S	R	S	S	S	R	S	
Kp_282		bla_{KPC}	S	S	S	S	R	S	S	S	S	S	
Kp_484	FIIK, FIB KQ	bla_{KPC}	R	R	R	S	S	S	R	R	R	R	
Kp_485		bla_{KPC}	S	S	S	S	S	S	S	S	S	S	
Kp_593		bla_{KPC}	S	S	R	R	R	R	S	S	R	S	
Kp_605		bla_{KPC}	S	S	S	S	R	S	R	R	S	S	
Kp_612		bla_{KPC}	R	R	R	S	S	S	R	R	R	R	
Kp_613		bla_{KPC}	S	S	S	S	R	R	R	R	R	S	
Kp_654		bla_{KPC}, bla_{VIM}	S	R	R	S	R	R	R	R	R	R	
Kp_660		bla_{KPC}	S	S	S	S	S	S	S	S	S	S	
Kp_672		bla_{KPC}	R	R	R	S	S	S	R	S	R	S	
Kp_506		bla_{KPC}	S	S	S	S	S	S	S	S	S	S	512
Kp_604		bla_{KPC}	S	R	R	S	S	S	R	R	R	S	
Kp_689		bla_{KPC}	S	S	S	R	R	R	R	R	S	S	
Kp_690	FIB KQ	bla_{KPC}	S	S	S	R	R	S	S	S	S	S	
Kp_691		bla_{KPC}	S	S	S	S	S	S	R	R	S	S	
Kp_696		bla_{KPC}	S	S	S	R	S	S	S	S	S	S	
Kp_712		bla_{KPC}	S	S	S	S	R	R	R	S	S	S	
Kp_714		bla_{KPC}	R	R	R	S	S	S	S	S	S	S	
Kp_176		bla_{KPC}	S	S	S	S	S	S	S	S	S	S	512
Kp_13		bla_{KPC}	R	S	S	R	R	R	R	R	R	S	
Kp_59		bla_{KPC}	S	S	S	R	R	R	S	S	S	S	
Kp_60		bla_{KPC}	S	S	S	S	R	R	S	S	R	S	
Kp_186	FIIK, FIB KQ, FIB KN	bla_{KPC}	S	S	S	S	S	S	S	S	S	S	
Kp_187		bla_{KPC}	S	S	S	S	S	R	S	S	R	S	
Kp_283		bla_{KPC}	S	S	S	R	S	R	S	S	S	S	
Kp_285		bla_{KPC}	S	S	S	R	S	R	S	S	S	S	
Kp_673		bla_{KPC}	S	S	S	S	S	S	S	S	R	S	
Kp_679		bla_{KPC}	S	S	S	S	S	S	S	S	S	S	
Kp_15	FIIK, FIB KQ, HI1	bla_{KPC}	R	S	S	S	S	S	R	S	R	S	512
Kp_245	FIIK, FIB KQ, FIB KN, X3	bla_{KPC}	S	S	S	S	S	R	S	S	S	S	512
Kp_709	FIB KN	bla_{KPC}	S	S	S	R	S	R	S	S	S	S	101
Kp_9		bla_{KPC}	S	S	S	R	S	R	S	R	S	S	101
Kp_16	FIIK, FIB KN	bla_{KPC}	R	S	S	S	S	R	R	S	R	S	
Kp_17		bla_{KPC}	S	R	R	S	R	S	R	R	R	S	
Kp_284		bla_{KPC}	S	S	S	S	S	S	S	S	S	S	
Kp_19	FIIK, FIB KN, A/C	bla_{KPC}	S	R	S	S	R	S	R	R	R	S	101
Kp_49	FIB KQ, FIB KN	bla_{KPC}	S	S	S	R	S	R	S	S	S	S	101
Kp_640	FIIK, X1,N, M	bla_{KPC}	S	R	R	S	S	S	R	R	R	S	307
Kp_707	FIB KQ, FIB KN, HI1	bla_{KPC}	S	S	S	R	R	R	S	R	R	S	307
Ec_136	FIA, FIB, FII	-	S	S	R	S	S	R	R	R	R	S	405
Ec_705	FIB KQ, FIB KN	bla_{KPC}	S	S	S	R	R	R	R	R	R	S	405
Ec_178	FIIK, FIB KQ, FIA, FIB, I1γ	bla_{KPC}	S	S	S	R	R	R	R	R	R	S	131

[a] Kp and Ec are abbreviations for the *K. pneumoniae* and *E. coli* strains, respectively; [b] PBRT, PCR-based replicon typing; [c] CR, carbapenem-resistance; not detected; [d] CN, gentamycin; AK, amikacin; TOB, tobramycin; TZP, piperacillin/tazobactam; FOS, fosfomycin; AMS, ampicillin-sulbactam; CIP, ciprofloxacin; LEV, levofloxacin; CXM, cefuroxime; SXT, co-trimoxazole; [e] ST, sequence type.

2.2. Plasmid Typing and Classification

Plasmid analysis showed that IncFIB KQ was the most predominant Incompatibility group observed in 81% of the strains. The combined results obtained from PBRT identified 14 profiles with a prevalence of FIIK, FIB KQ (33.3%; $n = 16$), FIB KQ (16.6%; $n = 8$), FIIK, FIB KQ, FIB KN (20.8%; $n = 10$), FIIK, FIB KN (8.3%; $n = 4$), whereas a single strain was found to be positive for the remaining profiles (Figure 1). Multireplicon status (two or more replicons) was recorded in 39 strains (81%), with a maximum of five replicons in one of the three strains of *E. coli*, which also showed resistance to six antibiotics. By contrast, the *K. pneumoniae* strain (reported in Table 1 as 245) showed a high number of replicons ($n = 4$) but low resistance. The only strain (654 in Table 1) carrying a combination of two carbapenem-resistance genes bla_{KPC}; bla_{VIM} was found among the most prevalent PBRT profile (FIIK, FIB KQ). Finally, FII was the only replicon recorded for the *E. coli* strain, which was the only strain found to be negative for carbapenemase-resistance genes.

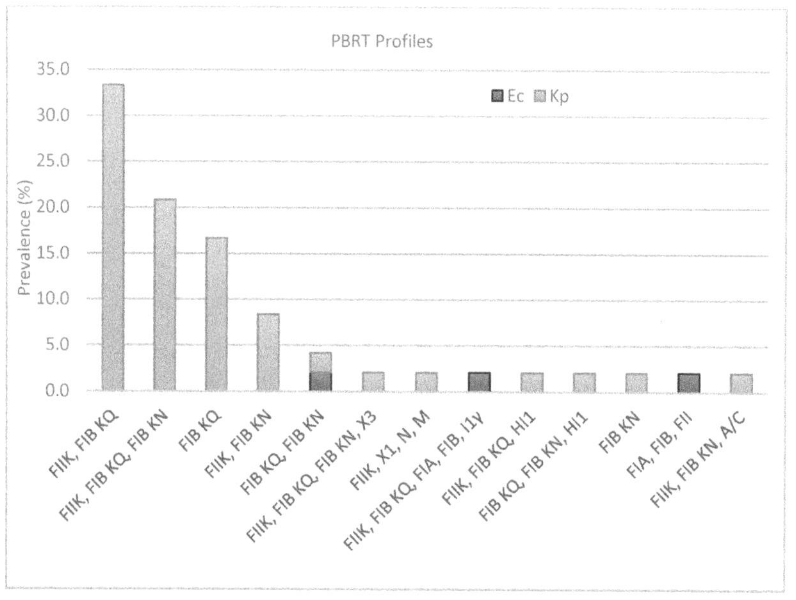

Figure 1. Prevalence of PBRT profiles in *K. pneumoniae* (Kp) and *E. coli* (Ec) strains.

2.3. Multilocus Sequence Typing Analysis

MLST analysis was performed on a representative isolate from each PBRT profile, and based on these results, 14 strains (11 *K. pneumoniae* and 3 *E. coli* strains) were characterized as shown in Table 1. Five different Sequence types (STs) were identified: ST512, ST101, ST405, ST307 and ST131. Specifically, among the 11 strains of *K. pneumoniae*, three STs were determined with a prevalence of ST512 ($n = 5$), followed by ST101 ($n = 4$) and ST307 ($n = 2$). All the STs detected in this study were found in strains showing high variability in terms of both antimicrobial resistance and PBRT patterns. The strains called Kp_506 and Kp_176, belonging to ST512, were susceptible to all of the tested antibiotics and associated with the FIB KQ and the FIIK, FIB KQ, FIB KN group, respectively. In addition, the Kp_245 strain was resistant to a single antibiotic (AMS) but associated with a multireplicon status showing four replicons (FIIK, FIB KQ, FIB KN, X3). By contrast, the MDR strain Kp_6 was resistant to five antibiotics (gentamycin (CN), amikacin (AK), tobramycin (TOB), ciprofloxacin (CIP), cefuroxime (CXM)) but only two replicons were detected (FIIK, FIB KQ). Finally, Kp_15, resistant to three antibiotics (CN, CIP, CXM), was the only strain showing a FIIK, FIB KQ, HI1 profile. The same variability was observed for ST10, which is shown in Table 1. Importantly, the two strains of *K. pneumoniae* classified as ST307 showed a high level of resistance with different antibiotic patterns.

On the other hand, two STs were detected among the three strains of *E. coli*: the well-known ST131 ($n = 1$) and the sporadic ST405 ($n = 2$). The former was associated with a strain showing a high resistance rate and five replicons, whereas the latter was associated with two strains showing different resistance patterns and PBRT profiles. In addition, one of those strains was negative for carbapenem-resistance genes, confirming the variability of the collection.

2.4. Epidemiological Data

The control patients did not differ from the cases in terms of age, sex, hospital admission rate during the previous 30 days, or comorbidities. However, there was a significant difference between the two groups in terms of origin upon admission, with 83.3% of the controls coming from the emergency department (ED) versus 52.1% of the cases. Moreover, none of the controls had previously resided in a long-term care facility (LTCF) (Table 2).

Table 2. Distribution of selected variables associated with the detection of CPE on admission.

Variables	Control	%	Cases	%	*p*
Age					
<44	2	4.2%	4	8.3%	NS
45–64	17	35.4%	17	35.4%	NS
>65	29	60.4%	27	56.3%	NS
Sex					
Male	30	62.5%	32	66.7%	NS
Origin of Patients					
Emergency department	40	83.3%	25	52.1%	<0.001
Home	3	6.3%	10	20.8%	NS
[1] LTCF	-		4	8.3%	0.041
Other hospital	5	10.4%	9	18.8%	NS
Hospital admission during the previous 30 days	11	22.9%	11	22.9%	NS
Comorbidities					
Diabetes	3	6.3%	8	16.7%	NS
Renal disease	13	27.1%	7	14.6%	NS
Cardiovascular disease	11	22.9%	6	12.5%	NS
[2] COPD	7	14.6%	3	6.3%	NS
Cancer	10	20.8%	10	20.8%	NS
Presence of disability	1	2.1%	5	10.4%	NS
Respiratory failure	14	29.2%	9	18.8%	NS
Urinary tract infection	9	18.8%	3	6.3%	NS
Administration of antibiotics	14	29.17%	25	52.08%	0.022

NS: >0.05; [1] LTCF, long-term care facility; [2] COPD, chronic obstructive pulmonary disease.

The results of multivariable logistic regression are shown in Table 3, highlighting a significant association between previous antibiotic use (OR 3.76, 95%CI 1.45–9.72) and hospital admission (OR 3.00, 95%CI 1.16–7.71) and CPE carriage on admission. On the other hand, admission to the ED was protective (OR 0.27, 95%CI 1.10–0.73).

Table 3. Result of logistic regression analysis to evaluate factors associated with CPE carriage on admission.

Variables	OR	*p* Value	95%CI
Previous use of antibiotics	3.76	0.006	1.45–9.72
Previous hospital admission 30 days	3.00	0.023	1.16–7.71
Emergency department (ED)	0.27	0.010	0.10–0.73

3. Discussion

In Italy, the rapid spread of CRE has become endemic, and it is a critical issue in the surveillance and treatment of infections [9].

CRE infection control and prevention require greater investments than other diseases across a range of areas, including patient screening, the management of long hospitalizations, as well as antimicrobials and patient isolation [3]. Once they are introduced by newly admitted patients, CRE strains rapidly spread through the hospital; hence, the prompt identification of colonized patients through active rectal surveillance can potentially reduce transmission. Incorporating rapid inexpensive methods into the clinical routine of hospitals to typify the plasmid conferring carbapenem-resistance can help to stem the spread of Enterobacterales [10].

We confirmed the efficacy of active screening for CRE through rectal swabs as an important component of any infection control program [11]. The observed high prevalence of *K. pneumoniae* was in agreement with other investigations, including the last European Survey [5,12]. Most of our CRE isolates were found to be resistant to many antibiotics commonly used in hospital settings, a phenomenon that has been widely reported [9,13,14]. These findings highlight the limited number of therapies that are available to treat CRE, which accounts for the high mortality rate currently associated with this type of infection [12]. Hence, the control of the spread of CRE infections is critically important, particularly in hospital settings.

Among the known groups of genes encoding for the carbapenemase enzymes, bla_{KPC} and bla_{NDM} are the most prevalent, and the co-occurrence of multiple resistance determinants is frequently reported [15,16]. However, in the present study, only the variant bla_{KPC} was detected in the isolates, and only one strain of *K. pneumoniae* was positive for two carbapenemase genes. To our knowledge, such a low incidence of the co-occurrence of bla_{KPC} and bla_{VIM} has rarely been described [17,18].

Considering the global spread of these determinants through plasmids, the characterization of incompatibility groups among CPE isolates makes it possible to track the dissemination of plasmids where antibiotic resistance genes can be located [16]. Not surprisingly, the most prevalent incompatibility group detected among our isolates was IncF, the common plasmid types largely associated with the spread of antibiotic resistance genes in Enterobacterales [19]. This is due to their advantageous intracellular adaptation supported by the regulatory sequences of replicons in constant and rapid evolution. Furthermore, the IncF are low-copy-number plasmids carrying more replicons to promote the initiation of replication. This feature was also observed in the present work, in which a high number of multiple replicons were detected with a prevalence of FII and FIB, typically found in a multireplicon status. Normally, the FII replicon is silent, whereas the activity of FIB, as well as that of FIA, is only related to enteric bacteria. Notably, the occurrence of multiple replicons allows plasmids to enlarge the host range replication, increasing the likelihood that they may be transferred between different species [20]. Our analysis thus highlights the usefulness of molecular typing to better understand the distribution of resistant strains in clinical settings monitored by surveillance programs. Furthermore, the proposed method for the amplicon analysis using the AATI Fragment Analyzer reduces detection time and simplifies the electrophoresis step with an automated workflow. The PBRT method therefore represents a valid tool that can also be used in hospital settings thanks to its rapidity.

The identification of ST through MLST allowed us to collect preliminary information about the clones circulating in the hospital. Of all of the sequence types obtained, ST512 and ST131 were the most frequent in *K. pneumoniae* and *E. coli*, respectively. ST512 belongs to the clonal complex 258 (CC258), in which ST258 is not only the dominant type but also the ancestor of all members. This finding confirms the endemic nature of ST512 in Italy, its association with several plasmids containing bla_{KPC} variants and its widespread rectal colonization [21,22]. In addition to the endemic ST512 clone, *K. pneumoniae* ST101 was also identified, and found to carry bla_{KPC}, pointing to the high potential for the spread of this clone, which is already found in Italy [23]. Notably, a comparative analysis carried out by Roe and colleagues (2019) revealed a similar resistome between emerging ST101 and the global ST258 of *K. pneumoniae*, strengthening the tendency of the former to become an epidemic dual-risk clone [24]. In agreement with other studies carried out in Italy, emerging ST307 was also identified among our isolates, confirming the pivotal role of this clone in clinical niches [25,26]. It is a novel distinctive lineage carrying KPC-plasmids acquired through horizontal transfer with particular virulence factors allowing an advantageous adaptation to the hospital environment [26].

Notwithstanding the limited number of *E. coli* strains collected, two relevant lineages were detected. ST405 has an international distribution and is associated with extended-spectrum beta-lactamases [27]. ST405, carrying bla_{NDM}, is usually responsible for urosepsis [28]; however, in our study this gene was found to be negative. Commonly, the acquisition of novel resistance determinants is essential for a successful and rapid spread of emerging clones. ST131 causes a wide range of extraintestinal infections with a high prevalence in patients in LTCF, who require regular health-care assistance [29]. The majority of the plasmids associated with ST131 belong to the IncF group containing FIA and FII replicons as was observed in the present study. The IncF plasmid carries multiple antibiotic determinants and virulence factors that ensure important advantages for ST131 throughout host colonization [21,29].

The identification of clones responsible for the spread of resistance determinants as well as the detection of emerging clones highlight the evolutionary dynamics of bacterial strains. Recent studies have identified the ED as an important reservoir for CRE colonization, highlighting the need to address infection control in the ED in order to better manage carbapenem resistance in other wards [30,31]. Our results, on the other hand, show entrance from ED as a protective factor. This discrepancy can probably be accounted for by the fact that in the present investigation, patients originating from the ED were admitted through that department but did not stay there for a prolonged period. In addition to being an established factor associated with colonization, previous admission to a LTCF was not independently associated with CPE carriage. However, we must underscore the role of LTCFs as a reservoir of CPE in the healthcare system and the specific clones belonging to patients originating from LTCF [32]. In this context, the patient's disability and need for assistance highlight the importance of contact precautions and infection control policies. The continuous collection of epidemiological and molecular information in healthcare facilities allows such facilities to enact prompt intervention to mitigate the risk of the colonization of patients during their stay.

4. Materials and Methods

4.1. Bacterial Isolates

A surveillance study involving the collection of isolates provided by the ongoing CRE screening on admission program at Ancona Associated Hospitals (Marche Region, Italy) was carried out from February to September 2018. A total of 2478 patients were screened on admission by rectal swab in the study period.

Rectal swabs from all adult patients were collected on admission to the hospital at the clinical microbiology laboratory in Ancona (Ancona Associated Hospitals, Italy). All samples were analyzed for the presence of carbapenem-resistant bacteria. The rectal swabs were transported to our laboratory on the day of collection and stored at 4 °C until processed. The swabs were then inoculated into 5 mL Tryptic Soy Broth (Liofilchem, Roseto degli Abruzzi, Italy) with a 10 µg meropenem disc (Becton Dickinson, Franklin Lakes, NJ, USA) and incubated overnight at 35 °C [33]. Subsequently, 100 µL of a 0.5 McFarland suspension of each swab sample were inoculated into chromogenic media (Brilliance CRE Agar-Thermo Fisher, Waltham, MA, USA). After 24 h of incubation at 37 °C, the plates were evaluated to verify the color of the colonies: pale pink colonies were considered presumptive carbapenem-resistant *E. coli*, while steel-blue colonies were assumed to be *K. pneumoniae*. Colorless or cream-colored colonies were assumed to be *Acinetobacter baumannii*. Subsequently, VITEK MS (bioMérieux, Marcy l'Etoile, France) was used to confirm the species.

The phenotypic assay for carbapenem resistance mechanisms was carried out using disk diffusion method with meropenem disks supplemented with phenylboronic acid, dipicolinic acid and cloxacillin according to the instruction manual (Biolife Italiana, Milan, Italy) [34].

4.2. Antimicrobial Susceptibility Testing

Antimicrobial resistance patterns were identified by the Molecular Epidemiology Laboratory at the Università Politecnica delle Marche, Ancona (Italy). The antibiotic susceptibility of all the collected strains was determined by the minimal inhibitory concentration (MIC) method using the SensiQuattro Gram-negative System (Liofilchem, Italy). This methodology is in accordance with the guidelines of the European Committee on Antimicrobial Susceptibility Testing (EUCAST) clinical breakpoint version 9.0 (https://www.eucast.org/fileadmin/src/media/PDFs/EUCAST_files/Breakpoint_tables/v_9.0_Breakpoint_Tables.pdf). The following antimicrobial agents were tested: gentamycin (CN), amikacin (AK), tobramycin (TOB), piperacillin/tazobactam (TZP), fosfomycin (FOS), ampicillin-sulbactam (AMS), ciprofloxacin (CIP), levofloxacin (LEV), cefuroxime (CXM), co-trimoxazole (SXT).

4.3. Molecular Detection of Resistance Determinants

All carbapenem-resistant strains were tested for the detection of associated determinants. DNA was obtained by boiling the lysis of isolated colonies for 10 min in distilled water. The samples were then centrifuged at $15{,}000 \times g$ for 10 min and the supernatant was transferred into a new 1.5 mL tube and used for the following reactions. Carbapenemase-resistance genes (bla_{IMP}, bla_{VIM}, $bla_{\mathrm{OXA\text{-}48}}$, bla_{NDM}, bla_{KPC}) were determined by PCR amplification using primers and conditions previously described by Poirel et al. [35]. Briefly, the thermal cycling settings were 30 cycles at 95 °C for 30 s, 55 °C for 1 min, 72 °C for 30 s and one cycle at 72 °C for 5 min. A positive control was used for the amplification of each gene. The PCR products were separated by electrophoresis on a 2.5% (w/v) agarose gel (Sigma-Aldrich, St. Louis, MO, USA) and visualized using UV transillumination.

4.4. PCR-Based Replicon Typing

Plasmid characterization was performed by PBRT [10] using the PBRT kit 2.0 (Diatheva, Fano, Italy). This system, consisting of eight multiplex PCR assays, allows the identification of the following 30 replicons found in the Enterobacteriaceae family: HI1, HI2, I1, I2, X1, X2, X3, X4, L, M, N, FIA, FIB, FIC, FII, FIIS, FIIK, FIB KN, FIB KQ, W, Y, P1, A/C, T, K, U, R, B/O, HIB-M and FIB-M. All PCR reactions were performed according to the manufacturer's instructions, including positive controls. The amplicons were detected through capillary electrophoresis on the AATI Fragment Analyzer (Agilent, Santa Clara, CA, USA) using the dsDNA 906 Reagent kit (Advanced Analytical, Ankeny, IA, USA). This amplicon analysis allows the combination of two multiplex PCRs in the same lane, resolving up to eight peaks. One µL of multiplex PCR 1 (M1) was combined with 1 µL of multiplex PCR 3 (M3), followed by M2 and M7, M6 and M8; the remaining M4 and M5 were loaded separately (2 µL each). The positive peaks were analyzed using the "PBRT plugin" developed in cooperation with the Advanced Analytical Company. This tool allows automatic peak calling and the recording of positive replicons.

4.5. Multilocus Sequence Typing (MLST)

MLST was performed on a representative strain for each PBRT pattern for both *K. pneumoniae* and *E. coli* strains. For the former, the amplification of the seven targeted housekeeping genes (*gapA, infB, mdh, pgi, phoE, rpoB,* and *tonB*) was carried out according to Protocol 2 of the MLST Institute Pasteur database (https://bigsdb.pasteur.fr/). This protocol uses primers with universal sequencing tails amplifying all genes at the same temperature and sequencing them with the same forward and reverse primers. Instead, MLST analysis for *E. coli* strains was performed using primers and conditions described by Wirth and colleagues [36] to detect the seven housekeeping genes *adk, icd, mdh, gyrB, purA, recA* and *fumC*. The same PCR primers were used for the sequencing carried out using the BigDye Terminator v. 1.1 Cycle Sequencing kit on the ABI PRISM® 310 Genetic Analyzer (Thermo Fisher Scientific, Waltham, MA, USA). For each locus, the obtained consensus sequences were submitted to the Pasteur online database and compared to assign the specific allele numbers. Based on the seven-allele combination found for each locus, the profile of the isolates corresponding to a specific sequence type (ST) was determined.

4.6. Epidemiological Analysis

A case control approach was used to assess factors associated with the isolation of resistant strains upon hospital admission. CPE carriage was defined as colonization by a strain with a confirmed carbapenem-resistance phenotype. A 1:1 matched case-control study adjusted for sex and age (five-year range) was used. It included 48 cases and 48 controls. A control was defined as a patient not carrying CPE on admission. The following factors were investigated through the analysis of hospital discharge records, including socio-demographic variables (age, sex), provenance on admission (home, other healthcare institution, LTCF, ED), comorbidities (diabetes, renal disease, cardiovascular disease, chronic obstructive pulmonary disease (COPD), cancer, the presence of a disability, respiratory failure, urinary tract infection). All variables were tested by bivariate analysis for their association with CPE carriage on admission, and a multivariate regression model was constructed to evaluate variables independently associated with CPE carriage using a stepwise approach. The goodness of fit of the model was evaluated with the Hosmer-Lemeshow test and its discriminative ability was assessed with the area under the ROC curve. The level of significance was set at $p < 0.05$. Data were analyzed by Stata 15 software (Stata Corp).

Ethical approval was granted by the Regional Ethics Committee of the Marches (Det. 816/DG, 11 October 2018); in accordance with the study protocol, the clinical isolates were collected and stored as a routine clinical procedure, and information concerning both the clinical isolates and patient records was anonymous. All patients provided written informed consent for their data to be used for surveillance and preventive purposes, and anonymized data were thus linked by means of an ID given to each patient at the time of hospital admission.

5. Conclusions

In conclusion, our study highlights the importance of screening patients upon their admission in order to limit the spread of CRE in hospitals. Moreover, this study provides a proof of concept for the introduction of rapid molecular typing methods such as the PBRT in routine clinical microbiological screening. We tested the feasibility of this approach for future epidemiological surveillance programs and believe that it could be a good starting point to screen patients on admission and could enhance infection control/screening programs in hospitals.

Author Contributions: Conceptualization, P.B., F.A., E.O. and M.M.; methodology, P.B., F.A., E.O.; software, F.A., E.O. and E.C.; validation, P.B., F.A. and E.O.; formal analysis, F.A., P.B., D.B. and E.C.; investigation, E.P., D.B., G.A. and E.C.; resources, E.O., E.C., P.B. and M.M.D.; data curation, R.M., A.L., F.O., A.M. and A.P.; writing—original draft preparation, F.A., D.B. and P.B.; writing—review and editing, F.A., D.B., P.B., E.O., E.C. and M.M.; supervision, M.M.; project administration, M.M.; funding acquisition, M.M. All authors have read and agreed to the published version of the manuscript.

Funding: This research was funded by the Programma Operativo Regionale MARCHE FESR 2014-2020-Asse 1-Os 3-Azione 3.1. Call: "Promuovere soluzioni innovative per affrontare le sfide delle comunità locali nell'ambito di specializzazione Salute e benessere". DDPF n. 1 del 20/01/2017, entitled: "Un approccio integrato per la promozione della salute degli anziani e delle persone fragili: prevenzione delle infezioni, integrazione nutrizionale e terapie personalizzate (PrInT-Age)", CUP: B98I17000000007.

Institutional Review Board Statement: The study was conducted according to the guidelines of the Declaration of Helsinki, and approved by Regional Ethics Committee of the Marches (protocol code Det. 816/DG and date of approval 11 October 2018).

Informed Consent Statement: Informed consent was obtained from all subjects involved in the study.

Data Availability Statement: Data is contained within the article.

Conflicts of Interest: The authors declare no conflict of interest.

References

1. Bonomo, R.A.; Burd, E.M.; Conly, J.; Limbago, B.M.; Poirel, L.; Segre, J.A.; Westblade, L.F. Carbapenemase-Producing Organisms: A Global Scourge. *Clin. Infect. Dis.* **2018**, *66*, 1290–1297. [CrossRef] [PubMed]
2. Queenan, A.M.; Bush, K. Carbapenemases: The versatile β-lactamases. *Clin. Microbiol. Rev.* **2007**, *20*, 440–458. [CrossRef] [PubMed]
3. Ambretti, S.; Bassetti, M.; Clerici, P.; Petrosillo, N.; Tumietto, F.; Viale, P.; Rossolini, G.M. Screening for carriage of carbapenem-resistant Enterobacteriaceae in settings of high endemicity: A position paper from an Italian working group on CRE infections. *Antimicrob. Resist. Infect. Control* **2019**, *8*, 1–11. [CrossRef]
4. Logan, L.K.; Weinstein, R.A. The Epidemiology of Carbapenem-Resistant Enterobacteriaceae: The Impact and Evolution of a Global Menace. *J. Infect. Dis.* **2017**, *215*, S28–S36. [CrossRef] [PubMed]
5. ECDC. *SURVEILLANCE REPORT. Surveillance of Antimicrobial Resistance in Europe 2018*; European Centre for Disease Prevention and Control (ECDC): Solna, Sweden, 2018.
6. Giannella, M.; Trecarichi, E.M.; De Rosa, F.G.; Del Bono, V.; Bassetti, M.; Lewis, R.E.; Losito, A.R.; Corcione, S.; Saffioti, C.; Bartoletti, M.; et al. Risk factors for carbapenem-resistant Klebsiella pneumoniae bloodstream infection among rectal carriers: A prospective observational multicentre study. *Clin. Microbiol. Infect.* **2014**, *20*, 1357–1362. [CrossRef]
7. Barbadoro, P.; Dichiara, A.; Arsego, D.; Ponzio, E.; Savini, S.; Manso, E.; D'errico, M.M. Spread of carbapenem-resistant Klebsiella pneumoniae in hub and spoke connected health-care networks: A case study from Italy. *Microorganisms* **2020**, *8*, 37. [CrossRef]
8. Magiorakos, A.P.; Srinivasan, A.; Carey, R.B.; Carmeli, Y.; Falagas, M.E.; Giske, C.G.; Harbarth, S.; Hindler, J.F.; Kahlmeter, G.; Olsson-Liljequist, B.; et al. Multidrug-resistant, extensively drug-resistant and pandrug-resistant bacteria: An international expert proposal for interim standard definitions for acquired resistance. *Clin. Microbiol. Infect.* **2012**, *18*, 268–281. [CrossRef]
9. Ridolfo, A.L.; Rimoldi, S.G.; Pagani, C.; Marino, A.F.; Piol, A.; Rimoldi, M.; Olivieri, P.; Galli, M.; Dolcetti, L.; Gismondo, M.R. Diffusion and transmission of carbapenem-resistant Klebsiella pneumoniae in the medical and surgical wards of a university hospital in Milan, Italy. *J. Infect. Public Health* **2016**, *9*, 24–33. [CrossRef]
10. Carloni, E.; Andreoni, F.; Omiccioli, E.; Villa, L.; Magnani, M.; Carattoli, A. Comparative analysis of the standard PCR-Based Replicon Typing (PBRT) with the commercial PBRT-KIT. *Plasmid* **2017**, *90*, 10–14. [CrossRef]
11. Singh, K.; Mangold, K.A.; Wyant, K.; Schora, D.M.; Voss, B.; Kaul, K.L.; Hayden, M.K.; Chundi, V.; Peterson, L.R. Rectal screening for Klebsiella pneumoniae carbapenemases: Comparison of real-time PCR and culture using two selective screening agar plates. *J. Clin. Microbiol.* **2012**, *50*, 2596–2600. [CrossRef]
12. Grundmann, H.; Glasner, C.; Albiger, B.; Aanensen, D.M.; Tomlinson, C.T.; Andrasević, A.T.; Cantón, R.; Carmeli, Y.; Friedrich, A.W.; Giske, C.G.; et al. Occurrence of carbapenemase-producing Klebsiella pneumoniae and Escherichia coli in the European survey of carbapenemase-producing Enterobacteriaceae (EuSCAPE): A prospective, multinational study. *Lancet Infect. Dis.* **2017**, *17*, 153–163. [CrossRef]
13. Sotgiu, G.; Are, B.M.; Pesapane, L.; Palmieri, A.; Muresu, N.; Cossu, A.; Dettori, M.; Azara, A.; Mura, I.I.; Cocuzza, C.; et al. Nosocomial transmission of carbapenem-resistant Klebsiella pneumoniae in an Italian university hospital: A molecular epidemiological study. *J. Hosp. Infect.* **2018**, *99*, 413–418. [CrossRef] [PubMed]

14. Wang, Q.; Wang, X.; Wang, J.; Ouyang, P.; Jin, C.; Wang, R.; Zhang, Y.; Jin, L.; Chen, H.; Wang, Z.; et al. Phenotypic and Genotypic Characterization of Carbapenem-resistant Enterobacteriaceae: Data from a Longitudinal Large-scale CRE Study in China (2012–2016). *Clin. Infect. Dis.* **2018**, *67*, S196–S205. [CrossRef] [PubMed]

15. Patel, G.; Bonomo, R.A. "Stormy waters ahead": Global emergence of carbapenemases. *Front. Microbiol.* **2013**, *4*, 48. [CrossRef] [PubMed]

16. Zhou, H.; Zhang, K.; Chen, W.; Chen, J.; Zheng, J.; Liu, C.; Cheng, L.; Zhou, W.; Shen, H.; Cao, X. Epidemiological characteristics of carbapenem-resistant Enterobacteriaceae collected from 17 hospitals in Nanjing district of China. *Antimicrob. Resist. Infect. Control* **2020**, *9*, 1–10. [CrossRef] [PubMed]

17. Papadimitriou-Olivgeris, M.; Bartzavali, C.; Lambropoulou, A.; Solomou, A.; Tsiata, E.; Anastassiou, E.D.; Fligou, F.; Marangos, M.; Spiliopoulou, I.; Christofidou, M. Reversal of carbapenemase-producing Klebsiella pneumoniae epidemiology from blaKPC-to blaVIM-harbouring isolates in a Greek ICU after introduction of ceftazidime/avibactam. *J. Antimicrob. Chemother.* **2019**, *74*, 2051–2054. [CrossRef]

18. Zagorianou, A.; Sianou, E.; Iosifidis, E.; Dimou, V.; Protonotariou, E.; Miyakis, S.; Roilides, E.; Sofianou, D. Microbiological and molecular characteristics of carbapenemase-producing klebsiella pneumonia endemic in a tertiary Greek hospital during 2004–2010. *Eurosurveillance* **2012**, *17*, 1–7. [CrossRef]

19. Carattoli, A. Plasmids and the spread of resistance. *Int. J. Med. Microbiol.* **2013**, *303*, 298–304. [CrossRef]

20. Villa, L.; García-Fernández, A.; Fortini, D.; Carattoli, A. Replicon sequence typing of IncF plasmids carrying virulence and resistance determinants. *J. Antimicrob. Chemother.* **2010**, *65*, 2518–2529. [CrossRef]

21. Mathers, A.J.; Peirano, G.; Pitout, J.D.D. The role of epidemic resistance plasmids and international high- risk clones in the spread of multidrug-resistant Enterobacteriaceae. *Clin. Microbiol. Rev.* **2015**, *28*, 565–591. [CrossRef]

22. Netikul, T.; Kiratisin, P. Genetic characterization of carbapenem-resistant enterobacteriaceae and the spread of carbapenem-resistant klebsiella pneumonia ST340 at a university hospital in Thailand. *PLoS ONE* **2015**, *10*, e0139116. [CrossRef] [PubMed]

23. Del Franco, M.; Paone, L.; Novati, R.; Giacomazzi, C.G.; Bagattini, M.; Galotto, C.; Montanera, P.G.; Triassi, M.; Zarrilli, R. Molecular epidemiology of carbapenem resistant Enterobacteriaceae in Valle d'Aosta region, Italy, shows the emergence of KPC-2 producing Klebsiella pneumoniae clonal complex 101 (ST101 and ST1789). *BMC Microbiol.* **2015**, *15*, 1–9. [CrossRef] [PubMed]

24. Roe, C.C.; Vazquez, A.J.; Esposito, E.P.; Zarrilli, R.; Sahl, J.W. Diversity, virulence, and antimicrobial resistance in isolates from the newly emerging klebsiella pneumoniaeST101 Lineage. *Front. Microbiol.* **2019**, *10*, 542. [CrossRef] [PubMed]

25. Loconsole, D.; Accogli, M.; De Robertis, A.L.; Capozzi, L.; Bianco, A.; Morea, A.; Mallamaci, R.; Quarto, M.; Parisi, A.; Chironna, M. Emerging high-risk ST101 and ST307 carbapenem-resistant Klebsiella pneumoniae clones from bloodstream infections in Southern Italy. *Ann. Clin. Microbiol. Antimicrob.* **2020**, *19*, 1–10. [CrossRef]

26. Villa, L.; Feudi, C.; Fortini, D.; Brisse, S.; Passet, V.; Bonura, C.; Endimiani, A.; Mammina, C.; Ocampo, A.M.; Jimenez, J.N.; et al. Diversity, virulence, and antimicrobial resistance of the KPCproducing klebsiella pneumoniae ST307 clone. *Microb. Genom.* **2017**, *3*, e000110. [CrossRef]

27. Coque, T.M.; Novais, A.; Carattoli, A.; Poirel, L.; Pitout, J.; Peixe, L.; Baquero, F.; Canton, R.; Nordmann, P. Dissemination of clonally related E.coli strains expressing ESBL CTX-M-15. *Emerg. Infect. Dis.* **2008**, *14*, 195–200. [CrossRef]

28. Chowdhury, P.R.; McKinnon, J.; Liu, M.; Djordjevic, S.P. Multidrug resistant uropathogenic Escherichia coli ST405 with a novel, composite IS26 transposon in a unique chromosomal location. *Front. Microbiol.* **2019**, *9*, 3212. [CrossRef]

29. Nicolas-Chanoine, M.H.; Bertrand, X.; Madec, J.Y. Escherichia coli st131, an intriguing clonal group. *Clin. Microbiol. Rev.* **2014**, *27*, 543–574. [CrossRef]

30. Salomão, M.C.; Freire, M.P.; Boszczowski, I.; Raymundo, S.F.; Guedes, A.R.; Levin, A.S. Increased risk for carbapenem-resistant enterobacteriaceae colonization in intensive care units after hospitalization in emergency department. *Emerg. Infect. Dis.* **2020**, *26*, 1156–1163. [CrossRef]

31. Salomão, M.C.; Guimarães, T.; Duailibi, D.F.; Perondi, M.B.M.; Letaif, L.S.H.; Montal, A.C.; Rossi, F.; Cury, A.P.; Duarte, A.J.S.; Levin, A.S.; et al. Carbapenem-resistant Enterobacteriaceae in patients admitted to the emergency department: Prevalence, risk factors, and acquisition rate. *J. Hosp. Infect.* **2017**, *97*, 241–246. [CrossRef]

32. Giufrè, M.; Ricchizzi, E.; Accogli, M.; Barbanti, F.; Monaco, M.; de Araujo, F.P.; Farina, C.; Fazii, P.; Mattei, R.; Sarti, M.; et al. Colonization by multidrug-resistant organisms in long-term care facilities in Italy: A point-prevalence study. *Clin. Microbiol. Infect.* **2017**, *23*, 961–967. [CrossRef]

33. Richter, S.S.; Marchaim, D. Screening for carbapenem-resistant Enterobacteriaceae: Who, When, and How? *Virulence* **2017**, *8*, 417–426. [CrossRef]

34. Giske, C.G.; Gezelius, L.; Samuelsen, Ø.; Warner, M.; Sundsfjord, A.; Woodford, N. A sensitive and specific phenotypic assay for detection of metallo-β-lactamases and KPC in Klebsiella pneumoniae with the use of meropenem disks supplemented with aminophenylboronic acid, dipicolinic acid and cloxacillin. *Clin. Microbiol. Infect.* **2011**, *17*, 552–556. [CrossRef] [PubMed]

35. Poirel, L.; Walsh, T.R.; Cuvillier, V.; Nordmann, P. Multiplex PCR for detection of acquired carbapenemase genes. *Diagn. Microbiol. Infect. Dis.* **2011**, *70*, 119–123. [CrossRef] [PubMed]

36. Wirth, T.; Falush, D.; Lan, R.; Colles, F.; Mensa, P.; Wieler, L.H.; Karch, H.; Reeves, P.R.; Maiden, M.C.J.; Ochman, H.; et al. Sex and virulence in Escherichia coli: An evolutionary perspective. *Mol. Microbiol.* **2006**, *60*, 1136–1151. [CrossRef] [PubMed]

Article

Phenotypic and Genotypic Features of *Klebsiella pneumoniae* Harboring Carbapenemases in Egypt: OXA-48-Like Carbapenemases as an Investigated Model

Suzan Mohammed Ragheb [1,†] ⓘ, Mahmoud Mohamed Tawfick [2,3,*,†] ⓘ, Amani Ali El-Kholy [4] ⓘ and Abeer Khairy Abdulall [5,*] ⓘ

1. Department of Microbiology and Immunology, Faculty of Pharmacy, Modern University for Technology and Information (MTI), Cairo 11571, Egypt; drsuzanragheb84@gmail.com
2. Department of Microbiology and Immunology, Faculty of Pharmacy (Boys), Al-Azhar University, Cairo 11884, Egypt
3. Department of Microbiology, Faculty of Pharmacy, Heliopolis University, Cairo 11785, Egypt
4. Department of Clinical and Chemical Pathology, Faculty of Medicine, Cairo University, Cairo 11562, Egypt; aaakholy@gmail.com
5. Department of Microbiology and Immunology, Faculty of Pharmacy (Girls), Al-Azhar University, Cairo 11884, Egypt
* Correspondence: mahmoud_tawfick@azhar.edu.eg (M.M.T.); Khairy.abeer@yahoo.com (A.K.A.)
† These authors contributed equally to this work.

Received: 25 October 2020; Accepted: 25 November 2020; Published: 28 November 2020

Abstract: This study aimed at the characterization of carbapenem-resistant *Klebsiella pneumoniae* isolates focusing on typing of the $bla_{OXA-48-like}$ genes. Additionally, the correlation between the resistance pattern and biofilm formation capacity of the carbapenem-resistant *K. pneumoniae* isolates was studied. The collected isolates were assessed for their antimicrobial resistance and carbapenemases production by a modified Hodge test and inhibitor-based tests. The carbapenemases encoding genes (bla_{KPC}, bla_{NDM}, bla_{VIM}, bla_{IMP}, and $bla_{OXA-48-like}$) were detected by PCR. Isolates harboring $bla_{OXA-48-like}$ genes were genotyped by Enterobacterial Repetitive Intergenic Consensus-Polymerase Chain Reaction (ERIC-PCR) and plasmid profile analysis. The discriminatory power of the three typing methods (antibiogram, ERIC-PCR, and plasmid profile analysis) was compared by calculation of Simpson's Diversity Index (SDI). The transferability of bla_{OXA-48} gene was tested by chemical transformation. The biofilm formation capacity and the prevalence of the genes encoding the fimbrial adhesins (*fimH-1* and *mrkD*) were investigated. The isolates showed remarkable resistance to β-lactams and non-β-lactams antimicrobials. The coexistence of the investigated carbapenemases encoding genes was prevalent except for only 15 isolates. The plasmid profile analysis had the highest discriminatory power (SDI = 0.98) in comparison with ERIC-PCR (SDI = 0.89) and antibiogram (SDI = 0.78). The transferability of bla_{OXA-48} gene was unsuccessful. All isolates were biofilm formers with the absence of a significant correlation between the biofilm formation capacity and resistance profile. The genes *fimH-1* and *mrkD* were prevalent among the isolates. The prevalence of carbapenemases encoding genes, especially $bla_{OXA-48-like}$ genes in Egyptian healthcare settings, is worrisome and necessitates further strict dissemination control measures.

Keywords: *Klebsiella pneumoniae*; carbapenemases; bla_{OXA-48}; ERIC-PCR; plasmid profile analysis; biofilm formation

1. Introduction

Antimicrobial resistance is among the serious problems that contribute to high morbidity and mortality rates, particularly in immunocompromised patients [1]. Several alarming reports released by international organizations have described the consequences of this crisis at health and economic levels [2–4]. *Klebsiella pneumoniae* is among the ESKAPE pathogens (*Enterococcus faecium, Staphylococcus aureus, Klebsiella pneumoniae, Acinetobacter baumannii, Pseudomonas aeruginosa,* and *Enterobacter* spp.) which adopt numerous mechanisms to "escape" from different antimicrobials actions [5]. Increased attention has emerged toward resistance to carbapenems, which are considered as the last resort β-lactam for treatment of life-threatening infections caused by multidrug-resistant Enterobacteriaceae. Thus, the WHO Global priority list of antibiotic-resistant bacteria classified carbapenem-resistant Enterobacteriaceae as pathogens of critical priority [6].

Carbapenem resistance may be attributed to porin mutations, efflux pumps, and/or carbapenemases production. Although there is an expansion in the number of emerged carbapenemases, five important members that belong to three Ambler classes are the most studied, namely, *Klebsiella pneumoniae* carbapenemases (*bla*KPC), New Delhi metallo-β-lactamases (*bla*NDM), Verona integron-encoded metallo-β-lactamases (*bla*VIM), Active on imipenem metallo-β-lactamases (*bla*IMP), and Oxacillinase-48-like carbapenemases (*bla*OXA-48-like) [7,8]. These carbapenemases have several variants according to the National Center for Biotechnology Information (NCBI) Antimicrobial Resistance Reference Gene database, available at https://www.ncbi.nlm.nih.gov/pathogens/isolates#/refgene. Additionally, and in the terms of the geographical dissemination, the "big five" carbapenemases differ in their spread globally, and their epidemiological status may be endemic, or just recorded cases [8]. KPC β-lactamases belong to the Ambler Class A carbapenemases which have serine at their active site. In addition to its activity against carbapenems, *bla*KPC show activity against a wide range of β-lactams as penicillins, cephalosporins, and monobactams, and also overcome the in-vitro activity of β-lactamase inhibitors such as clavulanic acid and sulbactam. Additionally, plasmids carrying *bla*KPC encoding gene may carry other genes that confer resistance to other antimicrobial classes such as fluoroquinolones and aminoglycosides [9]. NDM, VIM, and IMP metallo β-lactamases are the most common metallo-β-lactamases that belong to the Ambler Class B carbapenemases with zinc at their active site, and this leads to their inhibition by metal-chelating agents such as ethylenediaminetetraacetic acid (EDTA). They show resistance to penicillins, cephalosporins, and carbapenems and also to β-lactamase inhibitors. Furthermore, these carbapenemases are supported by genetic platforms that allow their dissemination with other resistance determinants of fluoroquinolones and aminoglycosides [10]. OXA-48-like carbapenemases belong to the Ambler Class D carbapenemases with serine at their active site. They are not inhibited by EDTA or clavulanic acid, and this makes their detection challenging [11].

In Egypt, *bla*OXA-48 is of a great concern due to its endemicity in healthcare settings, especially in a successful candidate such as *K. pneumoniae,* which is characterized by a great plasmid load and high ability of transfer of resistance determinants of different antimicrobial classes [12]. There are several features concerning the *bla*OXA-48 gene. It is located on pOXA-48a IncL plasmid which does not carry any further antimicrobial determinants other than the *bla*OXA-48 gene. The *bla*OXA-48 gene is surrounded by two copies of the insertion sequence, IS*1999* [13]. Additionally, the rapid dissemination of pOXA-48a may be attributed to the disruption of *tir* gene (encodes a protein that is responsible for transfer inhibition) by transposon Tn*1999* and its variants [14]. Furthermore, *bla*OXA-48 gene has several variants that differ by few amino acids, and consequently differ in their hydrolysis specificities [13].

The dissemination of resistance genes, especially those encoding carbapenemases, necessitates the application of typing methods that participate in the study of the features and clonality of isolates that harbor various antimicrobials determinants [15].

Another problematic feature associated with carbapenem resistance in *K. pneumoniae* is biofilm formation. The International Union of Pure and Applied Chemistry defines biofilm as "an aggregate of microorganisms in which cells that are frequently embedded within a self-produced matrix of extracellular polymeric substance (EPS) adhere to each other and/or to a surface" [16]. Biofilm formation participates in the aggravation of serious clinical manifestations and leads to reduced susceptibility to antimicrobial agents [17]. Type 1 and 3 fimbriae have a crucial role in adhesion, which is a key factor in *K. pneumoniae* biofilm formation besides capsule and lipopolysaccharide [18].

In this study, we aimed to characterize carbapenem-resistant *K. pneumoniae* isolated from patients admitted to an Egyptian tertiary hospital. Additionally, we focused on the typing of the detected $bla_{OXA-48-like}$ genes. Additionally, we shed light on the biofilm formation capacity of the carbapenem-resistant isolates and studied the possible correlation between the two traits

2. Results

2.1. Distribution and Identification of K. pneumoniae Isolates

A total of 117 isolates of carbapenem-non-susceptible *K. pneumoniae* were collected and recovered from diverse clinical specimens as shown in Figure 1. The confirmation of isolate identification using MALDI-TOF/MS had a high confidence ranging from 99.7 to 99.9%.

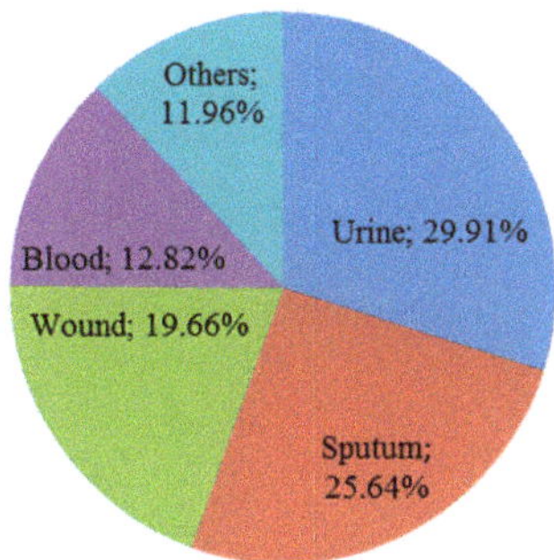

Figure 1. The distribution of collected isolates according to their origin: urine (29.91%), sputum (25.64%), wound (19.66%), blood (12.82%), and others (endotracheal aspirate (4.27%), central venous pressure tip (1.71%), cerebrospinal fluid (0.85%), pleural fluid (0.85%), drain culture (0.85%), bedsore (0.85%), bile culture (0.85%), central venous line culture (0.85%), and aspirate fluid (0.85%)).

2.2. Isolate Antimicrobial Susceptibility Pattern

According to interim standard definitions of acquired resistance by Magiorakos et al. [19], our isolates were considered as multidrug-resistant (MDR) with the possibility of being extensively drug-resistant (XDR). As shown in Table S1, there was remarkable resistance to most antimicrobial agents. All isolates showed a resistant or intermediate profile against most β-Lactams such as amoxicillin/clavulanic acid (AUG), piperacillin/tazobactam (TZP), cefoxitin (FOX), cefotaxime (CTX), cefotaxime/clavulanic acid (CTC), ceftazidime (CAZ), ceftazidime/clavulanic acid (CZC), cefepime (CPM), imipenem (IMI), meropenem (MEM), and ertapenem (ETR). The susceptibilities of isolates to aztreonam (ATM), gentamicin (GN), amikacin (AK), ciprofloxacin (CIP), sulfamethoxazole/trimethoprim (STX), and tigecycline (TGC) varied to be sensitive, intermediate, or resistant. The frequency of antimicrobial resistance is shown in Figure 2. Additionally, all isolates had minimum inhibitory concentration MIC > 1024 mg/L for imipenem, while MICs of meropenem ranged from 16 to > 1024 mg/L as shown in Figure 3.

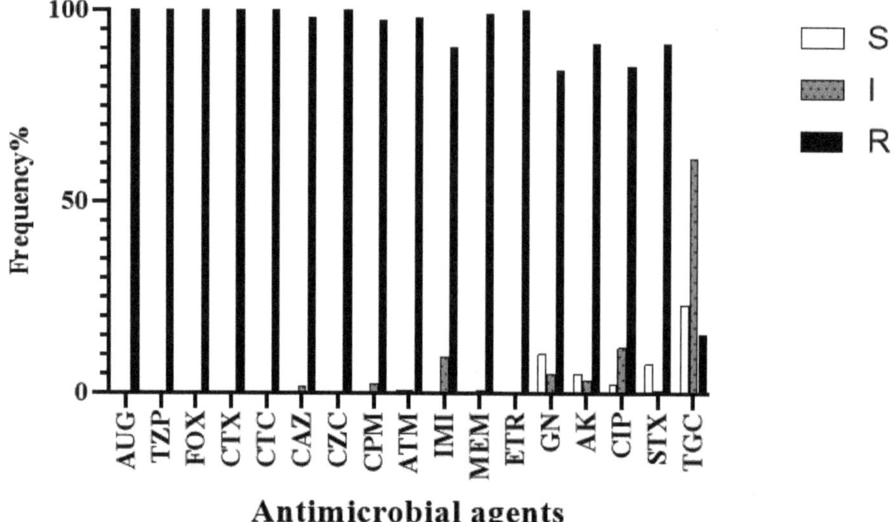

Figure 2. The frequencies of antimicrobial resistance of collected isolates to different antimicrobial classes. The frequencies of *K. pneumoniae* isolates resistance against β-lactams were very high (mostly 100%), while they were variable for non-β-lactam antimicrobials. S, Sensitive; I, Intermediate; R, Resistant. AUG, amoxicillin/clavulanic acid; TZP, piperacillin/tazobactam; FOX, cefoxitin; CTX, cefotaxime; CTC, cefotaxime/clavulanic acid; CAZ, ceftazidime; CZC, ceftazidime/clavulanic acid; CPM, cefepime; ATM, aztreonam; IMI, imipenem; MEM, meropenem; ETR, ertapenem; GN, gentamicin; AK, amikacin; CIP, ciprofloxacin; STX, sulfamethoxazole/trimethoprim, and (TGC) tigecycline. The interpretation of the results was according to the Clinical and Laboratory Standards Institute guidelines, CLSI (M100-S26, 2016). Interpretation of tigecycline results was according to the FDA recommendations available at http://www.accessdata.fda.gov/drugsatfda_docs/label/2009/021821s016lbl.pdf.

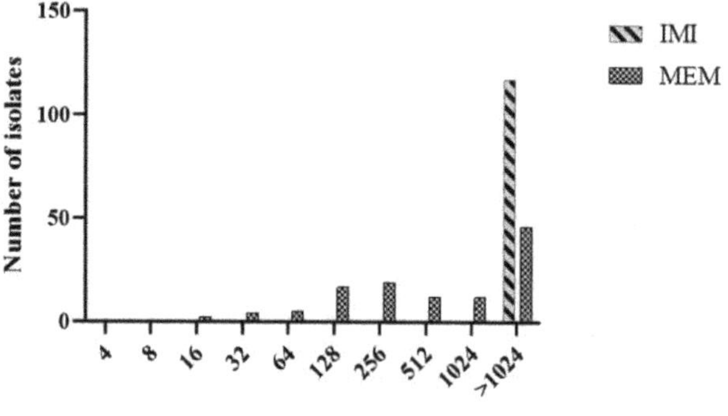

Figure 3. The distribution of the minimum inhibitory concentration MIC values of imipenem (IMI) and meropenem (MEM) among isolates. For both antimicrobial agents, the MICs for the resistant isolates were ≥4 mg/L according to CLSI (M100-S26, 2016) guidelines.

2.3. Phenotypic and Genotypic Characterization of Carbapenem-Resistant Isolates

Only 66 isolates (56.41%) were positive for non-specific detection of carbapenemases production by the modified Hodge test (MHT). In contrast, the inhibitor-based tests were more specific by confirmation of carbapenemases production and differentiation between their types. Seventy-nine isolates (67.52%) were positive for MβLs production, and no isolates showed positive results for KPC production or AmpC activity. The remaining isolates (n = 38) did not comply with the criteria of the previously mentioned enzymes but showed resistance to temocillin, giving a presumptive estimation of the presence of the $bla_{\text{OXA-48-like}}$ genes. The results of both phenotypic tests in comparison with the further molecular screening of carbapenemases genes are shown in Table S2.

The results of molecular screening of the investigated carbapenemases are shown in Table S2. VIM gene was the most prevalent gene amongst the studied genes as it was detected in 99 isolates (84.62%) followed by 88 isolates (75.21%) for NDM gene, 69 isolates (58.97%) for IMP gene, and 34 isolates (29.06%) for OXA-48-like genes. The presence of KPC gene was very limited as it was detected in only 6 isolates (5.13%).

Furthermore, all isolates showed coexistence of at least two carbapenemases under investigation except 4 isolates, namely, 33, 136, 147, and 396, which had VIM gene only; 2 isolates, namely, 16 and 417 had NDM gene only; and 9 isolates, namely, 8, 28,113,127, 151, 346, 356, 378, and 399 had OXA-48-like genes only. On the other hand, only 4 isolates, namely, 137, 142, 410, and 419 were carbapenem-resistant and showed positive MHT but did not show augmentation in the inhibitor-based tests or any positive result via PCR, and this gives an indication that carbapenem resistance in those isolates may be for mechanisms other than carbapenemases production.

The OXA-48-like genes amplicons of the isolates, namely, 151, 303, 311, 383, 399, 405, 409, and 415 were sequenced, and further sequence analysis showed 100% identity for the OXA-48 gene.

2.4. Genotyping of Isolates Harboring $bla_{\text{OXA-48-like}}$ Genes

Genotyping of the 34 isolates harboring $bla_{\text{OXA-48-like}}$ genes was performed by two molecular methods: Enterobacterial Repetitive Intergenic Consensus-Polymerase Chain Reaction (ERIC-PCR) and plasmid profile analysis. Figure 4 shows the ERIC-PCR dendrogram that divided the 34 isolates into 16 genotypes. Among them, there were three major groups: each represented genetic relatedness between its isolates despite their collection from different clinics at different periods. Additionally, 10 isolates showed unique patterns and were not related to each other. On the other hand, the plasmid profile analysis showed variation between isolates. Under experimental conditions, out of 34 isolates, 2 isolates (5.88%) harbored one plasmid, and other isolates harbored many plasmids ranging from 2 to 8 plasmids. The highest number of plasmids was 8 plasmids which were found in one isolate followed by 6 plasmids which were found in another one, representing 2.94% for each. Additionally, there were 11 isolates (32.35%) that had 3 plasmids, followed by 9 isolates (26.47%) that had 5 plasmids, 6 isolates (17.65%) that had 4 plasmids, and 4 isolates (11.76%) that had 2 plasmids. The results showed that the isolates belonging to the same cluster were mostly collected from different sources and at different times.

The Simpson's Diversity Index was calculated for antibiogram, ERIC-PCR, and plasmid profile analysis giving convergent values of 0.78, 0.89, and 0.98, respectively as shown in Figure 5. The SDI value of the plasmid profile analysis was very close to 1 indicating the maximum diversity that was achieved by this method. The criteria for SDI calculation of the applied typing methods—antibiogram, ERIC-PCR, and plasmid profile analysis—are shown in Tables S3–S5, respectively. A comparative summary of the SDI calculations of the three methods is shown in Table S6.

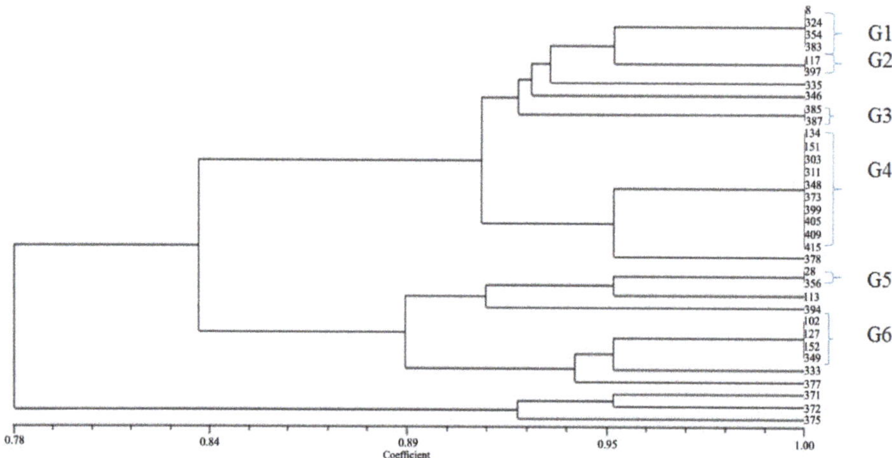

Figure 4. ERIC-PCR dendrogram for 34 isolates harboring $bla_{OXA-48-like}$ encoding genes; implemented using NTSYS 2.01 software. The investigated isolates were divided into 16 genotypes. Among them, there were many groups; each represented genetic relatedness. G1(Group 1; isolates: 8, 324, 354, and 383); G2 (Group 2; isolates: 117 and 397); G3 (Group 3; isolates: 385 and 387); G4 (Group 4; isolates: 134, 151, 303, 311, 348, 373, 399, 405, 409, and 415); G5 (Group 5; isolates: 28 and 356), and G6 (Group 6; isolates: 102, 127, 152, and 349). The remaining 10 isolates represent unique patterns (isolates: 113, 333, 335, 346, 371, 372, 375, 377, 378, and 394).

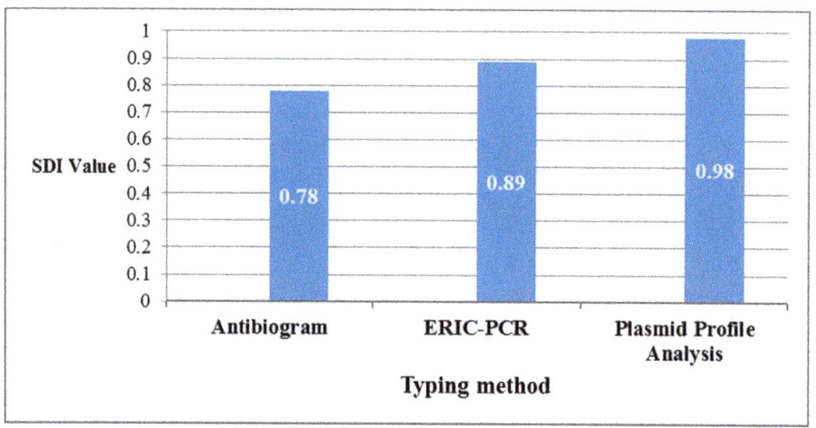

Figure 5. A comparative graph of calculated Simpson's Diversity Index (SDI) of the applied typing methods to study the relatedness among the 34 *K. pneumoniae* isolates harboring $bla_{OXA-48-like}$ genes. SDI values were convergent as they were 0.78, 0.89, and 0.98 for antibiogram, ERIC-PCR, and plasmid profile analysis, respectively.

2.5. Transferability of bla_{OXA-48} Gene

The plasmids of five selected isolates, namely, 311, 383, 399, 405, and 409, underwent genetic transfer by chemical transformation. PCR results of transformants showed the absence of bla_{OXA-48} gene. Further investigations were performed for two selected transformants, namely, 383 and 409, by an antimicrobial susceptibility test and PCR. The results showed reduced susceptibility to different β-lactams including carbapenems, sulfamethoxazole/trimethoprim, and tigecycline in comparison with *Escherichia coli* DH5α as shown in Table 1. Although the isolate 383 had bla_{NDM}, bla_{VIM}, and bla_{IMP} encoding genes and the isolate 409 had bla_{NDM} and bla_{VIM} encoding genes, only bla_{VIM} encoding gene was detected by colony PCR for both isolates transformants. This explains the successful transfer of such gene.

Table 1. Comparison of antimicrobial susceptibilities between *E. coli* DH5α and two transformants.

Isolate	AUG	TZP	FOX	CTX	CTC	CAZ	CZC	CPM	ATM	MEM	ETR	GN	AK	CIP	STX	TGC
DH5α	R	S	S	S	S	S	S	S	S	S	S	S	S	S	S	S
transformant 383	R	R	R	R	R	R	R	R	R	I	R	S	S	S	R	R
transformant 409	R	R	R	R	R	R	R	R	R	R	R	S	S	S	R	R

2.6. Assessment of Biofilm Formation and Detection of Adhesion Encoding Genes in Carbapenem-Resistant K. pneumoniae Isolates

The quantitative biofilm formation assay resulted in OD_{600} values ranging from 0.2 to 1.1. All isolates were classified as biofilm formers but with different capacities. The majority of the isolates were weak biofilm formers and were represented by 82 isolates (70.09%), followed by 34 isolates as moderate formers (29.06%), and only one isolate was classified as a strong biofilm former (0.85%). Furthermore, the PCR screening of the isolates showed the prevalence of the two adhesion encoding genes in most isolates. Out of 117 isolates, *fimH-1* gene was found in 108 isolates (92.31%) and *mrkD* gene was found in 115 isolates (98.29%). Additionally, 106 isolates (90.60%) represented the coexistence of both genes as shown in Table S7.

2.7. The Correlation between Non-Susceptibility Pattern and Biofilm Formation Category of Carbapenem-Resistant K. pneumoniae Isolates

As shown in Table 2, there is no significant correlation between non-susceptibility patterns and biofilm formation categories among carbapenem-resistant isolates. All *p*-values were > 0.05, and this was considered statistically insignificant.

Table 2. Correlation between non-susceptibility pattern and biofilm formation.

	Non-Susceptibility Pattern	Biofilm Pattern			*p*-Value (>0.05)
		Strong	Moderate	Weak	
A	Non-susceptible to all antimicrobials.	1	21	52	0.3333
B	Non-susceptible to all Except AK.	0	0	2	0.6667
C	Non-susceptible to all Except GN	0	0	4	0.6667
D	Non-susceptible to all Except CIP	0	0	1	0.6667
E	Non-susceptible to all Except STX	0	3	2	>0.9999
F	Non-susceptible to all Except TGC	0	7	13	0.3333
G	Non-susceptible to all Except ATM & AK	0	1	0	>0.9999
H	Non-susceptible to all Except GN & AK	0	0	2	0.6667
I	Non-susceptible to all Except TGC & AK	0	1	0	>0.9999
J	Non-susceptible to all Except STX & TGC	0	1	0	>0.9999
K	Non-susceptible to all Except GN & TGC	0	0	3	0.6667
L	Non-susceptible to all Except GN, CIP & STX	0	0	1	0.6667
M	Non-susceptible to all Except GN, STX & TGC	0	0	1	0.6667
N	Non-susceptible to all Except GN, CIP, STX & TGC	0	0	1	0.6667

3. Discussion

The dissemination of carbapenemases encoding genes in Egypt, especially in healthcare settings, is of great concern in several mapping studies [20–22]. The prevalence of such genes in Egypt could be attributed to considerable reasons that are well-identified globally. For example, hospital environments are considered as a pool for the dissemination of such genes through the horizontal gene transfer that enables their spread via mobile genetic elements such as plasmids [23,24]. Additionally, plasmids may carry more than one resistance gene, and this leads to resistance to diverse antimicrobial classes and as a result limited therapeutic options [25].

Although the transferability of the OXA-48 gene of our selected isolates failed and this was compliant with Skalova et al. trials [26], the reduced transformants susceptibility to most antimicrobials confirms the genetic transfer risk. Additionally, over-the-counter prescription and the misuse of antimicrobials participate in the exacerbation of the antimicrobial resistance problem, especially in Egypt [27,28].

In parallel, the investigated isolates in our study showed increased resistance to various antimicrobial agents and also high MIC values for imipenem and meropenem. This may be markedly attributed to various resistance mechanisms that led to resistance to most antimicrobial classes. Additionally, only 15 isolates (12.82%) of the total investigated isolates were of sole carbapenemase, indicating remarkable coexistence. This finding is most consistent with studies from Egyptian hospitals and unlike the study by Argente et al. in which bla_{OXA-48} gene was the only detected carbapenemase among the tested isolates [29]. This reflects the diverse regional distribution of carbapenemases.

In the current study, the detection and typing of carbapenemases depended on the merge between the phenotypic and molecular methods, and this is in consistency with the study by Karampatakis et al. [30]. The application of the molecular methods in the detection of resistance genes and further typing of isolates is considered as a gold standard tool in the field of epidemiology, and this appears on two occasions in this study; Firstly, the disagreement between phenotypic and genotypic methods for the detection of bla_{KPC} encoding gene which was detected in only 6 isolates by PCR. In this case, the genotypic methods act as an alarming tool for the silent dissemination of such gene which is absent phenotypically but likely to be successfully expressed under certain conditions [31]. Secondly, there is no phenotypic test for the direct detection of $bla_{OXA-48-like}$ genes. Resistance to temocillin is considered as a stepwise prediction of them, since other resistance mechanisms may confer temocillin resistance [32].

In our work, we applied ERIC-PCR and plasmid profile analysis, besides the previously determined antibiogram for the characterization of isolates harboring $bla_{OXA-48-like}$ genes. The discriminatory power of the applied typing methods (antibiogram, ERIC-PCR, and plasmid profile analysis) was approximated according to the SDI calculations, while the highest diversity was represented by plasmid profile analysis (SDI = 0.98). The three typing methods are different in their concepts, and this leads to variation in the clonality of the isolates among the three methods.

The interplay between the antimicrobial resistance and biofilm formation was studied from different sides. On the one hand, resistance genes encoded on some plasmids regulate the expression of fimbrial genes, which are among the essentials of biofilm formation. On the other hand, the plasmid horizontal gene transfer is enhanced in biofilms more than in planktonic cells [33]. In our work, all investigated isolates were carbapenem-resistant and showed biofilm formation ability which was various and independent on the resistance profile of those isolates. It is worth mentioning that despite the absence of such correlation, having 100% of isolates as biofilm formers with the high resistance pattern is considered a troublesome feature of the studied isolates.

Our results were in concordance with the study by El Fertas-Aissani et al. on the prevalence of *fimH-1* and *mrkD* adhesion encoding genes, giving an indication of their conservation in *K. pneumoniae* pathogen [34]. Furthermore, the prevalence of those two adhesion genes and the weak biofilm patterns among the isolates may suggest that the strong biofilm formation may need factors other than fimbrial adhesion genes.

Three isolates in our study, namely, 349, 354, and 375, represented unique findings, as they were resistant to all applied antimicrobials except tigecycline. Additionally, they showed positive results for the MHT, all carbapenemases (bla_{KPC}, bla_{NDM}, bla_{VIM}, bla_{IMP}, and $bla_{OXA-48-like}$) encoding genes, and for *fimH-1* and *mrkD* encoding genes. Although our applied typing methods are relatively simple and of low cost, the results of the three isolates encourage the adoption of other typing methods such as whole-genome sequencing (WGS) [35]. Whole-genome sequencing provides massive data for the determination of the sequence type, plasmid replicon, antimicrobial resistance determinants, and virulence genes. Additionally, *K. pneumoniae* was of special concern for the newly emerged bioinformatics interfaces such as Kaptive Web [36] and KlebNet [37]. Additionally, we are moving globally toward the extended drug resistance with some reported cases of alarmingly pan-drug-resistant isolates [38], research is directed toward finding alternatives for combating antibiotic resistance. The alternative strategies may include antibiotic combinations [39], phage therapy [40], vaccine development [41], natural products [42], revival of classical antibiotics [43], or repurposing non-antimicrobials for targeting pathogens [44].

Finally, the applied infection control measures and awareness campaigns may not be enough to combat the problem of antimicrobial resistance, especially in a developing country such as Egypt. Among the important strategies that are worth emphasizing is launching collaborative platforms between Egyptian healthcare institutes and other international ones for networking and gaining accurate knowledge about the possible patterns of antimicrobial resistance that could be easily disseminated globally [45,46].

4. Materials and Methods

4.1. Isolates Collection and Confirmation

A total of 117 archival non-duplicate and non-consecutive isolates of carbapenem-non-susceptible (resistant or intermediate) *K. pneumoniae* were recovered from clinical specimens. The specimens were collected and processed by dedicated healthcare workers. Furthermore, the specimens were collected from patients admitted to different departments of Kasr Al-Ainy Hospital in Cairo, Egypt, from September 2014 to December 2016. The study was approved by the Ethics Committee of the Faculty of Pharmacy, Al-Azhar University (Girls' branch). The ethical approval code is SMR/2015.

The collected isolates were obtained from the diagnostic routine work of the Clinical Pathology Department of Kasr Al-Ainy Hospital without direct contact with the patient or their related personal data. All laboratory techniques and procedures of isolation, identification, and storage of the included isolates in the current study were performed according to the standard microbiological techniques.

The isolates were previously identified using conventional microbiological methods which included Gram staining, cultural characteristics on MacCkonky's agar, and biochemical testing such as growth on Triple Sugar Iron agar (TSI) and Indole, Methyl red, Voges-Proskauer, and Citrate utilization (IMViC) test. The identification of *K. pneumoniae* isolates was confirmed using Matrix-Assisted Laser Desorption/Ionization Time-Of-Flight Mass Spectrometry (MALDI-TOF/MS) (Vitek MS; BioMérieux, Inc., Marcy-l'Etoile, France).

4.2. Antimicrobial Susceptibility Testing

Antimicrobial susceptibility testing was carried out by disc diffusion according to the Kirby-Bauer method. The isolates were tested against the following antimicrobial agents: amoxicillin/clavulanic acid (20/10 µg), piperacillin/tazobactam (100/10 µg), cefoxitin (30 µg), cefotaxime (30 µg), cefotaxime/clavulanic acid (30/10 µg), ceftazidime (30 µg), ceftazidime/clavulanic acid (30/10 µg), cefepime (30 µg), aztreonam (30 µg), imipenem (10 µg), meropenem (10 µg), ertapenem (10 µg), gentamicin (10 µg), amikacin (30 µg), ciprofloxacin (5 µg), sulfamethoxazole/trimethoprim (1.25/23.75 µg),and tigecycline (15 µg), (Oxoid, Basingstoke, UK). *E. coli* ATCC® 25922, *E. coli* ATCC® 35218, and *P.aeruginosa* ATCC® 27853 were used as quality control strains. The minimum inhibitory concentrations (MIC) of meropenem (commercially available as Meronem^{TM} (500 mg) IV, AstraZeneca, Macclesfield, UK, Ltd.) and imipenem (commercially available as Tienam ^{TM} (500 mg) Injection, Merck Sharp & Dohme BV, Haarlem, The Netherlands) were determined by a microdilution method. The antimicrobial-free media was used as a positive control and the un-inoculated media was used as a negative control. The results were interpreted according to the Clinical and Laboratory Standards Institute guidelines, CLSI (M100-S26, 2016) [47]. Tigecycline interpretive criteria were according to the FDA recommendations. http://www.accessdata.fda.gov/drugsatfda_docs/label/2009/021821s016lbl.pdf. The interpretation criteria of the different antimicrobial agents are mentioned in Table S8.

4.3. Phenotypic Confirmation of Carbapenemases Production

Non-specific screening of carbapenemases production was performed by the modified Hodge test (MHT) according to guidelines of CLSI [47]. The formation of cloverleaf-like indentation of grown *E. coli* ATCC® 25922 along the tested isolate is an indication of the positive result. *K. pneumoniae* ATCC® BAA-1705 was used as a positive control, and *K. pneumoniae* ATCC® BAA-1706 was used as a negative control. Moreover, the inhibitor-based tests were performed as another phenotypic approach for further differentiation between carbapenemases producing isolates. The inhibitor-based tests were based on the inactivation of carbapenemases by combining some substrates independently with meropenem disc. For example, EDTA as metallo-β-lactamases (MβLs) inhibitor, boronic acid as KPC inhibitor, and cloxacillin as AmpC inhibitor [48]. The inhibitor-based tests were performed using MASTDISCS^{TM} ID Carbapenemase Detection Disc Set D70C (MAST group, Merseyside, UK). The set consists of four different discs as shown in Table 3. The interpretation of the tests depends on the comparison of the inhibition zone of the meropenem disc with the inhibition zone of the disc containing meropenem combined with the carbapenemases inhibitor. An additional Temocillin disc (30 μg) was applied for obtaining a preliminary indication of the *bla*$_{OXA-48-like}$ production until performing further genotypic confirmatory tests.

Table 3. The interpretative criteria of applied inhibitor-based tests.

Disc	Content	Interpretative Criteria
	Meropenem (10 μg).	Confirmation of Carbapenem-non-susceptibility.
B	Meropenem + MβL inhibitor.	Confirmation of MβL production (if only B-A ≥ 5 mm).
C	Meropenem + KPC inhibitor.	Confirmation of KPC production (if only C-A ≥ 4 mm).
D	Meropenem + AmpC inhibitor.	Confirmation of AmpC+ porin loss (if both C-A ≥ 4 mm and D-A ≥ 5 mm)
TEM	Temocillin (30 μg).	Indicative for OXA-48 (if there was no synergy detected and the inhibition zone of Temocillin was <11 mm)

4.4. Molecular Detection of Carbapenemases Encoding Genes

Total DNA of the isolates was extracted by the boiling method according to Vaneechoutte et al. [49]. Uniplex PCR reactions were performed for detection of *bla*$_{KPC}$, *bla*$_{NDM}$, *bla*$_{VIM}$, *bla*$_{IMP}$, and *bla*$_{oxa-48-like}$ encoding genes with primers listed in Table S9 [50–52]. The primer specificity was checked against sequences retrieved from GenBank sequence databases using the National Center for Biotechnology Information/Basic Local Alignment Search Tool (NCBI/BLAST), allowing the detection of several variants of each carbapenemase. The primers were synthesized by Eurofins MWG Operon (Ebersberg, Germany). The applied profile was according to the criteria that were mentioned in primer references except for adjustment of annealing temperature to be 56 °C. The positive controls were obtained from the Clinical and Chemical Pathology Department laboratory collection (Kasr Al-Ainy hospital, Cairo, Egypt) and represented by positive strains that harbored investigated carbapenemases encoding genes and had been previously sequenced. The negative control was a template-free master mix.

The resulted PCR amplicons were electrophorized on 0.8% gel alongside a 1-kb ladder. Selected $bla_{OXA\text{-}48\text{-}like}$ gene amplicons, namely, 151, 303, 311, 383, 399, 405, 409, and 415, were sequenced for further investigation using pre-OXA-48 primers targeting the gene and its flanking regions [53].

PCR products were submitted for purification and sequencing in both directions via (Solgent Co. Ltd., Daejeon, Korea). The resulted DNA sequences were trimmed using Geneious Software v.8.1.6 (www.geneious.com) and compared to the Genbank sequence database using the BlASTn tool at https://blast.ncbi.nlm.nih.gov/Blast.cgi.

4.5. Genotyping of Isolates Harboring $bla_{OXA\text{-}48\text{-}like}$ Genes

Thirty-four isolates were confirmed either by inhibitor-based tests and/or PCR as harboring $bla_{OXA\text{-}48\text{-}like}$ genes. These isolates were genotyped by two methods: the first one is the Enterobacterial Repetitive Intergenic Consensus-Polymerase Chain Reaction (ERIC-PCR) which aimed to determine the clonal relatedness between the investigated isolates by detection of specific conserved sequences among them [54]. The second typing method is the plasmid profile analysis which depends on the investigation of the isolate plasmids and characterization of their profile [55].

4.5.1. Enterobacterial Repetitive Intergenic Consensus (ERIC-PCR)

The genomic DNA was extracted using a GeneJet Genomic DNA Purification Kit (K0721, Thermo Scientific, Rockford, IL, USA) according to the manufacturer's protocol. The investigated isolates were genotyped by ERIC-PCR using ERIC-2 primer (5′ AAGTAAGTGACTGGGGTGAGCG 3′) [56]. The PCR profile was according to Abdulall et al. [57] with slight modifications as follows: initial denaturation at 95 °C for 5 min, followed by 35 cycles of denaturation at 95 °C for 1 min, annealing at 45 °C for 1 min, and extension at 72 °C for 8 min, with a final extension at 72 °C for 10 min. Aliquots of PCR amplicons were electrophorized alongside a 10-kb Ladder. The interpretation of the results depended on binary data by which the positive and negative amplifications were assigned as 1/0, respectively. Cluster analysis was performed using NTSYS 2.01 software and the schematic dendrogram was built using the Unweighted Pair-Group Method using Arithmetic Mean (UPGMA).

4.5.2. Plasmid Profile Analysis

The isolate plasmids were extracted by a QIAprep® Spin Miniprep Kit according to the manufacturer's protocol. Extracted plasmids were electrophorized, and the generated patterns were judged visually and considered of distinct type if they showed a single band of different size, according to the ESGEM recommendations [15].

4.6. Calculation of the Discriminatory Power of Different Applied Typing Methods

The discriminatory power of the applied typing methods (antibiogram, ERIC-PCR, and plasmid profile analysis) among the 34 isolates was assessed using Simpson's Diversity Index (SDI). The concept of discrimination based on if there are two unrelated isolates, which were sampled from a test population, they will be probably found in different typing groups [58]. The SDI calculation depended on the following equation $D = 1 - \frac{\Sigma\ n(n-1)}{N(N-1)}$, where D is the diversity index, n is the number of individuals of each species, and N is the total number of individuals of all species. The SDI ranges from 0, which indicates the identical types of the tested isolates, to 1, which indicates the maximum diversity among them.

4.7. Genetic Transfer of $bla_{OXA\text{-}48}$ Gene by Transformation

The acquisition of resistance determinants harbored by the isolate plasmids was assessed by transformation. The preparation and transformation of competent *E. coli* DH5α cells were performed using the $CaCl_2$ method according to Cohen et al. [59]. The genetic transfer of $bla_{OXA\text{-}48}$ gene was assessed for the plasmids of five selected isolates, namely, 311, 383, 399, 405, and 409. The transformants were selected on LB agar containing ampicillin (50 mg/L) and also on LB agar containing meropenem (0.5 mg/L). The plasmids of transformants were extracted and underwent PCR for the detection of the bla_{OXA-48} gene.

4.8. Quantitative Biofilm Formation Assay for the Carbapenem-Resistant Isolates by Microtitre Plate Method

The quantitative assessment of biofilm formation for the 117 carbapenem-resistant isolates was performed by the crystal violet staining assay according to O'Toole protocol with slight modifications [60]. Briefly, overnight culture of each isolate was grown in Luria-Bertani broth, and then was diluted 1:100 into fresh LB medium. One hundred microliters of the isolate dilution was added per well in a flat-bottomed 96 well microtitre plate. After overnight incubation at 37 °C, the medium was discarded, and the wells were washed twice by adding 100 μL of distilled water in each well for further removal of unattached cells and media. The plate was allowed to dry, stained by adding 125 μL of 0.1% crystal violet to each well, and incubated at room temperature for 15 min. The crystal violet was discarded and the plate was washed four times for removing excess stain and allowed to dry. The stained biofilms were solubilized by adding 125 μL of 30% acetic acid to each well and incubated for 15 min at room temperature. The solubilized stain of each well was transferred to a new microtiter plate. The optical density of each well was measured at 600 nm (OD_{600}) using a microplate reader (Stat Fax® 2100, Awareness Technology Inc., Palm City, FL, USA). The experiments were carried out three times in triplicates; the readings of each isolate in each plate were averaged and compared for further interpretation. The positive control was *K. pneumoniae* ATCC® 700603 and the negative control was un-inoculated LB media. The interpretation of the results was according to the recommendations of Stepanović et al. recommendations [61].

4.9. Molecular Detection of Type 1 (fimH-1) and Type 3 (mrkD) Adhesion Encoding Genes

The carbapenem-resistant isolates were screened for the presence of *fimH-1* and *mrkD* encoding genes by uniplex PCR using primers listed in Table S9 [34,62]. The applied profile was as follows: initial denaturation at 94 °C for 4 min followed by 30 cycles of denaturation at 94 °C for 30 s, annealing at 55 °C for 40 s, extension at 72 °C for 1 min, and a final extension at 72 °C for 10 min. The positive controls were obtained from the MTI University Microbiology Laboratory archive, Egypt, and represented by positive strains that harbored investigated adhesion genes and had been previously sequenced. The negative control was a template-free master mix.

4.10. Statistical Analysis

Statistical analysis was performed using GraphPad Prism software (San Diego, CA, USA). The description of qualitative data was as a frequency percentage. The chi-square or Fisher's exact test was used for comparing categorical variables. The results were considered statistically significant when p-value < 0.05. Spearman's test was applied for the determination of the correlation between the biofilm formation category and antimicrobial resistance profile of all isolates.

5. Conclusions

In our work, we studied the prevalence of the "big five" carbapenemases encoding genes, (bla_{KPC}, bla_{NDM}, bla_{VIM}, bla_{IMP}, and $bla_{OXA-48-like}$) among carbapenem-resistant *K. pneumoniae* isolated from a tertiary hospital in Egypt. The tested carbapenem-resistant *K. pneumoniae* isolates showed a remarkable coexistence of different investigated carbapenemases. Additionally, we shed light on $bla_{OXA-48-like}$ genes which represented about one-third of our isolates. The assessment of the simple epidemiological typing methods—antibiogram, ERIC-PCR, and plasmid profile analysis—showed their convergent discriminatory power. Further investigations showed the biofilm formation ability of all isolates with the prevalence of adhesion genes (*fim-H* and *mrkD*). The presence of a high resistance pattern with biofilm formation ability represents a worrisome feature of a superbug such as *K. pneumoniae* in the Egyptian healthcare setting. Thus, there is an urgent need for further strategies for combating resistance, in-depth studies for the resistance genes and their transfer, and also other characteristics such as biofilm formation which all strongly improve the ability of any pathogen to cause high rates of morbidity and mortality.

Supplementary Materials: The following are available online http://www.mdpi.com/2079-6382/9/12/852/s1. Table S1: Antimicrobial susceptibility patterns of *Klebsiella pneumoniae* isolates by disc diffusion method; Table S2: Results of modified Hodge test, inhibitor-based tests, and genotypic detection of investigated carbapenemases; Table S3: Simpson's Diversity Index calculations for antibiogram; Table S4: Simpson's Diversity Index calculations for ERIC-PCR; Table S5: Simpson's Diversity Index calculations for plasmid profile analysis; Table S6: A comparative summary of SDI calculations of the three typing methods; Table S7: Distribution of *fimH-1* and *mrkD* adhesion genes among the carbapenem-resistant isolates; Table S8: The applied antimicrobial agents, their classes, and their interpretive categories; Table S9: Primer sequences and the expected amplicon sizes of investigated genes.

Author Contributions: Conceptualization, A.K.A., A.A.E.-K., and M.M.T.; methodology, S.M.R., M.M.T., and A.K.A.; software, M.M.T. and S.M.R.; formal analysis, S.M.R. and M.M.T.; investigation, A.K.A., A.A.E.-K., M.M.T., and S.M.R.; data curation, S.M.R. and M.M.T.; writing—original draft preparation, S.M.R.; writing—review and editing, S.M.R. and M.M.T.; supervision, A.K.A., A.A.E.-K., and M.M.T. All authors have read and agreed to the published version of the manuscript.

Funding: This research received no external funding.

Conflicts of Interest: The authors declare no conflict of interest.

References

1. O'Neill, J.I.M. Antimicrobial resistance: Tackling a crisis for the health and wealth of nations. *Rev. Antimicrob. Resist.* **2014**, *20*, 1–16.

2. World Health Organization. *Antimicrobial Resistance: Global Report on Surveillance*; WHO: Geneva, Switzerland, 2014; pp. 1–256. Available online: http://apps.who.int/iris/bitstream/10665/112642/1/9789241564748_eng.pdf?ua=1 (accessed on 18 June 2020).

3. Centers for Disease Control and Prevention. About Antimicrobial Resistance. 2018. Available online: https://www.cdc.gov/drugresistance/about.html (accessed on 18 June 2020).

4. World Bank. *Drug-Resistant Infections: A Threat to Our Economic Future*; World Bank: Washington, DC, USA, 2017; Available online: http://www.worldbank.org/en/topic/health/publication/drug-resistant-infections-a-threat-to-our-economic-future (accessed on 20 June 2020).

5. Rice, L.B. Progress and Challenges in Implementing the Research on ESKAPE Pathogens. *Infect. Control. Hosp. Epidemiol.* **2010**, *31*, S7–S10. [CrossRef]

6. World Health Organization. Global Priority List of Antibiotic-Resistant Bacteria to Guide Research, Discovery, and Development of New Antibiotics. 2017. Available online: http://www.who.int/medicines/publications/global-priority-list-antibiotic-resistant-bacteria/en/ (accessed on 22 June 2020).

7. Codjoe, F.S.; Donkor, E.S. Carbapenem Resistance: A Review. *Med. Sci.* **2017**, *6*, 1. [CrossRef]

8. Halat, D.H.; Moubareck, C.A. The Current Burden of Carbapenemases: Review of Significant Properties and Dissemination among Gram-Negative Bacteria. *Antibiotics* **2020**, *9*, 186. [CrossRef]

9. Bratu, S.; Landman, D.; Haag, R.; Recco, R.; Eramo, A.; Alam, M.; Quale, J. Rapid Spread of Carbapenem-Resistant Klebsiella pneumoniae in New York City. *Arch. Intern. Med.* **2005**, *165*, 1430–1435. [CrossRef]

10. Zhao, W.H.; Hu, Z.Q. Acquired metallo-β-lactamases and their genetic association with class 1 integrons and ISCRelements in Gram-negative bacteria. *Futur. Microbiol.* **2015**, *10*, 873–887. [CrossRef] [PubMed]

11. Poirel, L.; Potron, A.; Nordmann, P. OXA-48-like carbapenemases: The phantom menace. *J. Antimicrob. Chemother.* **2012**, *67*, 1597–1606. [CrossRef] [PubMed]

12. Wyres, K.L.; Lam, M.M.C.; Holt, K.E. Population genomics of Klebsiella pneumoniae. *Nat. Rev. Genet.* **2020**, *18*, 344–359. [CrossRef]

13. Pitout, J.; Peirano, G.; Kock, M.M.; Strydom, K.A.; Matsumura, Y. The Global Ascendency of OXA-48-Type Carbapenemases. *Clin. Microbiol. Rev.* **2019**, *33*, 1–48. [CrossRef]

14. Potron, A.; Poirel, L.; Nordmann, P. Derepressed Transfer Properties Leading to the Efficient Spread of the Plasmid Encoding Carbapenemase OXA-48. *Antimicrob. Agents Chemother.* **2013**, *58*, 467–471. [CrossRef]

15. Van Belkum, A.; Tassios, P.; Dijkshoorn, L.; Haeggman, S.; Cookson, B.; Fry, N.; Fussing, V.; Green, J.; Feil, E.; Gerner-Smidt, P.; et al. Guidelines for the validation and application of typing methods for use in bacterial epidemiology. *Clin. Microbiol. Infect.* **2007**, *13*, 1–46. [CrossRef] [PubMed]

16. Vert, M.; Doi, Y.; Hellwich, K.H.; Hess, M.; Hodge, P.E.; Kubisa, P.; Rinaudo, M.; Schué, F. Terminology for biorelated polymers and applications (IUPAC Recommendations 2012). *Pure Appl. Chem.* **2012**, *84*, 377–410. [CrossRef]

17. Vuotto, C.; Longo, F.; Balice, M.P.; Donelli, G.; Varaldo, P.E. Antibiotic Resistance Related to Biofilm Formation in *Klebsiella pneumoniae*. *Pathogens* **2014**, *3*, 743–758. [CrossRef] [PubMed]

18. Alcántar-Curiel, M.D.; Blackburn, D.; Saldaña, Z.; Gayosso-Vázquez, C.; Iovine, N.M.; De La Cruz, M.A.; Girón, J.A. Multi-functional analysis ofKlebsiella pneumoniaefimbrial types in adherence and biofilm formation. *Virulence* **2013**, *4*, 129–138. [CrossRef] [PubMed]

19. Magiorakos, A.P.; Srinivasan, A.; Carey, R.; Carmeli, Y.; Falagas, M.; Giske, C.; Harbarth, S.J.; Hindler, J.; Kahlmeter, G.; Olsson-Liljequist, B.; et al. Multidrug-resistant, extensively drug-resistant and pandrug-resistant bacteria: An international expert proposal for interim standard definitions for acquired resistance. *Clin. Microbiol. Infect.* **2012**, *18*, 268–281. [CrossRef]

20. Manenzhe, R.I.; Zar, H.J.; Nicol, M.P.; Kaba, M. The spread of carbapenemase-producing bacteria in Africa: A systematic review. *J. Antimicrob. Chemother.* **2015**, *70*, 23–40. [CrossRef]

21. Sekyere, J.O.; Govinden, U.; Essack, S. The Molecular Epidemiology and Genetic Environment of Carbapenemases Detected in Africa. *Microb. Drug Resist.* **2016**, *22*, 59–68. [CrossRef]

22. Touati, A.; Mairi, A. Epidemiology of Carbapenemase-producing Enterobacterales in the Middle East: A systematic review. *Expert Rev. Anti-Infect. Ther.* **2020**, *18*, 241–250. [CrossRef]

23. Yan, Z.; Zhou, Y.; Du, M.; Bai, Y.; Liu, B.; Gong, M.; Song, H.; Tong, Y.; Liu, Y. Prospective investigation of carbapenem-resistant *Klebsiella pneumonia* transmission among the staff, environment and patients in five major intensive care units, Beijing. *J. Hosp. Infect.* **2019**, *101*, 150–157. [CrossRef]

24. Lerminiaux, N.A.; Cameron, A.D. Horizontal transfer of antibiotic resistance genes in clinical environments. *Can. J. Microbiol.* **2019**, *65*, 34–44. [CrossRef]

25. Carattoli, A. Plasmids and the spread of resistance. *Int. J. Med. Microbiol.* **2013**, *303*, 298–304. [CrossRef] [PubMed]

26. Skalova, A.; Chudejova, K.; Rotova, V.; Medvecky, M.; Studentova, V.; Chudackova, E.; Lavicka, P.; Bergerova, T.; Jakubu, V.; Zemlickova, H.; et al. Molecular Characterization of OXA-48-Like-Producing Enterobacteriaceae in the Czech Republic and Evidence for Horizontal Transfer of pOXA-48-Like Plasmids. *Antimicrob. Agents Chemother.* **2016**, *61*. [CrossRef]

27. Laxminarayan, R.; Duse, A.; Wattal, C.; Zaidi, A.K.M.; Wertheim, H.; Sumpradit, N.; Vlieghe, E.; Hara, G.L.; Gould, I.M.; Goossens, H.; et al. Antibiotic Resistance—The Need for Global Solutions. *Lancet Infect. Dis.* **2013**, *13*, 1057–1098. [CrossRef]

28. Zakaa El-din, M.; Samy, F.; Mohamed, A.; Hamdy, F.; Yasser, S.; Ehab, M. Egyptian community pharmacists' attitudes and practices towards antibiotic dispensing and antibiotic resistance; a cross-sectional survey in Greater Cairo. *Curr. Med. Res. Opin.* **2019**, *35*, 939–946. [CrossRef] [PubMed]

29. Argente, M.; Miró, E.; Martí, C.; Vilamala, A.; Alonso-Tarrés, C.; Ballester, F.; Calderon, A.; Galles, C.; Gasós, A.; Mirelis, B.; et al. Molecular characterization of OXA-48 carbapenemase-producing *Klebsiella pneumoniae* strains after a carbapenem resistance increase in Catalonia. *Enferm. Infecc. Microbiol. Clínica* **2019**, *37*, 82–88. [CrossRef] [PubMed]

30. Karampatakis, T.; Geladari, A.; Politi, L.; Antachopoulos, C.; Iosifidis, E.; Tsiatsiou, O.; Karyoti, A.; Papanikolaou, V.; Tsakris, A.; Roilides, E. Cluster-distinguishing genotypic and phenotypic diversity of carbapenem-resistant Gram-negative bacteria in solid-organ transplantation patients: A comparative study. *J. Med. Microbiol.* **2017**, *66*, 1158–1169. [CrossRef] [PubMed]

31. Azimi, L.; Talebi, M.; Owlia, P.; Pourshafie, M.-R.; Najafi, M.; Lari, E.R.; Lari, A.R.; Najafi, M. Tracing of false negative results in phenotypic methods for identification of carbapenemase by Real-time PCR. *Gene* **2016**, *576*, 166–170. [CrossRef]

32. EUCAST Subcommittee for Detection of Resistance Mechanisms and Specific Resistances of Clinical and/or Epidemiological Importance. 2017. EUCAST Guidelines for Detection of Resistance Mechanisms and Specific Resistances of Clinical and/or Epidemiological Importance, Version 2.0. Available online: http://www.eucast.org/fileadmin/src/media/PDFs/EUCAST_files/Resistance_mechanisms/EUCAST_detection_of_resistance_mechanisms_170711.pdf (accessed on 25 June 2020).

33. Madsen, J.S.; Burmølle, M.; Hansen, L.H.; Sørensen, S.J. The interconnection between biofilm formation and horizontal gene transfer. *FEMS Immunol. Med. Microbiol.* **2012**, *65*, 183–195. [CrossRef]

34. El Fertas-Aissani, R.; Messai, Y.; Alouache, S.; Bakour, R. Virulence profiles and antibiotic susceptibility patterns of Klebsiella pneumoniae strains isolated from different clinical specimens. *Pathol. Biol.* **2013**, *61*, 209–216. [CrossRef]

35. Köser, C.U.; Ellington, M.J.; Cartwright, E.J.P.; Gillespie, S.H.; Brown, N.M.; Farrington, M.; Holden, M.T.G.; Dougan, G.; Bentley, S.D.; Parkhill, J.; et al. Routine Use of Microbial Whole Genome Sequencing in Diagnostic and Public Health Microbiology. *PLoS Pathog.* **2012**, *8*, e1002824. [CrossRef]

36. Wick, R.R.; Heinz, E.; Holt, K.E.; Wyres, K.L. Kaptive Web: User-Friendly Capsule and Lipopolysaccharide Serotype Prediction for Klebsiella Genomes. *J. Clin. Microbiol.* **2018**, *56*, 00197-18. [CrossRef] [PubMed]

37. Lee, M.; Pinto, N.A.; Kim, C.Y.; Yang, S.; D'Souza, R.; Yong, D.; Lee, I. Network Integrative Genomic and Transcriptomic Analysis of Carbapenem-Resistant Klebsiella pneumoniae Strains Identifies Genes for Antibiotic Resistance and Virulence. *mSystems* **2019**, *4*, e00202-19. [CrossRef] [PubMed]

38. Zowawi, H.M.; Forde, B.M.; Alfaresi, M.; Alzarouni, A.; Farahat, Y.; Chong, T.-M.; Yin, W.F.; Chan, K.-G.; Li, J.; Schembri, M.A.; et al. Stepwise evolution of pandrug-resistance in *Klebsiella pneumoniae*. *Sci. Rep.* **2015**, *5*, 15082. [CrossRef]

39. Daikos, G.L.; Tsaousi, S.; Tzouvelekis, L.S.; Anyfantis, I.; Psichogiou, M.; Argyropoulou, A.; Stefanou, I.; Sypsa, V.; Miriagou, V.; Nepka, M.; et al. Carbapenemase-Producing *Klebsiella pneumoniae* Bloodstream Infections: Lowering Mortality by Antibiotic Combination Schemes and the Role of Carbapenems. *Antimicrob. Agents Chemother.* **2014**, *58*, 2322–2328. [CrossRef]

40. Kęsik-Szeloch, A.; Drulis-Kawa, Z.; Weber-Dąbrowska, B.; Kassner, J.; Majkowska-Skrobek, G.; Augustyniak, D.; Łusiak-Szelachowska, M.; Żaczek, M.; Górski, A.; Kropinski, A.M. Characterising the biology of novel lytic bacteriophages infecting multidrug resistant *Klebsiella pneumoniae*. *Virol. J.* **2013**, *10*, 100. [CrossRef] [PubMed]

41. Ahmad, T.A.; El-Sayed, L.H.; Haroun, M.; Hussein, A.A.; El Ashry, E.S.H. Development of immunization trials against *Klebsiella pneumoniae*. *Vaccine* **2012**, *30*, 2411–2420. [CrossRef] [PubMed]

42. Somboro, A.M.; Sekyere, J.O.; Amoako, D.G.; Kumalo, H.M.; Khan, R.; Bester, L.A.; Essack, S. In vitro potentiation of carbapenems with tannic acid against carbapenemase-producing enterobacteriaceae: Exploring natural products as potential carbapenemase inhibitors. *J. Appl. Microbiol.* **2018**, *126*, 452–467. [CrossRef]

43. Cassir, N.; Rolain, J.-M.; Brouqui, P. A new strategy to fight antimicrobial resistance: The revival of old antibiotics. *Front. Microbiol.* **2014**, *5*, 551. [CrossRef]

44. Younis, W.; Abdelkhalek, A.; Mayhoub, A.S.; Seleem, M.N. In Vitro Screening of an FDA-Approved Library Against ESKAPE Pathogens. *Curr. Pharm. Des.* **2017**, *23*, 2147–2157. [CrossRef]

45. Rosenthal, V.D.; Bat-Erdene, I.; Gupta, D.; Belkebir, S.; Rajhans, P.; Zand, F.; Myatra, S.N.; Afeef, M.; Tanzi, V.L.; Muralidharan, S.; et al. Six-year multicenter study on short-term peripheral venous catheters-related bloodstream infection rates in 727 intensive care units of 268 hospitals in 141 cities of 42 countries of Africa, the Americas, Eastern Mediterranean, Europe, South East Asia, and Western Pacific Regions: International Nosocomial Infection Control Consortium (INICC) findings. *Infect. Control Hosp. Epidemiol.* **2020**, *41*, 553–563. [CrossRef]

46. Al-Hassan, L.; Roemer-Mahler, A.; Price, J.; Islam, J.; El-Mahallawy, H.; Higgins, P.G.; Hussein, A.F.A.; Roca, I.; Newport, M. The TACTIC experience: Establishing an international, interdisciplinary network to tackle antimicrobial resistance. *J. Med. Microbiol.* **2020**, *69*, 1213–1220. [CrossRef] [PubMed]

47. Clinical and Laboratory Standards Institute. Performance Standards for Antimicrobial Susceptibility Testing. In *CLSI Document M100-S26*; Clinical and Laboratory Standards Institute: Wayne, PA, USA, 2016.

48. Lutgring, J.D.; Limbago, B. The Problem of Carbapenemase-Producing-Carbapenem-Resistant-Enterobacteriaceae Detection. *J. Clin. Microbiol.* **2016**, *54*, 529–534. [CrossRef] [PubMed]

49. Vaneechoutte, M.; Dijkshoorn, L.; Tjernberg, I.; Elaichouni, A.; De Vos, P.; Claeys, G.; Verschraegen, G. Identification of Acinetobacter genomic species by amplified ribosomal DNA restriction analysis. *J. Clin. Microbiol.* **1995**, *33*, 11–15. [CrossRef] [PubMed]

50. Akpaka, P.E.; Swanston, W.H.; Ihemere, H.N.; Correa, A.; Torres, J.A.; Tafur, J.D.; Montealegre, M.C.; Quinn, J.P.; Villegas, M.V. Emergence of KPC-Producing Pseudomonas aeruginosa in Trinidad and Tobago. *J. Clin. Microbiol.* **2009**, *47*, 2670–2671. [CrossRef]

51. Poirel, L.; Walsh, T.R.; Cuvillier, V.; Nordmann, P. Multiplex PCR for detection of acquired carbapenemase genes. *Diagn. Microbiol. Infect. Dis.* **2011**, *70*, 119–123. [CrossRef]

52. Aktaş, Z.; Kayacan, C.B.; Schneider, I.; Can, B.; Midilli, K.; Bauernfeind, A. Carbapenem-Hydrolyzing Oxacillinase, OXA-48, Persists in *Klebsiella pneumoniae* in Istanbul, Turkey. *Chemotherapy* **2008**, *54*, 101–106. [CrossRef]

53. Potron, A.; Nordmann, P.; Lafeuille, E.; Al Maskari, Z.; Al Rashdi, F.; Poirel, L. Characterization of OXA-181, a Carbapenem-Hydrolyzing Class D β-Lactamase from *Klebsiella pneumoniae*. *Antimicrob. Agents Chemother.* **2011**, *55*, 4896–4899. [CrossRef]

54. Wilson, L.A. Sharp, P.M. Enterobacterial Repetitive Intergenic Consensus (ERIC) Sequences in Escherichia coli: Evolution and Implications for ERIC-PCR. *Mol. Biol. Evol.* **2006**, *23*, 1156–1168. [CrossRef]

55. Mayer, L.W. Use of plasmid profiles in epidemiologic surveillance of disease outbreaks and in tracing the transmission of antibiotic resistance. *Clin. Microbiol. Rev.* **1988**, *1*, 228–243. [CrossRef]

56. Versalovic, J.; Koeuth, T.; Lupski, J.R. Distribution of repetitive DNA sequences in eubacteria and application to finerpriting of bacterial enomes. *Nucleic Acids Res.* **1991**, *19*, 6823–6831. [CrossRef]

57. Abdulall, A.K.; Tawfick, M.M.; El-Manakhly, A.R.; El-Kholy, A. Carbapenem-resistant Gram-negative bacteria associated with catheter-related bloodstream infections in three intensive care units in Egypt. *Eur. J. Clin. Microbiol. Infect. Dis.* **2018**, *37*, 1647–1652. [CrossRef] [PubMed]

58. Hunter, P.R.; Gaston, M.A. Numerical index of the discriminatory ability of typing systems: An application of Simpson's index of diversity. *J. Clin. Microbiol.* **1988**, *26*, 2465–2466. [CrossRef] [PubMed]

59. Cohen, S.N.; Chang, A.C.Y.; Hsu, L. Nonchromosomal Antibiotic Resistance in Bacteria: Genetic Transformation of *Escherichia coli* by R-Factor DNA. *Proc. Natl. Acad. Sci. USA* **1972**, *69*, 2110–2114. [CrossRef]

60. O'Toole, G.A. Microtiter Dish Biofilm Formation Assay. *J. Vis. Exp.* **2011**, *47*, e2437. [CrossRef] [PubMed]

61. Stepanović, S.; Vuković, D.; Hola, V.; Bonaventura, G.D.; Djukić, S.; Ćirković, I.; Ruzicka, F. Quantification of biofilm in microtiter plates: Overview of testing conditions and practical recommendations for assessment of biofilm production by staphylococci. *APMIS* **2007**, *115*, 891–899. [CrossRef] [PubMed]

62. Hennequin, C.; Forestier, C. Influence of capsule and extended-spectrum beta-lactamases encoding plasmids upon Klebsiella pneumoniae adhesion. *Res. Microbiol.* **2007**, *158*, 339–347. [CrossRef]

Publisher's Note: MDPI stays neutral with regard to jurisdictional claims in published maps and institutional affiliations.

Article

Multiplicity of Carbapenemase-Producers Three Years after a KPC-3-Producing *K. pneumoniae* ST147-K64 Hospital Outbreak

Ana Margarida Guerra [1], Agostinho Lira [2], Angelina Lameirão [2], Aurélia Selaru [2], Gabriela Abreu [2], Paulo Lopes [2], Margarida Mota [2], Ângela Novais [1,*] and Luísa Peixe [1,*]

[1] UCIBIO, Faculdade de Farmácia, Universidade do Porto, 4050-313 Porto, Portugal; anamg.guerra@gmail.com

[2] Centro Hospitalar Vila Nova de Gaia/Espinho, 4434-502 Vila Nova de Gaia, Portugal; lira@chvng.min-saude.pt (A.L.); angelinalameirao@chvng.min-saude.pt (A.L.); aurelia.selaru@chvng.min-saude.pt (A.S.); gabriela.abreu@chvng.min-saude.pt (G.A.); microlopes@chvng.min-saude.pt (P.L.); mmota@chvng.min-saude.pt (M.M.)

* Correspondence: aamorim@ff.up.pt or angelasilvanovais@gmail.com (Â.N.); lpeixe@ff.up.pt (L.P.); Tel.: +351-220-428-586 (Â.N.); +351-220-428-580 (L.P.)

Received: 6 October 2020; Accepted: 7 November 2020; Published: 13 November 2020

Abstract: Carbapenem resistance rates increased exponentially between 2014 and 2017 in Portugal (~80%), especially in *Klebsiella pneumoniae*. We characterized the population of carbapanemase-producing Enterobacterales (CPE) infecting or colonizing hospitalized patients (2017–2018) in a central hospital from northern Portugal, where KPC-3-producing *K. pneumoniae* capsular type K64 has caused an initial outbreak. We gathered phenotypic (susceptibility data), molecular (population structure, carbapenemase, capsular type) and biochemical (FT-IR) data, together with patients' clinical and epidemiological information. A high diversity of Enterobacterales species, clones (including *E. coli* ST131) and carbapenemases (mainly KPC-3 but also OXA-48 and VIM) was identified three years after the onset of carbapenemases spread in the hospital studied. ST147-K64 *K. pneumoniae*, the initial outbreak clone, is still predominant though other high-risk clones have emerged (e.g., ST307, ST392, ST22), some of them with pandrug resistance profiles. Rectal carriage, previous hospitalization or antibiotherapy were presumptively identified as risk factors for subsequent infection. In addition, our previously described Fourier Transform infrared (FT-IR) spectroscopy method typed 94% of *K. pneumoniae* isolates with high accuracy (98%), and allowed to identify previously circulating clones. This work highlights an increasing diversity of CPE infecting or colonizing patients in Portugal, despite the infection control measures applied, and the need to improve the accuracy and speed of bacterial strain typing, a goal that can be met by simple and cost-effective FT-IR based typing.

Keywords: carbapenemase-producing Enterobacterales; KPC; carbapenem; multidrug resistance; nosocomial

---

1. Introduction

The dissemination of carbapenemase-producing *Enterobacterales* (CPE) has been causing serious concerns in many EU countries due to the limited therapeutic options and ineffective infection control policies [1,2]. Portugal is one of the EU countries where the rates of carbapenems resistance amongst invasive isolates has been alarmingly rising (~80% between 2014 and 2017), especially among *Klebsiella pneumoniae* for which reported rates in 2018 were 11.7% [3,4]. Reporting of CPE from national laboratories is mandatory since 2013 and recommendations to control CPE spread that include identification of rectal carriers and improvement of infection control policies in the hospital setting have been officially launched in 2017 [5].

Available data on molecular epidemiology and population structure of CPE populations in the country represent scattered snapshots on particular species and/or small institutions. The ST147 *K. pneumoniae* clone exhibiting K-type K64 has been one of the first responsible for KPC-3 spread in diverse healthcare settings in the north of Portugal [6,7]. It was responsible for the first large outbreak of CPE in a reference hospital from the north of Portugal in 2015 [8] and it has been described in other regions [9,10]. Analysis of recent CPE collections is still scarce and there is limited analysis on risk factors and the effect of infection control policies in a single institution.

Considering that precise and timely bacterial typing information can make the difference in the effectiveness of infection control measures, we have recently proposed a quick and cost-effective approach for accurate *K. pneumoniae* typing based on an in-house database of Fourier-Transform Infrared (FT-IR) discriminatory spectra [11].

In the present study, we aim to understand the evolution of CPE population in a central hospital from northern Portugal three years after the onset of carbapenemases spread [8] in order to identify critical factors for the increasing burden of CPE in hospitalized patients. Moreover, we use this collection to evaluate the performance of our method for accurate subtyping of *K. pneumoniae* contemporaneous isolates.

2. Results

2.1. The Increasing Diversity of CPE Population

Carbapenemase production was confirmed in 120 out of the 128 (94%) recovered isolates, that were subsequently characterized. The remaining ones were possible ESBL producers with permeability defects. These 120 isolates were identified in rectal swabs ($n = 86$) and clinical samples ($n = 34$) from 114 patients. Multiple *Enterobacterales* species were identified in either clinical samples or rectal swabs, though *K. pneumoniae* was by far the most frequent ($n = 98$; $n = 66$ from rectal swabs and $n = 32$ from clinical samples) (Table 1). While nearly all isolates from infection belonged to *K. pneumoniae* (n = 32/34; 94%), the isolates from fecal samples presented a greater diversity, including mostly *K. pneumoniae* ($n = 66/86$; 76.7%), but also *E. coli* ($n = 10/86$; 11.6%), different *E. cloacae* complex species including *E. asburiae* and *E. hormaechei* ($n = 5/86$; 5.8%), *K. oxytoca* ($n = 4/86$; 4.7%), and *Citrobacter freundii* complex ($n = 1/86$; 1.2%), the latter two appearing exclusively in colonization samples. *E. coli* was identified mainly from rectal swabs ($n = 10/11$; 91%). Most isolates produced KPC-3 ($n = 118/120$; 98.3%) including all *K. pneumoniae*, but OXA-48 and VIM-1 were also occasionally detected in 1 *E. coli* clinical isolate and 1 *E. hormaechei* identified as a gastrointestinal colonizer, respectively. *mcr* genes were not detected.

Table 1. Enterobacterales species identified as carbapenemase producers.

Microorganism	Colonization		Infection		Total	
	No.	%	No.	%	No.	%
Klebsiella pneumoniae	66	77	32	94	98	82
Escherichia coli	10	11	1	3	11	9
Enterobacter cloacae complex	5	6	1	3	6	5
Klebsiella oxytoca	4	5	-	-	4	3
Citrobacter freundii complex	1	1	-	-	1	
Total	86	100	34	100	120	100

Simultaneous rectal carriage (in the same sample) by two different species of CPE (*K. pneumoniae* and *E. coli*) or two different *K. pneumoniae* clones was found in four and one patients, respectively. In all cases, both species/clones were KPC-3 producers, which may indicate the interspecies transfer of a

KPC-3-encoding plasmid. In the only case where isolates from urine and blood were recovered from the same patient, they had identical phenotype and genotype.

2.2. Epidemiological and Clinical Data of Patients Carrying CPE Isolates

Patient's age ranged from 20 years to 96 years, with mean age of 74 ± 14 years (median 78 years) and female patients were predominant (59% female versus 41% male). They were mostly located in medical (*n* = 55, 48%) or surgical (*n* = 51; 45%) wards, whereas seven (6%) of them were patients attending the emergency room, and 1 (0.9%) was at intensive care unit. Clinical isolates (*n* = 34) were identified mostly at surgical units (*n* = 17/34; 50%), whereas isolates from rectal screenings were obtained mostly from medical wards (*n* = 50/86; 58%).

The clinical isolates were identified in urine (*n* = 20; 57%) and pus (*n* = 7; 21%) but also in blood (*n* = 3; 9%), biologic liquid (*n* = 2; 6%), or respiratory samples (*n* = 2; 6%). The majority of them were *K. pneumoniae* (*n* = 32/34; 94%), but also *E. cloacae* (*n* = 1/34; 3%) or *E. coli* (*n* = 1/34; 3%). Clinical isolates were identified in 4 medical units (internal medicine was the most frequent, 55%) and 7 surgical units (many from women's surgery department, 35%). Only ten of the patients with clinical isolates were primarily diagnosed with infectious diseases at ICU, emergency or medical units, whereas in many of the other patients, infection occurred during hospitalization for a median of 30 days (Table 2). Most infected patients (93%) were previous colonized (0–421 days, mean 44 days), most of them in the month previous to infection. Isolates from outpatients (*n* = 6; 74–94 years old) were collected from urine (*n* = 5/7; 71%) and blood (*n* = 2/7; 2.9%). Three of these outpatients were previous CPE carriers from previous hospitalizations, and two of the other three patients had been hospitalized in the last month, all of this suggesting nosocomial acquisition. Most of these patients (70%) received antibiotherapy in the three months prior to infection, even most (87%) of those that were not diagnosed with infectious diseases.

Isolates from rectal swabs were collected from patients at 9 medical (*n* = 56; 65%) or 7 surgical units (*n* = 29, 34%). Internal medicine (*n* = 24; 43%) and unit III (a satellite hospital at a 23Km distance, *n* = 10; 18%) were the most frequent units within the medical department, whereas orthopedics and women's surgery (*n* = 8; 23% each) were the most common units within surgical departments. Of highlight, a portion of isolates (*n* = 22, 26%) from both medical and surgical units were recovered from 20 patients identified as CPE carriers that were selected for decolonization protocol by gentamicin use. Eight of them were being followed for long periods of time (between 2016 and mid 2017). From these, most (87.5%) had at least one negative CPE screening, on average, 208 days after initial gentamicin treatment though 3 negative rectal screenings are needed to consider a successful decolonization.

In total, from the 81 patients screened for CPE in rectal swabs, 26 (32%) patients had positive screenings from previous hospitalizations at this hospital. From the 42 (52%) *de novo* CPE carriers, most had long (>14 days) hospitalization periods (*n* = 22; 52%) or had had previous antibiotherapy treatments (*n* = 17; 40%). The remaining 13 screening samples belonged to patients (16%) with no history of screenings.

In relation to antibiotic exposure, 39 from the 42 *de novo* colonized patients (93%) were exposed to antimicrobials in the previous three months. The main antibiotics administered were piperacillin plus tazobactam (27%), amoxicillin plus clavulanic acid (20%), and broad spectrum cephalosporins (10% ceftriaxone).

K. pneumoniae isolates were distributed among all origins (50% medical, 41% surgical, 8% emergency room, 1% ICU), whereas six out of the 11 *E. coli* isolates were recovered from patients under decolonization protocol (*n* = 6/11, 55%) and *E. cloacae* complex, *K. oxytoca*, and *C. freundii* complex were identified mainly or exclusively in surgical units (83%, 75%, and 100%, respectively).

Table 2. Epidemiological data on clinical CPE isolates.

Unit	Patient N° (Sex/Age)	Clone (MLST)	Species	Isolation Date	Diagnosis of Infectious Diseases	Product	Previous Antibiotherapy (3 months)	Previous Hospitalizations	Duration of Hospitalization	Previous Colonization [a]
Emergency	2 (M/79)	ST147	*K. pneumoniae*	19/11/17	N	U	Y	Y (this hospital, 2 months)	7 days	+ (2 months)
Emergency	3 (M/87)	ST147	*K. pneumoniae*	20/11/17	Y	U	Y	Y (this hospital, 1 month)	6 days	Not tested
Emergency	10 (F/94)	ST147	*K. pneumoniae*	07/12/17	N	U	N	Y (this hospital, >3 months)	13 days	+ (7 months)
Emergency	15 (F/79)	ST392	*K. pneumoniae*	18/12/17	Y	B	Y	Y (this hospital; 15 days)	6 days	+ (14 days)
Emergency	19 (F/79)	ST147	*K. pneumoniae*	29/12/17	Y	U	N	Y (this hospital; 1 month)	13 days	- (2 months)
Emergency	30 (M/74)	ST147	*K. pneumoniae*	12/02/18	Y	U	Y	N	10 days	Not tested
	30 (M/74)	ST147	*K. pneumoniae*	13/02/18						
Medical	11 (F/96)	ST307	*K. pneumoniae*	12/12/17	N	U	Y	Y (this hospital, 1 month)	16 days	+ (1 month)
Medical	18 (F/66)	ST15	*K. pneumoniae*	28/12/17	N	U	Y	N	45 days	+ (12 days)
Medical	20 (F/82)	ST22	*K. pneumoniae*	09/01/18	Y	U	Y	N	36 days	- (6 days)
Medical	22 (F/86)	ST147	*K. pneumoniae*	12/01/18	Y	U	Y	N	26 days	+ (1 day)
Medical	24 (M/85)	ST147	*K. pneumoniae*	22/01/18	Y	W	Y	N	28 days	+ (2 months)
Medical	31 (F/81)	ST392	*K. pneumoniae*	12/02/18	Y	U	Y	N	19 days	+ (2 days)
Medical	31 (F/75)	ST392	*K. pneumoniae*	21/02/18	N	U	Y	N	6 days	Not tested
Medical	6 (F/84)	ST392	*K. pneumoniae*	29/11/17	N	U	N	Y (this hospital, 1 month)	29 days	+ (1 month)
Medical	25 (M/42)	ST131	*E. coli*	25/01/18	N	W	Y	Y (another hospital)	50 days	+ (20 days)
Surgical	4 (F/60)	ST147	*K. pneumoniae*	21/11/17	N	W	Y	N	45 days	+ (7 days)
Surgical	5 (M/32)	ST147	*K. pneumoniae*	28/11/17	N	W	Y	Y (this hospital, 2 months)	22 days	+ (3 months)
Surgical	8 (F/91)	ST359	*K. pneumoniae*	04/12/17	N	R	Y	N	16 days	- (3 days)
Surgical	9 (F/48)	ST11	*K. pneumoniae*	05/12/17	N	U	Y	N	62 days	+ (5 days)
Surgical	12 (F/78)	ST147	*K. pneumoniae*	12/12/17	N	W	Y	Y (another hospital)	59 days	+ (1 month)
Surgical	13 (F/61)	ST147	*K. pneumoniae*	15/12/17	N	U	N	N	17 days	+ (1 day)
Surgical	14 (F/92)	ST147	*K. pneumoniae*	16/12/17	N	W	Y	N	36 days	+ (5 days)
Surgical	16 (M/69)	ST147	*K. pneumoniae*	19/12/17	N	B	Y	N	30 days	+ (1 day)
Surgical	17 (M/60)	ST147	*K. pneumoniae*	26/12/17	N	U	Y	N	25 days	+ (7 days)
Surgical	21 (F/82)	ST147	*K. pneumoniae*	11/01/18	N	U	Y	N	80 days	+ (1 month)
Surgical	23 (F/68)	ST147	*K. pneumoniae*	12/01/18	N	R	Y	N	40 days	+ (7 days)
Surgical	26 (M/73)	ND	*K. pneumoniae*	31/01/18	N	W	Y	N	30 days	+ (5 days)
Surgical	27 (F/82)	ST147	*K. pneumoniae*	03/02/18	N	W	Y	Y (this hospital, 2 months)	30 days	+ (1 month)
Surgical	28 (M/78)	-	*E. cloacae* complex	07/02/18	N	W	Y	Y (another hospital)	15 days	+ (1 day)
Surgical	29 (M/77)	ST147	*K. pneumoniae*	08/02/18	N	U	Y	N	39 days	+ (9 days)
Surgical	32 (M/78)	ST392	*K. pneumoniae*	12/02/18	Y	U	Y	N	57 days	+ (15 days)
Surgical	1 (M/79)	ST147	*K. pneumoniae*	04/11/17	N	U	Y	Y (another hospital)	41 days	+ (2 days)
UCI	7 (M/61)	ND	*K. pneumoniae*	03/12/17	Y	U	Y	Y (this hospital, 2 months)	8 days	+ (2 months)

[a] Date of last colonization (when tested). ND = Not Done. N = No; Y = Yes; B = Blood; U = Urine; W = Wound; R = Respiratory.

2.3. Multiplicity of Clonal Lineages and Emergence of Other International High-Risk Clones

Using our FT-IR spectral workflow and database (11), we were able to predict the capsular type of 94% of the *K. pneumoniae* isolates (6 isolates were not predicted by our model). From those, we grouped isolates in 9 clones. Four of them had identity with capsular types included in our model and identified as K64 (ST147), KL112 (ST15), K19 (ST15), and KL105 (ST11). The remaining five were predicted as not belonging to any of the 19 classes (K-types) included in our model. Indeed, they were subsequently identified as K9 (ST22), KL27 (ST392), KL102 (ST307), KL15/KL17/KL51/KL52 (not typed by MLST), or K53 (not typed by MLST). Only two isolates (2%) were incorrectly predicted (Table 3).

Table 3. Performance of Fourier Transform infrared (FT-IR) typing and epidemiological data of carbapenemase-producing *K. pneumoniae* isolates from the studied hospital.

n (%)	*wzi* Alelle	K-Type by *wzi* Sequencing	K-Type by FT-IR [a]	MLST	Source [b]	Hospital Ward
54 (55)	*wzi*64	K14/K64	KL64	ST147	S ($n = 34$) C ($n = 20$)	MU ($n = 25$) SU ($n = 23$) ER ($n = 6$)
20 (20.4)	*wzi*187	KL27	FT1, 1 NT	ST392	S ($n = 15$) C ($n = 5$)	MU ($n = 15$) SU ($n=3$) ER ($n = 2$)
5 (5.1)	*wzi*9	K9	FT2	ST22	S ($n = 4$) C ($n = 1$)	MU ($n = 4$) SU ($n = 1$)
4 (4.1)	*wzi*173	KL102	FT3	ST307	S ($n = 3$) C ($n = 1$)	MU ($n = 3$) SU ($n = 1$)
3 (3.1)	*wzi*19	K19	1 K19, KL107 *, 1 NT	ST15	S ($n = 3$)	SU ($n = 3$)
2 (3.1)	*wzi*93	K60/KL112	KL112	ST15	S ($n = 1$) C ($n = 1$)	MU ($n = 1$) SU ($n = 1$)
2 (2)	*wzi*53	K53	FT4	ND	S ($n = 1$) C ($n = 1$)	SU ($n = 2$)
2 (2)	*wzi*50	KL15/KL17/KL51/KL52	FT5	ND	S ($n = 1$) C ($n = 1$)	MU ($n = 1$) ICU ($n = 1$)
2 (2)	*wzi*461	NA	NT	ND	S ($n = 2$)	SU ($n = 2$)
2 (2)	*wzi*236	KL10	NT	ST359	C ($n = 1$) S ($n=1$)	SU ($n = 2$)
1 (1)	*wzi*75	KL105	NT	ST11	C ($n = 1$)	SU ($n = 1$)
1 (1)	*wzi*102	K31	KL62 *	ND	S ($n = 1$)	SU ($n = 1$)

[a] FT-IR types not recognized by the model were attributed arbitrary designations (FT1-FT5). [b] S = Rectal Screening; C = Clinical isolate; NT = Not Typeable; NA = Not attributable; ND = Not Done; * Incorrect prediction. MU = Medical Units; SU = Surgical Units; ER = Emergency Room; ICU = Intensive Care Unit.

Three years after the initial outbreak, ST147-K64 is still the predominant *K. pneumoniae* clone (55% overall) responsible for 62.5% of CPE infections and also the most frequent *K. pneumoniae* colonizer (52%). They were recovered from patients at several medical or surgical units (50% vs 39%, respectively) and also in patients at emergency room (11%). The second most represented clone was ST392-KL27 (20.4%) that was mainly identified in medical wards (85%) in February 2018 (Table 3). A diversity of other clones including ST22 K9 (5.1%), ST307-KL102 (4.1%), ST15-K19 (3%), ST15-KL112 (2%), or others were identified as colonizers before or after infections (Table 3).

The carbapenem resistant *E. coli* isolates belonged to distinct phylo-groups, being B2 the most prevalent one ($n = 5$; 46%), followed by groups A ($n = 2$; 18%), E ($n = 2$; 18%), B1 ($n = 1$, 9%), and F ($n = 1$, 9%). One of the B2 *E. coli* producing KPC-3 belonged to the worldwide disseminated clone ST131 (Table 4).

Table 4. Characterization of CPE producers other than *K. pneumoniae*.

Species	*E. coli* Phylogenetic Group (N°)	Carbapenemase (N°)	Source (N°)	Hospital Ward
E. coli	B2 (5) [a]	KPC-3 (3)	S (4)	*n* = 2 SU
		OXA-48 (1)	C (1)	*n* = 3 MU
	A (2)	KPC-3 (2)	S (2)	*n* = 2 MU
	E (2)	KPC-3 (2)	S (2)	*n* = 1 MU *n* = 1 SU
	B1 (1)	KPC-3 (1)	S (1)	*n* = 1 MU
	F (1)	KPC-3 (1)	S (1)	*n* = 1 MU
K. oxytoca	NA (4)	KPC-3 (4)	S (4)	*n* = 3 SU *n* = 1 MU
E. cloacae complex	NA (6)	KPC-3 (5)	S (4), C (1)	*n* = 5 SU
		VIM-1 (1)	S	*n* = 1 MU
C. freundii complex	NA (1)	KPC-3 (1)	S	*n* = 1 SU

[a] B2 *E. coli* producing KPC-3 belongs to ST131; S = Rectal Screening; C = Clinical Isolate; MU = Medical Units; SU = Surgical units; NA = Not applicable.

2.4. A Variety of Multidrug Resistance Profiles

All isolates showed multidrug resistance (MDR) phenotypes and were resistant to third generation cephalosporins (100%) and beta-lactam/beta-lactamase inhibitors (100%) and the vast majority was resistant to ciprofloxacin (90%). All carbapenemase producers tested were nonsusceptible to ertapenem (97% resistant, 3% intermediate) and demonstrated susceptible (S), intermediate (I), or resistance (R) phenotypes to imipenem (61% R, 32% I, 7% S) and meropenem (54% R, 31% I, 15% S). Conversely, amikacin (94%), as well as tigecycline (95%), tetracycline (80%), chloramphenicol (80%), or fosfomycin (78%) had the highest susceptibility rates.

Susceptibility patterns varied according to species and clone. Besides third generation cephalosporins, carbapenems and ciprofloxacin, ST307 and ST22 *K. pneumoniae* were resistant or intermediate to all aminoglycosides (except amikacin), and ST22 additionally to trimethoprim/trimethoprim-sulfamethoxazole and chloramphenicol. ST147 and ST392 were variably resistant to aminoglycosides (ST147-29%; ST392-57%) other than amikacin, trimethoprim-sulfamethoxazole (ST147-46%; ST392-43%). The clinical *E. cloacae* isolate was the most susceptible in this study (susceptibility to all aminoglycosides, ciprofloxacin, chloramphenicol, tetracycline, tigecycline, trimethoprim, and trimethoprim-sulfamethoxazole). On the contrary, the clinical OXA-48-producing *E. coli* was susceptible only to meropenem/imipenem, tetracycline, tigecycline, and chloramphenicol.

3. Discussion

This study highlights a high multiplicity of species, clones, and carbapenemases three years after the initial outbreak by KPC-3-producing *K. pneumoniae* ST147, irrespective of the infection control measures applied. This information, together with epidemiological and patients' data is critical to understand galloping carbapenems resistance rates in our country and support the revision of infection control measures. To our knowledge, this is the first study providing a comprehensive analysis of the molecular epidemiology of CPE in a single institution in our country in two different time-points of CPE spread, towards a comprehension of factors driving their extraordinary increase. The multiplicity of carbapenemase-producing clones and *Enterobacterales* species, together with data from previous studies in Portugal [7,9,10,12], highlights successful horizontal transfer of carbapenemases that might occur during gastrointestinal colonization, which represents an additional challenge for infection control.

In contrast to previous studies [9,10,12], ST147-K64 *K. pneumoniae* is still the predominant clone in different healthcare settings in our area. Its identification in patients that are colonized or infected in multiple occasions throughout time, and its dispersion in other community-based healthcare-associated settings suggests possible reintroductions of this highly transmissible clone in the hospital. The emergence of other KPC-3-producing high-risk *K. pneumoniae* clones with recognized worldwide expansion (ST392, ST307) and additional enhanced virulence and/or antibiotic resistance (ST307, ST22) is of concern.

The ST392 clone has been recently reported in other countries (China, Italy, Mexico) [13–15] but to our knowledge this is the first study to describe KPC-3-producing *K. pneumoniae* ST392 isolates in Portugal. Although detected in small numbers, the identification of 4 ST307 isolates producing KPC-3 among medical and surgical units is of great concern because of its higher resistance profile and the recognized virulence potential (high resistance to complement-mediated killing) of this high-risk clone reported worldwide [16,17]. Our data confirms the absence of ST258 *K. pneumoniae* in our country, a clone that is well-established in other neighboring European countries [18,19]. Furthermore, the detection of ST131 *E. coli* producing KPC-3 especially as rectal colonizers poses a major threat to public health, since it has the potential to cause widespread resistance to carbapenems in the community setting.

We demonstrated an excellent performance of our FT-IR based approach for subtyping multidrug resistant *K. pneumoniae* populations, that was corroborated by reference genotypic molecular methods. Isolates relationships' were correctly established for 98% of the isolates and, for some of them, it was possible to identify previously circulating *K. pneumoniae* lineages (e.g., ST147-K64, ST15-K19) for which epidemiological data and resistance patterns are well-known [7,11]. This information is useful for infection control teams in order to more quickly and effectively implement measures to control the spread of problematic lineages and eventually guide antibiotic therapy. Thus, the high accuracy of the method, together with its simplicity and extremely short time-to-response (we can provide results from 1-3h after standardized growth conditions) represent ideal features for routine implementation and a real-time support to infection control in the context of outbreak detection [8] or epidemiological surveillance (this study). New K-types and clones were introduced in our spectral database for further improvement of the model and correct predictions of future unknown isolates.

Besides establishing the potential for diversification of the CPE population, the data obtained in this study were critical to support more targeted infection approaches in the most affected units (medical and surgical wards) and to control the spread of ST392 in medical units (data now shown). The high number of CPE in internal medicine (22.5%) is in agreement with previous data [9], and occurs due to the higher burden of elderly patients, long hospitalization periods and the higher frequency of invasive procedures [20]. Besides, reinforcement of medical equipment, surfaces' cleaning and contact precautions between patients and healthcare professionals is critical to control the situation [21]. From the data presented, it can be concluded that nearly all *de novo* CPE carriers (n = 43; 96%) became colonized by CPE after hospital admission which strongly supports nosocomial acquisition. Previous colonization and exposure to antibiotics in the previous three months were also recognized as risk factors for development of CPE infections, as previously [20]. Patients maintained carrier status over 200 days, raising questions about the effectiveness of routine decolonization protocols that would need to be evaluated in a longer period and a larger sample. In any case, it is known that available data varies according to decolonization strategy and the benefits are still questionable, and for that reason routine decolonization of CPE is not recommended [22].

In conclusion, this work highlights an increasing diversity of CPE infecting or colonizing hospitalized patients in a central hospital from the north of Portugal, even after implementation of recommended infection control measures guided by clinical and epidemiological data and routine rectal screening results. These results alert for the need to improve the accuracy and speed of bacterial strain typing information, a goal that can be met by simple and cost-effective FT-IR based typing. We strongly believe that, together with epidemiological data and antibiotic resistance patterns, it can be an asset to support infection control and patient's management in real-time.

4. Materials and Methods

4.1. Study Design and CPE sample

The hospital studied is a central and reference hospital located in the north of Portugal that includes 3 units and 1 rehabilitation center that serves 700.000 inhabitants of that region. It contains capacity for 580-beds, and offers healthcare specialties within medical, surgical, and emergency (adults and pediatric) wards as well as attendance to outpatients.

In this hospital, the first noticed outbreak by CPE occurred in 2015 by a multidrug resistant KPC-3 producing *K. pneumoniae* ST147 clone with capsular (K)-type 64, that caused the death of 3 patients [8]. From then, the hospital implemented standard infection control measures (including active screening, isolation and contact precautions for infected or colonized patients) that are still in use; however, CPE increased throughout time.

At the hospital, suspected CPE isolates ($n = 128$) identified from either clinical samples ($n = 35$, November 2017-February 2018) or from rectal swabs ($n = 93$ in February 2018) from hospitalized patients were collected. Screenings of CPE in rectal swabs were performed: (i) at admission in high-risk patients, i.e. those with hospitalizations longer than 72 h in the last 6 months or transferred from another healthcare institution; (ii) at 48 h of hospitalization when the result is negative at admission; (iii) each 7 days of hospitalization if the prior result is negative.

At admission, CPE colonizers were detected by molecular biology using Xpert CarbaR (Cepheid), whereas subsequent screenings on hospitalized patients were performed using a cultural method. Rectal swabs (positive for Xpert CarbaR and subsequent screenings) were plated on the chromogenic agar chromID Carba Smart (bioMérieux) for recovery of carbapenem resistant isolates. Presumptive *Enterobacterales* isolates (pink colonies for *E. coli* and bluish-green colonies for *Klebsiella, Enterobacter, Serratia* or *Citrobacter* sp.) were selected for further characterization (>1 isolate per sample was studied when representing different morphotypes or species).

All patients with positive results were transferred to cohort areas and submitted to infection control measures. We collected patients' data (age, gender, underlying conditions), hospitalization, and rectal screening history and previous antibiotherapy.

4.2. CPE Identification and Antibiotic Susceptibility Testing

Isolates from clinical samples were identified and preliminarily tested for antibiotic susceptibility by VITEK®2 system (bioMérieux). When necessary, species identification was confirmed by MALDI-TOF MS (VITEK MS, bioMérieux) and further sequencing of *leuS* for speciation within *Enterobacter cloacae* species [23].

Extended antibiotic susceptibility profiles were subsequently obtained by disk diffusion for 20 antibiotics (cefotaxime, ceftazidime, cefepime, ertapenem, meropenem, imipenem, amoxicillin-clavulanic acid, piperacillin-tazobactam, amikacin, tobramycin, gentamicin, kanamycin, netilmicin, ciprofloxacin, tetracycline, tigecycline, fosfomycin, trimethoprim, trimethoprim-sulfamethoxazole, and chloramphenicol), according to EUCAST guidelines (www.eucast.org) and CLSI [24].

4.3. Molecular Characterization of CPE Producers

Carbapenemase production was confirmed by the Blue-Carba test [25] and the type of carbapenemase was identified by PCR directed to the most frequent carbapenemase gene families (bla_{KPC}, bla_{OXA-48}, bla_{IMP}; bla_{VIM}, bla_{NDM}) and further sequencing [7,26]. *mcr* genes (*mcr-1* to *mcr-5*) were sought by a multiplex PCR previously described [27] in representative isolates from different clones and species.

4.4. FT-IR for Subtyping of K. pneumoniae Carbapenemase Producers

The relationship between *K. pneumoniae* isolates was established by Fourier Transform Infrared (FT-IR) spectroscopy using our previously described workflow [11]. Briefly, bacterial spectra were acquired in standardized conditions and compared with those from our in-house *K. pneumoniae* database (including 19 international clones/K-types) for identification of capsular (K) types and because of established K-type and clone relationships, presumptive clonal identification. These FT-IR-based assignments were confirmed by PCR and sequencing of *wzi* gene [11,28] and multi-locus sequence typing (MLST) using the seven housekeeping genes (*gapA, infB, pgi, mdh, phoE, rpoB, tonB*) proposed in

Pasteur MLST scheme in representative isolates of clones with >3 isolates each (https://bigsdb.pasteur.fr/klebsiella/klebsiella.html).

E. coli phylogenetic groups were identified by PCR, according to the quadriplex method described by Clermont et al. [29]. *E. coli* ST131 was presumptively identified by a specific PCR in isolates belonging to the phylogenetic group B2 and *fumC* sequencing [30].

Author Contributions: A.M.G. was responsible for the experimental work and wrote the paper. A.L. (Agostinho Lira), A.L. (Angelina Lameirão), A.S., G.A., P.L. and M.M. were responsible for screening and preliminary identification of CPE isolates, and provided the epidemiological and clinical data. Â.N., together with L.P., designed the study, supervised the experimental work and revised the paper. All authors have read and agreed to the published version of the manuscript.

Funding: This work received financial support from Applied Molecular Biosciences Unit UCIBIO, which is supported by national funds from Fundação para a Ciência e a Tecnologia (FCT)/MCTES [UID/Multi/UIDB/04378/2020]. Ângela Novais is supported by national funds through FCT, I.P., in the context of the transitional norm [DL57/2016/CP1346/CT0032]. We are also grateful to BioMerieux for the support in microbiological media used in this study.

Conflicts of Interest: The authors declare no conflict of interest.

References

1. Cassini, A.; Högberg, L.D.; Plachouras, D.; Quattrocchi, A.; Hoxha, A.; Simonsen, G.S.; Colomb-Cotinat, M.; E Kretzschmar, M.; Devleesschauwer, B.; Cecchini, M.; et al. Attributable deaths and disability-adjusted life-years caused by infections with antibiotic-resistant bacteria in the EU and the European Economic Area in 2015: A population-level modelling analysis. *Lancet Infect. Dis.* **2019**, *19*, 56–66. [CrossRef]

2. A Bonomo, R.; Burd, E.M.; Conly, J.; Limbago, B.M.; Poirel, L.; A Segre, J.; Westblade, L.F. Carbapenemase-Producing Organisms: A Global Scourge. *Clin. Infect. Dis.* **2017**, *66*, 1290–1297. [CrossRef] [PubMed]

3. European Centre for Disease Prevention and Control. *Surveillance of Antimicrobial Resistance in Europe—Annual Report of the European Antimicrobial Resistance Surveillance Network (EARS-Net) 2017*; ECDC: Stockholm, Sweden, 2018.

4. Direção-Geral da Saúde. Infeções e Resistências aos Antimicrobianos: Relatório Anual do Programa Prioritário. 2018. Available online: https://www.dgs.pt/portal-da-estatistica-da-saude/diretorio-de-informacao/diretorio-de-informacao/por-serie-1003038-pdf.aspx?v=%3d%3dDwAAAB%2bLCAAAAAAABAArySzItzVUy81MsTU1MDAFAHzFEfkPAAAA (accessed on 21 June 2020).

5. Direção-Geral da Saúde. Vigilância Epidemiológica das Resistências aos Antimicrobianos; Norma n° 004/2013. 2013. Available online: https://www.dgs.pt/directrizes-da-dgs/normas-e-circulares-normativas/norma-n-0042013-de-21022013-jpg.aspx (accessed on 21 June 2020).

6. Rodrigues, C.; Machado, E.; Ramos, H.; Peixe, L.; Novais, Â. Expansion of ESBL-producing Klebsiella pneumoniae in hospitalized patients: A successful story of international clones (ST15, ST147, ST336) and epidemic plasmids (IncR, IncFIIK). *Int. J. Med. Microbiol.* **2014**, *304*, 1100–1108. [CrossRef] [PubMed]

7. Rodrigues, C.; Bavlovič, J.; Machado, E.; Amorim, J.; Peixe, L.; Novais, Â. KPC-3-Producing Klebsiella pneumoniae in Portugal Linked to Previously Circulating Non-CG258 Lineages and Uncommon Genetic Platforms (Tn4401d-IncFIA and Tn4401d-IncN). *Front. Microbiol.* **2016**, *7*, 1000. [CrossRef]

8. Silva, L.; Rodrigues, C.; Lira, A.; Leão, M.; Mota, M.; Lopes, P.; Novais, Â.; Peixe, L. Fourier transform infrared (FT-IR) spectroscopy typing: A real-time analysis of an outbreak by carbapenem-resistant Klebsiella pneumoniae. *Eur. J. Clin. Microbiol. Infect. Dis.* **2020**, 1–5. [CrossRef]

9. Aires-De-Sousa, M.; De La Rosa, J.M.O.; Gonçalves, M.L.; Pereira, A.L.; Nordmann, P.; Poirel, L. Epidemiology of Carbapenemase-Producing Klebsiella pneumoniae in a Hospital, Portugal. *Emerg. Infect. Dis.* **2019**, *25*, 1632–1638. [CrossRef]

10. Perdigão, J.; Modesto, A.; Pereira, A.L.; Neto, O.; Matos, V.; Godinho, A.; Phelan, J.; Charleston, J.; Spadar, A.; De Sessions, P.F.; et al. Whole-genome sequencing resolves a polyclonal outbreak by extended-spectrum beta-lactam and carbapenem-resistant Klebsiella pneumoniae in a Portuguese tertiary-care hospital. *Microb. Genom.* **2020**, *10*, mgen000349. [CrossRef]

11. Rodrigues, C.; Sousa, C.; Lopes, J.A.; Novais, Â.; Peixe, L. One the front line of Klebsiella pneumoniae surface structures understanding: Establishment of Fourier Transform Infrared (FT-IR) spectroscopy as a capsule typing method. *mSystems* **2020**. [CrossRef]

12. Manageiro, V.; Romão, R.; Moura, I.B.; Sampaio, D.A.; Vieira, L.; Ferreira, E.; Caniça, M.; EuSCAPE-Portugal, the N. Molecular Epidemiology and Risk Factors of Carbapenemase-Producing Enterobacteriaceae Isolates in Portuguese Hospitals: Results from European Survey on Carbapenemase-Producing Enterobacteriaceae (EuSCAPE). *Front. Microbiol.* **2018**, *9*, 2834. [CrossRef]

13. Bocanegra-Ibarias, P.; Garza-González, E.; Morfín-Otero, R.; Barrios, H.; Villarreal-Treviño, L.; Rodríguez-Noriega, E.; Garza-Ramos, U.; Petersen-Morfin, S.; Silva-Sanchez, J. Molecular and microbiological report of a hospital outbreak of NDM-1-carrying Enterobacteriaceae in Mexico. *PLoS ONE* **2017**, *12*, e0179651. [CrossRef]

14. Esposito, E.P.; Cervoni, M.; Bernardo, M.; Crivaro, V.; Cuccurullo, S.; Imperi, F.; Zarrilli, R. Molecular Epidemiology and Virulence Profiles of Colistin-Resistant Klebsiella pneumoniae Blood Isolates from the Hospital Agency "Ospedale dei Colli", Naples, Italy. *Front. Microbiol.* **2018**, *9*, 1463. [CrossRef] [PubMed]

15. Yang, J.; Ye, L.; Guo, L.; Zhao, Q.; Chen, R.; Luo, Y.; Chen, Y.; Tian, S.; Zhao, J.; Shen, D.; et al. A nosocomial outbreak of KPC-2-producing Klebsiella pneumoniae in a Chinese hospital: Dissemination of ST11 and emergence of ST37, ST392 and ST395. *Clin. Microbiol. Infect.* **2013**, *19*, E509–E515. [CrossRef] [PubMed]

16. Castanheira, M.; Farrell, S.E.; Wanger, A.; Rolston, K.V.; Jones, R.N.; Mendes, R.E. Rapid Expansion of KPC-2-Producing Klebsiella pneumoniae Isolates in Two Texas Hospitals due to Clonal Spread of ST258 and ST307 Lineages. *Microb. Drug Resist.* **2013**, *19*, 295–297. [CrossRef] [PubMed]

17. Villa, L.; Feudi, C.; Fortini, D.; Brisse, S.; Passet, V.; Bonura, C.; Endimiani, A.; Mammina, C.; Ocampo, A.M.; Jimenez, J.N.; et al. Diversity, virulence, and antimicrobial resistance of the KPC-producing Klebsiella pneumoniae ST307 clone. *Microb. Genom.* **2017**, *3*, e000110. [CrossRef]

18. Soria-Segarra, C.; González-Bustos, P.; López-Cerero, L.; Fernández-Cuenca, F.; Rojo-Martín, M.D.; Fernández-Sierra, M.A.; Gutiérrez-Fernández, J. Tracking KPC-3-producing ST-258 Klebsiella pneumoniae outbreak in a third-level hospital in Granada (Andalusia, Spain) by risk factors and molecular characteristics. *Mol. Biol. Rep.* **2019**, *47*, 1089–1097. [CrossRef]

19. David, S.; The EuSCAPE Working Group; Reuter, S.; Harris, S.R.; Glasner, C.; Feltwell, T.; Argimon, S.; AbuDahab, K.; Goater, R.; Giani, T.; et al. Epidemic of carbapenem-resistant Klebsiella pneumoniae in Europe is driven by nosocomial spread. *Nat. Microbiol.* **2019**, *4*, 1919–1929. [CrossRef]

20. Mathers, A.; Vegesana, K.; German-Mesner, I.; Ainsworth, J.; Pannone, A.; Crook, D.; Sifri, C.; Sheppard, A.; Stoesser, N.; Peto, T.; et al. Risk factors for Klebsiella pneumoniae carbapenemase (KPC) gene acquisition and clinical outcomes across multiple bacterial species. *J. Hosp. Infect.* **2020**, *104*, 456–468. [CrossRef]

21. De Rosa, F.G.; Corcione, S.; Cavallo, R.; Di Perri, G.; Bassetti, M. Critical issues for Klebsiella pneumoniae KPC-carbapenemase producing K. pneumoniae infections: A critical agenda. *Future Microbiol.* **2015**, *10*, 283–294. [CrossRef]

22. Tacconelli, E.; Mazzaferri, F.; De Smet, A.M.; Bragantini, D.; Eggimann, P.; Huttner, B.D.; Kuijper, E.J.; Lucet, J.-C.; Mutters, N.T.; Sanguinetti, M.; et al. ESCMID-EUCIC clinical guidelines on decolonization of multidrug-resistant Gram-negative bacteria carriers. *Clin. Microbiol. Infect.* **2019**, *25*, 807–817. [CrossRef]

23. Miyoshi-Akiyama, T.; Hayakawa, K.; Ohmagari, N.; Shimojima, M.; Kirikae, T. Multilocus Sequence Typing (MLST) for Characterization of Enterobacter cloacae. *PLoS ONE* **2013**, *8*, e66358. [CrossRef]

24. Clinical and Laboratory Standards Institute. *Performance Standards for Antimicrobial Susceptibility Testing: M100-S27*; Clinical and Laboratory Standards Institute: Wayne, PA, USA, 2017.

25. Pires, J.; Novais, A.; Peixe, L. Blue-Carba, an Easy Biochemical Test for Detection of Diverse Carbapenemase Producers Directly from Bacterial Cultures. *J. Clin. Microbiol.* **2013**, *51*, 4281–4283. [CrossRef] [PubMed]

26. Bogaerts, P.; Rezende de Castro, R.; de Mendonça, R.; Huang, T.D.; Denis, O.; Glupczynski, Y. Validation of carbapebemase and extended-spectrum β-lactamase multiplex endpoint PCR assays according to ISO 15189. *J. Antimicrob. Chemother.* **2013**, *68*, 1576–1582. [CrossRef] [PubMed]

27. Rebelo, A.R.; Bortolaia, V.; Kjeldgaard, J.S.; Pedersen, S.K.; Leekitcharoenphon, P.; Hansen, I.M.; Guerra, B.; Malorny, B.; Borowiak, M.; Hammerl, J.A.; et al. Multiplex PCR for detection of plasmid-mediated colistin resistance determinants, mcr-1, mcr-2, mcr-3, mcr-4 and mcr-5 for surveillance purposes. *Eurosurveillance* **2018**, *23*, 17-00672. [CrossRef] [PubMed]

28. Brisse, S.; Passet, V.; Haugaard, A.B.; Babosan, A.; Kassis-Chikhani, N.; Struve, C.; Decré, D. wzi Gene Sequencing, a Rapid Method for Determination of Capsular Type for Klebsiella Strains. *J. Clin. Microbiol.* **2013**, *51*, 4073–4078. [CrossRef] [PubMed]

29. Clermont, O.; Christenson, J.K.; Denamur, E.; Gordon, D.M. The ClermontEscherichia coliphylo-typing method revisited: Improvement of specificity and detection of new phylo-groups. *Environ. Microbiol. Rep.* **2012**, *5*, 58–65. [CrossRef]

30. Rodrigues, C.; Machado, E.; Fernandes, S.; Peixe, L.; Novais, Â. An update on faecal carriage of ESBL-producing Enterobacteriaceae by Portuguese healthy humans: Detection of theH30 subclone of B2-ST131Escherichia coliproducing CTX-M-27. *J. Antimicrob. Chemother.* **2016**, *71*, 1120–1122. [CrossRef]

Publisher's Note: MDPI stays neutral with regard to jurisdictional claims in published maps and institutional affiliations.

2020, 9, 675

Article

Antibiotic Resistance and Mobile Genetic Elements in Extensively Drug-Resistant *Klebsiella pneumoniae* Sequence Type 147 Recovered from Germany

Kyriaki Xanthopoulou [1,2], Alessandra Carattoli [3], Julia Wille [1,2], Lena M. Biehl [2,4], Holger Rohde [5,6], Fedja Farowski [2,4,7], Oleg Krut [8], Laura Villa [9], Claudia Feudi [10], Harald Seifert [1,2] and Paul G. Higgins [1,2,*]

[1] Institute for Medical Microbiology, Immunology and Hygiene, University of Cologne, 50935 Cologne, Germany; kyriaki.xanthopoulou@uk-koeln.de (K.X.); julia.wille_@uk-koeln.de (J.W.); harald.seifert@uni-koeln.de (H.S.)

[2] German Centre for Infection Research (DZIF), Partner site Bonn-Cologne, 50935 Cologne, Germany; lena.biehl@uk-koeln.de (L.M.B.); fedja.farowski@uk-koeln.de (F.F.)

[3] Department of Molecular Medicine, Sapienza University of Rome, 00185 Rome, Italy; alessandra.carattoli@uniroma1.it

[4] Department I of Internal Medicine, Faculty of Medicine and University Hospital of Cologne, University of Cologne, 50937 Cologne, Germany

[5] Institute for Medical Microbiology, Virology and Hygiene, University Medical Centre Hamburg-Eppendorf, 20246 Hamburg, Germany; rohde@uke.de

[6] German Centre for Infection Research, Partner Site Hamburg-Lübeck-Borstel, 20246 Hamburg, Germany

[7] Department of Internal Medicine II, Infectious Diseases, University Hospital Frankfurt, Goethe University Frankfurt, 60590 Frankfurt am Main, Germany

[8] Paul-Ehrlich Institute, Federal Institute for Vaccines and Biomedicine, 63225 Langen, Germany; oleg.krut@pei.de

[9] Department of Infectious Diseases, Istituto Superiore di Sanità, 00161 Rome, Italy; laura.villa@iss.it

[10] Institute of Microbiology and Epizootics, Centre for Infection Medicine, Department of Veterinary Medicine, Freie Universität Berlin, 14163 Berlin, Germany; claudia.feudi@fu-berlin.de

* Correspondence: paul.higgins@uni-koeln.de; Tel.: +49-221-47832011

Received: 17 September 2020; Accepted: 2 October 2020; Published: 5 October 2020

Abstract: Mobile genetic elements (MGEs), especially multidrug-resistance plasmids, are major vehicles for the dissemination of antimicrobial resistance determinants. Herein, we analyse the MGEs in three extensively drug-resistant (XDR) *Klebsiella pneumoniae* isolates from Germany. Whole genome sequencing (WGS) is performed using Illumina and MinION platforms followed by core-genome multi-locus sequence typing (MLST). The plasmid content is analysed by conjugation, S1-pulsed-field gel electrophoresis (S1-PFGE) and Southern blot experiments. The *K. pneumoniae* isolates belong to the international high-risk clone ST147 and form a cluster of closely related isolates. They harbour the $bla_{OXA-181}$ carbapenemase on a ColKP3 plasmid, and 12 antibiotic resistance determinants on an multidrug-resistant (MDR) IncR plasmid with a recombinogenic nature and encoding a large number of insertion elements. The IncR plasmids within the three isolates share a high degree of homology, but present also genetic variations, such as inversion or deletion of genetic regions in close proximity to MGEs. In addition, six plasmids not harbouring any antibiotic resistance determinants are present in each isolate. Our study indicates that genetic variations can be observed within a cluster of closely related isolates, due to the dynamic nature of MGEs. The mobilome of the *K. pneumoniae* isolates combined with the emergence of the XDR ST147 high-risk clone have the potential to become a major challenge for global healthcare.

Keywords: carbapenem resistance; carbapenemase; whole genome sequencing; long reads, plasmid; *Klebsiella pneumoniae*; extensively drug-resistant; molecular typing

1. Introduction

The evolution and spread of antibiotic-resistant pathogens has emerged as one of the most important public health problems worldwide over the last decades (https://www.who.int/en/news-room/fact-sheets/detail/antibiotic-resistance). In bacterial genomes, capture, accumulation and dissemination of antibiotic resistance determinants are often associated with mobile genetic elements (MGEs) like plasmids, transposons and insertion sequences (ISs) [1]. Plasmids are often assemblies of different MGE modules and are the most efficient intra- and interspecies DNA transfer mechanism among prokaryotes [2]. This is well exemplified by the global spread of the KPC carbapenemase involving the incompatibility group FIIk (IncFIIk) plasmids in *Klebsiella pneumoniae* [3]. Moreover, bla_{NDM-1} in *K. pneumoniae* has been mainly associated with broad host range IncA/C2, IncHI1, IncX3 and IncN2 plasmids [4]. In *Acinetobacter baumannii*, the transposon Tn*125*, harbouring the insertion element IS*Aba125*, is considered as the main vehicle for the dissemination of NDM-1 enzymes [5,6].

K. pneumoniae, belonging to the Enterobacterales family, is a natural inhabitant of the gastrointestinal tract of humans and animals. Nevertheless, it is also encountered as a nosocomial pathogen causing various infections such as pneumonia, urinary tract infection and bloodstream infection [4]. Of concern is the rapid expansion of carbapenem-resistant *K. pneumoniae*, mainly associated with those carbapenemases which are endemic in certain countries, such as KPC-positive *K. pneumoniae* in Greece and Italy [7,8]. OXA-48-like is the most common carbapenemase in Enterobacterales in some regions of the world including Germany. Other frequently encountered carbapenemases in Germany include VIM-1 and NDM-1 [9,10]. The successful propagation of OXA-48-positive Enterobacterales is reinforced by the global distribution of certain high-risk clones (e.g., *K. pneumoniae* sequence type (ST) 307 or *Escherichia coli* ST38) as also its association with MGEs, e.g., OXA-48 linked with different Tn*1999* variants on highly transferable IncL plasmids [10,11]. The expansion of high-risk *K. pneumoniae* clones with a multidrug-resistant (MDR) or extensive drug-resistant (XDR) phenotype has been observed in recent years [4]. *K. pneumoniae* ST147 has been reported as an emerging high-risk clone associated with plasmid-encoded extended-spectrum β-lactamases (ESBLs) like $bla_{CTX-M-15}$, or carbapenemases such as bla_{OXA-48} and bla_{NDM-1} [4,12–17].

In the present study, we characterise the content and genetic structure of MGEs and the clonal relatedness of three OXA-181-producing *K. pneumoniae* ST147 clinical isolates recovered in Germany.

2. Results and Discussion

Dissemination of antibiotic resistance is driven by clonal expansion or horizontal gene transfer, including mainly MGEs [1,2]. In the present study, all three isolates colonising haematology/oncology patients were identified as *K. pneumoniae* ST147 and were the only representatives of this ST among 40 in total collected *K. pneumoniae* isolates. The three isolates were also characterised by their capsular type KL64 (*wzi* allele 64). MDR *K. pneumoniae* ST147 isolates represent a successful clone with a global spread and these isolates are often armed with carbapenemases and ESBLs [15,18]. The German National Reference Centre for Multidrug-Resistant Gram-negative Bacteria and the Robert Koch Institute reported, between 2008 and 2014, 13 carbapenemase-producing ST147 *K. pneumoniae* isolates in Germany. In particular, 9/42 OXA-48-, 3/34 KPC-2- and 1/5 NDM-1-producing isolates were assigned to ST147 [19].

In the present study, the isolates HKP0018, HKP0064 and HKP0067 were analysed by whole genome sequencing (WGS) and harboured on the chromosome a gene encoding the intrinsic SHV-11, as well as *oqxAB* and *fosA* genes, belonging to the core genome of the KpI–III phylogroups [20]. The plasmid-encoded resistome of the investigated isolates, summarised in Table 1, was identical and included beta-lactam, aminoglycoside, fluoroquinolone, tetracycline and other antimicrobial resistance determinants. Antimicrobial susceptibility testing showed that all three *K. pneumoniae* isolates exhibited an XDR phenotype; resistant to ampicillin, aztreonam, ceftazidime, chloramphenicol, ciprofloxacin, gentamicin, imipenem, meropenem, minocycline, tetracycline, ticarcillin, tigecycline, and trimethoprim and susceptible only to amikacin and colistin (Table 2). MDR and XDR *K. pneumoniae* isolates involved in nosocomial outbreaks have been widely reported [4,21,22]. Between June and October 2109, an outbreak of XDR *K. pneumoniae* producing NDM-1 and OXA-48 was reported in four medical facilities in Mecklenburg-Western Pomerania, Germany [23]. Molecular characterisation using core genome multi-locus sequence typing (cgMLST) analysis revealed that the three investigated isolates were closely related and formed a cluster with 0–1 allelic differences (data not shown). One could speculate that the closely related isolates were likely transmitted within the hospital. All three patients had been hospitalised in the same department (Table 3) and two of the patients had an overlapping hospitalisation at the same ward (C5A). However, a direct connection to HKP0018 could not be established within the study.

Table 1. Plasmid encoded antimicrobial resistance determinants, plasmid content and plasmid size of the isolates.

Plasmid	Replicon	Size (bp)	Antimicrobial Resistance Determinants	HKP0018	Isolate No. HKP0064	HKP0067
pHKP0018.1	ColKP3	6103	blaOXA$_{-181}$	+	+	+
pHKP0018.2	IncR	66,330	$bla_{\text{CTX-M-15}}{}^{b}$, $bla_{\text{OXA-1}}$, $bla_{\text{TEM-1B}}$, $aac(6')Ib\text{-}cr$, $aac(3)\text{-}IIa$, $strA$, $strB$, $qnrS1$, $sul1$, $dfrA1$, $tet(A)$, $catB3$-like	+	-	-
pHKP0064.2	IncR	70,762	$bla_{\text{CTX-M-15}}{}^{b}$, $bla_{\text{OXA-1}}$, $bla_{\text{TEM-1B}}$, $aac(6')Ib\text{-}cr$, $aac(3)\text{-}IIa$, $strA$, $strB$, $qnrS1$, $sul1$, $dfrA1$, $tet(A)$, $catB3$-like	-	+	+
pHKP0018.3	IncFIB	113,014	-	+	+	+
pHKP0018.4	NT [a]	57,450	-	+	+	+
pHKP0018.5	Col-like	8428	-	+	+	+
pHKP0018.6	Col-like	5499	-	+	+	+
pHKP0018.7	NT [a]	2044	-	+	+	+
pHKP0018.8	Col-like	1459	-	+	+	+

[a] NT, not typeable; [b] gene present in two copies.

Table 2. Antimicrobial susceptibility of the three *K. pneumoniae* isolates.

Antimicrobial Agent	MIC (mg/L)			Susceptibility [a]
	HKP0018	**HKP0064**	**HKP0067**	
Amikacin	8	8	8	S
Ampicillin	>128	>128	>128	R
Aztreonam	>128	>128	>128	R
Ceftazidime	128	128	128	R
Chloramphenicol	32	32	32	R
Ciprofloxacin	128	128	128	R
Colistin [b]	1	2	1	S
Gentamicin	128	128	128	R
Imipenem	8	8	8	R
Levofloxacin	64	64	64	R
Meropenem	32	32	32	R
Minocycline [c]	64	64	64	R
Rifampicin [d]	64	64	64	-
Tetracycline [c]	128	128	128	R
Ticarcillin	>128	>128	>128	R
Tigecycline [b]	2	2	2	R
Trimethoprim	128	128	128	R

[a] R, resistant; S, susceptible; [b] tested by broth microdilution method; [c] only CLSI breakpoint available; [d] no breakpoint available.

Table 3. *K. pneumoniae* clinical isolates information.

Isolate	Date of Isolation	Source	Department	Ward	ST
HKP0018	16.02.2015	Rectal swab	Haematology/Oncology	C5A	147
HKP0064	08.05.2015	Throat swab	Haematology/Oncology	1G	147
HKP0067	19.05.2015	Rectal swab	Haematology/Oncology	C5A	147

Phylogenetic analysis of 30 ST147 *K. pneumoniae* isolates from different countries showed several branches (Figure 1). The isolates HKP0018, HKP0064 and HKP0067 were on the same branch with ST147 isolates from different countries, such as Switzerland, USA, United Kingdom and Singapore, illustrating the worldwide spread of this clone. In addition, the three investigated isolates clustered together with 7 ST147 *K. pneumoniae* isolates recovered between 2013 and 2014 in Göttingen, Germany. The latter MDR isolates harboured the carbapenemase OXA-48 on a 63.6 kb IncL plasmid [15]. The close genetic relatedness observed between the German isolates suggests that an OXA-48-like producing ST147 clone is circulating in the country.

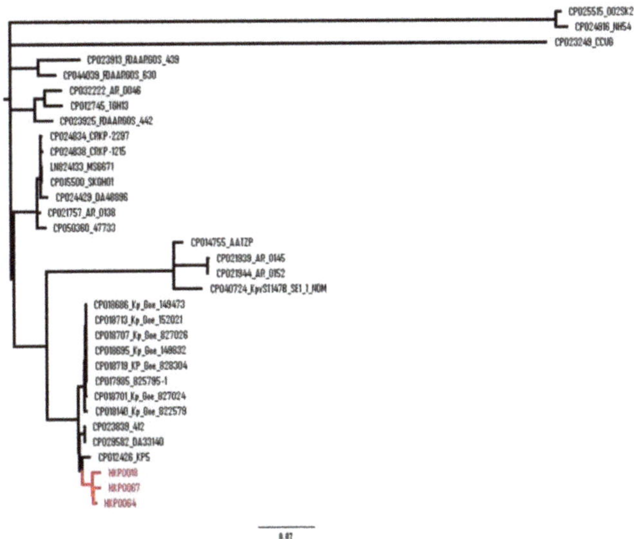

Figure 1. Phylogenetic analysis of HKP0018, HKP0064, HKP0067 and 30 ST147 *K. pneumoniae* isolates. Phylogenetic maximum-likelihood tree was generated using the FigTree v1.4.3 software of the SNP analysis performed using the kSNP3 tool (Galaxy version 3.1) software at the ARIES Galaxy server (https://aries.iss.it/).

In the present study, plasmid analysis revealed eight closed plasmids for each individual *K. pneumoniae* isolate.

OXA-48-like is the most prevalent carbapenem-hydrolysing β-lactamase in Enterobacterales isolates from Germany [9,24]. MDR *K. pneumoniae* ST147 encoding OXA-48 on a conjugative IncL plasmid have been recently reported in Germany [15]. In the present study, all three investigated isolates harboured OXA-181 on an identical 6103 bp ColKP3 plasmid, pHKP0018.1. This plasmid also encoded the mobilisation genes *mobA*, *mobB*, *mobC* and *mobD*, and, upstream of $bla_{OXA-181}$ gene, 170 bp of a disrupted IS*Ecp1* was present. A blastn analysis to compare pHKP0018.1 to sequences available in the GenBank database revealed high similarities mainly with three groups of plasmids, of which Carbapenemase OXA-232_ColKP3 (Acc. No CP050165), pKP3-A (Acc. No JN205800) and p50595_OXA_181 (Acc. No CP050375) were chosen as exemplars for a more detailed comparison. The first one, with a size of 6141 bp, showed an identity of 99.98% to our plasmid, has a longer fragment of the interrupted IS*Ecp1* (208 bp) and carries the $bla_{OXA-232}$ gene, a $bla_{OXA-181}$ variant from which it differs by a single nucleotide, leading to the Arg214-Ser amino acid substitution, and from which it probably originated (Figure 2) [25]. Plasmid pKP3-A, obtained from a clinical *K. pneumoniae* isolate in 2010, is a ColKP3 plasmid carrying $bla_{OXA-181}$, proved to be mobilisable but not self-transmissible. It showed 99.95% similarity when compared to pHKP0018.1, from which it differs by the presence of the complete IS*Ecp1* element. In this plasmid, the carbapenemase gene was described as part of the Tn*2013* transposon, made up by the 3139 bp module IS*Ecp1*-$bla_{OXA-181}$-Δ*lysR*-Δ*ereA* [26]. In plasmid pHKP0018.1 this transposon was disrupted, with only the two right inverted repeats (IRR1 and IRR2) and the 3' target site duplication (ATATA) still identifiable (Figure 2) [26]. Lastly, p50595_OXA_181 plasmid depicts the group of X3-ColKP3 plasmids of approximately 51 kb in size, which held 50% of the pHKP0018.1 plasmid, with an identity of 100%. This portion contained the interrupted Tn*2013* (ΔIS*Ecp1*-$bla_{OXA-181}$-Δ*lysR*-Δ*ereA*) and an almost complete *repA* gene of ColKP3, inserted between the two insertion sequences IS*3000* and IS*Kpn19* (Figure 2). The sequence comparative analysis also showed that $bla_{OXA-181}$ seems to be almost uniquely located on X3-ColKP3 plasmids, frequently harboured by *E. coli* isolates, while its variant *bla*OXA-232 is primarily located on ColKP3 plasmids harboured predominantly by *K. pneumoniae*. Nevertheless, both variants are distributed on a global scale, including not only clinical isolates but also animal and environmental ones. Indeed, OXA-181 and OXA-232 represent, respectively, the second and third most common and widespread OXA-48-like enzyme and both are described as part of the Tn*2013* transposon, which, together with its localisation on plasmids like ColE-type, IncX3, IncN1 and IncT, is responsible for their dissemination [10].

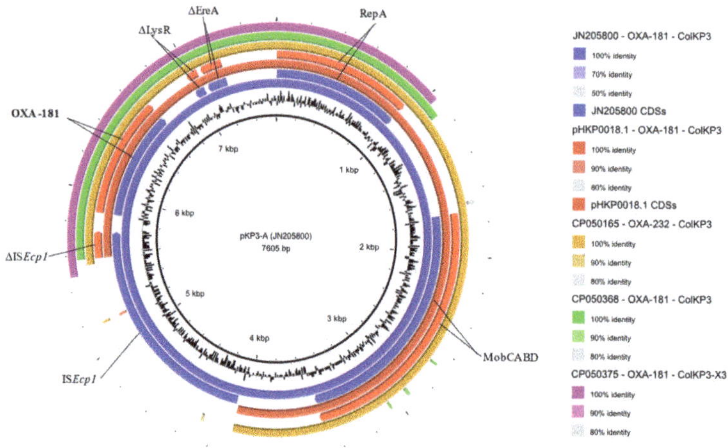

Figure 2. Graphical representation of *bla*OXA-181/*bla*OXA-232-carrying plasmids sequence comparison. Starting from the inner ring: GC content of pKP3-A plasmid sequenced (here used as reference), *bla*OXA-181-positive pKP3-A plasmid sequence (JN205800), pKP3-A CDSs, *bla*OXA-181-positive pHKP0018.1 plasmid sequence (CP061063.1), pHKP0018.1 CDSs, *bla*OXA-232-positive Carbapenemase (OXA-232)_ColKP3 plasmid sequence (CP050165), *bla*OXA-181-positive p47733_OXA_181 plasmid sequence (CP050368), *bla*OXA-181-positive p50595_OXA_181 plasmid sequence (CP050375). CDS's arrows indicate their transcription direction. Hypothetical proteins are not displayed. The figure was generated with BRIG v0.95.

S1-PFGE, Southern blot and WGS analysis revealed that all three isolates harboured an IncR plasmid, pHKP0018.2, pHKP0064.2 and pHKP0067.2, presenting only the *repB* gene and lacking the *repE* and *repA* genes and encoding the same antibiotic resistance determinants (Figure 3). The MDR region included a mosaic structure of 12 antibiotic resistance genes, including β-lactamases *bla*CTX-M-15 (present in two copies on each IncR plasmid), *bla*OXA-1, *bla*TEM-1B, aminoglycoside modifying enzymes *aac(6')Ib-cr*, *aac(3)-IIa*, *strA*, *strB*, as well as the resistance determinants *qnrS1*, *sul1*, *dfrA1*, *tet*(A) and *catB3*-like (Table 1). The MDR region was highly recombinogenic and encoded several copies of different ISs (n = 9). Furthermore, pHKP0018.2, pHKP0064.2 and pHKP0067.2 encoded a *higB/higA* toxin-antitoxin (TA) module and *parA/parB* partitioning genes, contributing to plasmid stabilisation and inheritance. As many others previously described, containing only the *repB* gene alone, the IncR plasmid of this study did not harbour known conjugative loci, and consequently attempts to transfer by conjugation IncR and to mobilise the ColKP3-OXA-181 into *E. coli* J53 were not successful. The IncR plasmids showed high sequence homology to IncR plasmids pKp_Goe_304-4 (Acc. No CP018724.1), pKp_Goe_021-4 (Acc. No CP018718.1), pKp_Goe_024-4 (Acc. No CP018705.1), and CP017989.1 of a ST147 *K. pneumoniae* isolate collected in Germany in 2014, and to the IncR plasmid pSg1-NDM (Acc. No CP011839.1) identified in a ST147 *K. pneumoniae* isolate from Singapore [18].

Sequence analysis revealed that pHKP0064.2 and pHKP0067.2 were identical and 70,762 bp in size. Nevertheless, comparative analysis revealed a rearrangement of a composite transposon flanked by two inverted copies of IS*26* and containing *catB3*-like, *aac(6′)Ib-cr* and *bla*_{OXA-1} genes. This 3826 bp region was inserted in the same position in the two IncR plasmids but in opposite orientation. Similarly, another reshuffling of a 13,957 bp region was observed for pHKP0064.2 and pHKP0067.2. This genomic region was flanked by two copies of IS*Ecp1* in inverse orientation and harboured a truncated transposase, Tn3 resolvase, *bla*_{TEM-1B}, *qnrS1*, recombinase, IS*Kpn19*, *umuC*, HAMP-domain and IS*26* (Figure 3). In the isolate HKP0018 an IncR plasmid, pHKP0018.2, with a size of 66,330 bp was identified. The plasmids pHKP0064.2 and pHKP0067.2 shared a high degree of sequence homology with pHKP0018.2, apart from a 4432 bp region which was missing from the latter plasmid. The missing region was part of the 13,957 bp genomic region involved in the rearrangement in pHKP0064.2 and pHKP0067.2. This subregion was comprised of genes encoding for the error-prone DNA polymerase V subunit (*umuC*) and a sensor histidine kinase (HAMP-domain) followed by the MGE IS*Kpn19* (Figure 3). These results indicate that within a group of clonal isolates, diversity can still be observed. Genetic variation within clonal bacterial groups caused by homologous recombination has been described in *E. coli* [27]. While genetic rearrangement, such as inversion or duplication, caused by MGEs have been confirmed by diverse studies [28–30].

An identical 113,014 bp IncFIB-like plasmid, pHKP0018.3, was identified in the *K. pneumoniae* isolates and did not encode any known antibiotic resistance determinants (Figure S1). However, this plasmid harboured a tellurite/colicin resistance determinant, phage-related genes, and was lacking known conjugative transfer genes. Furthermore, the plasmid harboured two members of the IS3 family, IS*Kpn1* and IS*2*. The IncFIB-like plasmid showed high homology (coverage 97%, identity 100%) to pSG1.1 (Acc. No CP012427.1) from an NDM-1 positive ST147 *K. pneumoniae* isolate from Singapore and also with other ST147 IncFIB plasmids (Acc. No CP021940.1, CP021945.1 and CP014756.1), indicating that this plasmid might be intrinsic to this ST [18].

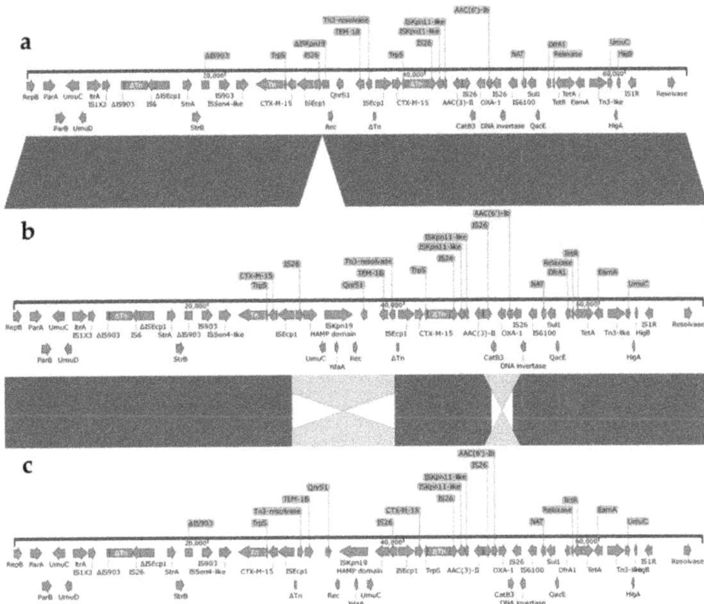

Figure 3. Major structural features of the IncR plasmids, pHKP0018.2 (**a**), pHKP0064.2 (**b**) and pHKP0067.2 (**c**), identified in *K. pneumoniae* isolates HKP0018, HKP0064 and HKP0067, respectively. Arrows indicate the deduced open reading frames (ORFs) and their orientations. Hypothetical proteins are not shown. The figure was generated with EasyFig 2.1 [31].

Southern blot and WGS revealed that the ST147 isolates carried also an identical 54,750 bp plasmid, pHKP0018.4 (Figure S2). The plasmid showed similarity (coverage 78%, identity 99%) to *K. pneumoniae* ST147 plasmids recovered from Singapore, pSg1-3 (Acc. No CP012429) [18]. pHKP0018.4 exhibited also similarity (coverage 62%, identity 83%) to phiKO2 of a *Klebsiella oxytoca* isolate which was described as a prophage able to replicate as linear plasmids with covalently closed ends [32].

Small plasmids, often present in high copy numbers, can serve as an important reservoir for antibiotic resistance determinants, such as small ColE plasmid derivatives encoding *qnrS1* in *Salmonella enterica* [33–35]. In the present study, apart from the ColKP3 *bla*$_{OXA-181}$-encoding plasmid, the ST147 *K. pneumoniae* isolates harboured in addition four identical small plasmids, which varied in size from 1.4 kb to 8.4 kb and did not encode known antimicrobial resistance determinants. An identical 8428 bp plasmid, pHKP0018.5, was identified in the three isolates. The plasmid carried two Col-like replication initiation proteins and showed similarity to pKpvST147B_4 (Acc. No CP040727.1, coverage 56%, identity 100%) from a ST147 *K. pneumoniae* isolate recovered at a hospital in south-east England (Figure S3). Another plasmid, 5499 bp in size and identical for the investigated ST147 isolates (pHKP0018.6), was detected and typed as a Col-like plasmid (Figure S4).

Moreover, a 2044 bp plasmid identical for the three *K. pneumoniae* plasmids, pHKP0018.7, was detected and encoded two hypothetical proteins, with no conserved domains. This plasmid could not be assigned to a replicon type and was identical (coverage 100%, identity 100%) to plasmids p4_1_2.4 (Acc. No CP023843.1) and pDA33140-2 (Acc. No CP029584.1) both from ST147 *K. pneumoniae* isolates recovered in Sweden (plasmid map not shown). Finally, an identical 1459 bp plasmid, pHKP0018.8, replicon typed as Col-like and bearing a hypothetical protein was identified. The Col-like plasmid was identical (coverage 100%, identity 100%) to plasmids found in *E. coli*, such as pEC881_8 (Acc. No CP019021.1) and pEC648_7 (Acc. No CP008721.1), which can be a result of interspecies plasmid transfer (plasmid map not shown).

3. Materials and Methods

3.1. Bacterial Isolates and Transformants

The isolates HKP0018, HKP0064 and HKP0067 were recovered in 2015 from throat and rectal swabs of three patients on admission to a university hospital in northern Germany (Table 3). The isolates were collected as part of the CONTAIN multicentre cohort study of the German Centre for Infection Research (DZIF) on the efficiency of infection control measures to prevent the transmission of ESBL producing Enterobacterales in haematology/oncology units [36]. The selection of the three isolates for further investigation was based on their clonal relatedness, ST, and acquired resistome. The surveillance swabs were plated on selective media (chromID® ESBL; bioMérieux, Nürtingen, Germany) and incubated for 18–24 h. The species identification was performed with MALDI-TOF mass spectrometry. Additionally, plasmid DNA was extracted from the isolate HKP0018 with the PureYield Plasmid Midiprep System (Promega, Madison, WI, USA) and then used to transform One Shot MAX Efficiency DH5α-T1R Competent Cells (Thermo Fisher Scientific, Waltham MA, USA). Selection of transformants was performed using ampicillin (40 mg/L) and tetracycline (30 mg/L) and was confirmed by PCR (Supplementary data).

3.2. Antimicrobial Susceptibility Testing

MICs for ampicillin, tetracycline, trimethoprim, gentamicin (Sigma–Aldrich, Steinheim, Germany), amikacin, aztreonam, imipenem, meropenem, minocycline, rifampicin, (Molekula, Newcastle-upon-Tyne, UK), levofloxacin (Sanofi Aventis, Frankfurt, Germany), ciprofloxacin (Bayer Pharma AG, Berlin, Germany) and ticarcillin (Carl Roth GmbH, Karlsruhe, Germany) were determined using the agar dilution method [37]. MICs for colistin and tigecycline were determined by broth microdilution method (Merlin Diagnostika GmbH, Bornheim, Germany). *E. coli* ATCC 25922, *Pseudomonas aeruginosa* ATCC 27853, and *Staphylococcus aureus* ATCC 25923 were used as quality control strains. MICs were interpreted using the resistance breakpoints for Enterobacterales from EUCAST (Version 10.0, January 2020, http://www.eucast.org/clinical_breakpoints/) and CLSI (https://clsi.org/standards/products/microbiology/documents/m100/).

3.3. S1-Pulsed-Field Gel Electrophoresis (S1-PFGE) and Southern Blot Hybridisation

Plasmid linearisation by S1 nuclease followed by PFGE was used to determine the size and total number of plasmids. Bacterial DNA embedded in agarose plugs was digested using 50 Units S1 nuclease (Thermo Fisher Scientific, Waltham, MA, USA) per plug slice and incubated according to the manufacturer's instructions. Samples were run on a CHEF-DR II system (Bio-Rad, Munich, Germany) for 17 h at 6 V/cm and 14 °C while initial and final pulses were conducted at 4 and 16 s, respectively. The Lambda PFG Ladder and λ DNA-Mono Cut Mix (New England Biolabs, Frankfurt, Germany) were used as markers. The approximate plasmid size was calculated using Image Lab^{TM} software (Bio-Rad, Munich, Germany).

Southern blot hybridisation was performed to determine the plasmid/chromosomal gene location by hybridisation with digoxigenin (DIG)-labelled probes (Roche, Mannheim, Germany). For the IncR replicon and *strA* of pHKP0018.2 and for the terminase pHKP0018.4 specific probes were used respectively (Table S1). Signal detection was performed according to the manufacturer's instructions using CDP-Star®ready-to-use (Roche, Mannheim, Germany) chemiluminescent substrate by autoradiography on a X-ray film (GE Healthcare, Buckinghamshire, United Kingdom). Chromosomal location was shown by colocalisation with a *rpoB* probe.

3.4. Whole Genome Sequencing (WGS) and Bioinformatics

Total DNA from the bacterial isolates and transformants was extracted using the MagAttract HMW DNA Kit (Qiagen, Hilden, Germany) and plasmid DNA was extracted using PureYield Plasmid Midiprep System according to manufacturer's instructions and used for short-read sequencing. Sequencing libraries were prepared using a Nextera XT library prep kit (Illumina GmbH, Munich, Germany) for a 250 bp paired-end sequencing run on an Illumina MiSeq platform. The obtained reads were de novo assembled with the Velvet assembler integrated in the Ridom SeqSphere+ v. 7.2.1 software, and SPAdes 3.11 [38]. Finally, where necessary, overlapping assembly contigs and predicted gaps were filled and confirmed by PCR-based gap closure as described previously [39].

DNA extraction for long-read sequencing was performed using the Genomic-Tips 100/G kit and Genomic DNA Buffers kit (Qiagen, Hilden, Germany) according to the manufacturer's instructions. Libraries were prepared using the 1D Ligation Sequencing Kit (SQK-LSK108) in combination with Native Barcoding Kit (EXP-NBD103) and Rapid Barcoding Kit (SQK-RBK004) in accordance with the manufacturer's instructions (Oxford Nanopore Technologies, Oxford, United Kingdom) and were loaded onto a R9.4 flow cell (Oxford Nanopore Technologies, Oxford, United Kingdom). The run was performed on a MinION MK1b device (Oxford Nanopore Technologies, Oxford, United Kingdom). Collection of raw electronic signal data and live base-calling was performed using the MinKNOW software and Albacore (Oxford Nanopore Technologies, Oxford, United Kingdom). *De novo* assembly of the MinION long-reads was performed using Canu [40]. The Illumina short-reads were assembled with the MinION long-reads using hybridSPAdes and Unicycler [41,42]. Additionally, plasmidSPAdes was implemented to identify plasmid sequences [43].

The assembled genomes generated in this project have been deposited in the NCBI under the BioProject ID PRJNA660340 (BioSample accessions: HKP0018, SAMN15946735; HKP00164, SAMN15946736; HKP0067, SAMN15946737).

3.5. Molecular Epidemiology, Resistome, Mobilome and Genome Annotation

The Pasteur multi-locus sequence typing (MLST) scheme was used to assign the ST (https://bigsdb.pasteur.fr/index.html). The molecular epidemiology was investigated with a validated cgMLST scheme, including 2358 target alleles, using the Ridom SeqSphere+ v. 7.2.1 software [44]. Capsular type

(KL-type) were assigned using Kaptive Web [45]. The resistome and plasmidome were analysed using ResFinder v.3.2.0 (https://cge.cbs.dtu.dk/services/ResFinder/) and PlasmidFinder v.2.0.1 [46,47]. Genome sequences were annotated using the RAST server (http://rast.nmpdr.org/) and partially manually edited. Plasmids were graphically depicted using SnapGene (http://www.snapgene.com/).

3.6. Conjugation Experiments

Broth mate conjugation experiments were performed using the sodium azide-resistant *E. coli* J53 as recipient. Selection of transconjugants was performed using sodium azide (200 mg/L) and ampicillin (40 mg/L), or tetracycline (30 mg/L). Transconjugants were tested by PCR for the presence of the *bla*$_{OXA-181}$ and *tet*(A) genes, while their susceptibility to meropenem (10 µg) and tetracycline (30 µg) was tested using the disk diffusion method, according to EUCAST recommendations (Version 10.0, January 2020, http://www.eucast.org/clinical_breakpoints/).

4. Conlusions

In conclusion, the present study describes a complex variety of plasmids within three clonal ST147 *K. pneumoniae* isolates recovered from haematology/oncology patients hospitalised in the same German hospital. The ST147 *K. pneumoniae* isolates harboured the *bla*$_{OXA-181}$ carbapenemase gene on a small ColKP3 plasmid, but also a complex array of 12 antibiotic resistance determinants on an MDR IncR plasmid, severely limiting treatment options. The recombinogenic nature of the MDR IncR plasmid encoding a large number of ISs can serve as genome plasticity mediators. The IncR plasmids of the studied isolates differed overall in a 4 kb region which could be attributed to an IS transposition event, as also in the opposite orientation of two composite transposons (3.8 kb and 13 kb). These results indicate that within a cluster of closely related isolates, variation can be observed due to the dynamic nature of MGEs. The abundant mobilome and resistome of the *K. pneumoniae* isolates combined with the emergence of ST147 as an international high-risk clone has the potential to become a major challenge for the healthcare setting and requires special attention and vigilance.

Supplementary Materials: The following are available online at http://www.mdpi.com/2079-6382/9/10/675/s1. Table S1: List of oligonucleotides used in the present study, Figure S1: Major structural features of the IncFIB plasmid pHKP0018.3 identified in K. pneumoniae isolates HKP0018, HKP0064 and HKP0067. Arrows represent predicted open reading frames (ORFs) and their direction represents the direction of transcription. Hypothetical proteins are not shown, Figure S2: Major structural features of the pHKP0018.4 plasmid identified in K. pneumoniae isolates HKP0018, HKP0064 and HKP0067. Arrows represent predicted ORFs and their direction represents the direction of transcription. Hypothetical proteins are not shown, Figure S3: Major structural features of pHKP0018.5 identified in K. pneumoniae isolates HKP0018, HKP0064 and HKP0067. Arrows represent predicted ORFs and their direction represents the direction of transcription. Hypothetical proteins are assigned as hp, Figure S4: Major structural features of pHKP0018.6 identified in K. pneumoniae isolates HKP0018, HKP0064 and HKP0067, respectively. Arrows represent predicted ORFs and the direction of the arrow represents the direction of transcription. Hypothetical proteins are assigned as hp.

Author Contributions: Conceptualization, K.X., A.C, H.S. and P.G.H.; methodology, K.X, A.C. and P.G.H.; software, F.F., O.K. and C.F; formal analysis, K.X., A.C., C.F., L.V., P.G.H.; investigation, K.X., J.W., P.G.H.; data curation, K.X.; writing—original draft preparation, K.X. and P.G.H.; writing—review and editing, K.X., A.C., J.W., L.M.B., H.R., F.F., O.K., L.V., C.F., H.S., P.G.H.; supervision, A.C. and P.G.H.; funding acquisition, H.S. All authors have read and agreed to the published version of the manuscript.

Funding: This work was supported by the German Center for Infection Research (DZIF).

Acknowledgments: We would like to thank Yvonne Pfeifer for providing the conjugation protocol.

Conflicts of Interest: The authors declare no conflict of interest.

References

1.   Partridge, S.R.; Kwong, S.M.; Firth, N.; Jensen, S.O. Mobile genetic elements associated with antimicrobial resistance. *Clin Microbiol Rev.* **2018**, *31*, e00088-17. [CrossRef] [PubMed]
2.   Carattoli, A. Plasmids and the spread of resistance. *Int. J. Med. Microbiol.* **2013**, *303*, 298–304. [CrossRef]

3. Peirano, G.; Bradford, P.A.; Kazmierczak, K.M.; Chen, L.; Kreiswirth, B.N.; Pitout, J.D.D. Importance of clonal complex 258 and IncF(K2-like) plasmids among a global collection of *Klebsiella pneumoniae* with *bla*(KPC). *Antimicrob. Agents Chemother.* **2017**, *61*, 5. [CrossRef]

4. Navon-Venezia, S.; Kondratyeva, K.; Carattoli, A. Klebsiella pneumoniae: A major worldwide source and shuttle for antibiotic resistance. *FEMS Microbiol. Rev.* **2017**, *41*, 252–275. [CrossRef]

5. Khan, A.U.; Maryam, L.; Zarrilli, R. Structure, genetics and worldwide spread of New Delhi Metallo-beta-lactamase (NDM): A threat to public health. *BMC Microbiol.* **2017**, *17*, 12. [CrossRef]

6. Bontron, S.; Nordmann, P.; Poirel, L. Transposition of Tn125 Encoding the NDM-1 Carbapenemase in *Acinetobacter baumannii. Antimicrob. Agents Chemother.* **2016**, *60*, 7245–7251. [CrossRef] [PubMed]

7. EARS-Net. Surveillance of Antimicrobial Resistance in Europe 2018. Available online: https://www.ecdc. europa.eu/sites/default/files/documents/surveillance-antimicrobial-resistance-Europe-2018.pdf (accessed on 18 November 2019).

8. David, S.; Reuter, S.; Harris, S.R.; Glasner, C.; Feltwell, T.; Argimon, S.; Abudahab, K.; Goater, R.; Giani, T.; Errico, G.; et al. Epidemic of carbapenem-resistant *Klebsiella pneumoniae* in Europe is driven by nosocomial spread. *Nat. Microbiol.* **2019**, *4*, 1919–1929. [CrossRef] [PubMed]

9. Robert Koch-Institut. Epidemiologisches Bulletin 31/2019. Available online: https://edoc.rki.de/handle/ 176904/6733 (accessed on 1 August 2019).

10. Pitout, J.D.D.; Peirano, G.; Kock, M.M.; Strydom, K.A.; Matsumura, Y. The global ascendency of OXA-48-Type carbapenemases. *Clin Microbiol Rev.* **2020**, *33*, 48. [CrossRef]

11. Poirel, L.; Bonnin, R.A.; Nordmann, P. Genetic Features of the Widespread Plasmid Coding for the Carbapenemase OXA-48. *Antimicrob. Agents Chemother.* **2012**, *56*, 559–562. [CrossRef]

12. Simner, P.J.; Antar, A.A.R.; Hao, S.; Gurtowski, J.; Tamma, P.D.; Rock, C.; Opene, B.N.A.; Tekle, T.; Carroll, K.C.; Schatz, M.C.; et al. Antibiotic pressure on the acquisition and loss of antibiotic resistance genes in Klebsiella pneumoniae. *J. Antimicrob. Chemother.* **2018**, *73*, 1796–1803. [CrossRef]

13. Nahid, F.; Zahra, R.; Sandegren, L. A *bla*$_{OXA-181}$-harbouring multi-resistant ST147 *Klebsiella pneumoniae* isolate from Pakistan that represent an intermediate stage towards pan-drug resistance. *PLoS ONE* **2017**, *12*, e0189438. [CrossRef]

14. Avgoulea, K.; Di Pilato, V.; Zarkotou, O.; Sennati, S.; Politi, L.; Cannatelli, A.; Themeli-Digalaki, K.; Giani, T.; Tsakris, A.; Rossolini, G.M.; et al. Characterization of extensively drug-resistant or pandrug-resistant sequence type 147 and 101 OXA-48-producing *Klebsiella pneumoniae* causing bloodstream infections in patients in an intensive care unit. *Antimicrob. Agents Chemother.* **2018**, *62*, e02457-17. [CrossRef]

15. Zautner, A.E.; Bunk, B.; Pfeifer, Y.; Sproer, C.; Reichard, U.; Eiffert, H.; Scheithauer, S.; Groß, U.; Overmann, J.; Bohne, W. Monitoring microevolution of OXA-48-producing *Klebsiella pneumoniae* ST147 in a hospital setting by SMRT sequencing. *J. Antimicrob. Chemother.* **2017**, *72*, 2737–2744. [CrossRef]

16. Rojas, L.J.; Hujer, A.M.; Rudin, S.D.; Wright, M.S.; Domitrovic, T.N.; Marshall, S.H.; Hujer, K.M.; Richter, S.S.; Cober, E.; Perez, F.; et al. NDM-5 and OXA-181 beta-lactamases, a significant threat continues to spread in the Americas. *Antimicrob. Agents Chemother.* **2017**, *61*, e00454-17. [CrossRef]

17. Peirano, G.; Chen, L.; Kreiswirth, B.N.; Pitout, J.D.D. Emerging antimicrobial-resistant high-risk *Klebsiella pneumoniae* clones ST307 and ST147. *Antimicrob. Agents Chemother.* **2020**, *64*, e01148-20. [CrossRef] [PubMed]

18. Khong, W.X.; Marimuthu, K.; Teo, J.; Ding, Y.; Xia, E.; Lee, J.J.; Ong, R.T.-H.; Venkatachalam, I.; Cherng, B.; Pada, S.K.; et al. Tracking inter-institutional spread of NDM and identification of a novel NDM-positive plasmid, pSg1-NDM, using next-generation sequencing approaches. *J. Antimicrob. Chemother.* **2016**, *71*, 3081–3089. [CrossRef] [PubMed]

19. Becker, L.; Kaase, M.; Pfeifer, Y.; Fuchs, S.; Reuss, A.; von Laer, A.; Sin, M.A.; Korte-Berwanger, M.; Gatermann, S.; Werner, G. Genome-based analysis of carbapenemase-producing *Klebsiella pneumoniae* isolates from German hospital patients, 2008–2014. *Antimicrob. Resist. Infect. Control* **2018**, *7*, 62. [CrossRef]

20. Holt, K.E.; Wertheim, H.; Zadoks, R.N.; Baker, S.; Whitehouse, C.A.; Dance, D.; Jenney, A.; Connor, T.R.; Hsu, L.Y.; Severin, J.; et al. Genomic analysis of diversity, population structure, virulence, and antimicrobial resistance in *Klebsiella pneumoniae*, an urgent threat to public health. *Proc. Natl. Acad. Sci. USA* **2015**, *112*, E3574–E3581. [CrossRef]

21. Du, J.; Cao, J.M.; Shen, L.Z.; Bi, W.Z.; Zhang, X.X.; Liu, H.Y.; Lu, H.; Zhou, T. Molecular epidemiology of extensively drug-resistant *Klebsiella pneumoniae* outbreak in Wenzhou, Southern China. *J. Med. Microbiol.* **2016**, *65*, 1111–1118. [CrossRef]

22. Zhou, T.L.; Zhang, Y.P.; Li, M.M.; Yu, X.; Sun, Y.; Xu, J.R. An outbreak of infections caused by extensively drug-resistant *Klebsiella pneumoniae* strains during a short period of time in a Chinese teaching hospital: Epidemiology study and molecular characteristics. *Diagn. Microbiol. Infect. Dis.* **2015**, *82*, 240–244. [CrossRef] [PubMed]

23. Haller, S.; Kramer, R.; Becker, K.; Bohnert, J.A.; Eckmanns, T.; Hans, J.B.; Hecht, J.; Heidecke, C.-D.; Hübner, N.-O.; Kramer, A.; et al. Extensively drug-resistant *Klebsiella pneumoniae* ST307 outbreak, north-eastern Germany, June to October 2019. *Eurosurveillance* **2019**, *24*, 1900734. [CrossRef]

24. Albiger, B.; Glasner, C.; Struelens, M.J.; Grundmann, H.; Monnet, D.L.; the European Survey of Carbapenemase-Producing Enterobacteriaceae (EuSCAPE) working group. Carbapenemase-producing Enterobacteriaceae in Europe: Assessment by national experts from 38 countries, May 2015. *Eurosurveillance* **2015**, *20*, 17–34. [CrossRef]

25. Potron, A.; Rondinaud, E.; Poirel, L.; Belmonte, O.; Boyer, S.; Camiade, S.; Nordmann, P. Genetic and biochemical characterisation of OXA-232, a carbapenem-hydrolysing class D beta-lactamase from Enterobacteriaceae. *Int. J. Antimicrob. Agents.* **2013**, *41*, 325–359. [CrossRef] [PubMed]

26. Potron, A.; Nordmann, P.; Lafeuille, E.; Al Maskari, Z.; Al Rashdi, F.; Poirel, L. Characterization of OXA-181, a carbapenem-Hydrolyzing Class D beta-lactamase from *Klebsiella pneumoniae*. *Antimicrob. Agents Chemother.* **2011**, *55*, 4896–4899. [CrossRef] [PubMed]

27. Hao, W. Extensive genomic variation within clonal bacterial groups resulted from homologous recombination. *Mob. Genet. Elements.* **2013**, *3*, e23463. [CrossRef]

28. Darmon, E.; Leach, D.R.F. Bacterial Genome Instability. *Microbiol. Mol. Biol. Rev.* **2014**, *78*, 1–39. [CrossRef] [PubMed]

29. Smet, A.; Van Nieuwerburgh, F.; Vandekerckhove, T.T.M.; Martel, A.; Deforce, D.; Butaye, P.; Haesebrouck, F. Complete nucleotide sequence of CTX-M-15-plasmids from clinical *Escherichia coli* isolates: Insertional events of transposons and insertion sequences. *PLoS ONE* **2010**, *5*, e11202. [CrossRef]

30. Miriagou, V.; Carattoli, A.; Tzelepi, E.; Villa, L.; Tzouvelekis, L.S. IS26-Associated In4-type integrons forming multiresistance loci in enterobacterial plasmids. *Antimicrob. Agents Chemother.* **2005**, *49*, 3541–3543. [CrossRef]

31. Sullivan, M.J.; Petty, N.K.; Beatson, S.A. Easyfig: A genome comparison visualizer. *Bioinformatics* **2011**, *27*, 1009–1010. [CrossRef]

32. Casjens, S.R.; Gilcrease, E.B.; Huang, W.M.; Bunny, K.L.; Pedulla, M.L.; Ford, M.E.; Houtz, J.M.; Hatfull, G.F.; Hendrix, R.W. The pKO2 linear plasmid prophage of *Klebsiella oxytoca*. *J. Bacteriol.* **2004**, *186*, 1818–1832.

33. Garcia-Fernandez, A.; Fortini, D.; Veldman, K.; Mevius, D.; Carattoli, A. Characterization of plasmids harbouring *qnrS1*, *qnrB2* and *qnrB19* genes in Salmonella. *J. Antimicrob. Chemother.* **2009**, *63*, 274–281. [CrossRef] [PubMed]

34. Hopkins, K.L.; Wootton, L.; Day, M.R.; Threlfall, E.J. Plasmid-mediated quinolone resistance determinant qnrS1 found in *Salmonella enterica* strains isolated in the UK. *J. Antimicrob. Chemother.* **2007**, *59*, 1071–1075. [CrossRef]

35. Branger, C.; Ledda, A.; Billard-Pomares, T.; Doublet, B.; Barbe, V.; Roche, D.; Médigue, C.; Arlet, G.; Denamur, E. Specialization of small non-conjugative plasmids in *Escherichia coli* according to their family types. *Microb. Genom.* **2019**, *5*, e000281. [CrossRef]

36. Biehl, L.M.; Higgins, P.; Wille, T.; Peter, K.; Hamprecht, A.; Peter, S.; Dörfel, D.; Vogel, W.; Häfner, H.; Lemmen, S.; et al. Impact of single room contact precautions on hospital-acquisition and transmission of multidrug-resistant *Escherichia coli*: A prospective multicentre cohort-study in haematological and oncological wards. *Clin. Microbiol. Infect.* **2019**, *25*, 1013–1020. [CrossRef]

37. CLSI. *Methods for Dilution Antimicrobial Susceptibility Tests for Bacteria That Grow Aerobically. Approved Standard-Tenth Edition: M07. 10th ed.* 2015. Available online: https://clsi.org/media/1632/m07a10_sample.pdf (accessed on 9 January 2020).

38. Bankevich, A.; Nurk, S.; Antipov, D.; Gurevich, A.A.; Dvorkin, M.; Kulikov, A.S.; Lesin, V.M.; Nikolenko, S.I.; Pham, S.; Prjibelski, A.D.; et al. SPAdes: A new genome assembly algorithm and its applications to single-cell sequencing. *J. Comput. Biol.* **2012**, *19*, 455–477. [CrossRef] [PubMed]

39. Dolejska, M.; Villa, L.; Poirel, L.; Nordmann, P.; Carattoli, A. Complete sequencing of an IncHI1 plasmid encoding the carbapenemase NDM-1, the ArmA 16S RNA methylase and a resistance nodulation cell division/multidrug efflux pump. *J. Antimicrob. Chemother.* **2013**, *68*, 34–39. [CrossRef] [PubMed]

40. Koren, S.; Walenz, B.P.; Berlin, K.; Miller, J.R.; Bergman, N.H.; Phillippy, A.M. Canu: Scalable and accurate long-read assembly via adaptive k-mer weighting and repeat separation. *Genome Res.* **2017**, *27*, 722–736. [CrossRef]

41. Antipov, D.; Korobeynikov, A.; McLean, J.S.; Pevzner, P.A. HYBRIDSPADES: An algorithm for hybrid assembly of short and long reads. *Bioinformatics* **2016**, *32*, 1009–1015. [CrossRef]

42. Wick, R.R.; Judd, L.M.; Gorrie, C.L.; Holt, K.E. Unicycler: Resolving bacterial genome assemblies from short and long sequencing reads. *PLoS Comput. Biol.* **2017**, *13*, e1005595. [CrossRef]

43. Antipov, D.; Hartwick, N.; Shen, M.; Raiko, M.; Lapidus, A.; Pevzner, P.A. plasmidSPAdes: Assembling plasmids from whole genome sequencing data. *Bioinformatics* **2016**, *32*, 3380–3387. [CrossRef] [PubMed]

44. de Been, M.; Pinholt, M.; Top, J.; Bletz, S.; Mellmann, A.; van Schaik, W.; Brouwer, E.; Rogers, M.; Kraat, Y.; Bonten, M.; et al. Core genome multilocus sequence typing scheme for high- resolution typing of *Enterococcus faecium*. *J. Clin. Microbiol.* **2015**, *53*, 3788–3797. [CrossRef] [PubMed]

45. Wick, R.R.; Heinz, E.; Holt, K.E.; Wyres, K.L. Kaptive Web: User-Friendly Capsule and Lipopolysaccharide Serotype Prediction for *Klebsiella* Genomes. *J. Clin. Microbiol.* **2018**, *56*, e00197-18. [CrossRef] [PubMed]

46. Zankari, E.; Hasman, H.; Cosentino, S.; Vestergaard, M.; Rasmussen, S.; Lund, O.; Aarestrup, F.M.; Larsen, M.V. Identification of acquired antimicrobial resistance genes. *J. Antimicrob. Chemother.* **2012**, *67*, 2640–2644. [CrossRef]

47. Carattoli, A.; Zankari, E.; Garcia-Fernandez, A.; Larsen, M.V.; Lund, O.; Villa, L.; Frank Møller, A.; Hasman, H. In silico detection and typing of plasmids using PlasmidFinder and plasmid multilocus sequence typing. *Antimicrob. Agents Chemother.* **2014**, *58*, 3895–3903. [CrossRef] [PubMed]

 antibiotics

Article

Genetic Characterization of Carbapenem-Resistant *Klebsiella* spp. from Municipal and Slaughterhouse Wastewater

Mykhailo Savin [1,2,*], Gabriele Bierbaum [3], Nico T. Mutters [1], Ricarda Maria Schmithausen [4], Judith Kreyenschmidt [2,5], Isidro García-Meniño [6,7], Silvia Schmoger [6], Annemarie Käsbohrer [6,8] and Jens Andre Hammerl [6,*]

1 Institute for Hygiene and Public Health, University Hospital Bonn, 53127 Bonn, Germany; nico.mutters@ukbonn.de
2 Institute of Animal Sciences, University of Bonn, 53115 Bonn, Germany; j.kreyenschmidt@uni-bonn.de
3 Institute for Medical Microbiology, Immunology and Parasitology, Medical Faculty, University of Bonn, 53115 Bonn, Germany; g.bierbaum@uni-bonn.de
4 Department of Hygiene and Environmental Medicine, University Hospital Essen, University of Duisburg-Essen, 45147 Essen, Germany; ricarda.schmithausen@uk-essen.de
5 Department of Fresh Produce Logistics, Hochschule Geisenheim University, 65366 Geisenheim, Germany
6 Department for Biological Safety, German Federal Institute for Risk Assessment, 10589 Berlin, Germany; isidro.garcia-menino@bfr.bund.de (I.G.-M.); silvia.schmoger@bfr.bund.de (S.S.); annemarie.kaesbohrer@bfr.bund.de (A.K.)
7 Laboratorio de Referencia de *Escherichia coli* (LREC), Departamento de Microbioloxía e Parasitoloxía, Facultade de Veterinaria, Universidade de Santiago de Compostela (USC), 27002 Lugo, Spain
8 Unit for Veterinary Public Health and Epidemiology, University of Veterinary Medicine, AT-1210 Vienna, Austria
* Correspondence: m.savin@uni-bonn.de (M.S.); jens-andre.hammerl@bfr.bund.de (J.A.H.)

Citation: Savin, M.; Bierbaum, G.; Mutters, N.T.; Schmithausen, R.M.; Kreyenschmidt, J.; García-Meniño, I.; Schmoger, S.; Käsbohrer, A.; Hammerl, J.A. Genetic Characterization of Carbapenem-Resistant *Klebsiella* spp. from Municipal and Slaughterhouse Wastewater. *Antibiotics* 2022, 11, 435. https://doi.org/10.3390/antibiotics11040435

Academic Editor: Francesca Andreoni

Received: 22 February 2022
Accepted: 22 March 2022
Published: 24 March 2022

Publisher's Note: MDPI stays neutral with regard to jurisdictional claims in published maps and institutional affiliations.

Abstract: Currently, human and veterinary medicine are threatened worldwide by an increasing resistance to carbapenems, particularly present in opportunistic *Enterobacterales* pathogens (e.g., *Klebsiella* spp.). However, there is a lack of comprehensive and comparable data on their occurrence in wastewater, as well as on the phenotypic and genotypic characteristics for various countries including Germany. Thus, this study aims to characterize carbapenem-resistant *Klebsiella* spp. isolated from municipal wastewater treatment plants (mWWTPs) and their receiving water bodies, as well as from wastewater and process waters from poultry and pig slaughterhouses. After isolation using selective media and determination of carbapenem (i.e., ertapenem) resistance using broth microdilution to apply epidemiological breakpoints, the selected isolates ($n = 30$) were subjected to WGS. The vast majority of the isolates (80.0%) originated from the mWWTPs and their receiving water bodies. In addition to ertapenem, *Klebsiella* spp. isolates exhibited resistance to meropenem (40.0%) and imipenem (16.7%), as well as to piperacillin-tazobactam (50.0%) and ceftolozan-tazobactam (50.0%). A high diversity of antibiotic-resistance genes ($n = 68$), in particular those encoding β-lactamases, was revealed. However, with the exception of $bla_{GES-5-like}$, no acquired carbapenemase-resistance genes were detected. Virulence factors such as siderophores (e.g., enterobactin) and fimbriae type 1 were present in almost all isolates. A wide genetic diversity was indicated by assigning 66.7% of the isolates to 12 different sequence types (STs), including clinically relevant ones (e.g., ST16, ST252, ST219, ST268, ST307, ST789, ST873, and ST2459). Our study provides information on the occurrence of carbapenem-resistant, ESBL-producing *Klebsiella* spp., which is of clinical importance in wastewater and surface water in Germany. These findings indicate their possible dissemination in the environment and the potential risk of colonization and/or infection of humans, livestock and wildlife associated with exposure to contaminated water sources.

Keywords: *Klebsiella pneumoniae*; wastewater; antimicrobial resistance; carbapenem resistance; virulence

1. Introduction

Antimicrobial resistance (AMR) is currently considered one of the major threats to public health and modern healthcare worldwide [1]. In 2015, more than 670,000 infections in the European Union (EU) and European Economic Area (EEA) countries were caused by bacteria resistant to antibiotics, resulting in an estimated 33,000 deaths [2]. Of those, *K. pneumoniae* resistant to third-generation cephalosporins and carbapenems accounted for 84,500 infections in healthcare settings with approx. 6,000 attributable deaths, where this pathogen was most frequently associated with bloodstream and ventilator-associated pneumonia infections [2]. Furthermore, *K. pneumoniae* and *K. oxytoca/K. michiganensis*, which represent the most important clinical species, have been associated with community-acquired infections such as UTIs, meningitis, pneumonia and bacteraemia [3]. In addition, *Klebsiella* spp. are ubiquitous in the environment and have been recovered from surface water, soil, and plants [4].

At 11.3%, *K. pneumoniae* was one of the most commonly reported bacterial species in EU/EEA countries in 2019 among invasive isolates originating from blood or cerebrospinal fluid [5]. In Germany in 2019, less than 1% of clinical *K. pneumoniae* isolates exhibited resistance to carbapenems (i.e., imipenem and/or meropenem); meanwhile, several Southern and East European countries reported rates of more than 10% [6]. However, an increase in carbapenem resistance in *K. pneumoniae* isolates in Germany is clearly noticeable, with a rise from 0.1% in 2015 to 0.9% in 2019 [6]. Furthermore, significantly increasing trends are seen in EU/EEA population-weighted mean percentages of carbapenem resistance among *K. pneumoniae* isolates, from 6.8% in 2015 to 7.9% in 2019 [6].

Currently, the most clinically used carbapenems are meropenem and imipenem [7]. They are the sole, or one of the few, safe and efficacious therapies available for people affected by severe and polymicrobial infections caused by critical priority Gram-negative pathogens, such as multidrug-resistant (MDR) *A. baumannii*, *P. aeruginosa*, and various bacteria of the *Enterobacteriaceae* family [8]. However, amid increasing rates of resistance to carbapenems, e.g., due to the production of carbapenemases, the effectiveness of most β-lactam antimicrobials is compromised. Genes encoding clinically relevant carbapenemases (i.e., KPC, NDM, IMP, OXA-48-like, and VIM), are often located on mobile genetic elements such as plasmids, transposons and intergrons and can be exchanged between *Enterobacteriaceae* and other Gram-negative bacteria, contributing to their spread [9].

While the incidence of carbapenem-resistant *Enterobacteriaceae* (CRE) in the general population is still low (0.3–2.93 infections per 100,000 person-years in USA), they show a high potential to cause outbreaks in healthcare settings [10]. Through clinical wastewater, such high-risk bacterial pathogens are introduced into municipal wastewater systems and are discharged into surface water due to inadequate wastewater treatment. In Germany, a study by Kehl et al. (2021) demonstrated the discharge of a high-risk *K. pneumoniae* clone, ST147, carrying *bla*$_{NDM}$ and *bla*$_{OXA-48}$ from a hospital into surface water [11]. Similar findings of carbapenem-resistant *K. pneumoniae* in rivers with high genetic concordance to clinical isolates have also been reported in other European countries [12–15].

Carbapenems are restricted to human use only and are not approved for use in veterinary medicine [8,16]. However, the risk of co-resistance to carbapenems, through the use of other antimicrobials in livestock or through horizontal gene transfer from human pathogens, cannot be ruled out [16]. CRE have been sporadically reported in the food chain in various European countries [17]. Carbapenem-resistant, and carbapenemase-producing *K. pneumoniae* have already been detected in poultry, chicken meat, cows and fish in countries that lack strict antimicrobial stewardship in livestock production [18–21]. However, the transmission of CRE from non-human sources is still limited. Nevertheless, antibiotic resistance remains a notable One Health problem, since not only animals and humans but also the environment is affected by CRE. The aquatic environment is of particular importance, since it provides a basic resource for all ecosystems, including agroecosystems, and holds a crucial role in the dissemination of AMR and their propagation between the natural environment, humans and other animals.

In Germany, the data are still lacking regarding the occurrence of the most clinically relevant species of *Klebsiella* spp. (i.e., *K. pneumoniae* and *K. oxytoca*) with resistance to carbapenems in wastewater, as well as their phenotypic and genotypic characteristics. Thus, the aim of this study is to evaluate the occurrence of carbapenem-resistant *Klebsiella* spp. in municipal wastewater treatment plants (mWWTPs) and their receiving water bodies, as well as in wastewater and process waters from poultry and pig slaughterhouses. In order to better assess their clinical relevance to public and environmental health, we also aim to characterize the recovered *Klebsiella* spp. isolates by applying phenotypic and genotypic methods.

2. Results

An overview of the phenotypic antimicrobial resistance of the investigated *Klebsiella* spp. isolates is presented in Figure 1. The isolates exhibit various phenotypic resistances to antimicrobials, inter alia to those highly and critically important for humans (Figure 1).

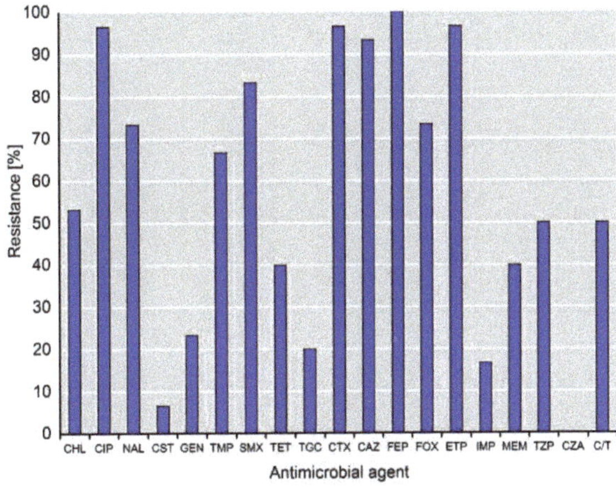

Figure 1. Phenotypical resistance to antimicrobial agents detected among isolates of *Klebsiella* spp. (*n* = 30). Abbreviations for antimicrobial agents: CHL, chloramphenicol; CIP, ciprofloxacin; NAL, nalidixic acid; CST, colistin; GEN, gentamicin; TMP, trimethoprim; SMX, sulfamethoxazole; TET, tetracycline; TGC, tigecycline; CTX, cefotaxime; CAZ, ceftazidime; FEP, cefepime; FOX, cefoxitin; ETP, ertapenem; IMI, imipenem; MEM, meropenem; TZP, piperacillin-tazobactam; CZA, ceftazidime-avibactam; C/T, ceftolozane-tazobactam.

As expected, the resistance rates to third- and fourth-generation cephalosporins (i.e., cefotaxime, ceftazidime, cefepime) were high and ranged from 93.3% (28/30) to 100%, whereas the rate of resistance to cefoxitin was lower at 73.3% (22/30). In addition to ertapenem resistance (96.7%, 29/30), as a selection criterion for this study, 16.7% (5/30) and 40.0% (12/30) of the isolates were resistant to imipenem and meropenem, respectively. Notably, 50% (15/30) of the isolates showed resistance to piperacillin-tazobactam and ceftolozan-tazobactam, whereas all of the isolates were susceptible to ceftazidime-avibactam. Almost all of the isolates (96.7%, 29/30) exhibited resistance to fluoroquinolones (i.e., ciprofloxacin), whereas only two isolates (6.7%) were resistant to colistin. Of note, the resistance rates to tigecycline and gentamicin were 20.0% (6/30) and 23.3% (7/30), respectively. The phenotypic resistance patterns of individual isolates are shown in Table 1.

Table 1. Selected phenotypic and genotypic characteristics of carbapenem-resistant *Klebsiella* spp. isolates recovered from municipal WWTPs and their receiving water bodies as well as from process waters of poultry and pig slaughterhouses.

Isolate	Species	Origin	Resistance Phenotype [a]	Combinations of β-Lactam–β-Lactamase Inhibitor	Antimicrobial Resistance Genes to Β-Lactams	MLST
05/11-30	*K. oxytoca*	Effluent mWWTP	CIP, NAL, TMP, SMX, CTX, CAZ, FEP, ETP	TZP, C/T	$bla_{OXY-2-8-like}$ [c]	_ [d]
05/11-32	*K. oxytoca*	Effluent mWWTP	CIP, NAL, TMP, SMX, CTX, CAZ, FEP, FOX, ETP	TZP, C/T	$bla_{OXY-2-8-like}$	-
05/10-58	*K. oxytoca*	Influent mWWTP	CIP, NAL, TMP, SMX, CTX, CAZ, FEP, FOX, ETP	TZP, C/T	$bla_{OXY-2-8-like}$	-
05/10-60	*K. oxytoca*	Influent mWWTP	CIP, NAL, TMP, SMX, CTX, CAZ, FEP, FOX, ETP	TZP, C/T	$bla_{OXY-2-8-like}$	-
03/12-04Bki	*K. oxytoca*	On-site preflooder downstream	CIP, NAL, GEN, SMX, CTX, CAZ, FEP, FOX, ETP	TZP, C/T	$bla_{CTX-M-9}$, bla_{OXA-4}, $bla_{OXY-2-8-like}$	-
05/13-23	*K. oxytoca*	On-site preflooder upstream	CIP, TMP, SMX, TET, CTX, CAZ, FEP, FOX, ETP	_ [b]	$bla_{CTX-M-15}$, $bla_{OXY-2-5}$ [c]	-
05/13-25	*K. oxytoca*	On-site preflooder upstream	CIP, NAL, TMP, SMX, CTX, CAZ, FEP, FOX, ETP	TZP, C/T	$bla_{OXY-2-8-like}$	-
03/11-12	*K. pneumoniae*	Effluent mWWTP	CIP, NAL, TET, CTX, CAZ, FEP, FOX, ETP, MEM	-	$bla_{OKP-B-3-like}$ [c]	-
03/11-28	*K. pneumoniae*	Effluent mWWTP	CIP, NAL, TMP, SMX, CTX, CAZ, FEP, ETP	-	$bla_{CTX-M-15}$, bla_{OXA-1}, bla_{SHV-28}, bla_{TEM-1B}	ST307
03/11-38	*K. pneumoniae*	Effluent mWWTP	CHL, CIP, NAL, CTX, CAZ, FEP, FOX, ETP, IMI, MEM	-	$bla_{GES-5-like}$, $bla_{SHV-2-like}$	-
05/11-29	*K. pneumoniae*	Effluent mWWTP	CHL, CIP, NAL, CST, CTX, CAZ, FEP, FOX, ETP, IMI	TZP	$bla_{CTX-M-15}$, bla_{OXA-1}, bla_{SHV-1} [c], $bla_{SHV-148-like}$	ST16
05/11-43	*K. pneumoniae*	Effluent mWWTP	CHL, CIP, NAL, GEN, TMP, SMX, TET, TGC, CTX, CAZ, FEP, FOX, ETP	TZP	$bla_{CTX-M-15}$, $bla_{OXY-2-2-like}$ [c], bla_{TEM-1B}	-
04/08-35	*K. pneumoniae*	Poultry Eviscerators	CIP, CTX, CAZ, FEP, FOX, ETP	-	bla_{SHV-25}	ST789
03/06-23	*K. pneumoniae*	Pig Holding Pens	CHL, CIP, TMP, SMX, TET, CTX, CAZ, FEP, ETP	TZP, C/T	$bla_{CTX-M-1}$, $bla_{SHV-27-like}$, bla_{TEM-1B}	ST873
03/01-52	*K. pneumoniae*	Influent in-house chemical-physical WWTP	CIP, TMP, SMX, CTX, FEP, FOX, ETP, MEM	-	bla_{SHV-33}	ST1948
03/10-26	*K. pneumoniae*	Influent mWWTP	CHL, CIP, NAL, TMP, SMX, CTX, CAZ, FEP, FOX, ETP, MEM	TZP, C/T	$bla_{CTX-M-15}$, bla_{OXA-1}, bla_{SHV-1}	ST2459
03/10-27	*K. pneumoniae*	Influent mWWTP	CHL, CIP, NAL, TMP, SMX, CTX, CAZ, FEP, FOX, ETP, MEM	TZP, C/T	$bla_{CTX-M-15}$, bla_{OXA-1}, bla_{SHV-1}	ST2459
03/10-46	*K. pneumoniae*	Influent mWWTP	CIP, TMP, SMX, CTX, CAZ, FEP, IMI	-	$bla_{CTX-M-15}$, $bla_{SHV-1-like}$	ST219
05/10-20	*K. pneumoniae*	Influent mWWTP	CHL, CIP, NAL, GEN, SMX, TET, CTX, CAZ, FEP, FOX, ETP, IMI, MEM	TZP, C/T	bla_{OXA-10}, bla_{SHV-31}	ST252
05/10-21	*K. pneumoniae*	Influent mWWTP	CHL, CIP, NAL, GEN, SMX, TET, CTX, CAZ, FEP, FOX, ETP, IMI, MEM	TZP, C/T	bla_{OXA-10}, bla_{SHV-31}	ST252
05/10-59	*K. pneumoniae*	Influent mWWTP	CHL, CIP, NAL, GEN, TMP, SMX, TET, TGC, CTX, CAZ, FEP, FOX, ETP	C/T	$bla_{CTX-M-15}$, bla_{SHV-11}, bla_{TEM-1A}	ST268
05/10-69A	*K. pneumoniae*	Influent mWWTP	CHL, CIP, NAL, SMX, CTX, CAZ, FEP, FOX, ETP, MEM	TZP, C/T	bla_{OXA-10}, $bla_{SHV-69-like}$, bla_{TEM-1B}	ST503
05/10-69B	*K. pneumoniae*	Influent mWWTP	CHL, CIP, NAL, SMX, FEP, ETP, MEM	TZP, C/T	bla_{OXA-10}, $bla_{SHV-69-like}$, bla_{TEM-1B}	ST503
05/10-71	*K. pneumoniae*	Influent mWWTP	CHL, CIP, NAL, CST, SMX, CTX, CAZ, FEP, ETP, MEM	TZP, C/T	bla_{OXA-10}, $bla_{SHV-69-like}$	ST503
05/10-83	*K. pneumoniae*	Influent mWWTP	CHL, CIP, NAL, GEN, TMP, SMX, TET, TGC, CTX, CAZ, FEP, FOX, ETP	TZP	$bla_{CTX-M-15}$, bla_{OXA-1}, $bla_{SHV-38-like}$, bla_{TEM-1B}	ST441
03/13-21	*K. pneumoniae*	On-site preflooder upstream	CHL, CIP, NAL, TMP, CTX, CAZ, FEP, ETP	-	$bla_{CTX-M-15}$, bla_{SHV-1}	ST2459

Table 1. *Cont.*

Isolate	Species	Origin	Resistance Phenotype [a]	Combinations of β-Lactam–β-Lactamase Inhibitor	Antimicrobial Resistance Genes to B-Lactams	MLST
05/13-31	*K. pneumoniae*	On-site preflooder upstream	CHL, CIP, NAL, GEN, TMP, SMX, TET, TGC, CTX, CAZ, FEP, FOX, ETP	-	$bla_{\text{CTX-M-15}}$, $bla_{\text{SHV-11}}$, $bla_{\text{TEM-1A}}$	ST268
03/05-22	*K. pneumoniae*	Pig Transporters	CHL, TMP, SMX, TET, CTX, CAZ, FEP, ETP	-	$bla_{\text{CTX-M-1}}$, $bla_{\text{SHV-27-like}}$, $bla_{\text{TEM-1B}}$	ST873
01/07-40	*K. pneumoniae*	Poultry Stunning Facilities	CIP, TMP, SMX, TET, TGC, CTX, CAZ, FEP, FOX, ETP, MEM	TZP	$bla_{\text{SHV-28-like}}$, $bla_{\text{TEM-1B}}$	ST458
01/07-41	*K. pneumoniae*	Poultry Stunning Facilities	CIP, TMP, SMX, TET, TGC, CTX, CAZ, FEP, FOX, ETP, MEM	-	$bla_{\text{SHV-28-like}}$, $bla_{\text{TEM-1B}}$	ST458

[a] Abbreviations for antimicrobial agents: CHL, chloramphenicol; CIP, ciprofloxacin; NAL, nalidixic acid; CST, colistin; GEN, gentamicin; TMP, trimethoprim; SMX, sulfamethoxazole; TET, tetracycline; TGC, tigecycline; CTX, cefotaxime; CAZ, ceftazidime; FEP, cefepime; FOX, cefoxitin; ETP, ertapenem; IMI, imipenem; MEM, meropenem; TZP, piperacillin-tazobactam; C/T, ceftolozane-tazobactam.; [b] susceptible to the combinations of β-lactam–β-lactamase inhibitor TZP, C/T and CZA; [c] Intrinsic chromosomally encoded β-lactamases; [d] The ST could not be determined using the prevailing scheme.

Klebsiella spp. isolates represented a reservoir for 68 different ARGs (antimicrobial resistance genes) conferring resistance to antimicrobials belonging to 11 different classes (Table 2).

Table 2. Prevalence of antibiotic-resistance genes detected in carbapenem-resistant *Klebsiella* spp. isolates recovered from municipal WWTPs and their receiving water bodies as well as from process waters of poultry and pig slaughterhouses.

Antimicrobial Class	Genes	Percentage [%]
	$bla_{CTX-M-15}$	36.7
	bla_{TEM-1B}	30.0
	$bla_{OXY-2-8-like}$ [a]	20.0
	bla_{OXA-1}	16.7
β-lactams	bla_{SHV-1} [a]	16.7
	bla_{OXA-10}	16.7
	$bla_{SHV-69-like}$	10.0
	bla_{TEM-1B}, $bla_{SHV-27-like}$, $bla_{SHV-27-like}$, bla_{SHV-11}, bla_{TEM-1A}	each 6.7
	$bla_{OKP-B-3-like}$ [a], bla_{SHV-28}, $bla_{GES-5-like}$, $bla_{SHV-2-like}$, $bla_{SHV-148-like}$, $bla_{OXY-2-2-like}$ [a], bla_{SHV-25}, bla_{SHV-33}, $bla_{SHV-38-like}$, $bla_{CTX-M-9}$, bla_{OXA-4}, $bla_{OXY-2-5}$ [a], $bla_{SHV-28-like}$	each 3.3
	strB	40.0
	strA	36.7
	aadA5	23.3
	aac(3)-I-like	16.7
	aadA1	16.7
Aminoglycosides	*aadA2*	16.7
	strA-like	13.3
	strB-like	13.3
	aadB	10.0
	aac(3)-IId-like, aacA4, aadA24-like, aph(3′)-Ia	each 6.7
	aac(3)-IIa-like, aacA4-like, aadA22, aph(3′)-XV	each 3.3
	catB3-like	20.0
Phenicols	*floR-like*	6.7
	catB2	3.3
Fluoroquinolones and aminoglycosides	*aac(6′)Ib-cr*	20.0
	aac(6′)Ib-cr-like	10.0
	dfrA14-like	26.7
Diaminopyrimidines (Trimethoprim)	*dfrA17*	23.3
	dfrA1	13.3
	dfrA12	6.7
	sul1	56.7
Sulfonamides	*sul2*	33.3
	sul2-like	16.7
Phosphonic Acid (Fosfomycin)	*fosA-like* [a]	56.7
	fosA [a]	13.3

Table 2. *Cont.*

Antimicrobial Class	Genes	Percentage [%]
Quinolones	*oqxA*-like [a]	70.0
	oqxB-like [a]	70.0
	qnrB66-like	13.3
	qnrS1	13.3
	qnrA1-like	10.0
	qnrB1	6.7
Tetracyclines	*tet(A)*	13.3
	tet(A)-like	6.7
	tet(B)	6.7
Macrolides	*mph(A)*	30.0
	erm(B)-like	6.7
Lincosamides	*lnu(F)*	3.3

[a] Intrinsic chromosomally encoded ARGs.

Of the detected ARGs, 25 encoded β-lactamases which, as expected, were found in all isolates. They encoded enzymes of seven families: bla_{CTX-M}, bla_{TEM}, bla_{SHV}, bla_{OXA}, bla_{OXY}, bla_{GES}, and bla_{OKP}. The most abundant were $bla_{CTX-M-15}$, bla_{TEM-1B} and $bla_{OXY-2-8-like}$, accounting for 36.7% (11/30), 30.0% (9/30) and 20.0% (6/30) of the isolates, respectively; and bla_{OXA-1}, bla_{SHV-1} and bla_{OXA-10} were each detected in 16.7% (5/30) of the isolates. Of note, combinations of up to four β-lactamases were detected in *K. pneumoniae* isolates from both in- and effluent of mWWTP. Interestingly, bla_{OXA-1} and bla_{OXA-10}, in combination with other extended spectrum β-lactamases of SHV and TEM families, were found to a large extent in isolates with resistance to piperacillin-tazobactam and ceftolozan-tazobactam (Table 1). No carbapenemases were detected, with the exception of $bla_{GES-5-like}$ carried by a *K. pneumoniae* isolate recovered from the effluent of mWWTP. Thus, resistance could be mediated by chromosomal alterations.

Among the *Klebsiella* spp. isolates of this study, only some of the genes could be backed to mobile genetic elements. Using the MobileElementFinder tool (version 1.0) from the Center for Genomic Epidemiology, some of the genes could be associated with plasmid or insertion sequences (IS) (Table S1). Interestingly, IncR plasmids often comprise a combination of the genes *aph(3″)-Ib-aph(6)-Id*, leading to a streptomycin resistance phenotype, while some other resistance genes coding for resistances against extended-spectrum β-lactamase antibiotics, tetracycline and sulphonamides were found on other plasmid types (based on the Inc-groups). Overall, the majority of the isolates exhibit a broad diversity of different IS, but mainly IS elements such as ISVsa3, ISAhy2, ISEc9, IS6100, ISKpn19 were found to be associated with resistance determinants. Based on the prevailing data, it cannot be excluded that further genes will be associated with plasmid or IS sequences, due to the use of short-read sequencing data for in silico analysis. Further information about the impact of the plasmids or IS elements on the spread of resistances needs to be determined in detail in another study.

The results of the multilocus sequence-typing (MLST), performed to identify high-risk clones of public health importance, showed that 66.7% (20/30) of the isolates could be assigned to 12 different sequence types (STs). Three isolates each belonged to ST503 and ST2459. ST873, ST458, ST268 and ST252 each accounted for two isolates, whereas the remaining six isolates were identified as ST789, ST441, ST307, ST219, ST1948 and ST16. The STs of seven *K. oxytoca* and three *K. pneumoniae* isolates (Table 1) could not be determined using the prevailing genotyping schemes, possibly indicating new STs.

Among known virulence factors, genes encoding various siderophores, and fimbriae were detected (Table 3). Almost all isolates (96.7%, 29/30) carried genes coding for the enterobactin siderophore system, whereas yersiniabactin, salmochelin, and aerobactin were less prevalent and accounted for 40.0% (12/30) and 33.3% (10/30) of the isolates, respectively. Fimbriae type 1 and fimbriae type 3 were detected in 90.0% (27/30) and 63.3% (19/30) of the isolates, respectively.

Table 3. Virulence factors detected in carbapenem-resistant *Klebsiella* spp. isolates recovered from municipal WWTPs and their receiving water bodies as well as from process waters of poultry and pig slaughterhouses.

Virulence Factor	Genes	Percentage [a]
Enterobactin	*ent*	96.7
Yersiniabactin	*ybt, irp1, irp2, fyuA*	40.0
Salmochelin	*iroN, iroBCD*	40.0
Aerobactin	*iucABCD, iutA*	33.3
Colibactin	*clbA-R*	0
Regulators of mucoid phenotype	*rmpA, rmpA2, rmpB*	0
K1 capsule synthesis	*magA*	0
Chromosomal capsule production	*cps*	0
Fimbriae type 1	*fim*	90.0
Fimbriae type 3	*mrk*	63.3

[a] Percentage of isolates carrying particular virulence factor.

Surface polysaccharide locus typing, performed in order to determine the capsule (K antigen) serotypes, showed that the capsule polysaccharide (CPS) types of the vast majority of the isolates (80.0%, 24/30) could be assigned to 14 different types. Three isolates each accounted for KL9 and KL60. KL81, KL74, KL62, KL52, KL21 and KL20 were each represented by two isolates, whereas the remaining six isolates were assigned to KL51, KL24, KL18, KL151, KL114 and KL102.

3. Discussion

This study provides novel data on antimicrobial resistance, genetic lineages, virulence factors and CPS-types of carbapenem-resistant (CR) *Klebsiella* spp. from municipal WWTPs as well as process waters and wastewater from German poultry and pig slaughterhouses.

The occurrence of ESBL-producing and CR *Klebsiella* spp. in municipal WWTPs indicates its possible dissemination in the general population and the impact of clinical effluents on the municipal sewer system. Wielders and colleagues (2017) reported an overall prevalence of ESBL-producing *K. pneumoniae* in the general population in the Netherlands of 4.3%, with seasonal differences ranging from 2.6% to 7.4% [22]. Meijs and colleagues (2021) reported even higher levels of ESBL-producing *K. pneumoniae* carriage for veterinary healthcare workers in the Netherlands of 9.8%, emphasizing occupational contact with animals as a potential source of ESBL-producing *K. pneumoniae* in the general population [23]. Several studies have reported high abundances of carbapenemase-producing *Klebsiella* spp. in clinical wastewater and its discharge into the municipal sewer system [11,24–26]. The subsequent incidence of such bacteria in surface waters suggests that conventional biological treatment is insufficient in terms of eliminating microbial loads, and shows the negative impact of inadequately treated wastewater on surface waters. Similar findings on ESBL, and on carbapenemase-producing *K. pneumoniae* in Austrian and Swiss rivers mostly within urbanized areas, also highlight the anthropological pollution in aquatic environments [13,14]. *Klebsiella* spp. are known for their ability to survive under adverse conditions and are widely distributed in nature, including in surface water and nutrient rich wastewater [27]. Lepuschitz and colleagues (2019) recovered two multidrug-resistant *K. pneumoniae* ST985 isolates, which share the same cgMLST profile, from sampling sites on a river 200 km apart, demonstrating the possible survival distance of *K. pneumoniae* in river water [13]. A study by Rocha and colleagues (2022) suggests that *Klebsiella* spp. isolates in wastewater retain clinically relevant features, including those acquired through HGT, even after treatment. Thus, further dissemination of the CR isolates recovered in our study among animals and humans, and their colonization and/or infection cannot be ruled out [28]. The application of state-of-the-art wastewater treatment techniques based on oxidative, adsorptive, and membrane-based technologies, as well as the establishment of a surveillance system for clinically relevant antimicrobial-resistant bacteria in surface water should be encouraged.

In this study, almost all isolates exhibited resistance to ciprofloxacin, which is considered to be critically important in human medicine and is often administered to outpatients [29]. Thus, the use of (fluoro)quinolones may contribute to the selection of CR *Klebsiella* spp. in the general community. This finding is in line with the EDCD report indicating that resistance to carbapenems is almost always combined with resistance to other antimicrobial classes, severely narrowing the treatment options for invasive infections caused by "critical pathogens" (i.e., CR *A. baumannii*, *P. aeruginosa* and *Enterobacteriaceae*) and decreasing the likelihood of a positive outcome [6].

Resistance to carbapenems at a clinical level is most frequently caused by the production of carbapenemases, however, other mechanisms may also be involved in the development of such phenotypes [30]. In this study, the results of antimicrobial susceptibility testing were interpreted based on epidemiological cut-off values. This allows the detection of early changes in resistance patterns that could possibly lead to resistance at a clinical level. Nevertheless, no clinically relevant carbapenemases were detected, with only one *K. pneumoniae* isolate from the effluent of mWWTP carrying bla_{GES-5}. Since no carbapenemases and AmpC β-lactamases were detected, possible mechanisms of resistance to carbapenems, and combinations of β-lactam–β-lactamase inhibitor (i.e., piperacillin-tazobactam and ceftolozane-tazobactam), could be: changes in membrane permeability due to mutations in the genes encoding efflux pump, alterations in the expression and function of porins, and the association of impermeability with the production of ESBL [30]. Furthermore, bla_{OXA-1}, which encodes a penicillinase with weak affinity for inhibitors such as tazobactam, and hyperproduction of bla_{TEM-1} could also be responsible for resistance to piperacillin-tazobactam [31,32]. Narrow-spectrum oxacillinases, e.g., OXA-10-type class D β-lactamases, were previously shown to exhibit weak carbapenemase activity at a level comparable with that of OXA-58 [33,34]. Genes encoding class D β-lactamases are commonly found in *P. aeruginosa*; however, they are also detected in *Enterobacteriaceae*, albeit with lower abundance than in *Pseudomonas* spp., underlying the important role of horizontal gene transfer (HGT) in the spread of AMRs [35]. Resistance to ceftolozane-tazobactam, and ceftazidime-avibactam have been reported in clinical MDR/extensively-drug resistant (XDR) *P. aeruginosa* isolates due to mutations in bla_{OXA-10} which developed during antimicrobial treatment [36].

K. pneumoniae isolates belonging to the sequence types (STs) determined in this study (ST16, ST252, ST219, ST268, ST307, ST789, ST873, and ST2459) have been detected in various clinical settings across Europe and Asia, causing urinary and respiratory tract infections [37–42]. All of them carried different carbapenemases belonging to NDM, OXA, IMP, and KPC families, with some of them exhibiting an extensively drug-resistant (XDR) phenotype. In Europe, the spread of carbapenemases among *K. pneumoniae* is frequently linked to specific clonal lineages such as ST11, ST15, ST101, ST258/512 and their derivatives [43]. However, novel high-risk CR *K. pneumoniae* lineages are continuously emerging. Wyres and colleagues (2018) showed that some *K. pneumoniae* clones are generally better than others at acquiring genetic material via HGT [44]. Currently, comprehensive data on the mechanisms underlying this phenomenon are still lacking, and experimental studies are needed for its further investigation. Interestingly, *K. pneumoniae* ST789 carrying bla_{NDM-5} has been reported in neonates in China, and was classified as a novel high-risk CR lineage [45]. In our study *K. pneumoniae* ST789 was detected in wastewater from poultry eviscerators and was carrying bla_{SHV-25}. However, given the potential of some *Klebsiella* spp. to become high-risk clonal lineage, the possible factors contributing to increased virulence and antimicrobial resistance that might occur in livestock production need to be investigated. This would help develop mitigation strategies in order to interrupt possible dissemination of CRE from livestock to humans. ESBL-producing *Enterobacteriaceae* can serve as a basic model for its spread, since the bacterial species involved are the same and the antimicrobial resistance genes are located on plasmids as well.

The contamination of food of animal and vegetable origin with ESBL-producing *Enterobacteriaceae* is already well described [46–50]. However, considering the risks of CRE to human health, there have been appeals for a zero-tolerance policy and an international ban on the sale of food contaminated with CRE [46]. The fact that carbapenems are not approved for use in veterinary medicine and are predominantly used in human hospital settings can explain the low incidence of carbapenem resistance among the isolates recovered from poultry and pig slaughterhouses. These findings are in line with other reports indicating the absence of or single cases of CRE in European livestock, in particular pigs and broilers [51]. Nevertheless, the risk of AMR transmission through horizontal gene transfer from human pathogens or of co-resistance through the use of other antimicrobials in agriculture cannot be ruled out.

According to epidemiological studies, the first step in the majority of *K. pneumoniae* infections is the colonization of the host's gastrointestinal tract [52]. The recovered isolates carried genes encoding fimbrial adhesins (type 1 and type 3 fimbriae), which play an essential role in adhesion to the host's mucosal surfaces and in biofilm formation, as well as genes encoding components of siderophore systems that mediate the uptake of ferric iron [53]. The presence of these virulence factors increases the probability of adherence to the host, colonization, and invasive infections. These data reinforce the recent trend of increasing occurrence of community-acquired *K. pneumoniae* infections in young and healthy individuals, rather than primarily nosocomial infections in immunocompromised patients [54]. However, in comparison to clinical isolates, all of the recovered isolates lacked the factors responsible for the hypermucoviscous phenotype that protects the bacteria against opsonization and phagocytosis.

4. Materials and Methods

The sampling sites, procedures and preparation of the samples have been previously described [55,56]. Briefly, the process waters and wastewater ($n = 87$) arising during the operation and cleaning of production facilities were collected from the delivery areas (transport trucks, transport crates, and holding pens) and unclean areas (stunning facilities, scalders, eviscerators, and aggregate wastewater from production facilities) of two poultry and two pig slaughterhouses. In- and effluents ($n = 62$) from their in-house wastewater treatment plants (WWTPs) were also sampled. Further samples ($n = 36$) were taken at two municipal WWTPs (mWWTPs) receiving pretreated wastewater from the pig slaughterhouses, including their on-site preflooders upstream and downstream from the discharge points [55]. At each sampling site, 1 L of water was collected using sterile Nalgene Wide Mouth Environmental Sample Bottles (Thermo Fisher Scientific, Waltham, MA, USA). For further information on selected characteristics of the sampled slaughterhouses, sampling sites and number of samples taken at each sampling site, please see [55,56].

Klebsiella spp. isolates with resistance to third-generation cephalosporins, and carbapenems were recovered from water samples by selective cultivation on CHROMagar ESBL and CHROMagar mSuperCarba plates (MAST Diagnostica, Reinfeld, Germany), as previously described [56]. Presumptive colonies of *Klebsiella* spp. were unselectively subcultured on Columbia Agar supplemented with 5% sheep blood (*v/v*) (Mast Diagnostics, Reinfeld, Germany). Species identification for the individual isolates was conducted using MALDI-ToF MS (bioMérieux, Marcy-l'Étoile, France) equipped with the Myla software.

Antimicrobial susceptibility testing was performed according to CLSI guidelines (M07-A10), using broth microdilution and applying the epidemiological cut-off values (ECOFFs) from the European Committee on Antimicrobial Susceptibility Testing (EU-

CAST). In order to assess the clinical relevance of the presumptive ESBL-producing and carbapenem-resistant *Klebsiella* spp. isolates for human medicine, they were tested against the newly approved β-lactam/β-lactamase inhibitor combinations ceftazidime-avibactam, ceftolozan-tazobactam, and piperacillin-tazobactam, by a microdilution method using the clinical cut-off values as previously described [56,57].

A total of 185 *Klebsiella* spp. (155 *K. pneumoniae*, 30 *K. oxytoca*) were isolated, of which 30 (16.2%), comprising 23 *K. pneumoniae* and 7 *K. oxytoca*, showed resistance to at least one of the tested carbapenems (i.e., ertapenem, imipenem, meropenem), and were further investigated in detail. The vast majority (80%, 24/30) originated from mWWTPs (influent, $n = 12$; effluent, $n = 7$) and their on-site preflooders upstream ($n = 4$) and downstream ($n = 1$) from the discharge points. Further isolates were recovered from the process waters and wastewater accruing in poultry (stunning facilities, $n = 2$; eviscerators, $n = 1$) and pig slaughterhouses (pig transporters, $n = 1$; holding pens, $n = 1$; influent in-house chemical-physical WWTP, $n = 1$).

Extraction of genomic DNA (gDNA) from the individual colonies of *Klebsiella* spp., DNA library preparation, and whole-genome sequencing (WGS) were performed as previously described [58]. Briefly, gDNA was extracted using PureLink® Genomic DNA Mini Kit (Invitrogen, Darmstadt, Germany) following the manufacturer's instructions. Commercial DNA library preparation and WGS were conducted using LGC Genomics GmbH (Berlin, Germany) on an Illumina NextSeq 500/550 V2 (Illumina, CA, USA). De novo assembly of high-quality ~150 bp paired-end sequencing reads was conducted using the SPAdes algorithm of the PATRIC database (v. 3.5.27) [59]. ResFinder v 3.0 and MLST v 2.0 under default values, as well as MyDbFinder (release 1.1; parameters: 90% sequence identity, 60% sequence coverage) of the Center for Genomic Epidemiology (https://cge.cbs.dtu.dk/services/ (accessed on 10 December 2020)) were used for bioinformatics analysis of ARGs, sequence types (STs) and virulence factors, respectively [60]. The captive tool (https://kaptive-web.erc.monash.edu/ (accessed on 10 December 2020)) was used for surface polysaccharide locus typing and variant evaluation. The tool MobileElementFinder (Center for Genomic Epidemiology, version 1.0, default parameters; https://cge.cbs.dtu.dk/services/MobileElementFinder/ (accessed on 10 March 2022) was used for the prediction of mobile genetic elements (MGEs) in combination with plasmid or insertion sequences.

Supplementary Materials: The following supporting information can be downloaded at https://www.mdpi.com/article/10.3390/antibiotics11040435/s1, Table S1: Mobile genetic elements detected in carbapenem-resistant *Klebsiella* spp. isolates recovered from municipal WWTPs and their receiving water bodies as well as from process waters of poultry and pig slaughterhouses.

Author Contributions: Conceptualization, M.S.; methodology, M.S. and J.A.H.; formal analysis, M.S., I.G.-M., S.S. and J.A.H.; investigation, M.S., I.G.-M., S.S. and J.A.H.; data curation, M.S.; writing—original draft preparation, M.S.; writing—review and editing, G.B., N.T.M., R.M.S., J.K., A.K. and J.A.H.; project administration, M.S. and J.A.H.; funding acquisition, J.K. and J.A.H. All authors have read and agreed to the published version of the manuscript.

Funding: This work was financed by the BMBF (Federal Ministry of Education and Research) funding measure HyReKA [02WRS1377], internal grants from BfR (43-001, 43-002 and 1322-648), the European Joint Projects (EJP) "ARDIG" and "FULL_FORCE" funded by the European Union's Horizon 2020 research and innovation programme (Grant Agreement No 773830), as well as by the project "GÜCCI" funded by the German Federal Ministry of Health. The funders had no role in the study design, data collection and interpretation, nor in the decision to submit the work for publication. I. García-Meniño acknowledges the Consellería de Cultura, Educación e Ordenación Universitaria, Xunta de Galicia for the post-doctoral grant (Grant Number: ED481B-2021-006).

Institutional Review Board Statement: Not applicable.

Informed Consent Statement: Not applicable.

Data Availability Statement: The data for this study have been uploaded to Sequence Read Archive (SRA). The accession number for the bioproject is PRJNA816413 and the individual biosamples are aslo available there. The temporary number will change as soon as the raw reads are processed by NCBI.

Acknowledgments: We thank the staff of the participating municipal WWTPs and slaughterhouses for their kind cooperation.

Conflicts of Interest: The authors declare no conflict of interest.

References

1. De Kraker, M.E.A.; Stewardson, A.J.; Harbarth, S. Will 10 Million People Die a Year due to Antimicrobial Resistance by 2050? *PLoS Med.* **2016**, *13*, e1002184. [CrossRef] [PubMed]
2. Cassini, A.; Högberg, L.D.; Plachouras, D.; Quattrocchi, A.; Hoxha, A.; Simonsen, G.S.; Colomb-Cotinat, M.; Kretzschmar, M.E.; Devleesschauwer, B.; Cecchini, M.; et al. Attributable deaths and disability-adjusted life-years caused by infections with antibiotic-resistant bacteria in the EU and the European Economic Area in 2015: A population-level modelling analysis. *Lancet Infect. Dis.* **2019**, *19*, 56–66. [CrossRef]
3. Weiner, L.M.; Webb, A.K.; Limbago, B.; Dudeck, M.A.; Patel, J.; Kallen, A.J.; Edwards, J.R.; Sievert, D.M. Antimicrobial-Resistant Pathogens Associated With Healthcare-Associated Infections: Summary of Data Reported to the National Healthcare Safety Network at the Centers for Disease Control and Prevention, 2011–2014. *Infect. Control Hosp. Epidemiol.* **2016**, *37*, 1288–1301. [CrossRef]
4. Zumla, A. Mandell, Douglas, and Bennett's principles and practice of infectious diseases. *Lancet Infect. Dis.* **2010**, *10*, 303–304. [CrossRef]
5. European Centre for Disease Prevention and Control. Carbapenem-Resistant Enterobacteriaceae, Second Update—26 September 2019. Available online: https://www.ecdc.europa.eu/sites/default/files/documents/carbapenem-resistant-enterobacteriaceae-risk-assessment-rev-2.pdf (accessed on 10 February 2022).
6. European Centre for Disease Prevention and Control. Antimicrobial Resistance in the EU/EEA (EARS-Net)—Annual Epidemiological Report 2019. Available online: https://www.ecdc.europa.eu/sites/default/files/documents/surveillance-antimicrobial-resistance-Europe-2019.pdf (accessed on 10 February 2022).
7. Baughman, R.P. The use of carbapenems in the treatment of serious infections. *J. Intensive Care Med.* **2009**, *24*, 230–241. [CrossRef] [PubMed]
8. World Health Organization. Guidelines for the Prevention and Control of Carbapenem-Resistant Enterobacteriaceae, *Acinetobacter baumannii* and *Pseudomonas aeruginosa* in Health Care Facilities. Available online: https://apps.who.int/iris/bitstream/handle/10665/259462/9789241550178-eng.pdf?sequence=1&isAllowed=y (accessed on 10 February 2022).
9. Tzouvelekis, L.S.; Markogiannakis, A.; Psichogiou, M.; Tassios, P.T.; Daikos, G.L. Carbapenemases in Klebsiella pneumoniae and other Enterobacteriaceae: An evolving crisis of global dimensions. *Clin. Microbiol. Rev.* **2012**, *25*, 682–707. [CrossRef]
10. Livorsi, D.J.; Chorazy, M.L.; Schweizer, M.L.; Balkenende, E.C.; Blevins, A.E.; Nair, R.; Samore, M.H.; Nelson, R.E.; Khader, K.; Perencevich, E.N. A systematic review of the epidemiology of carbapenem-resistant Enterobacteriaceae in the United States. *Antimicrob. Resist. Infect. Control* **2018**, *7*, 55. [CrossRef]
11. Kehl, K.; Schallenberg, A.; Szekat, C.; Albert, C.; Sib, E.; Exner, M.; Zacharias, N.; Schreiber, C.; Parčina, M.; Bierbaum, G. Dissemination of carbapenem resistant bacteria from hospital wastewater into the environment. *Sci. Total Environ.* **2022**, *806*, 151339. [CrossRef] [PubMed]
12. Khan, F.A.; Hellmark, B.; Ehricht, R.; Söderquist, B.; Jass, J. Related carbapenemase-producing Klebsiella isolates detected in both a hospital and associated aquatic environment in Sweden. *Eur. J. Clin. Microbiol. Infect. Dis.* **2018**, *37*, 2241–2251. [CrossRef]
13. Lepuschitz, S.; Schill, S.; Stoeger, A.; Pekard-Amenitsch, S.; Huhulescu, S.; Inreiter, N.; Hartl, R.; Kerschner, H.; Sorschag, S.; Springer, B.; et al. Whole genome sequencing reveals resemblance between ESBL-producing and carbapenem resistant Klebsiella pneumoniae isolates from Austrian rivers and clinical isolates from hospitals. *Sci. Total Environ.* **2019**, *662*, 227–235. [CrossRef]
14. Bleichenbacher, S.; Stevens, M.J.A.; Zurfluh, K.; Perreten, V.; Endimiani, A.; Stephan, R.; Nüesch-Inderbinen, M. Environmental dissemination of carbapenemase-producing Enterobacteriaceae in rivers in Switzerland. *Environ. Pollut.* **2020**, *265*, 115081. [CrossRef]
15. Teban-Man, A.; Farkas, A.; Baricz, A.; Hegedus, A.; Szekeres, E.; Pârvu, M.; Coman, C. Wastewaters, with or without Hospital Contribution, Harbour MDR, Carbapenemase-Producing, but Not Hypervirulent Klebsiella pneumoniae. *Antibiotics* **2021**, *10*, 361. [CrossRef] [PubMed]
16. Food and Agriculture Organization of the United Nations. Drivers, Dynamics and Epidemiology of Antimicrobial Resistance in Animal Production. Available online: https://www.fao.org/3/i6209e/i6209e.pdf (accessed on 10 February 2022).
17. The European Union summary report on antimicrobial resistance in zoonotic and indicator bacteria from humans, animals and food in 2017. *EFSA J.* **2019**, *17*, e05598. [CrossRef]
18. Hamza, E.; Dorgham, S.M.; Hamza, D.A. Carbapenemase-producing Klebsiella pneumoniae in broiler poultry farming in Egypt. *J. Glob. Antimicrob. Resist.* **2016**, *7*, 8–10. [CrossRef] [PubMed]

19. He, T.; Wang, Y.; Sun, L.; Pang, M.; Zhang, L.; Wang, R. Occurrence and characterization of blaNDM-5-positive Klebsiella pneumoniae isolates from dairy cows in Jiangsu, China. *J. Antimicrob. Chemother.* **2017**, *72*, 90–94. [CrossRef] [PubMed]
20. Diab, M.; Hamze, M.; Bonnet, R.; Saras, E.; Madec, J.-Y.; Haenni, M. OXA-48 and CTX-M-15 extended-spectrum beta-lactamases in raw milk in Lebanon: Epidemic spread of dominant Klebsiella pneumoniae clones. *J. Med. Microbiol.* **2017**, *66*, 1688–1691. [CrossRef] [PubMed]
21. Abdallah, H.M.; Reuland, E.A.; Wintermans, B.B.; Al Naiemi, N.; Koek, A.; Abdelwahab, A.M.; Ammar, A.M.; Mohamed, A.A.; Vandenbroucke-Grauls, C.M.J.E. Extended-Spectrum β-Lactamases and/or Carbapenemases-Producing Enterobacteriaceae Isolated from Retail Chicken Meat in Zagazig, Egypt. *PLoS ONE* **2015**, *10*, e0136052. [CrossRef]
22. Wielders, C.C.H.; van Hoek, A.H.A.M.; Hengeveld, P.D.; Veenman, C.; Dierikx, C.M.; Zomer, T.P.; Smit, L.A.M.; van der Hoek, W.; Heederik, D.J.; de Greeff, S.C.; et al. Extended-spectrum β-lactamase- and pAmpC-producing Enterobacteriaceae among the general population in a livestock-dense area. *Clin. Microbiol. Infect.* **2017**, *23*, 120.e1–120.e8. [CrossRef]
23. Meijs, A.P.; Gijsbers, E.F.; Hengeveld, P.D.; Dierikx, C.M.; De Greeff, S.C.; van Duijkeren, E. ESBL/pAmpC-producing Escherichia coli and Klebsiella pneumoniae carriage among veterinary healthcare workers in the Netherlands. *Antimicrob. Resist. Infect. Control* **2021**, *10*, 147. [CrossRef]
24. Müller, H.; Sib, E.; Gajdiss, M.; Klanke, U.; Lenz-Plet, F.; Barabasch, V.; Albert, C.; Schallenberg, A.; Timm, C.; Zacharias, N.; et al. Dissemination of multi-resistant Gram-negative bacteria into German wastewater and surface waters. *FEMS Microbiol. Ecol.* **2018**, *94*, fiy057. [CrossRef] [PubMed]
25. Kizny Gordon, A.E.; Mathers, A.J.; Cheong, E.Y.L.; Gottlieb, T.; Kotay, S.; Walker, A.S.; Peto, T.E.A.; Crook, D.W.; Stoesser, N. The Hospital Water Environment as a Reservoir for Carbapenem-Resistant Organisms Causing Hospital-Acquired Infections-A Systematic Review of the Literature. *Clin. Infect. Dis.* **2017**, *64*, 1435–1444. [CrossRef] [PubMed]
26. Surleac, M.; Czobor Barbu, I.; Paraschiv, S.; Popa, L.I.; Gheorghe, I.; Marutescu, L.; Popa, M.; Sarbu, I.; Talapan, D.; Nita, M.; et al. Whole genome sequencing snapshot of multi-drug resistant Klebsiella pneumoniae strains from hospitals and receiving wastewater treatment plants in Southern Romania. *PLoS ONE* **2020**, *15*, e0228079. [CrossRef] [PubMed]
27. Cherak, Z.; Loucif, L.; Moussi, A.; Rolain, J.-M. Carbapenemase-producing Gram-negative bacteria in aquatic environments: A review. *J. Glob. Antimicrob. Resist.* **2021**, *25*, 287–309. [CrossRef] [PubMed]
28. Rocha, J.; Ferreira, C.; Mil-Homens, D.; Busquets, A.; Fialho, A.M.; Henriques, I.; Gomila, M.; Manaia, C.M. Third generation cephalosporin-resistant Klebsiella pneumoniae thriving in patients and in wastewater: What do they have in common? *BMC Genom.* **2022**, *23*, 72. [CrossRef]
29. WHO Advisory Group on Integrated Surveillance. Critically Important Antimicrobials for Human Medicine: 6th Revision 2018. Ranking of Medically Important Antimicrobials for Risk Management of Antimicrobial Resistance Due to Non-Human Use. Available online: https://www.who.int/publications/i/item/9789241515528 (accessed on 10 February 2022).
30. Nordmann, P.; Poirel, L. Epidemiology and Diagnostics of Carbapenem Resistance in Gram-negative Bacteria. *Clin. Infect. Dis.* **2019**, *69*, S521–S528. [CrossRef]
31. Hubbard, A.T.M.; Mason, J.; Roberts, P.; Parry, C.M.; Corless, C.; van Aartsen, J.; Howard, A.; Bulgasim, I.; Fraser, A.J.; Adams, E.R.; et al. Piperacillin/tazobactam resistance in a clinical isolate of Escherichia coli due to IS26-mediated amplification of blaTEM-1B. *Nat. Commun.* **2020**, *11*, 4915. [CrossRef]
32. Livermore, D.M.; Day, M.; Cleary, P.; Hopkins, K.L.; Toleman, M.A.; Wareham, D.W.; Wiuff, C.; Doumith, M.; Woodford, N. OXA-1 β-lactamase and non-susceptibility to penicillin/β-lactamase inhibitor combinations among ESBL-producing Escherichia coli. *J. Antimicrob. Chemother.* **2019**, *74*, 326–333. [CrossRef]
33. Antunes, N.T.; Fisher, J.F. Acquired Class D β-Lactamases. *Antibiotics* **2014**, *3*, 398–434. [CrossRef]
34. Kotsakis, S.D.; Flach, C.-F.; Razavi, M.; Larsson, D.G.J. Characterization of the First OXA-10 Natural Variant with Increased Carbapenemase Activity. *Antimicrob. Agents Chemother.* **2019**, *63*, 4–6. [CrossRef]
35. Maurya, A.P.; Dhar, D.; Basumatary, M.K.; Paul, D.; Ingti, B.; Choudhury, D.; Talukdar, A.D.; Chakravarty, A.; Mishra, S.; Bhattacharjee, A. Expansion of highly stable bla OXA-10 β-lactamase family within diverse host range among nosocomial isolates of Gram-negative bacilli within a tertiary referral hospital of Northeast India. *BMC Res. Notes* **2017**, *10*, 145. [CrossRef]
36. Arca-Suárez, J.; Lasarte-Monterrubio, C.; Rodiño-Janeiro, B.-K.; Cabot, G.; Vázquez-Ucha, J.C.; Rodríguez-Iglesias, M.; Galán-Sánchez, F.; Beceiro, A.; González-Bello, C.; Oliver, A.; et al. Molecular mechanisms driving the in vivo development of OXA-10-mediated resistance to ceftolozane/tazobactam and ceftazidime/avibactam during treatment of XDR Pseudomonas aeruginosa infections. *J. Antimicrob. Chemother.* **2021**, *76*, 91–100. [CrossRef] [PubMed]
37. Heiden, S.E.; Hübner, N.-O.; Bohnert, J.A.; Heidecke, C.-D.; Kramer, A.; Balau, V.; Gierer, W.; Schaefer, S.; Eckmanns, T.; Gatermann, S.; et al. A Klebsiella pneumoniae ST307 outbreak clone from Germany demonstrates features of extensive drug resistance, hypermucoviscosity, and enhanced iron acquisition. *Genome Med.* **2020**, *12*, 113. [CrossRef] [PubMed]
38. Bathoorn, E.; Rossen, J.W.; Lokate, M.; Friedrich, A.W.; Hammerum, A.M. Isolation of an NDM-5-producing ST16 Klebsiella pneumoniae from a Dutch patient without travel history abroad, August 2015. *Euro Surveill.* **2015**, *20*, 30040. [CrossRef] [PubMed]
39. Papagiannitsis, C.C.; Dolejska, M.; Izdebski, R.; Dobiasova, H.; Studentova, V.; Esteves, F.J.; Derde, L.P.G.; Bonten, M.J.M.; Hrabák, J.; Gniadkowski, M. Characterization of pKP-M1144, a Novel ColE1-Like Plasmid Encoding IMP-8, GES-5, and BEL-1 β-Lactamases, from a Klebsiella pneumoniae Sequence Type 252 Isolate. *Antimicrob. Agents Chemother.* **2015**, *59*, 5065–5068. [CrossRef]

40. Kiaei, S.; Moradi, M.; Hosseini-Nave, H.; Ziasistani, M.; Kalantar-Neyestanaki, D. Endemic dissemination of different sequence types of carbapenem-resistant Klebsiella pneumoniae strains harboring blaNDM and 16S rRNA methylase genes in Kerman hospitals, Iran, from 2015 to 2017. *Infect. Drug Resist.* **2019**, *12*, 45–54. [CrossRef]
41. Van Duijkeren, E.; Wielders, C.C.H.; Dierikx, C.M.; van Hoek, A.H.A.M.; Hengeveld, P.; Veenman, C.; Florijn, A.; Lotterman, A.; Smit, L.A.M.; van Dissel, J.T.; et al. Long-term Carriage of Extended-Spectrum β-Lactamase-Producing Escherichia coli and Klebsiella pneumoniae in the General Population in The Netherlands. *Clin. Infect. Dis.* **2018**, *66*, 1368–1376. [CrossRef]
42. Yoon, E.-J.; Kim, J.O.; Kim, D.; Lee, H.; Yang, J.W.; Lee, K.J.; Jeong, S.H. Klebsiella pneumoniae Carbapenemase Producers in South Korea between 2013 and 2015. *Front. Microbiol.* **2018**, *9*, 56. [CrossRef]
43. David, S.; Reuter, S.; Harris, S.R.; Glasner, C.; Feltwell, T.; Argimon, S.; Abudahab, K.; Goater, R.; Giani, T.; Errico, G.; et al. Epidemic of carbapenem-resistant Klebsiella pneumoniae in Europe is driven by nosocomial spread. *Nat. Microbiol.* **2019**, *4*, 1919–1929. [CrossRef] [PubMed]
44. Wyres, K.L.; Holt, K.E. Klebsiella pneumoniae as a key trafficker of drug resistance genes from environmental to clinically important bacteria. *Curr. Opin. Microbiol.* **2018**, *45*, 131–139. [CrossRef]
45. Wei, L.; Feng, Y.; Wen, H.; Ya, H.; Qiao, F.; Zong, Z. NDM-5-producing carbapenem-resistant Klebsiella pneumoniae of sequence type 789 emerged as a threat for neonates: A multicentre, genome-based study. *Int. J. Antimicrob. Agents* **2021**, *59*, 106508. [CrossRef]
46. Kluytmans, J.A.J.W.; Overdevest, I.T.M.A.; Willemsen, I.; Kluytmans-van den Bergh, M.F.Q.; van der Zwaluw, K.; Heck, M.; Rijnsburger, M.; Vandenbroucke-Grauls, C.M.J.E.; Savelkoul, P.H.M.; Johnston, B.D.; et al. Extended-spectrum β-lactamase-producing Escherichia coli from retail chicken meat and humans: Comparison of strains, plasmids, resistance genes, and virulence factors. *Clin. Infect. Dis.* **2013**, *56*, 478–487. [CrossRef]
47. Kola, A.; Kohler, C.; Pfeifer, Y.; Schwab, F.; Kühn, K.; Schulz, K.; Balau, V.; Breitbach, K.; Bast, A.; Witte, W.; et al. High prevalence of extended-spectrum-β-lactamase-producing Enterobacteriaceae in organic and conventional retail chicken meat, Germany. *J. Antimicrob. Chemother.* **2012**, *67*, 2631–2634. [CrossRef] [PubMed]
48. Zarfel, G.; Galler, H.; Luxner, J.; Petternel, C.; Reinthaler, F.F.; Haas, D.; Kittinger, C.; Grisold, A.J.; Pless, P.; Feierl, G. Multiresistant bacteria isolated from chicken meat in Austria. *Int. J. Environ. Res. Public Health* **2014**, *11*, 12582–12593. [CrossRef] [PubMed]
49. Ghodousi, A.; Bonura, C.; Di Noto, A.M.; Mammina, C. Extended-Spectrum ß-Lactamase, AmpC-Producing, and Fluoroquinolone-Resistant Escherichia coli in Retail Broiler Chicken Meat, Italy. *Foodborne Pathog. Dis.* **2015**, *12*, 619–625. [CrossRef]
50. Egea, P.; López-Cerero, L.; Torres, E.; Del Gómez-Sánchez, M.C.; Serrano, L.; Navarro Sánchez-Ortiz, M.D.; Rodriguez-Baño, J.; Pascual, A. Increased raw poultry meat colonization by extended spectrum beta-lactamase-producing Escherichia coli in the south of Spain. *Int. J. Food Microbiol.* **2012**, *159*, 69–73. [CrossRef] [PubMed]
51. The European Union Summary Report on Antimicrobial Resistance in zoonotic and indicator bacteria from humans, animals and food in 2017/2018. *EFSA J.* **2020**, *18*, e06007. [CrossRef]
52. Russo, T.A.; Marr, C.M. Hypervirulent Klebsiella pneumoniae. *Clin. Microbiol. Rev.* **2019**, *32*, e00001-19. [CrossRef]
53. Clegg, S.; Murphy, C.N. Epidemiology and Virulence of Klebsiella pneumoniae. *Microbiol. Spectr.* **2016**, *4*. [CrossRef] [PubMed]
54. Paczosa, M.K.; Mecsas, J. Klebsiella pneumoniae: Going on the Offense with a Strong Defense. *Microbiol. Mol. Biol. Rev.* **2016**, *80*, 629–661. [CrossRef]
55. Savin, M.; Bierbaum, G.; Hammerl, J.A.; Heinemann, C.; Parcina, M.; Sib, E.; Voigt, A.; Kreyenschmidt, J. Antibiotic-resistant bacteria and antimicrobial residues in wastewater and process water from German pig slaughterhouses and their receiving municipal wastewater treatment plants. *Sci. Total Environ.* **2020**, *727*, 138788. [CrossRef]
56. Savin, M.; Bierbaum, G.; Hammerl, J.A.; Heinemann, C.; Parcina, M.; Sib, E.; Voigt, A.; Kreyenschmidt, J. ESKAPE Bacteria and Extended-Spectrum-β-Lactamase-Producing Escherichia coli Isolated from Wastewater and Process Water from German Poultry Slaughterhouses. *Appl. Environ. Microbiol.* **2020**, *86*, e02748-19. [CrossRef] [PubMed]
57. The European Committee on Antimicrobial Susceptibility Testing. Breakpoint Tables for Interpretation of MICs and Zone Diameters. Version 11.0, 2021. Available online: https://www.eucast.org/fileadmin/src/media/PDFs/EUCAST_files/Breakpoint_tables/v_11.0_Breakpoint_Tables.pdf (accessed on 10 February 2022).
58. Savin, M.; Bierbaum, G.; Schmithausen, R.M.; Heinemann, C.; Kreyenschmidt, J.; Schmoger, S.; Akbaba, I.; Käsbohrer, A.; Hammerl, J.A. Slaughterhouse wastewater as a reservoir for extended-spectrum β-lactamase (ESBL)-producing, and colistin-resistant Klebsiella spp. and their impact in a "One Health" perspective. *Sci. Total Environ.* **2021**, *804*, 150000. [CrossRef] [PubMed]
59. Wattam, A.R.; Abraham, D.; Dalay, O.; Disz, T.L.; Driscoll, T.; Gabbard, J.L.; Gillespie, J.J.; Gough, R.; Hix, D.; Kenyon, R.; et al. PATRIC, the bacterial bioinformatics database and analysis resource. *Nucleic Acids Res.* **2014**, *42*, D581–D591. [CrossRef] [PubMed]
60. Chen, L.; Zheng, D.; Liu, B.; Yang, J.; Jin, Q. VFDB 2016: Hierarchical and refined dataset for big data analysis—10 years on. *Nucleic Acids Res.* **2016**, *44*, D694–D697. [CrossRef]

Article

Persistence and Dissemination Capacities of a bla_{NDM-5}-Harboring IncX-3 Plasmid in *Escherichia coli* Isolated from an Urban River in Montpellier, France

Florence Hammer-Dedet [1], Fabien Aujoulat [1], Estelle Jumas-Bilak [2] and Patricia Licznar-Fajardo [2,*]

1   HSM, University Montpellier, CNRS, IRD, 34090 Montpellier, France; florencehammer@hotmail.fr (F.H.-D.); fabien.aujoulat@umontpellier.fr (F.A.)

2   HSM, University of Montpellier, CNRS, IRD, CHU Montpellier, 34090 Montpellier, France; estelle.bilak@umontpellier.fr

*   Correspondence: patricia.licznar-fajardo@umontpellier.fr

Abstract: To investigate the capacities of persistence and dissemination of bla_{NDM-5} within *Escherichia coli* and in aquatic environment, we characterized *E. coli* (sequence type 636) strains B26 and B28 isolated one month apart from the same urban river in Montpellier, France. The two isolates carried a pTsB26 plasmid, which sized 45,495 Kb, harbored bla_{NDM-5} gene and belonged to IncX-3 incompatibility group. pTsB26 was conjugative in vitro at high frequency, it was highly stable after 400 generations and it exerted no fitness cost on its host. bla_{NDM-5}harboring plasmids are widely dispersed in *E. coli* all around the world, with no lineage specialization. The genomic comparison between B26 and B28 stated that the two isolates probably originated from the same clone, suggesting the persistence of pTsB26 in an *E. coli* host in aquatic environment.

Keywords: Carbapenemase producing Enterobacterales; plasmid; IncX-3; NDM; carbapenem resistance; one health; water; environment

Citation: Hammer-Dedet, F.; Aujoulat, F.; Jumas-Bilak, E.; Licznar-Fajardo, P. Persistence and Dissemination Capacities of a bla_{NDM-5}-Harboring IncX-3 Plasmid in *Escherichia coli* Isolated from an Urban River in Montpellier, France. *Antibiotics* **2022**, *11*, 196. https://doi.org/10.3390/antibiotics11020196

Academic Editor: Francesca Andreoni

Received: 31 December 2021
Accepted: 28 January 2022
Published: 2 February 2022

Publisher's Note: MDPI stays neutral with regard to jurisdictional claims in published maps and institutional affiliations.

1. Introduction

Antimicrobial resistance (AMR) occurs worldwide and the World Health Organization has identified it as one of the three main threats to human health [1,2]. Due to the overuse and misuse of antimicrobials, selective pressure exerted on bacteria significantly enhances AMR, with consequences for antimicrobial treatments failures [3]. At first neglected, the environment is now considered as a main player in the emergence and diffusion of AMR. Water plays a major role in interconnecting different ecosystems such as humans, animals, soils and hydrosystems [4–8]. Urban and rural surface waters constitute hotspots for exchanges among microorganisms of human and environmental origin, which are all subject to strong selection pressures due to diverse pollutions [3,5]. β-lactams are by far the most widely consumed antibiotics worldwide [9], and among β-lactams, carbapenems are last resort treatment for multidrug-resistant bacterial infections [10–12]. Several studies reported the occurrence of Carbapenemase Encoding Genes (CEGs) in aquatic environments [6–8]. However, there is a lack of studies on the persistence and dissemination capacities of CEGs and bacteria carrying CEGs in waters.

In 2009, a New Delhi Metallo-β-lactamase-1 (NDM-1) encoding gene was first identified in a carbapenem-resistant *Klebsiella pneumoniae* involved in a urinary tract infection [13]. Since then, NDM-producing *Enterobacterales* have spread rapidly in humans and animals, and in various environments [14–17]. NDM enzymes belong to class B β-lactamases, which displays a broad lysis spectrum, hydrolyzing almost all β-lactams [7]. Until now, 29 variants of NDM have been described [18]. Several plasmids harboring bla_{NDM} genes have been identified. Among them, plasmids of the incompatibility group IncX-3 are frequently associated with bla_{NDM-5} [14,19–24]. The IncX-3 plasmids group gathers self-transmissible plasmids with a narrow host spectrum restricted to *Enterobacterales* [25]. Characterizing these plasmids would be contributive for understanding their role in environmental AMR issue.

Several clinical, animal and environmental strains containing IncX-3 plasmids carrying bla_{NDM-5} have been isolated in China, South East Asian countries [26–28] and French urban waters [29]. Here, we described pTsB26, an IncX-3 plasmid encoding bla_{NDM-5} gene carried by two strains of *E. coli* isolated from an urban river one month apart. We studied the capacities of dissemination and persistence of pTsB26 by in vitro assays and the fitness cost associated with pTsB26 carriage. Metadata analysis was performed in order to identify common themes in bla_{NDM-5}-harboring IncX-3 plasmids and to gain insights in IncX-3 success in *E. coli* population. The worldwide circulation of bla_{NDM-5}-harboring IncX-3 plasmids in *E. coli* and in various environments as well as the high stability in host bacteria observed herein could explain the contribution of this type of plasmid in the global dissemination of carbapenemase genes.

2. Results

2.1. Characteristics of the Two NDM-5 Producing E. coli Isolates B26 and B28

B26 and B28 *E. coli* were isolated in the urban river of Font d'Aurelle, in August and September 2015, respectively. They both displayed high level of carbapenem resistance, with minimum inhibitory concentrations (MIC) over 15 mg/L for ertapenem, meropenem and imipenem.

The B26 genome is 4927 Mb and includes four plasmids belonging to IncFIA/FIB, IncFII, IncQ1 and IncX-3 incompatibility groups. The inventory of antimicrobial resistance genes identified bla_{NDM-5}, bla_{TEM-1B} and bla_{SHV-12} (previously identified as bla_{SHV} by multiplex PCR GeneXpert Cepheid). Genome of B28 is 4858 Mb, including a unique IncX-3 plasmid and bla_{NDM-5}, was the only antimicrobial resistance gene detected.

Both isolates were affiliated to the sequence type (ST) 636 (B2 phylogroup) by in silico multilocus sequence typing. Whole-genome alignment and comparison showed no large indels. Only 21 variations between B26 and B28 genomes were identified, including 19 single nucleotide polymorphisms (SNPs) and two multi-nucleotide polymorphisms (four nucleotides). Among the 21 variations, six variations were related to hypothetic encoding sequences, four of them corresponding to synonymous mutations and two to missense mutations.

2.2. Characteristics of the bla_{NDM-5}-Harboring IncX-3 Plasmids

Transformation assays in *E. coli* TOP10 with plasmids extracted from B26 and B28 were successful and gave two transformants, TsB26 and TsB28. These two transformants displayed high level of resistance to carbapenems (MIC $\geq$ 7.5 mg/L whatever the carbapenem). They were positive to specific PCRs anchored in *pir* gene (i.e., specific of IncX-3 plasmid) and bla_{NDM}. PCR tests indicated that both transconjugants did not contain any other plasmid.

Genomic sequences of TsB26 and TsB28 were aligned and compared with those of *E. coli* TOP10. TsB26 presented a unique additional sequence onto a single contig corresponding to IncX-3 plasmid. The extremities of this contig were bla_{NDM-5} and IS*Aba125*, corresponding to a plasmid region already sequenced [29]. Complete identity was observed between pTsB28 and pTsB26 encoding bla_{NDM-5} plasmids. So, the plasmid was called pTsB26 from then on. pTsB26 sizes 45,495 kb with a GC content of 46.5%. It belongs to IncX-3 plasmid incompatibility group. Nucleotide sequence analysis revealed 57 predicted open reading frames corresponding to 57 encoding genes (Figure 1). Alignment of pTsB26 with IncX3-plasmid conserved backbone, described by Liakopoulos [30], showed that pTsB26 backbone was typical of IncX-3 group. It is approximately 25 kb and includes encoding genes for replication (*pir* and *bis*), entry exclusion (*eex*), plasmid stability (*par*AB, *topB* and *hns*) and conjugative transfer (*pilX1–11* and *taxA-C*) [30]. The accessory module of about 20 kb contains bla_{NDM-5}, which is preceded by IS*3000* and IS*Aba125* and followed by

ble_{MBL} (bleomycin resistance gene), *trp*F (N-5′phosphoribosylanthranilate isomerase), *dsb*D (disulfide oxidoreductase) and *umu*D (encoding a protein implicated in the SOS system).

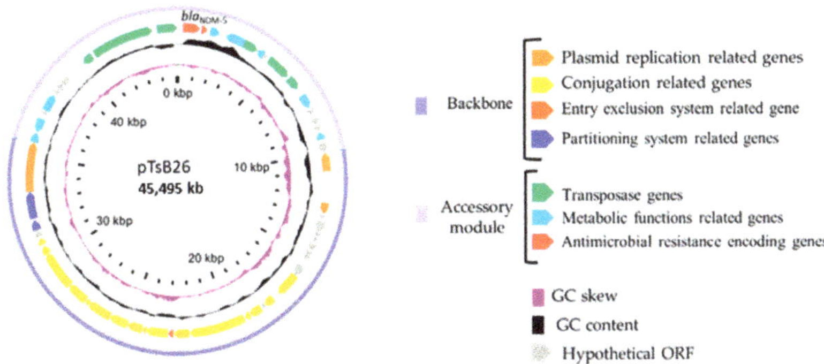

Figure 1. pTsB26 plasmid representation.

BLASTn analysis showed that pTsB26 displayed nearly the same sequence (45494/45495 bp) as the plasmid pEC7-NDM-5 (accession number: MH347484) found in a *E. coli* strain isolated from dog in South Korea. A high homology (45475/45495 bp) was also observed with the well characterized IncX-3-blaNDM5 pEC463-NDM5 plasmid (accession number: MG545911) in an *E. coli* clinical strain from China [31].

2.3. Conjugative Transfer Success of pTsB26 In Vitro

Conjugative transfer rate of pTsB26 was studied by mating assays, with B26, B28, TsB26 and TsB28 as donor strains and XL1-Blue *E. coli* as receptor. All strains successfully transferred pTsB26, with high transfer rates (Table 1), and all transconjugant strains displayed carbapenem resistance (MIC $\geq$ 7.5 mg/L).

Table 1. Conjugative transfer rates of pTsB26 from different donor strains to *Escherichia coli* XL1-Blue receptor.

Donor Strains	Conjugative Frequency
B26	2.09×10^{-3}
B28	4.81×10^{-4}
TsB26	3.52×10^{-3}
TsB28	3.76×10^{-3}

2.4. In Silico Population Study of bla$_{NDM-5}$-Harboring IncX-3 Plasmids among E. coli Species

A genomes dataset was constructed with 28 complete genomes of *E. coli*-carrying $bla_{\text{NDM-5}}$ on an IncX-3 plasmid and the genomes of B26 and B28 strains. In silico sequence types, phylotypes and metadata associated with the genomes are presented in Table 2. More than 96% of the genomes belong to A, B1 and C phylogroups. B26 and B28 were the only genomes of the B2 phylogroup in the dataset. The 31 genomes corresponded to strains isolated from diverse origins: humans (*n* = 20), environment (*n* = 6) and animals (*n* = 4).

Table 2. Characteristics of *E. coli* genomes carrying a *bla*$_{NDM-5}$ IncX-3 plasmid.

Strain	Accession Number	ST	Phylotype	Country	Source Type	Source Niche	Year
CRE1493	CP019071	167	A	China	rectal swab	*homo sapiens*	2013
165	CP020509	101	B1	USA	abdominal	*homo sapiens*	2015
CREC-591	CP024821	101	B1	South Korea	peritoneal fluid	*homo sapiens*	2015
WCHEC025943	CP027205	410	C	China	wastewater	environment	2017
WCHEC005784	CP028578	617	A	China	rectal swab	*homo sapiens*	2014
135	CP028632	11	E	Canada	NA	livestock	2006
ECCRA-119	CP029242	156	B1	China	stools	dog	2017
CH613	MCRE01000001	10	A	China	urine	*homo sapiens*	2015
GSH8M-2	NZ_AP019675	542	A	Japan	wastewater treatment plant	environment	2018
WP8-S18-CRE-02	NZ_AP022245	542	A	Japan	wastewater treatment plant	environment	2018
TUM18781	NZ_AP023205	2040	B1	Japan	NA	*homo sapiens*	2018
YJ3	NZ_AP023226	10	A	Myanma	stools	*homo sapiens*	2018
WCHEC005237	NZ_CP026580	167	A	China	rectal swab	*homo sapiens*	2014
SCEC020001	NZ_CP032426	410	C	China	urinary tract	*homo sapiens*	2016
SCEC020022	NZ_CP032892	156	B1	China	stools	*homo sapiens*	2016
WCHEC020031	NZ_CP033401	410	C	China	NA	*homo sapiens*	2016
L37	NZ_CP034589	48	A	China	rectal swab	*homo sapiens*	2018
L65	NZ_CP034738	3076	B1	China	NA	*homo sapiens*	2018
SCEC020026	NZ_CP034958	410	C	China	NA	*homo sapiens*	2016
WCHEC020032	NZ_CP034966	410	C	China	NA	*homo sapiens*	2016
WCHEC025970	NZ_CP036177	167	A	China	NA	*homo sapiens*	2017
L725	NZ_CP036202	2161	B1	China	stools	*homo sapiens*	2018
EC-129	NZ_CP038453	167	A	Japan	sputum	*homo sapiens*	2018
GZ04-0086	NZ_CP042336	44	A	China	stools	*homo sapiens*	2018
GZEC065	NZ_CP048025	156	B1	China	blood	*homo sapiens*	2017
pV11-19-E11-025-038	NZ_CP049050	1721	A	South Korea	NA	dog	2019
3R	NZ_CP049348	156	B1	China	NA	poultry	2015
SFE8	NZ_CP051219	533	B1	China	stools	pork	2019
B26	B26	636	B2	France	urban water	environment	2015
B28	B28	636	B2	France	urban water	environment	2015

ST, Sequence Type; NA, Not Available.

In order to study the distribution of *E. coli* encoding *bla*$_{NDM-5}$ on IncX-3 plasmids within the whole *E. coli* populations, we reconstructed genetic links by goeBURST analysis. The dataset of 30 genomes was matched with 178 776 available genomes of *E. coli* (11 058 STs) in the dataset EnteroBase (13 July 2021) (Figure 2). The genomes of *E. coli*-carrying *bla*$_{NDM-5}$ IncX-3 plasmid spread out in 16 STs with no obvious lineage specialization. However, about half of the genomes (41.9%) belong to the CC10. This CC is a major sub-population in *E. coli* because it gathers 11.85% of the STs available in EnteroBase. The other genomes are scattered in the overall *E. coli* population structure. B26 and B28 belonged to ST636, which forms the CC636 together with 4 related STs. CC636 was relatively isolated in the *E. coli* population and gathers 451 strains (0.25%) of the 178,776 strains of EnteroBase.

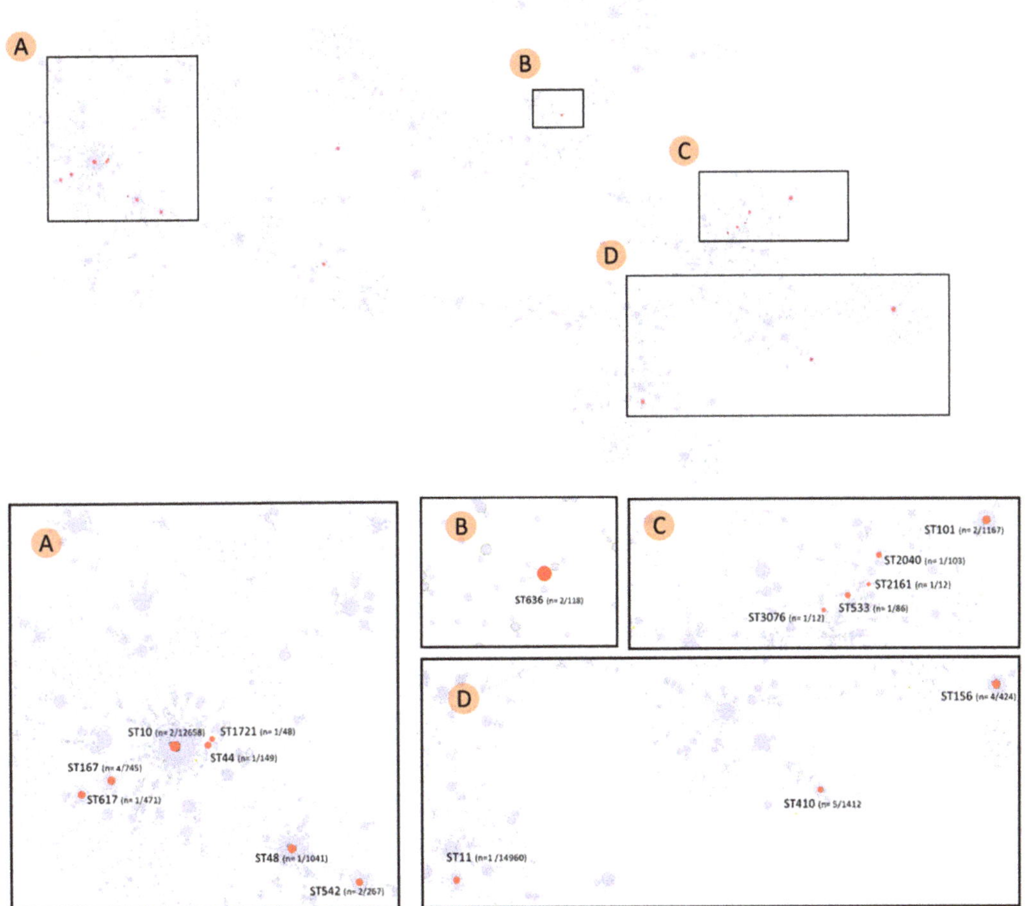

Figure 2. goeBURST diagram of *E. coli* genomes carrying a *bla*$_{NDM-5}$ encoding IncX-3 plasmid within a global population of *E. coli* established on 178 776 strains available in the EnteroBase database. Each node corresponds to a Sequence Type (ST). The size of the node is scaled to the number of genomes of that ST. Nodes linked between them present one allele in common among the 7 genes considered in the MultiLocus Sequence Type scheme. Red nodes correspond to STs for which genomes with a *bla*$_{NDM-5}$ encoding IncX-3 plasmid was identified; the proportion in the ST of genomes containing the plasmid is noted in parentheses.

2.5. Stability and Fitness Cost of pTsB26 on B26 and B28 E. coli

To evaluate the stability of the pTsB26 plasmid in B26 and B28, strains were passaged daily for 40 days without antibiotic selection. pTsB26 is highly stable, with more than 96% of plasmid containing cells after approximately 400 generations (Figure 3).

These assays allowed one to isolate strains without pTsB26: B26ΔpJ40 and B28ΔpJ19, respectively, isolated at day 40 and day 19 of the experiment. In parallel, two strains containing pTsB26 plasmid (B26J40a and B28J19a) were isolated the same day as B26ΔpJ40 and B28ΔpJ19. These strains, used as control, were thereafter called "LB-adapted" strains, with the letter "a" at the end of the strain name.

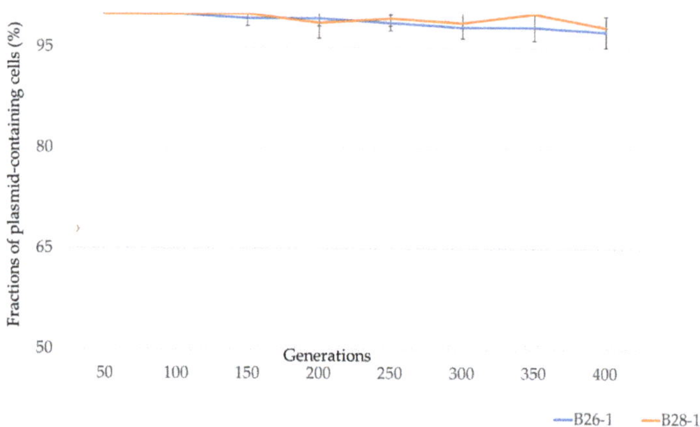

Figure 3. Stability of pTsB26 in B26 and B28.

Growth kinetics assays were performed for the strains with (ancestral strains and LB-adapted strains) and without pTsB26 plasmid (Table 3). Growth rates of B26, B26ΔpJ40 and B26J40a did not vary significantly ($p > 0.05$), suggesting that the carriage of pTsB26 did not produce fitness lost for B26. Of note, B26ΔpJ19 and B26J19a contained the other plasmids IncFIA/FIB, IncFII and IncQ1 as the ancestral B26. On the other hand, growth rates of ancestral B28 or LB-adapted B28 (B28J19a) were equivalent and significantly better than observed for B28 cured for pTsB26 ($p < 0.05$). This last observation suggests that pTsB26 could provide fitness advantage even to its host, without selective pressure.

Table 3. Growth rates of strains with and without the pTsB26 plasmid.

Strain	μmax (h^{-1}) ($\pm$sd)
B26	0.43351667 ($\pm$0.02560949) [a]
B26ΔpJ40	0.4441375 ($\pm$0.02775884) [a]
B26J40a	0.44380714 ($\pm$0.03785335) [a]
B28	0.43683889 ($\pm$0.0217273) [a, b]
B28ΔpJ19	0.40048824 ($\pm$0.0333677) [c]
B28J19a	0.43095 ($\pm$0.03460016) [a, b]

sd, standard deviation; values with a different letter (a,b,c) are significantly different at $p < 0.05$ (Student's t-test).

3. Discussion

The spread of NDM-5 variant has been extensively described in hospital environments, including hospital sewage water [32,33] and in wastewater [34]. In France, in 2020, about 20% of carbapenemase-producing enterobacteria isolated from clinical samples were NDM-producers [35]. Among them, more than 35% were NDM-5 variant. Its emergence and successful spread are worth emphasizing as this variant represented only 5% of NDMs reported in 2013, 15% in 2016 and more than 30% in 2017, with stabilization since this date [35].

Focusing on hydric environments, this variant has been found in various environments such as urban waters [29,36]; rivers and lakes [8,37]; sediments and soil [38]; and seawater [8]. Various plasmid incompatibility groups carrying bla_{NDM-5} have been reported, with IncX-3 and IncF being among the most prevalent [8,27,34,37–40].

Here, we report the description of the bla_{NDM-5} carrying plasmid pTsB26 from two *E. coli* ST636 isolated from an urban river in Montpellier, France [29]. IncX-3 plasmids have a narrow host spectrum restricted to *Enterobacterales* species [25,28,41]. Their association with antimicrobial resistance has been documented worldwide [28,41–49].

The involvement of IncX-3 plasmids in the dissemination of *bla*NDM genes was first described in the 2000's [43]. It became quickly prevalent within *Enterobacterales* all around the world, with predominance in Asia [27,28,41,42,44–47,49,50]. Here, we show that *bla*NDM-5-harboring IncX-3 plasmids are widely disseminated in the species *E. coli* in all the continents, with no lineage specialization (Figure 3). This strongly suggests horizontal spread within *E. coli* and proves the important implication of IncX-3 plasmid in the success of *bla*NDM-5 gene in this species.

Only 21 variations (19 SNPs and 2 multi-nucleotide polymorphisms) were detected between B26 and B28 isolated in the same urban aquatic environment a month apart. The scarce genomic differences strongly suggest their clonal origin. Clonal strains have certainly persisted for at least one month in the river, with iterative input being highly unlikely but not excluded. This hypothesis is strongly supported by the fact that the two isolates carried the same plasmid pTsB26 (100% nucleotide identity). Interestingly, the IncFIA/FIB, IncFII and IncQ1 plasmids carried by B26 (isolated in August 2015) were absent from the genome of B28 (isolated in September 2015). Only pTsB26 has persisted in the *E. coli* strain isolated in September, showing that pTsB26 is stable in B26 and B28 isolates, relative to other plasmids in the same strains. This in situ observation was verified by in vitro evolution experiments, demonstrating the longtime persistence of the plasmid after 400 generations in B26 and B28 isolates (Figure 3). Several factors could explain the stability of pTsB26 in the cell lineage. First, like the other IncX-3 plasmids, pTsB26 harbors the widespread partitioning system ParAB. This system limits the number of segregant cells during cell division, ensuring the correct inheritance of the plasmid to the daughter cells [51]. Moreover, the conjugative traits of pTsB26 allow for infection of segregant cells and thus limit their number. Thus, harboring pTsB26 does not reduce the fitness of B26 and B28 (Table 3) and enhances the growth of B28. It is generally admitted that plasmids cause a fitness burden on their bacterial host [52–55] and that the plasmid could be quickly eliminated from its host. High stability of IncX-3 plasmids has already been reported in *Enterobacterales* transconjugant strains [28,56]. The absence of fitness cost of IncX-3 plasmids could be explained by their small size [25] and by the presence of transcriptional regulator H-NS like protein [28,30,57–60]. The observed high stability of pTsB26 has potential due to its high conjugation frequency, to a low rate of segregational loss and due to the fact it does not pass a fitness cost onto the bacterial host.

The isolates B26 and B28 are the unique representants of ST636 in the studied dataset of *E. coli*-carrying *bla*NDM-5-encoding IncX-3 plasmid (Table 2). However, in their study, Kumwenda et al. reported two *E. coli* ST636 clinical isolates that carried *bla*NDM-5 onto a plasmid not affiliated to an incompatibility group [61]. EnteroBase reported 118 isolates belonging to ST636 (Figure 3). They were isolated in different countries (all continents are represented) and in environmental and clinical samples. Other studies reported the occurrence of ST636-producing ESBLs in clinical and environmental samples [62–64], suggesting a generalist trait for this ST.

Persistence of strains or STs carrying CEGs on self-transmissible plasmid such as pTsB26 in aquatic environment is of concern. Aquatic environment contains diverse autochtonous bacteria, including *Enterobacterales* (e.g., *Enterobacter* sp. [65] or *Raoultella* sp. [66]), which can exchange and receive IncX-3 plasmids [27,28,41]. These autochtonous bacteria can constitute an environmental reservoir and shuttles for *bla*NDM genes. Aquatic environments are strongly linked with anthropic activities, and during recreational activities, after flood episodes or by alimentation, humans can become exposed to bacteria from aquatic environments [7,67–70]. Thus, if water contains carbapenemase-producing-bacteria such as B26 or B28 or other autochtonous bacteria, and has acquired the plasmid by horizontal gene transfer, it represents a risk for human health (i) directly by causing antimicrobial-resistant bacterial infections [71] and (ii) indirectly by participating in the dissemination of *bla*NDM-5 on the occasion of gut colonization [71,72], or transit. These resistant bacteria can transfer the plasmid to host microbiota bacteria, making a "shuttle" between aquatic environment and humans [7,71].

We described for the first time the in situ persistence of a $bla_{\text{NDM-5}}$harboring IncX-3 self-conjugative plasmid in an *E. coli* lineage in an aquatic environment. This study underlines, once again, the importance of investigations into environmentally emerging, resistant bacteria. Beside genomics, testing genetic transfer and resistance stability by in vitro evolution is a proxy for diffusion and persistence in natural environment. In addition to the strategy proposed in this study, experiments of resistance genes transfer to waterborne autochthonous bacteria in microcosm would be interesting to conduct for a better description of the resistance reservoirs and the conditions influencing these reservoirs. On another hand, rapid alerts on environmentally emerging antimicrobial resistance are needed for rapid responses to resistance with public health concern. For this, efficient surveillance of AMR in environment should be undertaken. This is one of the challenges of the current national and international projects aiming to limit global AMR outbreak.

4. Materials and Methods

4.1. Escherichia coli Strains

E. coli strains B26 and B28 were isolated from water sampled at the same site of the urban river Font d'Aurelle in the city of Montpellier (N43.62711 E003.85316), France. They were isolated in August and September 2015, respectively [29].

Transformant strains, TsB26 and TsB28, were obtained by transformation experiments (see Section 4.2.).

Strains B26ΔpJ40, B28ΔpJ19, B26J40a and B28J19a were obtained during plasmid stability assays (see Section 4.3.).

4.2. Transformation and Conjugation Assays

Plasmid DNA extraction was done using the NucleoSpin Plasmid Kit (Macherey-Nagel, Allentown, PA, USA). Plasmid extracts were used for transformation assays using One Shot TOP10 chemically competent *E. coli* (Invitrogen, ThermoFisher Scientific, Paisley, UK) as recipient cell.

Conjugation experiments were performed using non-competent XL1-Blue *E. coli* MRF', a recipient strain resistant to tetracycline and sensitive to meropenem. Briefly, donor (B26, B28 and transformants TsB26 and TsB28) and recipient strains were grown overnight at 37 °C in Luria Bertani (LB) broth supplemented (donor strains) or not (recipient strain) with ertapenem (4 mg/L). Cells were washed from antibiotic and resuspended in LB broth, and each donor strain suspension was mixed (1:1 ratio) with the receptor strain. 200 μL of each mix was deposited onto nitrocellulose membrane, itself stuck on LB agar media and incubated at 37 °C during 24 h. Transconjugants were selected by plating the bacteria from the nitrocellulose membrane onto LB agar plates supplemented with ertapenem (4 mg/L) and tetracycline (12 mg/L). The conjugative frequencies were determined by calculating the transfer rate (ratio transconjugant/donor).

The presence of bla_{NDM} and *pir* (encoding the IncX-3 plasmid-specific Pir protein) in selected transconjugants and transformants was assessed by specific PCRs [14,73].

4.3. Evaluation of Plasmid Stability

Strains B26 and B28 were grown overnight at 37 °C in 10 mL of LB broth supplemented with ertapenem (4 mg/L). Bacterial cells were washed from antibiotic by centrifugation, the pellet was resuspended in 1 mL of LB broth and 10 mL of fresh LB broth without antibiotic was spiked with 10 μL of the bacterial suspension and incubated 24 h at 37 °C in a shaking water bath. Serial passages of 10 μL of overnight culture to 10 mL of fresh LB broth were done daily. One passage corresponded approximatively to 10 generations of growth. Every 50 generations, samples were diluted and plated on LB agar plates. Then, 50 colonies from each lineage were screened on LB agar plates supplemented or not with ertapenem (4mg/L) to determine the fraction of plasmid-containing cells. The lack of plasmid was confirmed by the absence of bla_{NDM} and *pir* genes with specific PCRs, and these strains (B26ΔpJ40 and B28ΔpJ19) were harvested for fitness cost assays. Parallelly, strains from the same generation, carrying pTsB26 plasmid, were harvested as controls (B26J40a and B28J19a). Experiments were done in triplicate.

4.4. Fitness Cost of Plasmid Carriage

Growth of strains carrying (B26, B28, B26J40a and B28J19a) or not (B26ΔpJ40 and B28ΔpJ19) pTsB26 plasmid were measured at 37 °C in LB broth without antibiotics using a CLARIOstar Plus microplate reader (BMG, Labtech). Every 15 min, the microplate was shacked at 200 rpm during 20 s, and optical density was measured at 600 nm. Growth rates were calculated according to Sandegren et al. [74].

4.5. Carbapenem Susceptibility Testing

Susceptibility to carbapenems of B26, B28, TsB26, TsB28, B26J40a, B28J19a, B26ΔpJ40 and B281ΔpJ19 was assessed by determining the Minimal Inhibitory Concentration (MIC) in liquid media [75] for ertapenem, meropenem and imipenem. *E. coli* strain ATCC 25922 was used as control strain, as recommended by the CA-SFM (https://www.sfm-microbiologie.org/2021/04/23/casfm-avril-2021-v1-0/, accessed on 28 September 2021).

4.6. In Silico Analysis

4.6.1. DNA Extraction and Whole-Genome Sequencing

Genomic DNA of B26, B28, TsB26, TsB28 and *E. coli* TOP10 was extracted using the MasterPure™ purification kit (Lucigen, Middleton, WI, USA). High-throughput genome sequencing was carried out at the Plateforme de Microbiologie Mutualisée (P2M, Institut Pasteur, Paris, France). DNAs were processed for sequencing with Illumina systems (libraries using the Nextera XT DNA Library Prep kit and sequencing with the NextSeq 500 system). Paired-end reads were submitted to pre-processing using fqCleaner and to de novo assembly using SPAdes v3.12.0 [76] with k-mer lengths 21, 33, 55 and 77. The raw data and assemblies have been deposited in GenBank under the BioProject accession number PRJNA796954.

4.6.2. Genotyping Methods

B26 and B28 strains were genotyped by Clermont typing [77] using the in silico tool EzClermont [78] and by MultiLocus Sequence Typing (MLST) using the Achtman scheme (https://pubmlst.org/data, accessed on 28 September 2021).

4.6.3. Plasmid Sequence and Annotation

The sequence of the plasmid pTsB26 was deduced from the alignment of the genomes of TsB26 and TsB28 with that of *E. coli* TOP10 using the ProgressiveMauve algorithm of Mauve software [79]. Identification of plasmid incompatibility group was done with PlasmidFinder [80]. Plasmid annotation was performed with Prokka v1.14.6 [81] with default parameters through the Galaxy platform (v4.6.0 + galaxy0) [82], and the plasmid sequence was manually curated; plasmid-specific genes were named according to Thomas et al. recommendations [83].

For ORF annotation as hypothetical protein, functional prediction was performed with default settings using NCBI BLASTp [84], InterProScan [85] and Pfam [86] servers. Identification of antimicrobial resistance genes was done using ResFinder 4.1 [87]. pTsB26 representation was done using genomeVx [88].

4.6.4. Comparative Genomics of B26 and B28 Genomes

Genomes of B26 and B28 were first aligned with the ProgressiveMauve algorithm of Mauve software. Snippy v4.6.0 (https://github.com/tseemann/snippy, accessed on 20 November 2021) was used to detect both substitutions and insertions/deletions (indels) between B26 and B28 genomes. Snippy was run on the Galaxy platform (v4.6.0 + galaxy0) [82] with the default parameters.

4.6.5. Distribution of IncX3-bla_{NDM} Plasmids in *E. coli* Population

In order to select *E. coli* genomes carrying bla_{NDM-5} on an IncX3 plasmid, complete genomes available on NCBI database (6 April 2021) were investigated using BLASTn tool [84]. bla_{NDM-5} and *pir* gene of pTsB26 sequences were used, and genomes carrying the 2 genes (minimum homology 100% and 92%, respectively) on the same replicon were selected. Metadata associated with each genome were collected, and ST were determined in silico using the MLST 2.0 [89]. STs (Achtman MLST scheme) associated with selected genomes and B26 and B28 strains were compared to general *E. coli* population using EnteroBase database [90] (178,776 strains the 13 July 2021) and the goesBURST algorithm (single locus variant level) of PHYLOViZ 2.0 software [91]. Plasmid homologies were determined by BLASTn analysis [84] against the NCBI nr/nt database with default parameters.

4.7. Statistical Assays

Statistical analysis of growth rates was performed using Student's t-test. All statistics were made using the GraphPad Prism software V 5.03. Test results were considered as statistically significant when the associated *p*-value was less than 0.05.

Author Contributions: F.H.-D. conceptualized and conducted the in vitro and in silico studies, carried out the statistical analysis, wrote the manuscript and designed all tables and Figures. F.A. brought his expertise in bioinformatics tools and participated in the in silico study. P.L.-F. and E.J.-B. reviewed the manuscript. All authors have read and agreed to the published version of the manuscript.

Funding: This research received no external funding.

Institutional Review Board Statement: Not applicable.

Informed Consent Statement: Not applicable.

Acknowledgments: We acknowledge the Institute of Biomolecules Max Mousseron for the loan of their equipment, particularly Morgan Pellerano for his welcoming manner, his availability and his explanations. We acknowledge the UNESCO water-related Centre ICIReWARD 'International Center for Interdisciplinary Research on Water Systems Dynamics' that supports the study.

Conflicts of Interest: The authors declare no conflict of interest.

References

1. Infectious Diseases Society of America The 10 x '20 Initiative: Pursuing a Global Commitment to Develop 10 New Antibacterial Drugs by 2020. *Clin. Infect. Dis.* **2010**, *50*, 1081–1083. [CrossRef] [PubMed]
2. Zellweger, R.M.; Carrique-Mas, J.; Limmathurotsakul, D.; Day, N.P.J.; Thwaites, G.E.; Baker, S. Southeast Asia Antimicrobial Resistance Network A Current Perspective on Antimicrobial Resistance in Southeast Asia. *J. Antimicrob. Chemother.* **2017**, *72*, 2963–2972. [CrossRef] [PubMed]
3. Harbarth, S.; Balkhy, H.H.; Goossens, H.; Jarlier, V.; Kluytmans, J.; Laxminarayan, R.; Saam, M.; Van Belkum, A.; Pittet, D. Antimicrobial Resistance: One World, One Fight! *Antimicrob. Resist. Infect. Control* **2015**, *4*. [CrossRef]
4. Stewardson, A.J.; Marimuthu, K.; Sengupta, S.; Allignol, A.; El-Bouseary, M.; Carvalho, M.J.; Hassan, B.; Delgado-Ramirez, M.A.; Arora, A.; Bagga, R.; et al. Effect of Carbapenem Resistance on Outcomes of Bloodstream Infection Caused by *Enterobacteriaceae* in Low-Income and Middle-Income Countries (PANORAMA): A Multinational Prospective Cohort Study. *Lancet Infect. Dis.* **2019**, *19*, 601–610. [CrossRef]
5. Almakki, A.; Jumas-Bilak, E.; Marchandin, H.; Licznar-Fajardo, P. Antibiotic Resistance in Urban Runoff. *Sci. Total Environ.* **2019**, *667*, 64–76. [CrossRef]
6. Cherak, Z.; Loucif, L.; Moussi, A.; Rolain, J.-M. Carbapenemase-Producing Gram-Negative Bacteria in Aquatic Environments: A Review. *J. Glob. Antimicrob. Resist.* **2021**, *25*, 287–309. [CrossRef]
7. Hammer-Dedet, F.; Jumas-Bilak, E.; Licznar-Fajardo, P. The Hydric Environment: A Hub for Clinically Relevant Carbapenemase Encoding Genes. *Antibiotics* **2020**, *9*, E699. [CrossRef]
8. Hooban, B.; Joyce, A.; Fitzhenry, K.; Chique, C.; Morris, D. The Role of the Natural Aquatic Environment in the Dissemination of Extended Spectrum Beta-Lactamase and Carbapenemase Encoding Genes: A Scoping Review. *Water Res.* **2020**, *180*, 115880. [CrossRef]
9. WHO Report on Surveillance of Antibiotic Consumption. Available online: https://www.who.int/publications-detail-redirect/who-report-on-surveillance-of-antibiotic-consumption (accessed on 18 January 2022).
10. Codjoe, F.S.; Donkor, E.S. Carbapenem Resistance: A Review. *Med. Sci.* **2017**, *6*, E1. [CrossRef]

11. Bonomo, R.A.; Burd, E.M.; Conly, J.; Limbago, B.M.; Poirel, L.; Segre, J.A.; Westblade, L.F. Carbapenemase-Producing Organisms: A Global Scourge. *Clin. Infect. Dis.* **2018**, *66*, 1290–1297. [CrossRef]
12. Papp-Wallace, K.M.; Endimiani, A.; Taracila, M.A.; Bonomo, R.A. Carbapenems: Past, Present, and Future. *Antimicrob. Agents Chemother.* **2011**, *55*, 4943–4960. [CrossRef]
13. Yong, D.; Toleman, M.A.; Giske, C.G.; Cho, H.S.; Sundman, K.; Lee, K.; Walsh, T.R. Characterization of a New Metallo-Beta-Lactamase Gene, bla(NDM-1), and a Novel Erythromycin Esterase Gene Carried on a Unique Genetic Structure in *Klebsiella pneumoniae* Sequence Type 14 from India. *Antimicrob. Agents Chemother.* **2009**, *53*, 5046–5054. [CrossRef] [PubMed]
14. Poirel, L.; Dortet, L.; Bernabeu, S.; Nordmann, P. Genetic Features of blaNDM-1-Positive *Enterobacteriaceae*. *Antimicrob. Agents Chemother.* **2011**, *55*, 5403–5407. [CrossRef]
15. Nordmann, P.; Poirel, L.; Walsh, T.R.; Livermore, D.M. The Emerging NDM Carbapenemases. *Trends Microbiol.* **2011**, *19*, 588–595. [CrossRef] [PubMed]
16. Nordmann, P.; Naas, T.; Poirel, L. Global Spread of Carbapenemase-Producing *Enterobacteriaceae*. *Emerg. Infect. Dis.* **2011**, *17*, 1791–1798. [CrossRef] [PubMed]
17. Walsh, T.R.; Weeks, J.; Livermore, D.M.; Toleman, M.A. Dissemination of NDM-1 Positive Bacteria in the New Delhi Environment and Its Implications for Human Health: An Environmental Point Prevalence Study. *Lancet Infect. Dis.* **2011**, *11*, 355–362. [CrossRef]
18. Basu, S. Variants of the New Delhi Metallo-β-Lactamase: New Kids on the Block. *Future Microbiol.* **2020**, *15*, 465–467. [CrossRef]
19. Wu, W.; Feng, Y.; Tang, G.; Qiao, F.; McNally, A.; Zong, Z. NDM Metallo-β-Lactamases and Their Bacterial Producers in Health Care Settings. *Clin. Microbiol. Rev.* **2019**, *32*, e00115-18. [CrossRef]
20. Bontron, S.; Nordmann, P.; Poirel, L. Transposition of Tn125 Encoding the NDM-1 Carbapenemase in *Acinetobacter baumannii*. *Antimicrob. Agents Chemother.* **2016**, *60*, 7245–7251. [CrossRef]
21. Rolain, J.M.; Parola, P.; Cornaglia, G. New Delhi Metallo-Beta-Lactamase (NDM-1): Towards a New Pandemia? *Clin. Microbiol. Infect.* **2010**, *16*, 1699–1701. [CrossRef]
22. Kumarasamy, K.K.; Toleman, M.A.; Walsh, T.R.; Bagaria, J.; Butt, F.; Balakrishnan, R.; Chaudhary, U.; Doumith, M.; Giske, C.G.; Irfan, S.; et al. Emergence of a New Antibiotic Resistance Mechanism in India, Pakistan, and the UK: A Molecular, Biological, and Epidemiological Study. *Lancet Infect. Dis.* **2010**, *10*, 597–602. [CrossRef]
23. Khan, N.H.; Ishii, Y.; Kimata-Kino, N.; Esaki, H.; Nishino, T.; Nishimura, M.; Kogure, K. Isolation of *Pseudomonas aeruginosa* from Open Ocean and Comparison with Freshwater, Clinical, and Animal Isolates. *Microb. Ecol.* **2007**, *53*, 173–186. [CrossRef] [PubMed]
24. Datta, S.; Mitra, S.; Chattopadhyay, P.; Som, T.; Mukherjee, S.; Basu, S. Spread and Exchange of bla NDM-1 in Hospitalized Neonates: Role of Mobilizable Genetic Elements. *Eur. J. Clin. Microbiol. Infect. Dis.* **2017**, *36*, 255–265. [CrossRef]
25. Johnson, T.J.; Bielak, E.M.; Fortini, D.; Hansen, L.H.; Hasman, H.; Debroy, C.; Nolan, L.K.; Carattoli, A. Expansion of the IncX Plasmid Family for Improved Identification and Typing of Novel Plasmids in Drug-Resistant *Enterobacteriaceae*. *Plasmid* **2012**, *68*, 43–50. [CrossRef] [PubMed]
26. An, J.; Guo, L.; Zhou, L.; Ma, Y.; Luo, Y.; Tao, C.; Yang, J. NDM-Producing *Enterobacteriaceae* in a Chinese Hospital, 2014–2015: Identification of NDM-Producing *Citrobacter werkmanii* and Acquisition of blaNDM-1-Carrying Plasmid in Vivo in a Clinical *Escherichia coli* Isolate. *J. Med. Microbiol.* **2016**, *65*, 1253–1259. [CrossRef] [PubMed]
27. Zhang, F.; Xie, L.; Wang, X.; Han, L.; Guo, X.; Ni, Y.; Qu, H.; Sun, J. Further Spread of bla NDM-5 in *Enterobacteriaceae* via IncX3 Plasmids in Shanghai, China. *Front. Microbiol.* **2016**, *7*, 424. [CrossRef] [PubMed]
28. Ma, T.; Fu, J.; Xie, N.; Ma, S.; Lei, L.; Zhai, W.; Shen, Y.; Sun, C.; Wang, S.; Shen, Z.; et al. Fitness Cost of blaNDM-5-Carrying P3R-IncX3 Plasmids in Wild-Type NDM-Free *Enterobacteriaceae*. *Microorganisms* **2020**, *8*, 377. [CrossRef] [PubMed]
29. Almakki, A.; Maure, A.; Pantel, A.; Romano-Bertrand, S.; Masnou, A.; Marchandin, H.; Jumas-Bilak, E.; Licznar-Fajardo, P. NDM-5-Producing *Escherichia coli* in an Urban River in Montpellier, France. *Int. J. Antimicrob. Agents* **2017**, *50*, 123–124. [CrossRef]
30. Liakopoulos, A.; van der Goot, J.; Bossers, A.; Betts, J.; Brouwer, M.S.M.; Kant, A.; Smith, H.; Ceccarelli, D.; Mevius, D. Genomic and Functional Characterisation of IncX3 Plasmids Encoding blaSHV-12 in *Escherichia coli* from Human and Animal Origin. *Sci. Rep.* **2018**, *8*, 7674. [CrossRef]
31. Li, X.; Fu, Y.; Shen, M.; Huang, D.; Du, X.; Hu, Q.; Zhou, Y.; Wang, D.; Yu, Y. Dissemination of blaNDM-5 Gene via an IncX3-Type Plasmid among Non-Clonal *Escherichia coli* in China. *Antimicrob. Resist. Infect. Control* **2018**, *7*, 59. [CrossRef]
32. Wang, Z.; Li, M.; Shen, X.; Wang, L.; Liu, L.; Hao, Z.; Duan, J.; Yu, F. Outbreak of blaNDM-5-Harboring *Klebsiella pneumoniae* ST290 in a Tertiary Hospital in China. *Microb. Drug Resist.* **2019**, *25*, 1443–1448. [CrossRef] [PubMed]
33. Parvez, S.; Khan, A.U. Hospital Sewage Water: A Reservoir for Variants of New Delhi Metallo-β-Lactamase (NDM)- and Extended-Spectrum β-Lactamase (ESBL)-Producing *Enterobacteriaceae*. *Int. J. Antimicrob. Agents* **2018**, *51*, 82–88. [CrossRef] [PubMed]
34. Zurfluh, K.; Stevens, M.J.A.; Stephan, R.; Nüesch-Inderbinen, M. Complete and Assembled Genome Sequence of an NDM-5- and CTX-M-15-Producing *Escherichia coli* Sequence Type 617 Isolated from Wastewater in Switzerland. *J. Glob. Antimicrob. Resist.* **2018**, *15*, 105–106. [CrossRef]
35. Caractéristiques et Évolution Des Souches D'entérobactéries Productrices de Carbapénémases (EPC) Isolées En France, 2012–2020. Available online: https://www.santepubliquefrance.fr/maladies-et-traumatismes/infections-associees-aux-soins-et-resistance-aux-antibiotiques/resistance-aux-antibiotiques/documents/article/caracteristiques-et-evolution-des-souches-d-enterobacteries-productrices-de-carbapenemases-epc-isolees-en-france-2012-2020 (accessed on 21 January 2022).

36. Li, Y.; Tang, M.; Dai, X.; Zhou, Y.; Zhang, Z.; Qiu, Y.; Li, C.; Zhang, L. Whole-Genomic Analysis of NDM-5-Producing *Enterobacteriaceae* Recovered from an Urban River in China. *Infect. Drug Resist.* **2021**, *14*, 4427–4440. [CrossRef]
37. Bleichenbacher, S.; Stevens, M.J.A.; Zurfluh, K.; Perreten, V.; Endimiani, A.; Stephan, R.; Nüesch-Inderbinen, M. Environmental Dissemination of Carbapenemase-Producing *Enterobacteriaceae* in Rivers in Switzerland. *Environ. Pollut.* **2020**, *265*, 115081. [CrossRef] [PubMed]
38. Zhao, Q.; Berglund, B.; Zou, H.; Zhou, Z.; Xia, H.; Zhao, L.; Nilsson, L.E.; Li, X. Dissemination of *bla*NDM-5 via IncX3 Plasmids in Carbapenem-Resistant *Enterobacteriaceae* among Humans and in the Environment in an Intensive Vegetable Cultivation Area in Eastern China. *Environ. Pollut.* **2021**, *273*, 116370. [CrossRef]
39. Zhai, R.; Fu, B.; Shi, X.; Sun, C.; Liu, Z.; Wang, S.; Shen, Z.; Walsh, T.R.; Cai, C.; Wang, Y.; et al. Contaminated In-House Environment Contributes to the Persistence and Transmission of NDM-Producing Bacteria in a Chinese Poultry Farm. *Environ. Int.* **2020**, *139*, 105715. [CrossRef]
40. Zou, H.; Jia, X.; Liu, H.; Li, S.; Wu, X.; Huang, S. Emergence of NDM-5-Producing *Escherichia coli* in a Teaching Hospital in Chongqing, China: IncF-Type Plasmids May Contribute to the Prevalence of Bla NDM-5. *Front. Microbiol.* **2020**, *11*, 334. [CrossRef]
41. Wang, Y.; Tong, M.-K.; Chow, K.-H.; Cheng, V.C.-C.; Tse, C.W.-S.; Wu, A.K.-L.; Lai, R.W.-M.; Luk, W.-K.; Tsang, D.N.-C.; Ho, P.-L. Occurrence of Highly Conjugative IncX3 Epidemic Plasmid Carrying *bla* NDM in *Enterobacteriaceae* Isolates in Geographically Widespread Areas. *Front. Microbiol.* **2018**, *9*, 2272. [CrossRef]
42. Sonnevend, A.; Al Baloushi, A.; Ghazawi, A.; Hashmey, R.; Girgis, S.; Hamadeh, M.B.; Al Haj, M.; Pál, T. Emergence and Spread of NDM-1 Producer *Enterobacteriaceae* with Contribution of IncX3 Plasmids in the United Arab Emirates. *J. Med. Microbiol.* **2013**, *62*, 1044–1050. [CrossRef]
43. Ho, P.-L.; Li, Z.; Lai, E.L.; Chiu, S.S.; Cheng, V.C.C. Emergence of NDM-1-Producing *Enterobacteriaceae* in China. *J. Antimicrob. Chemother.* **2012**, *67*, 1553–1555. [CrossRef] [PubMed]
44. Fortini, D.; Villa, L.; Feudi, C.; Pires, J.; Bonura, C.; Mammina, C.; Endimiani, A.; Carattoli, A. Double Copies of Bla(KPC-3)::Tn4401a on an IncX3 Plasmid in *Klebsiella pneumoniae* Successful Clone ST512 from Italy. *Antimicrob. Agents Chemother.* **2016**, *60*, 646–649. [CrossRef] [PubMed]
45. Espedido, B.A.; Dimitrijovski, B.; van Hal, S.J.; Jensen, S.O. The Use of Whole-Genome Sequencing for Molecular Epidemiology and Antimicrobial Surveillance: Identifying the Role of IncX3 Plasmids and the Spread of BlaNDM-4-like Genes in the *Enterobacteriaceae*. *J. Clin. Pathol.* **2015**, *68*, 835–838. [CrossRef] [PubMed]
46. Wailan, A.M.; Paterson, D.L.; Kennedy, K.; Ingram, P.R.; Bursle, E.; Sidjabat, H.E. Genomic Characteristics of NDM-Producing *Enterobacteriaceae* Isolates in Australia and Their *bla*NDM Genetic Contexts. *Antimicrob. Agents Chemother.* **2016**, *60*, 136–141. [CrossRef]
47. Liu, Z.; Wang, Y.; Walsh, T.R.; Liu, D.; Shen, Z.; Zhang, R.; Yin, W.; Yao, H.; Li, J.; Shen, J. Plasmid-Mediated Novel *bla*NDM-17 Gene Encoding a Carbapenemase with Enhanced Activity in a Sequence Type 48 *Escherichia coli* Strain. *Antimicrob. Agents Chemother.* **2017**, *61*, e02233-16. [CrossRef]
48. Kieffer, N.; Nordmann, P.; Aires-de-Sousa, M.; Poirel, L. High Prevalence of Carbapenemase-Producing *Enterobacteriaceae* among Hospitalized Children in Luanda, Angola. *Antimicrob. Agents Chemother.* **2016**, *60*, 6189–6192. [CrossRef]
49. Petrosillo, N.; Vranić-Ladavac, M.; Feudi, C.; Villa, L.; Fortini, D.; Barišić, N.; Bedenić, B.; Ladavac, R.; D'Arezzo, S.; Andrašević, A.T.; et al. Spread of *Enterobacter cloacae* Carrying *bla*NDM-1, *bla*CTX-M-15, *bla*SHV-12 and Plasmid-Mediated Quinolone Resistance Genes in a Surgical Intensive Care Unit in Croatia. *J. Glob. Antimicrob. Resist.* **2016**, *4*, 44–48. [CrossRef]
50. Ho, P.-L.; Wang, Y.; Liu, M.C.-J.; Lai, E.L.-Y.; Law, P.Y.-T.; Cao, H.; Chow, K.-H. IncX3 Epidemic Plasmid Carrying *bla*NDM-5 in *Escherichia coli* from Swine in Multiple Geographic Areas in China. *Antimicrob. Agents Chemother.* **2018**, *62*, e02295-17. [CrossRef]
51. Volante, A.; Alonso, J.C. Molecular Anatomy of ParA-ParA and ParA-ParB Interactions during Plasmid Partitioning. *J. Biol. Chem.* **2015**, *290*, 18782–18795. [CrossRef]
52. De Gelder, L.; Ponciano, J.M.; Joyce, P.; Top, E.M. Stability of a Promiscuous Plasmid in Different Hosts: No Guarantee for a Long-Term Relationship. *Microbiology* **2007**, *153*, 452–463. [CrossRef]
53. Andersson, D.I.; Patin, S.M.; Nilsson, A.I.; Kugelberg, E. The Biological Cost of Antibiotic Resistance. In *Enzyme-Mediated Resistance to Antibiotics*; John Wiley & Sons, Ltd.: Hoboken, NJ, USA, 2007; pp. 339–348. ISBN 978-1-68367-166-4.
54. Göttig, S.; Pfeifer, Y.; Wichelhaus, T.A.; Zacharowski, K.; Bingold, T.; Averhoff, B.; Brandt, C.; Kempf, V.A. Global Spread of New Delhi Metallo-β-Lactamase 1. *Lancet Infect. Dis.* **2010**, *10*, 828–829. [CrossRef]
55. Baltrus, D.A. Exploring the Costs of Horizontal Gene Transfer. *Trends Ecol. Evol.* **2013**, *28*, 489–495. [CrossRef] [PubMed]
56. Long, D.; Zhu, L.-L.; Du, F.-L.; Xiang, T.-X.; Wan, L.-G.; Wei, D.-D.; Zhang, W.; Liu, Y. Phenotypic Profile and Global Transcriptomic Profile of Hypervirulent *Klebsiella pneumoniae* Due to Carbapenemase-Encoding Plasmid Acquisition. *BMC Genom.* **2019**, *20*, 480. [CrossRef]
57. Doyle, M.; Fookes, M.; Ivens, A.; Mangan, M.W.; Wain, J.; Dorman, C.J. An H-NS-like Stealth Protein Aids Horizontal DNA Transmission in Bacteria. *Science* **2007**, *315*, 251–252. [CrossRef] [PubMed]
58. Lucchini, S.; Rowley, G.; Goldberg, M.D.; Hurd, D.; Harrison, M.; Hinton, J.C.D. H-NS Mediates the Silencing of Laterally Acquired Genes in Bacteria. *PLoS Pathog.* **2006**, *2*, e81. [CrossRef]
59. Navarre, W.W.; Porwollik, S.; Wang, Y.; McClelland, M.; Rosen, H.; Libby, S.J.; Fang, F.C. Selective Silencing of Foreign DNA with Low GC Content by the H-NS Protein in *Salmonella*. *Science* **2006**, *313*, 236–238. [CrossRef] [PubMed]
60. San Millan, A.; MacLean, R.C. Fitness Costs of Plasmids: A Limit to Plasmid Transmission. *Microbiol. Spectr.* **2017**, *5*. [CrossRef]

61. Kumwenda, G.P.; Sugawara, Y.; Abe, R.; Akeda, Y.; Kasambara, W.; Chizani, K.; Takeuchi, D.; Sakamoto, N.; Tomono, K.; Hamada, S. First Identification and Genomic Characterization of Multidrug-Resistant Carbapenemase-Producing *Enterobacteriaceae* Clinical Isolates in Malawi, Africa. *J. Med. Microbiol.* **2019**, *68*, 1707–1715. [CrossRef]

62. van Hoek, A.H.A.M.; Veenman, C.; Florijn, A.; Huijbers, P.M.C.; Graat, E.A.M.; de Greeff, S.; Dierikx, C.M.; van Duijkeren, E. Longitudinal Study of ESBL *Escherichia coli* Carriage on an Organic Broiler Farm. *J. Antimicrob. Chemother.* **2018**, *73*, 3298–3304. [CrossRef]

63. Baniga, Z.; Hounmanou, Y.M.G.; Kudirkiene, E.; Kusiluka, L.J.M.; Mdegela, R.H.; Dalsgaard, A. Genome-Based Analysis of Extended-Spectrum β-Lactamase-Producing *Escherichia coli* in the Aquatic Environment and Nile Perch (Lates Niloticus) of Lake Victoria, Tanzania. *Front. Microbiol.* **2020**, *11*, 108. [CrossRef]

64. Büdel, T.; Kuenzli, E.; Clément, M.; Bernasconi, O.J.; Fehr, J.; Mohammed, A.H.; Hassan, N.K.; Zinsstag, J.; Hatz, C.; Endimiani, A. Polyclonal Gut Colonization with Extended-Spectrum Cephalosporin- and/or Colistin-Resistant *Enterobacteriaceae*: A Normal Status for Hotel Employees on the Island of Zanzibar, Tanzania. *J. Antimicrob. Chemother.* **2019**, *74*, 2880–2890. [CrossRef] [PubMed]

65. Grimont, F.; Grimont, P.A.D. The Genus *Enterobacter*. In *The Prokaryotes: A Handbook on the Biology of Bacteria: Proteobacteria: Gamma Subclass*; Dworkin, M., Falkow, S., Rosenberg, E., Schleifer, K.-H., Stackebrandt, E., Eds.; Springer: New York, NY, USA, 2006; Volume 6, pp. 197–214. ISBN 978-0-387-30746-6.

66. Drancourt, M.; Bollet, C.; Carta, A.; Rousselier, P. Phylogenetic Analyses of *Klebsiella* Species Delineate *Klebsiella* and *Raoultella* Gen. Nov., with Description of *Raoultella ornithinolytica* Comb. Nov., *Raoultella terrigena* Comb. Nov. and *Raoultella planticola* Comb. Nov. *Int. J. Syst. Evol. Microbiol.* **2001**, *51*, 925–932. [CrossRef] [PubMed]

67. Mills, M.C.; Lee, J. The Threat of Carbapenem-Resistant Bacteria in the Environment: Evidence of Widespread Contamination of Reservoirs at a Global Scale. *Environ. Pollut.* **2019**, *255*, 113143. [CrossRef] [PubMed]

68. Zhang, J.; Li, W.; Chen, J.; Qi, W.; Wang, F.; Zhou, Y. Impact of Biofilm Formation and Detachment on the Transmission of Bacterial Antibiotic Resistance in Drinking Water Distribution Systems. *Chemosphere* **2018**, *203*, 368–380. [CrossRef] [PubMed]

69. Sanganyado, E.; Gwenzi, W. Antibiotic Resistance in Drinking Water Systems: Occurrence, Removal, and Human Health Risks. *Sci. Total Environ.* **2019**, *669*, 785–797. [CrossRef] [PubMed]

70. Hölzel, C.S.; Tetens, J.L.; Schwaiger, K. Unraveling the Role of Vegetables in Spreading Antimicrobial-Resistant Bacteria: A Need for Quantitative Risk Assessment. *Foodborne Pathog. Dis.* **2018**, *15*, 671–688. [CrossRef] [PubMed]

71. Laurens, C.; Jean-Pierre, H.; Licznar-Fajardo, P.; Hantova, S.; Godreuil, S.; Martinez, O.; Jumas-Bilak, E. Transmission of IMI-2 Carbapenemase-Producing *Enterobacteriaceae* from River Water to Human. *J. Glob. Antimicrob. Resist.* **2018**, *15*, 88–92. [CrossRef]

72. Bosch, T.; Schade, R.; Landman, F.; Schouls, L.; Dijk, K. van A *bla*VIM-1 Positive *Aeromonas hydrophila* Strain in a near-Drowning Patient: Evidence for Interspecies Plasmid Transfer within the Patient. *Future Microbiol.* **2019**, *14*, 1191–1197. [CrossRef]

73. Carattoli, A.; Bertini, A.; Villa, L.; Falbo, V.; Hopkins, K.L.; Threlfall, E.J. Identification of Plasmids by PCR-Based Replicon Typing. *J. Microbiol. Methods* **2005**, *63*, 219–228. [CrossRef]

74. Sandegren, L.; Linkevicius, M.; Lytsy, B.; Melhus, Å.; Andersson, D.I. Transfer of an Escherichia coli ST131 Multiresistance Cassette Has Created a *Klebsiella pneumoniae*-Specific Plasmid Associated with a Major Nosocomial Outbreak. *J. Antimicrob. Chemother.* **2012**, *67*, 74–83. [CrossRef]

75. Bonnel, C.; Legrand, B.; Simon, M.; Clavié, M.; Masnou, A.; Jumas-Bilak, E.; Kang, Y.K.; Licznar-Fajardo, P.; Maillard, L.T.; Masurier, N. Tailoring the Physicochemical Properties of Antimicrobial Peptides onto a Thiazole-Based γ-Peptide Foldamer. *J. Med. Chem.* **2020**, *63*, 9168–9180. [CrossRef]

76. Bankevich, A.; Nurk, S.; Antipov, D.; Gurevich, A.A.; Dvorkin, M.; Kulikov, A.S.; Lesin, V.M.; Nikolenko, S.I.; Pham, S.; Prjibelski, A.D.; et al. SPAdes: A New Genome Assembly Algorithm and Its Applications to Single-Cell Sequencing. *J. Comput. Biol.* **2012**, *19*, 455–477. [CrossRef] [PubMed]

77. Clermont, O.; Gordon, D.; Denamur, E. Guide to the Various Phylogenetic Classification Schemes for *Escherichia coli* and the Correspondence among Schemes. *Microbiology* **2015**, *161*, 980–988. [CrossRef] [PubMed]

78. Waters, N.R.; Abram, F.; Brennan, F.; Holmes, A.; Pritchard, L. Easy Phylotyping of *Escherichia coli* via the EzClermont Web App and Command-Line Tool. *Access Microbiol.* **2020**, *2*. [CrossRef]

79. Darling, A.C.E.; Mau, B.; Blattner, F.R.; Perna, N.T. Mauve: Multiple Alignment of Conserved Genomic Sequence with Rearrangements. *Genome Res.* **2004**, *14*, 1394–1403. [CrossRef]

80. Carattoli, A.; Zankari, E.; García-Fernández, A.; Voldby Larsen, M.; Lund, O.; Villa, L.; Møller Aarestrup, F.; Hasman, H. In Silico Detection and Typing of Plasmids Using PlasmidFinder and Plasmid Multilocus Sequence Typing. *Antimicrob. Agents Chemother.* **2014**, *58*, 3895–3903. [CrossRef] [PubMed]

81. Seemann, T. Prokka: Rapid Prokaryotic Genome Annotation. *Bioinformatics* **2014**, *30*, 2068–2069. [CrossRef]

82. Afgan, E.; Baker, D.; van den Beek, M.; Blankenberg, D.; Bouvier, D.; Čech, M.; Chilton, J.; Clements, D.; Coraor, N.; Eberhard, C.; et al. The Galaxy Platform for Accessible, Reproducible and Collaborative Biomedical Analyses: 2016 Update. *Nucleic Acids Res.* **2016**, *44*, W3–W10. [CrossRef]

83. Thomas, C.M.; Thomson, N.R.; Cerdeño-Tárraga, A.M.; Brown, C.J.; Top, E.M.; Frost, L.S. Annotation of Plasmid Genes. *Plasmid* **2017**, *91*, 61–67. [CrossRef]

84. Madden, T.L.; Tatusov, R.L.; Zhang, J. Applications of Network BLAST Server. *Methods Enzymol.* **1996**, *266*, 131–141. [CrossRef]

85. Blum, M.; Chang, H.-Y.; Chuguransky, S.; Grego, T.; Kandasaamy, S.; Mitchell, A.; Nuka, G.; Paysan-Lafosse, T.; Qureshi, M.; Raj, S.; et al. The InterPro Protein Families and Domains Database: 20 Years On. *Nucleic Acids Res.* **2021**, *49*, D344–D354. [CrossRef] [PubMed]
86. Mistry, J.; Chuguransky, S.; Williams, L.; Qureshi, M.; Salazar, G.A.; Sonnhammer, E.L.L.; Tosatto, S.C.E.; Paladin, L.; Raj, S.; Richardson, L.J.; et al. Pfam: The Protein Families Database in 2021. *Nucleic Acids Res.* **2021**, *49*, D412–D419. [CrossRef] [PubMed]
87. Bortolaia, V.; Kaas, R.S.; Ruppe, E.; Roberts, M.C.; Schwarz, S.; Cattoir, V.; Philippon, A.; Allesoe, R.L.; Rebelo, A.R.; Florensa, A.F.; et al. ResFinder 4.0 for Predictions of Phenotypes from Genotypes. *J. Antimicrob. Chemother.* **2020**, *75*, 3491–3500. [CrossRef] [PubMed]
88. Conant, G.C.; Wolfe, K.H. GenomeVx: Simple Web-Based Creation of Editable Circular Chromosome Maps. *Bioinformatics* **2008**, *24*, 861–862. [CrossRef]
89. Larsen, M.V.; Cosentino, S.; Rasmussen, S.; Friis, C.; Hasman, H.; Marvig, R.L.; Jelsbak, L.; Sicheritz-Pontén, T.; Ussery, D.W.; Aarestrup, F.M.; et al. Multilocus Sequence Typing of Total-Genome-Sequenced Bacteria. *J. Clin. Microbiol.* **2012**, *50*, 1355–1361. [CrossRef]
90. Zhou, Z.; Alikhan, N.-F.; Mohamed, K.; Fan, Y.; Agama Study Group; Achtman, M. The EnteroBase User's Guide, with Case Studies on *Salmonella* Transmissions, *Yersinia pestis* Phylogeny, and *Escherichia* Core Genomic Diversity. *Genome Res.* **2020**, *30*, 138–152. [CrossRef]
91. Nascimento, M.; Sousa, A.; Ramirez, M.; Francisco, A.P.; Carriço, J.A.; Vaz, C. PHYLOViZ 2.0: Providing Scalable Data Integration and Visualization for Multiple Phylogenetic Inference Methods. *Bioinformatics* **2017**, *33*, 128–129. [CrossRef]

Article

Extended Spectrum Beta-Lactamase-Resistant Determinants among Carbapenem-Resistant *Enterobacteriaceae* from Beef Cattle in the North West Province, South Africa: A Critical Assessment of Their Possible Public Health Implications

Lungisile Tshitshi [1,2], Madira Coutlyne Manganyi [3], Peter Kotsoana Montso [4], Moses Mbewe [2] and Collins Njie Ateba [1,4,*]

1 Antimicrobial Resistance and Phage Biocontrol Research Group, Department of Microbiology, School of Biological Sciences, Faculty of Natural and Agricultural Sciences, North West University, Private Bag X2046, Mmabatho 2735, South Africa; Lungisile.Tshitshi@ump.ac.za
2 Faculty of Agriculture and Natural Sciences, University of Mpumalanga, Private Bag X11283, Mbombela 1200, South Africa; Moses.Mbewe@ump.ac.za
3 Unit for Environmental Sciences and Management, North-West University, Potchefstroom 2520, South Africa; madira.manganyi@nwu.ac.za
4 Food Security and Safety Niche Area, Faculty of Natural and Agricultural Sciences, North West University, Private Bag X2046, Mmabatho 2735, South Africa; montsokp@gmail.com
* Correspondence: collins.ateba@nwu.ac.za

Received: 23 October 2020; Accepted: 12 November 2020; Published: 17 November 2020

Abstract: Carbapenems are considered to be the last resort antibiotics for the treatment of infections caused by extended-spectrum beta-lactamase (ESBL)-producing strains. The purpose of this study was to assess antimicrobial resistance profile of Carbapenem-resistant *Enterobacteriaceae* (CRE) isolated from cattle faeces and determine the presence of carbapenemase and ESBL encoding genes. A total of 233 faecal samples were collected from cattle and analysed for the presence of CRE. The CRE isolates revealed resistance phenotypes against imipenem (42%), ertapenem (35%), doripenem (30%), meropenem (28%), cefotaxime, (59.6%) aztreonam (54.3%) and cefuroxime (47.7%). Multidrug resistance phenotypes ranged from 1.4 to 27% while multi antibiotic resistance (MAR) index value ranged from 0.23 to 0.69, with an average of 0.40. *Escherichia coli* (*E. coli*), *Klebsiella pneumoniae* (*K. pneumoniae*), *Proteus mirabilis* (*P. mirabilis*) and *Salmonella* (34.4, 43.7, 1.3 and 4.6%, respectively) were the most frequented detected species through genus specific PCR analysis. Detection of genes encoding carbapenemase ranged from 3.3% to 35% (*bla*KPC, *bla*NDM, *bla*GES, *bla*OXA-48, *bla*VIM and *bla*OXA-23). Furthermore, CRE isolates harboured ESBL genes (*bla*SHV (33.1%), *bla*TEM (22.5%), *bla*CTX-M (20.5%) and *bla*OXA (11.3%)). In conclusion, these findings indicate that cattle harbour CRE carrying ESBL determinants and thus, proper hygiene measures must be enforced to mitigate the spread of CRE strains to food products.

Keywords: *Enterobacteriaceae*; multidrug resistance; carbapenemase; ESBL; resistance genes; cattle

---

1. Introduction

Carbapenem-resistant *Enterobacteriaceae* (CRE) strains pose a serious threat, especially in public health worldwide [1,2]. These strains cause severe infections such as bloodstream, pneumonia and complicated urinary tract infections in debilitated immunocompromised patients, thus leading to prolonged hospital stay as well as increased healthcare costs and mortality rates [2,3]. According to the Centers for Disease Control and Prevention (CDC), direct healthcare costs associated with antimicrobial resistance infections are estimated at \$20 billion per annum in developed countries [4]. Carbapenem-resistant *Enterobacteriaceae* strains have been commonly reported in hospital settings and patients in intensive care units [1,5,6]. Although *Escherichia coli* (*E. coli*) and *Klebsiella pnuemoniae* (*K. pneumoniae*) are the most frequently detected CRE species, other clinical pathogens such *Citrobacter freundii, Enterobacter aerogenes* (*E. aerogenes*), *Enterobacter cloacae* (*E. cloacae*), *P. mirabillis, Salmonella* and *Serratia marcescens* species have been detected from environmental samples [7,8]. In addition, several studies have reported that these species harbours clinically important carbapenemase encoding genes such as *blaKPC, blaVIM, blaIMP, blaNDM* and *blaOXA-48* [9]. These genes are usually plasmid-borne, thereby accelerating horizontal transfer of resistance determinants between the same and/or different species [10]. Worrisome is the fact that Carbapenem-resistant strains may carry extended-spectrum beta-lactamase (ESBL) resistance genes.

Several studies have detected major ESBL genes (*blaTEM, blaSHV, blaCTX-M* and *blaOXA*) in CRE strains [11]. Given the fact that carbapenem antibiotics are considered as the last resort for treating infections caused by ESBL-producing strains, presence of carbapenemase and ESBL genes in CRE strains coupled with lack new therapeutic option is cause for concern [9,10]. Against this background, the World Health Organisation (WHO) has classified CRE strains (*Acinetobacter baumannii* (*A. baumannii*), *E. coli, K. pneumoniae, Pseudomonas aeruginosa* (*P. aeruginosa*) and *Enterobacter* species) as "critical priority pathogens", which require urgent research and development novel and effective antibiotics [12]. Furthermore, the WHO has published an "action plan", which encourages all countries to establish their own action plan to curb antibiotic resistance phenomenon [13]. This could be achieved via extensive research as well as surveillance and epidemiological studies both nationally and internationally. Data obtained from such studies may assist to strengthen the existing departmental and national policies to mitigate the spread of antimicrobial resistance pathogens.

Despite the fact that most of the studies conducted on CRE have been confined within hospital environment and admitted patients [5,6,14], there is a new evidence indicating that food producing animals, especially cattle, pigs and poultry, may harbour CRE carrying ESBL genes [15–18]. Nevertheless, the number of studies investigating cattle as potential reservoir of CRE harbouring ESBL genes is limited. Hence, the current study was undertaken to determine the occurrence of CRE carrying ESBL determinants in cattle.

2. Results

2.1. Enterobacteriaceae Isolated from Cattle Faeces

Out of the 233 faecal samples collected from four farms, a total of 280 presumptive isolates belonging to the *Enterobacteriaceae* family were obtained. *Enterobacteriaceae* isolates were mostly obtained from Farm_R-B and Farm_N-D (101 and 92, respectively). Only 59 and 28 of the isolates were obtained from Farm_K-A and Farm_L-C, respectively.

2.2. Carbapenem Resistance Isolates

All 280 isolates revealed various antimicrobial susceptibility profiles against four carbapenem antimicrobial agents tested; Figure 1. Of these, 69.3% of the isolates showed intermediate or resistance to one or more of the four carbapenem antibiotics. The isolates revealed resistance to imipenem (42%), ertapenem (35%), doripenem (30%) and meropenem (28%), while intermediate resistance was shown by 26, 31, 33 and 41%, respectively. All 194 isolates that exhibited carbapenem resistance phenotypes were positive for the modified Hodge test. A total of 151 isolates were capable of growing on Brilliance™ ESBL agar and produced colonies with different colours (pink, green, colourless and brown halo), indicating that the isolates could be composed of different species such as *E. coli, Klebsiella, Enterobacter, Serratia, Citrobacter* and *Salmonella, Proteus*.

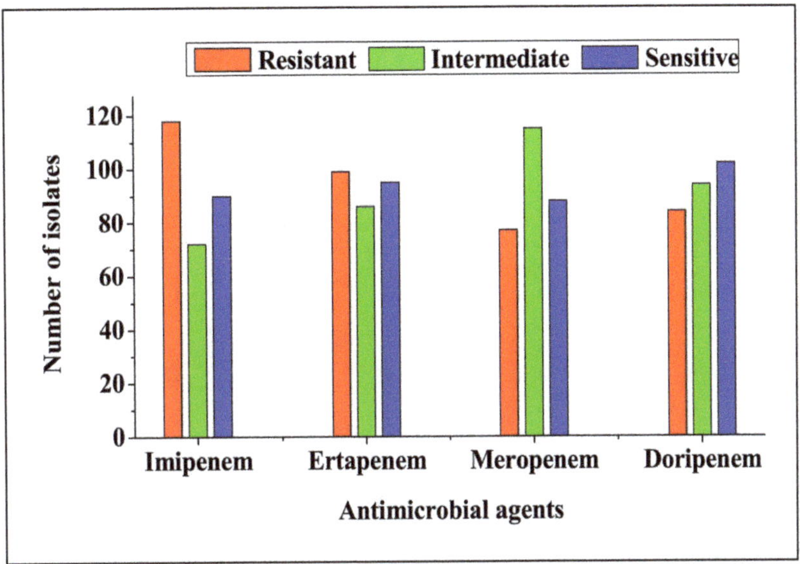

Figure 1. Antimicrobial susceptibility profile of *Enterobacteriaceae* isolated from cattle faeces.

2.3. Antimicrobial Resistance Profile of CRE

The 151 isolates that were Carbapenem-resistant and also positive for ESBL production revealed various antimicrobial resistance profiles against 13 different antibiotics tested, Figure 2. The isolates were resistant to at least two or more antimicrobial agents tested. Most of the isolates were resistant to cefotaxime, aztreonam and cefuroxime (59.6, 54.3 and 47.7%, respectively). Resistance to other antimicrobial agents (ceftiofur, amoxicillin, piperacillin, ticarcillin, cephalothin, ceftazidime and cefoxitin) ranged from 39.1 to 43.7%. Low resistance (12.6 and 9.3%) was observed for ciprofloxacin and amoxicillin-clavulanate, respectively. Large proportion (93.4%) of CRE isolates were resistant to three or more antimicrobial agents and were defined as multidrug resistant strains. Multidrug resistance phenotypes ranged from 1.4 to 27%, with the most common MDR pattern being observed against four, five and six different antimicrobial agents (27, 27 and 22%, respectively); Figure 3. Multi antibiotic resistance (MAR) index value ranged from 0.23 to 0.69, with an average of 0.40. The MAR indices 0.31, 0.38 and 0.46 were the most frequently observed among CRE isolates, Figure 4 and Table S1.

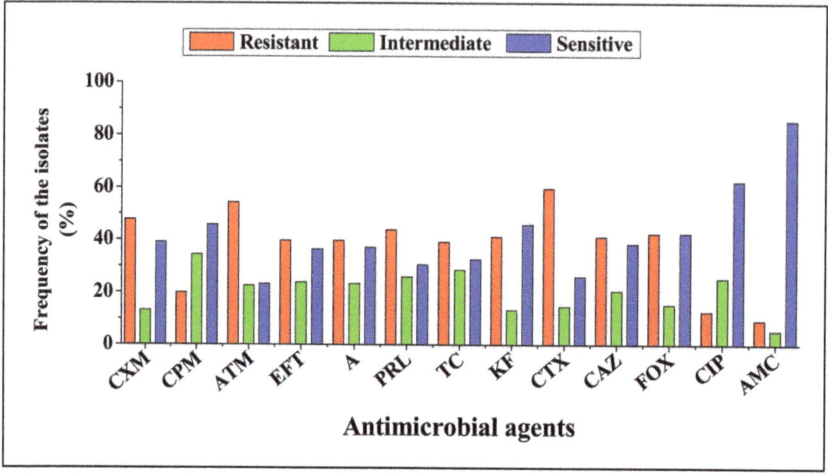

Figure 2. Antimicrobial susceptibility pattern of carbapenem-resistant *Enterobacteriaceae*. A = Amoxicillin, AMC = Amoxicillin-clavulanate, ATM = Aztreonam, CPM = Cefepime, CTX = Cefotaxime, FOX = Cefoxitin, CAZ = Ceftazidime, CXM = Cefuroxime, EFT = Ceftiofur, KF = Cephalothin, CIP = Ciprofloxacin, PRL = Piperacillin and TC = Ticarcillin.

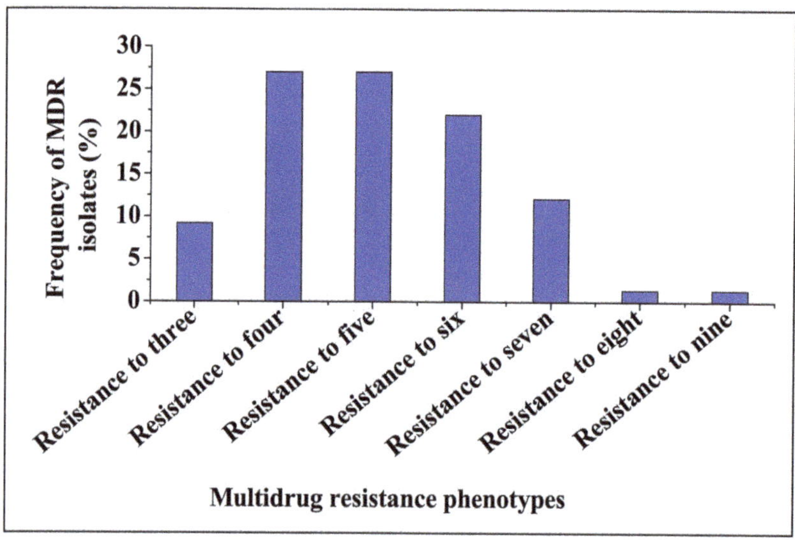

Figure 3. Frequency of multidrug resistance pattern in CRE isolated from cattle faeces.

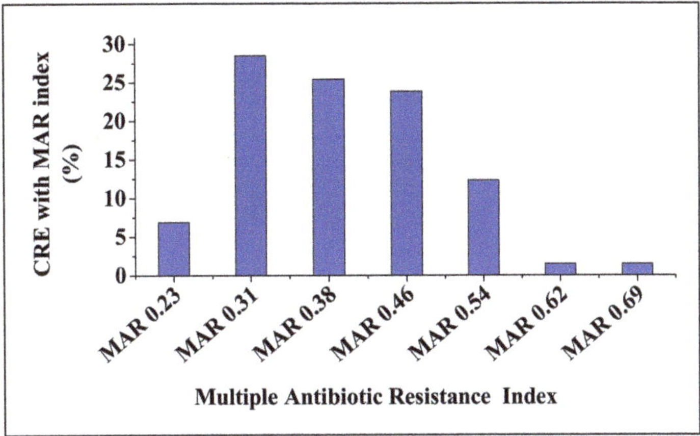

Figure 4. Mean frequency of multiple antibiotic resistance profiles of CRE strains.

2.4. Molecular Identification of CRE Species

A total of 151 CRE isolates were confirmed through amplifying the 16S rRNA conserved region. Of these, 84% were confirmed at species level (*E. coli* (34.4%), *K. pneumoniae* (43.7%), *P. mirabilis* (1.3%) and *Salmonella* (4.6%)) through genus-specific PCR analysis. The remaining 16% of the isolates that could not be identified as either one of these four species were classified as unspecified CRE species.

2.5. Detection of Genes Encoding Carbapenemases in CRE

All six carbapenemase-encoding genes screened were detected in CRE isolated from cattle faeces. The *blaKPC* (35.8%), *blaNDM* (20.5%) and *blaGES* (17.9%) were the most frequently detected genes, Figure 5. The *blaOXA-48*, *blaVIM* and *blaOXA-23* were detected in low proportions (10.6, 6.6 and 3.3%, respectively). Simultaneous detection of *blaKPC_blaOXA-23* (2.6%) *blaKPC_blaNDM* (1.3%) and *blaGES_blaOXA-48* (1.3%) genes in some of the isolates were observed. In general, large proportion (94.7%) of CRE species carried carbapenem resistance genes. Carbapenemase encoding genes were commonly detected in *E. coli* and *K. pneumoniae* species (96.2 and 92.4%, respectively), while all *Salmonella* and *Proteus mirabilis* species detected in this study possessed CR determinants; Table 1.

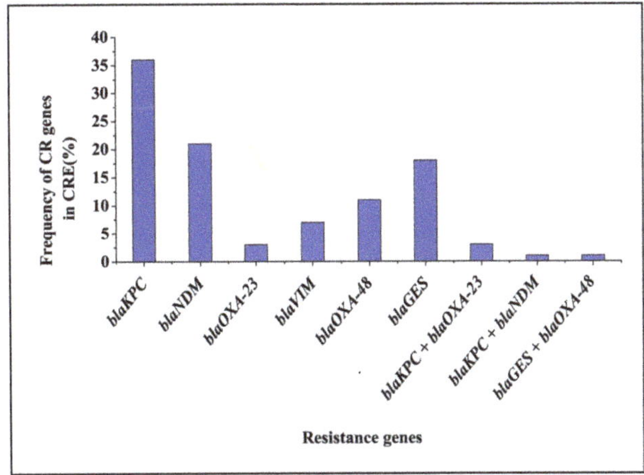

Figure 5. Number of carbapenemase-encoding genes detected in CRE isolated from cattle faeces.

Table 1. Proportion of CRE species harbouring carbapenemase genes.

CRE Species	No. of Isolates	Carbapenemase Encoding Genes (%)								
		bla_{KPC}	bla_{NDM}	$bla_{OXA\text{-}23}$	bla_{VIM}	$bla_{OXA\text{-}48}$	bla_{GES}	$bla_{KPC}\_bla_{OXA\text{-}23}$	$bla_{KPC}\_bla_{NDM}$	$bla_{GES}\_bla_{OXA\text{-}48}$
E. coli	52	34.6	32.7	3.8	1.9	13.5	9.6	0.0	3.8	0.0
K. pneumoniae	66	42.4	12.1	1.5	3.0	6.1	27.3	6.1	0.0	1.5
P. mirabilis	2	0.0	0.0	50.0	0.0	50.0	0.0	0.0	0.0	0.0
Salmonella species	7	28.6	42.9	14.3	0.0	14.3	0.0	0.0	0.0	0.0
Unspecified CRE species	24	25.0	12.5	0.0	29.2	12.5	16.7	0.0	0.0	4.2
Total	151	35.8	20.5	3.3	6.6	10.6	17.9	2.6	1.3	1.3

Large proportion (42.4 and 27.3%) of *K. pneumoniae* species harboured *blaKPC* and *blaGES*, followed by 34.6 and 32.7% of *E. coli* species carrying *blaKPC* and *blaNDM*, respectively. *Salmonella* species (42.9 and 28.6%) carried *blaNDM* and *blaKPC*, while unspecified CRE species (29.2 and 25.0%) of possessed of *blaVIM* and *blaKPC* and 50% of *P. mirabilis* species harboured either *blaOXA-23* or *blaOXA-48*. Some of *E. coli* species (3.8%) also possessed *blaKPC_blaNDM*, while *K. pneumoniae* species (6.1 and 1.5%) harboured *blaKPC_blaOXA-23* and *blaGES_blaOXA-48*, followed by unspecified CRE species (4.2%) possessing *blaGES_blaOXA-48*.

2.6. Detection of ESBL Determinants in CRE

All four major ESBL-encoding genes screened were detected in CRE isolates. The *blaSHV* (33.1%), *blaTEM* (22.5%) and *blaCTX-M* (20.5%) were most frequently detected genes while *blaOXA* (11.3%) detection was very low, Figure 6. Concurrent detection by *blaOXA_blaCTX-M* (7.3%) and *blaSHV_blaTEM* (5.3%) genes in CRE isolates were observed. Generally, 87.4% CRE species harboured ESBL genes. As indicated in Table 2, most of *E. coli* (86.5%) and *K. pneumoniae* (84.8%) and *Salmonella* (85.7%) species were positive for ESBL genes. All *P. mirabilis* and 95.8% unspecified CRE species carried ESBL genes. Large proportion (45.5%) of *K. pneumoniae* species harboured *blaSHV*. *E. coli* species (34.6 and 26.9%) and *Salmonella* species (57.1 and 28.6%) carried *blaCTX-M* and *blaTEM*, respectively). The 33.3% of unspecified CRE species harboured *blaSHV* while 50% of *P. mirabilis* species carried either *blaTEM* or *blaOXA*. Some *E. coli* (9.6 and 3.8%) and *K. pneumoniae* species (3.0 and 12.1%) possessed *blaSHV_blaTEM* and *blaOXA_blaCTX-M*, respectively. *Salmonella* (14.3%) and *P. mirabilis* species (9.1%) possessed *blaOXA_blaCTX-M*, while unspecified CRE species (4.2%) harboured *blaSHV_blaTEM*.

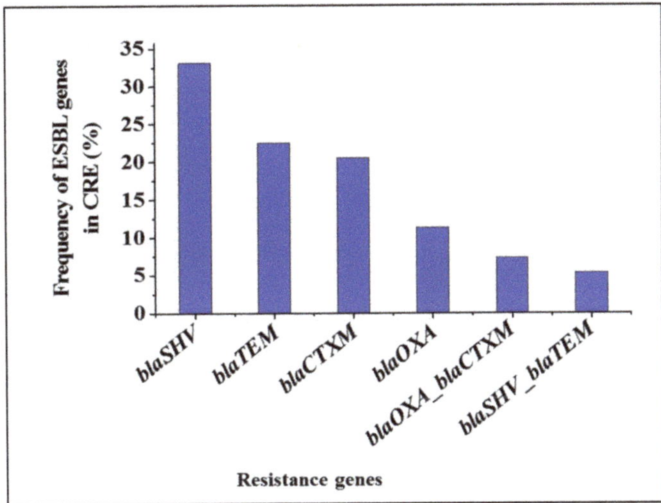

Figure 6. Frequency of ESBL-encoding genes detected in CRE isolated from cattle faeces.

Table 2. Proportion of CRE species carrying ESBL genes.

CRE Species	No. of Isolates	ESBL Encoding Genes (%)					
		bla_{SHV}	bla_{TEM}	$bla_{CTX\text{-}M}$	bla_{OXA}	$bla_{OXA}\_bla_{CTX\text{-}M}$	$bla_{SHV}\_bla_{TEM}$
E. coli	52	23.1	26.9	34.6	1.9	3.8	9.6
K. pneumoniae	66	45.5	16.7	9.1	13.6	12.1	3.0
P. mirabilis	2	0.0	50.0	0.0	50.0	9.1	0.0
Salmonella species	7	0.0	28.6	57.1	0.0	14.3	0.0
Unspecified CRE species	24	33.3	25.0	12.5	25.0	0.0	4.2
Total	151	33.1	22.5	20.5	11.3	7.3	5.3

3. Discussion

This study reports the occurrence of Carbapenem-resistant *Enterobacteriaceae* from cattle faeces obtained from different commercial farms in the North West province, South Africa. Overall, 280 *Enterobacteriaceae* were successfully isolated. Notably, *Enterobacteriaceae* isolates revealed intermediate or resistance to carbapenem antibiotics (imipenem, ertapenem, doripenem and meropenem) tested. Moreover, resistance to carbapenem antibiotics ranged from 28 to 42%. High resistance was observed against imipenem (42%). This finding was lower than 98% imipenem resistance reported in the previous study [19]. Interestingly, all *Enterobacteriaceae* isolates that revealed intermediate and/or resistant phenotypes against carbapenem antibiotics tested positive for the modified Hodge test and were thus regarded as CRE strains [15,20]. In addition, the isolates were capable of growing on Brilliance™ ESBL media, and this suggests that those isolates carry ESBL determinants [15]. Since carbapenem antibiotics are regarded as being the last resort for the treatment of infections caused by ESBL-producing *Enterobacteriaceae* or multidrug resistant pathogens, resistance to this antibiotic group may exacerbate morbidity and mortality rates in humans [14,21–23]. Therefore, monitoring of CRE and the genes associated with CR in food producing animals is essential to determine prevalence of Carbapenem-resistant pathogens in cattle.

Extensive use of antibiotics in agriculture, especially in food producing exacerbate the emergence antibiotic resistant pathogens in food producing animals. Several studies have reported the occurrence of antimicrobial resistant pathogens in foodborne pathogens [24–26]. In this study, Carbapenem-resistant *Enterobacteriaceae* isolates revealed various resistance patterns against beta-lactam, 1st, 2nd and 3rd cephalosporins antibiotics. Most of the isolates were highly resistant to cefotaxime, aztreonam and cefuroxime (59.6, 54.3 and 47.7%, respectively). These results corroborate the previous study [26]. In addition, large proportion (93.4%) of CRE isolates revealed MDR phenotypes, which is higher than the 83.78% MRD phenotype previously reported in spinach in South Africa [27]. Moreover, resistance against a maximum of nine antimicrobial agents was observed, suggesting that some of the CRE isolates obtained in this study could possibly be considered as extensively drug resistant (XDR) strains [28]. Interestingly, the MAR index value of 129 MDR strains ranged from 0.23 to 0.69, with an average of 0.40. Although these findings were lower than that of the other study [29], the MAR indices observed in this study were > 0.2. A MAR index of 0.2 or higher indicates high-risk sources of contamination [30,31]. This implies that continuous monitoring of carbapenem resistance in food producing animals, especially cattle is crucial to ensure the safety of food.

Increasing emergence of Carbapenem-resistant species poses a severe threat in public health [32]. Several studies reported have reported the occurrence of CRE in food, animals, water, hospitalised patients and hospitals and/or clinic environments [2,5,33,34]. *E. coli* and *K. pneumoniae* are the most predominant species associated with carbapenem resistance [32]. Overall, four CR species (*E. coli*, *K. pneumoniae*, *Salmonella* and *P. mirabilis*) were identified using genus-specific PCR analysis in this study. Similar to other studies [5,34,35], *K. pneumoniae* (43.7%) and *E. coli* (34.4%) species were the most commonly detected species. Moreover, other CR species, *Salmonella* (4.6%) and *P. mirabilis* (1.3%) were detected in low quantity. However, 16% of CRE isolates could not be identified at the specie level and were classified as 'unspecified CRE species'.

Carbapenem resistance determinants (*blaKPC*, *blaNDM*, *blaOXA-23*, *blaVIM*, *blaOXA-48* and *blaGES*) have been detected in CRE isolated from different sources such as hospitals, drinking and recreational water, agricultural environments, food producing animals and food products [23]. In this study, 97.4% of CRE isolates possessed carbapenem resistance genes were detected in CRE species. These findings were higher than the 14.3% detection of carbapenem resistance determinants reported in the previous study in from vegetables [33]. This variation could be attributed to different sample sources and isolation methods used per study. Furthermore, high detection of CR genes in this study indicates that the use of carbapenem antibiotics in agriculture may increase the occurrence of CR pathogens in food producing animal, especially cattle. As a result, this may accelerate dissemination of CR pathogens to humans through consumption of contaminated food [22,33,36]. Based on each species, *E. coli* (96.2%), *K. pneumoniae* (92.4%), *Salmonella* (100%), *P. mirabilis* (100%) and unspecified CRE species (95.8%) carried carbapenem resistance determinants. However, these findings were higher than that of the other studies [14,33,37]. However, these findings were higher than those of the other studies [14,33,37]. A possible explanation for this variation could be attributed to differences in the source of samples and geographical location and management practices per farm, which may have different selective pressure for the antimicrobial resistance levels [31]. Most *K. pneumoniae* species and *E. coli* carried *blaKPC*, while *Salmonella* species possessed the *blaNDM* gene. In addition, the *P. mirabilis* species possessed either *blaOXA-23* or *blaOXA-48*, whereas unspecified CRE species possessed of *blaVIM*. Some *E. coli* species possessed a combination of *blaKPC_blaNDM*, while *K. pneumoniae* species harboured *blaKPC_blaOXA-23* and *blaGES_blaOXA-48*, followed by unspecified CRE species possessing *blaGES_blaOXA-48*. Given that *E. coli* and *K. pnuemoniae* species cause severe infection in humans, detection of Carbapenem-resistant determinants in these species cannot be overemphasised.

In *Enterobacteriaceae*, ESBL determinants (*blaSHV*, *blaTEM*, *blaCTX-M* and *blaOXA*) are considered as the primary mechanism for mediating beta-lactam antibiotics resistance in *Enterobacteriaceae* [38,39]. Given that extended spectrum cephalosporins antibiotics are used in veterinary medicine, this has resulted in the emergence of ESBL resistance genes in food producing animals, especially cattle [40]. Numerous studies have detected ESBL determinants in *Enterobacteriaceae* isolated from food, cattle and pigs [26,37,41]. Likewise, ESBL-encoding genes (*blaSHV*, *blaTEM*, *blaCTX-M* and *blaOXA*) were detected in this study. Overall, large proportion (87.4%) of CRE isolates harboured ESBL genes. This finding was higher than that of the previous studies, which reported the occurrence of ESBL-producing *Enterobacteriaceae* in cattle and food [26,42,43]. The *blaSHV*, *blaTEM* and *blaCTX-M* were the most frequently (20.5–33.1%) detected genes. However, these findings were lower than those of the previous studies, which reported the prevalence of ESBL-producing strains obtained from hospital environment [27,44,45]. The variation could be attributed to the source of samples, geographical location and the number of isolates analysed per study. Furthermore, *blaOXA* was detected at low rate. Based on each species, large proportion (84.8–100%) of CRE species (*E. coli* (86.5%), *K. pneumoniae* (84.8%), *Salmonella* (85.7%) and *P. mirabilis* (100%) unspecified CRE species (95.8%)) harboured ESBL genes. Large proportions of *K. pneumoniae* (45.5%) and unspecified CRE (33.3%) species harboured *blaSHV*. This finding is consistent with the other study, in which *blaSHV* was predominantly detected in *K. pnuemoniae* [44]. Moreover, *E. coli* (34.6%) and *Salmonella* species (57.1%) possessed *blaCTX-M*, while *P. mirabilis* (50%) species carried either *blaTEM* or *blaOXA*. Notably, some *E. coli* and *K. pneumoniae* species harboured *blaSHV_blaTEM*, which is similar to another study from Eastern Cape Province, South Africa [44]. Another ESBL combination observed was *blaOXA_blaCTX-M*. These findings suggest that strict measures must be implemented throughout the food chain to mitigate transmission of antimicrobial resistant pathogens to the environment, as well as humans.

4. Materials and Methods

4.1. Ethical Consideration

Ethical clearance and approval for the study was obtained from the Animal Care Research Ethics Committee (AnimCare REC), of the North-West University, South Africa (Reference number: NWU-00066-15-S9).

4.2. Study Area, Sample Collection and Processing

This is a cross-sectional study, and it was conducted from July 2016 to July 1017 in the North-West province, South Africa. A total of 233 faecal samples were collected from four cattle farms (Farm_K-A, Farm_R-B, Farm_L-C and Farm_N-D) located in the Ngaka Modiri Molema district. The selection of the sampling sites was based on the accessibility and the wiliness of the farm owners to participate in the study. The farms had 100 to 400 head of cattle. Sampling was done between July 2016 and August 2017 and all the ethical procedures were followed during handling of the animals. Faecal samples were collected directly from the rectum of individual animals using sterile arm-length gloves and in order to avoid duplication of sampling, the cattle were locked into their respective handling pens. Samples were placed in sterile sample collection bottles, labelled appropriately and immediately transported on ice packs to the Antimicrobial Resistance and Phage Biocontrol Research Group (ARPHBRG) laboratory, North-West University for microbial analysis. For bacteria isolation, 1 g of each sample was dissolved in 2% (*w/v*) sterile buffered peptone water (Biolab, Lawrenceville, GA, USA). Aliquot of 5 µL of each sample (mixture) was transferred into 10 mL buffered peptone water. Ten-fold serial dilutions were prepared and aliquots of 100 µL from each dilution was spread-plated on MacConkey agar supplemented with crystal violet and salt (Biolab, Lawrenceville, GA, USA) using a standard procedure. The plates were incubated aerobically at 37 °C for 24 h. The colonies depicting different colours (pale, pink or red) were selected and purified by streaking on MacConkey agar and the plates were at 37 °C for 24 h. Pure colonies were preserved in 20% glycerol and the stock cultures were stored at −80 °C for future use.

4.3. Culture-Based Methods for Identification of Carbapenem Resistance Enterobacteriaceae Colonies

A total of 280 isolates were screened for carbapenem resistance using Kirby–Bauer disc diffusion method [46], the isolates were revived on MacConkey agar and the plates were at 37 °C for 24 h. After incubation, one colony was transferred into 50 mL falcon tubes containing 10 mL nutrient broth. The tubes were incubated at 37 °C for 20 h. The turbidity of the cultures was adjusted to 1×10^8 CFU/mL (equivalent to 0.5 McFarlan standard) using Thermo Spectronic (Model, Helios Epsilon) [Thermo-Fisher Scientific, Waltham, MA, USA]. Aliquot of 100 µL of the culture was spread on Muellerf–Hinton agar (Biolab, Lawrenceville, GA, USA). Four carbapenem antibiotic discs: imipenem (IPM, 10 µg), ertapenem (ETP, 10 µg), meropenem (MEM, 10 µg and doripenem (DOR, 10 µg) (Mast Diagnostics, Randburg, South Africa) were placed on inoculated plates. The plates were incubated at 37 °C for 24 h. The results were interpreted using both Clinical Laboratory Standards Institute and European Committee on Antimicrobial Susceptibility Testing guidelines [20,47]. Colonies of all the isolates showing intermediate or resistant phenotypes to one of the three tested antibiotics were further subjected to Modified Hodge Test to confirm their ability to produce carbapenemase [48]. Carbapenemase-negative *E. coli* (ATCC 25922) and Carbapenemase-positive *K. pneumoniae* (ATCC BAA-1705) were used as quality control.

4.4. Phenotypic Screening for Identification of ESBL-Producing Enterobacteriaceae

The isolates that showed intermediate or resistant phenotypes to at least one of the carbapenem antibiotics were screened for the presence of extended spectrum beta-lactamase (ESBL) traits. Briefly, the isolates were culture on chromogenic Brilliance™ ESBL agar (Thermo-Fisher Scientific, Waltham, MA, USA). *E. coli* (ATCC 25922) and *K. pneumoniae* (ATCC 700603) were used as the quality control organisms. Growth on Brilliance™ ESBL agar indicated the ability of the isolates to produce extended spectrum beta-lactamase, and the results were interpreted according to manufacturer's instructions.

4.5. Antimicrobial Susceptibility Test

All the CRE isolates revealing ESBL characteristics on Brilliance™ ESBL agar were subjected to antibiotic sensitivity test to determine their antimicrobial resistance profile according to Kirby–Bauer disc diffusion method [46], following Clinical Laboratory Standard Institute guidelines [20]. The thirteen antibiotics used were: Amoxicillin (A, 25 µg), Amoxicillin-clavulanate (AMC, 30 µg), Aztreonam (ATM, 30 µg), Cefepime (CPM, 30 µg), Cefotaxime (CTX, 30 µg), Cefoxitin (FOX, 30 µg), Ceftazidime (CAZ, 30 µg), Cefuroxime (CXM, 30 µg), Ceftiofur (EFT, 30 µg), Cephalothin (KF, 30 µg), Ciprofloxacin (CIP, 5 µg), Piperacillin (PRL, 30 µg) and Ticarcillin (TC, 75 µg) (Mast Diagnostics, Randburg, South Africa). *E. coli* (ATCC 25922) and *K. pneumoniae* (ATCC 700603) were used as the quality control organisms. The isolates were classified as sensitive, intermediate resistant or resistant based on standard reference values [20]. Any isolate revealing resistance to at least three or more different antibiotics tested was considered multidrug resistant. Multiple antibiotic resistance (MAR) index was determined as the ratio of the number of antibiotics to which CRE isolate showed resistance to the number of antibiotics to which the isolate was exposed [30], using the following formula:

$$MARI = X/Y \tag{1}$$

where 'X' is the number of antimicrobial agents which bacteria revealed resistance while 'Y' is the total number of antimicrobial agents tested.

4.6. DNA Extraction from CRE Isolates

Genomic DNA was extracted from all CRE/ESBL-producing isolates using the Zymo Research Genomic DNA™–Tissue MiniPrep Kit (Zymo Research Corp, Irvine, CA, USA) according to the manufacturer's instructions. The quality and quantity of DNA was determined using Nanodrop™-Lite spectrophotometer (Thermo Scientific, Walton, MA, USA). Pure DNA samples were stored at −20 °C for future use.

4.7. Genus-Specific Identification of CRE Isolates

Bacterial universal primer was used to amplify bacterial 16S rRNA gene fragments from the genomic DNA. The identities of the isolates were confirmed by genus-specific PCR, targeting the *uidA*, *ntrA*, *tuf* and *invA* genes specific for *E. coli*, *K. pneumoniae*, *P. mirabilis* and *Salmonella* species, respectively. Details of oligonucleotide primer sequence and PCR conditions are listed in Table 3. PCR reactions were prepared in standard 25 µL (comprising 12.5 µL 2 × DreamTaq Green Master Mix, 0.25 µL of each oligonucleotides primer, 11 µL RNase-DNase free water and 1 µL DNA template). A no DNA template (nuclease-free water) reaction tube served as a negative control while a DNA sample from *E. coli* (ATCC 25922), *K. pneumoniae* (ATCC 700603), *S. enterica* (ATCC 14028 and 12325) were used as quality control organisms. All the PCR reagents were New England Biolabs (Ipswich, MA, USA, supplied by Inqaba, Pretoria, South Africa) products supplied by Inqaba Biotechnical Industry Ltd., Pretoria, South Africa. Amplifications were performed using a DNA thermal cycler (model- Bio-Rad C1000 Touch™ Thermal Cycler). All PCR products were resolved by agarose gel electrophoresis and the rest were stored at 4 °C.

4.8. Detection of Carbapenemase-Encoding Genes Using Multiplex PCR

All confirmed CRE isolates were subjected to Multiplex PCR analysis for detection of carbapenemase resistance genes (*blaNDM*, *blaKPC*, *blaVIM*, *blaOXA-23*, *blaOXA-48* and *blaGES*). The oligonucleotide primer sequence and PCR conditions are listed in Table 4. PCR reactions were prepared in standard 25 µL (comprising of 12.5µL 2 × DreamTaq Green Master Mix, 0.25 µL of each oligonucleotides primer, 11 µL RNase-DNase free water and 1 µL DNA template). A non-DNA template (nuclease-free water) reaction tube served as a negative control. DNA sample from *K. pneumoniae* (ATCC BAA-1705), *S. enterica* (ATCC 14028 and 12325) was used as quality control. All the PCR reagents were New England Biolabs (Ipswitch, MA, USA supplied by Inqaba, Pretoria, South Africa) products supplied by Inqaba Biotechnical Industry Ltd., Pretoria, South Africa. Amplifications were performed using a DNA thermal cycler (model-Bio-Rad C1000 Touch™ Thermal Cycler). All PCR products were resolved by agarose gel electrophoresis and the rest were stored at 4 °C.

Table 3. List of oligonucleotide primer sequences and PCR conditions used in this study.

Primers	Oligonucleotide Sequence (5′–3′)	Genes	Amplicon Size (bp)	Annealing Tm (°C)	References
16S rRNA					
27F	AGAGTTTGATCATGGCTCAG	16S rRNA	1420	55	[49]
1492R	GGTACCTTGTTACGACTT				
Genus Specific Genes					
ntrA-F	CATCTCGATCTGCTGGCCAA	*ntrA*	90	52	[50]
ntrA-R	GCGCGGATCCAGCGATTGGA				
uidA-F	CTGGTATCAGCGCGAAGTC	*uidA*	556	52	
uidA-R	AGCGGGTAGATATCACACTC				
Tuf-F	TCTACTTCACACGTAG	*tuf*	240	58	[51]
Tuf-R	TTCTAACAGCTCTTCA				
invA-F	GTGAAATTATCGCCACGTGGCAA	*invA*	284	64	[52]
invA-R	TCATCGCACCGTCAAAGGAACC				

Table 4. List of oligonucleotide primer sequences and PCR conditions used in this study.

Primers	Oligonucleotide Sequence (5′–3′)	Genes	Amplicon Size (bp)	Annealing Tm (°C)	References
CRE Genes					
KPC-F	CGTCTAGTTCTGCTGTCTTG	bla_{KPC}	798	52	[53]
KPC-R	CTTGTCATCCTTGTTAGGCG				
NDM-F	GGTTTGGCGATCTGGTTTTC	bla_{NDM}	621		
NDM-R	CGGAATGGCTCATCACGATC				
OXA-23-F	ATGAGTTATCTATTTTTGTC	$bla_{OXA\text{-}23}$	501		
OXA-23-R	TGTCAAGCTCTTAAATAATA				
GES-C-F	GTTTTGCAATGTGCTCAACG	bla_{GES}	371		
GES-D-R	TGCCATAGCAATAGGCGTAG				
VIM-F	GATGGTGTTTGGTCGCATA	bla_{VIM}	390		
VIM-R	CGAATGCGCAGCACCAG				
OXA-48-F	TTCGGCCACGGAGCAAATCAG	$bla_{OXA\text{-}48}$	438		
OXA-48-R	GATGTGGGCATATCCATATTCATCGCA				
ESBL Genes					
blaTEM-F	AAACGCTGGTGAAAGTA	bla_{TEM}	822	45	[54]
blaTEM-R	AGCGATCTGTCTAT				
blaSHV-F	ATGCGTTATATCGCCTGTG	bla_{SHV}	753		
blaSHV-R	TGCTTTGTTATTCGGGCCAA				
blaCTX-M-F	CGCTTTGCGATGTGCAG	$bla_{CTX\text{-}M}$	550		
blaCTX-M-R	ACCGCGATATCGTTGGT				
blaOXA-F	ATATCTCTACTGTTGCATCTCC	bla_{OXA}	619		
blaOXA-R	AAACCCTTCAAACCATCC				

4.9. Detection of Extended Spectrum Beta-Lactamase-Encoding Genes in CRE

The CRE isolates were further screened for the presence of the major ESBL genes (*blaCTX*-M, *blaOXA*, *blaSHV* and *blaTEM*). Details of oligonucleotide primer sequences and PCR conditions are listed in Table 4. PCR reactions were prepared in standard 25 μL (comprising of 12.5 μL 2 × DreamTaq Green Master Mix, 0.25 μL of each oligonucleotides primer, 11 μL RNase-DNase free water and 1 μL DNA template). A no DNA template (nuclease-free water) reaction tube served as a negative control. All the PCR reagents were New England Biolabs (Ipswitch, MA, USA supplied by Inqaba, Pretoria, South Africa) products supplied by Inqaba Biotechnical Industry Ltd., Pretoria, South Africa. Amplifications were performed using a DNA thermal cycler (model: Bio-Rad C1000 Touch™ Thermal Cycler). All PCR products were resolved by agarose gel electrophoresis and the rest were stored at 4 °C.

4.10. Agarose Gel Electrophoresis and Visualization

A 2% (*w/v*) agarose gel containing 0.1 μg/mL Ethidium bromide was used to separate all PCR products by electrophoresis. A 1 Kb or 100 bp DNA molecular weight marker (Fermentas, Foster City, CA, USA) was included in each gel. A horizontal Pharmacia Biotech equipment system (Model Hoefer HE 99X, Amersham Pharmacia Biotech, Stockholm, Sweden) was used to perform the electrophoresis, each cycle ran for 1 h at 80V. Visualization and image capturing was performed using a ChemiDoc Imaging System (Bio-Rad ChemiDoc™ MP Imaging System, Hercules, CA, USA).

4.11. Statistical Analysis

The data generated in this study was entered into an Excel spreadsheet. Descriptive analysis was conducted using SAS, 2010 (v 9.4, SAS Institute, Cary, NC, USA). The proportions of positive for *E. coli* O177 serotype, antimicrobial resistance and virulence genes across the farms were determined. The analysed data were used to draw tables and bar graphs.

5. Conclusions

To the best of our knowledge, this is the first study reporting the occurrence of CR- and ESBL-producing *Enterobacteriaceae* in South African beef cattle. The results contained herein revealed high MDR phenotypes among CR- and ESBL-producing *Enterobacteriaceae*. Furthermore, the most frequently detected species were *E. coli* and *K. pneumoniae*. Notably, high detection of carbapenemase and/or ESBL resistance determinants in CRE was alarming given the significant importance of carbapenem antibiotics in public health. Therefore, stringent hygiene measures must be enforced to mitigate the spread and transmission of CRE- and ESBL-producing pathogens in beef farms.

Supplementary Materials: The following are available online at http://www.mdpi.com/2079-6382/9/11/820/s1, Table S1: Multidrug resistance phenotypes and MAR index of CRE isolates.

Author Contributions: Conceptualization, C.N.A. and M.M.; methodology, L.T.; software, P.K.M.; validation, L.T., M.C.M. and P.K.M.; formal analysis, L.T.; investigation, L.T.; resources, C.N.A. and M.M.; data curation L.T., M.C.M. and P.K.M.; writing—original draft preparation, L.T., M.C.M. and P.K.M.; writing—review and editing, C.N.A. and M.M.; visualization, L.T., M.C.M. and P.K.M.; supervision, C.N.A. and M.M.; project administration, C.N.A. and M.M.; funding acquisition, C.N.A. and M.M. All authors have read and agreed to the published version of the manuscript.

Funding: This research was funded by HWSETA awarded to L.T. This work was also supported in part by University of Mpumalanga and North West University, South Africa.

Conflicts of Interest: The authors declare no conflict of interest.

References

1. Pitout, J.D.D.; Nordmann, P.; Poirel, L. Carbapenemase-Producing Klebsiella pneumoniae, a Key Pathogen Set for Global Nosocomial Dominance. *Antimicrob. Agents Chemother.* **2015**, *59*, 5873–5884. [CrossRef] [PubMed]
2. Suay-Garcia, B.; Pérez-Gracia, M.T. Present and Future of Carbapenem-resistant Enterobacteriaceae (CRE) Infections. *Antibiotics* **2019**, *8*, 122. [CrossRef] [PubMed]
3. Bartsch, S.; McKinnell, J.; Mueller, L.; Miller, L.; Gohil, S.; Huang, S.; Lee, B.Y. Potential economic burden of carbapenem-resistant Enterobacteriaceae (CRE) in the United States. *Clin. Microbiol. Infect.* **2017**, *23*, 48.e9–48.e16. [CrossRef] [PubMed]
4. Antibiotic Resistance Threats in the United States. 2013. Available online: https://www.cdc.gov/drugresistance/pdf/ar-threats-2013-508.pdf (accessed on 28 September 2020).
5. Ballot, D.E.; Bandini, R.; Nana, T.; Bosman, N.; Thomas, T.; Davies, V.A.; Cooper, P.A.; Mer, M.; Lipman, J. A review of -multidrug-resistant Enterobacteriaceae in a neonatal unit in Johannesburg, South Africa. *BMC Pediatr.* **2019**, *19*, 320. [CrossRef]
6. Ramsamy, Y.; Mlisana, K.; Allam, M.; Amoako, D.G.; Abia, A.L.K.; Ismail, A.; Singh, R.; Kisten, T.; Han, K.S.S.; Muckart, D.J.J.; et al. Genomic Analysis of Carbapenemase-Producing Extensively Drug-Resistant Klebsiella pneumoniae Isolates Reveals the Horizontal Spread of p18-43_01 Plasmid Encoding blaNDM-1 in South Africa. *Microorganisms* **2020**, *8*, 137. [CrossRef]
7. Boutal, H.; Vogel, A.; Bernabeu, S.; Devilliers, K.; Creton, E.; Cotellon, G.; Plaisance, M.; Oueslati, S.; Dortet, L.; Jousset, A.; et al. A multiplex lateral flow immunoassay for the rapid identification of NDM-, KPC-, IMP- and VIM-type and OXA-48-like carbapenemase-producing Enterobacteriaceae. *J. Antimicrob. Chemother.* **2018**, *73*, 909–915. [CrossRef]
8. Fernández, J.; Guerra, B.; Rodicio, M.R. Resistance to Carbapenems in Non-Typhoidal Salmonella enterica Serovars from Humans, Animals and Food. *Vet. Sci.* **2018**, *5*, 40. [CrossRef]
9. Solgi, H.; Nematzadeh, S.; Giske, C.G.; Badmasti, F.; Westerlund, F.; Lin, Y.-L.; Goyal, G.; Nikbin, V.S.; Nemati, A.H.; Shahcheraghi, F. Molecular Epidemiology of OXA-48 and NDM-1 Producing Enterobacterales Species at a University Hospital in Tehran, Iran, Between 2015 and 2016. *Front. Microbiol.* **2020**, *11*, 936. [CrossRef]
10. Teixeira, P.; Tacão, M.; Pureza, L.; Gonçalves, J.; Silva, A.; Cruz-Schneider, M.P.; Henriques, I. Occurrence of carbapenemase-producing Enterobacteriaceae in a Portuguese river: blaNDM, blaKPC and blaGES among the detected genes. *Environ. Pollut.* **2020**, *260*, 113913. [CrossRef]
11. Li, M.; Guo, M.; Chen, L.; Zhu, C.; Xiao, Y.; Li, P.; Guo, H.; Chen, L.; Zhang, W.; Du, H. Isolation and Characterization of Novel Lytic Bacteriophages Infecting Epidemic Carbapenem-Resistant Klebsiella pneumoniae Strains. *Front. Microbiol.* **2020**, *11*, 1554. [CrossRef]
12. Shrivastava, S.R.; Shrivastava, P.S.; Ramasamy, J. World health organization releases global priority list of antibiotic-resistant bacteria to guide research, discovery, and development of new antibiotics. *J. Med Soc.* **2018**, *32*, 76. [CrossRef]
13. Zhen, X.; Lundborg, C.S.; Sun, X.; Hu, X.; Dong, H. Economic burden of antibiotic resistance in ESKAPE organisms: A systematic review. *Antimicrob. Resist. Infect. Control.* **2019**, *8*, 1–23. [CrossRef] [PubMed]
14. Han, R.; Shi, Q.; Wu, S.; Yin, D.; Peng, M.; Dong, D.; Zheng, Y.; Guo, Y.; Zhang, R.; Hu, F.; et al. Dissemination of Carbapenemases (KPC, NDM, OXA-48, IMP, and VIM) Among Carbapenem-Resistant Enterobacteriaceae Isolated From Adult and Children Patients in China. *Front. Cell. Infect. Microbiol.* **2020**, *10*, 314. [CrossRef] [PubMed]
15. Gazin, M.; Paasch, F.; Goossens, H.; Malhotra-Kumar, S. Current Trends in Culture-Based and Molecular Detection of Extended-Spectrum- -Lactamase-Harboring and Carbapenem-Resistant Enterobacteriaceae. *J. Clin. Microbiol.* **2012**, *50*, 1140–1146. [CrossRef]
16. Ben Said, L.; Jouini, A.; Klibi, N.; Dziri, R.; Alonso, C.A.; Boudabous, A.; Ben Slama, K.; Torres, C. Detection of extended-spectrum beta-lactamase (ESBL)-producing Enterobacteriaceae in vegetables, soil and water of the farm environment in Tunisia. *Int. J. Food Microbiol.* **2015**, *203*, 86–92. [CrossRef]

17. Köck, R.; Daniels-Haardt, I.; Becker, K.; Mellmann, A.; Friedrich, A.W.; Mevius, D.; Schwarz, S.; Jurke, A. Carbapenem-resistant Enterobacteriaceae in wildlife, food-producing, and companion animals: A systematic review. *Clin. Microbiol. Infect.* **2018**, *24*, 1241–1250. [CrossRef]

18. Zhang, W.; Zhu, Y.; Wang, C.; Liu, W.; Li, R.; Chen, F.; Luan, T.; Zhang, Y.; Schwarz, S.; Liu, S. Characterization of a Multidrug-Resistant Porcine Klebsiella pneumoniae Sequence Type 11 Strain Coharboring blaKPC-2 and fosA3 on Two Novel Hybrid Plasmids. *mSphere* **2019**, *4*, 00590-19. [CrossRef]

19. Gondal, A.J.; Saleem, S.; Jahan, S.; Choudhry, N.; Yasmin, N. Novel Carbapenem-Resistant Klebsiella pneumoniae ST147 Coharboring blaNDM-1, blaOXA-48 and Extended-Spectrum β-Lactamases from Pakistan. *Infect. Drug Resist.* **2020**, *13*, 2105–2115. [CrossRef]

20. CLSI. *Performance Standards for Antimicrobial Susceptibility Testing*, 28th ed.; Supplement M100; Clinical and Laboratory Standards Institute: Wayne, PA, USA, 2018.

21. Kanj, S.S.; Kanafani, Z.A. Current Concepts in Antimicrobial Therapy Against Resistant Gram-Negative Organisms: Extended-Spectrum β-Lactamase–Producing Enterobacteriaceae, Carbapenem-Resistant Enterobacteriaceae, and Multidrug-Resistant Pseudomonas aeruginosa. *Mayo Clin. Proc* **2011**, *86*, 250–259. [CrossRef]

22. Potter, R.F.; D'Souza, A.W.; Dantas, G. The rapid spread of carbapenem-resistant Enterobacteriaceae. *Drug Resist. Updates* **2016**, *29*, 30–46. [CrossRef]

23. Mills, M.C.; Lee, J. The threat of carbapenem-resistant bacteria in the environment: Evidence of widespread contamination of reservoirs at a global scale. *Environ. Pollut.* **2019**, *255*, 113143. [CrossRef] [PubMed]

24. Ateba, C.N.; Bezuidenhout, C. Characterisation of Escherichia coli O157 strains from humans, cattle and pigs in the North-West Province, South Africa. *Int. J. Food Microbiol.* **2008**, *128*, 181–188. [CrossRef] [PubMed]

25. Dlamini, B.S.; Montso, P.K.; Kumar, A.; Ateba, C.N. Distribution of virulence factors, determinants of antibiotic resistance and molecular fingerprinting of Salmonella species isolated from cattle and beef samples: Suggestive evidence of animal-to-meat contamination. *Environ. Sci. Pollut. Res.* **2018**, *25*, 32694–32708. [CrossRef] [PubMed]

26. Montso, K.P.; Dlamini, S.B.; Kumar, A.; Ateba, C.N. Antimicrobial Resistance Factors of Extended-Spectrum Beta-Lactamases Producing Escherichia coli and Klebsiella pneumoniae Isolated from Cattle Farms and Raw Beef in North-West Province, South Africa. *BioMed Res. Int.* **2019**, *2019*, 1–13. [CrossRef] [PubMed]

27. Richter, L.; Du Plessis, E.M.; Duvenage, S.; Korsten, L. Occurrence, Identification, and Antimicrobial Resistance Profiles of Extended-Spectrum and AmpC β-Lactamase-Producing Enterobacteriaceae from Fresh Vegetables Retailed in Gauteng Province, South Africa. *Foodborne Pathog. Dis.* **2019**, *16*, 421–427. [CrossRef]

28. Magiorakos, A.-P.; Srinivasan, A.; Carey, R.; Carmeli, Y.; Falagas, M.; Giske, C.; Harbarth, S.J.; Hindler, J.; Kahlmeter, G.; Olsson-Liljequist, B.; et al. Multidrug-resistant, extensively drug-resistant and pandrug-resistant bacteria: An international expert proposal for interim standard definitions for acquired resistance. *Clin. Microbiol. Infect.* **2012**, *18*, 268–281. [CrossRef]

29. Fadare, F.T.; Adefisoye, M.A.; Okoh, A.I. Occurrence, identification and antibiogram signatures of selected Enterobacteriaceae from Tsomo and Tyhume rivers in the Eastern Cape Province, Republic of South Africa. *bioRxiv* **2020**, 2–46. [CrossRef]

30. Krumperman, P.H. Multiple antibiotic resistance indexing of Escherichia coli to identify high-risk sources of faecal contamination of foods. *Appl. Environ. Microbiol.* **1983**, *46*, 165–170. [CrossRef]

31. Ahmed, H.A.; El Bayomi, R.M.; Hussein, M.A.; Khedr, M.H.; Remela, E.M.A.; El-Ashram, A.M. Molecular characterization, antibiotic resistance pattern and biofilm formation of Vibrio parahaemolyticus and V. cholerae isolated from crustaceans and humans. *Int. J. Food Microbiol.* **2018**, *274*, 31–37. [CrossRef]

32. Sheu, C.-C.; Chang, Y.-T.; Lin, S.-Y.; Chen, Y.-H.; Hsueh, P.-R. Infections Caused by Carbapenem-Resistant Enterobacteriaceae: An Update on Therapeutic Options. *Front. Microbiol.* **2019**, *10*, 80. [CrossRef]

33. Liu, B.-T.; Zhang, X.-Y.; Wan, S.-W.; Hao, J.-J.; Jiang, R.-D.; Song, F.-J. Characteristics of Carbapenem-Resistant Enterobacteriaceae in Ready-to-Eat Vegetables in China. *Front. Microbiol.* **2018**, *9*, 1147. [CrossRef]

34. Perovic, O.; Germs-Sa, F.; Ismail, H.; Quan, V.; Bamford, C.; Nana, T.; Chibabhai, V.; Bhola, P.; Ramjathan, P.; Swe-Han, K.S.; et al. Carbapenem-resistant Enterobacteriaceae in patients with bacteraemia at tertiary hospitals in South Africa, 2015 to 2018. *Eur. J. Clin. Microbiol. Infect. Dis.* **2020**, *39*, 1287–1294. [CrossRef]

35. Grundmann, H.; Glasner, C.; Albiger, B.; Aanensen, D.M.; Tomlinson, C.T.; Andrasević, A.T.; Cantón, R.; Carmeli, Y.; Friedrich, A.W.; Giske, C.G.; et al. Occurrence of carbapenemase-producing Klebsiella pneumoniae and Escherichia coli in the European survey of carbapenemase-producing Enterobacteriaceae (EuSCAPE): A prospective, multinational study. *Lancet Infect. Dis.* **2017**, *17*, 153–163. [CrossRef]

36. Bleichenbacher, S.; Stevens, M.J.; Zurfluh, K.; Perreten, V.; Endimiani, A.; Stephan, R.; Nüesch-Inderbinen, M. Environmental dissemination of carbapenemase-producing Enterobacteriaceae in rivers in Switzerland. *Environ. Pollut.* **2020**, *265*, 115081. [CrossRef] [PubMed]

37. Braun, S.D.; Ahmed, M.F.E.; El-Adawy, H.; Hotzel, H.; Engelmann, I.; Weiß, D.; Monecke, S.; Ehricht, R. Surveillance of Extended-Spectrum Beta-Lactamase-Producing Escherichia coli in Dairy Cattle Farms in the Nile Delta, Egypt. *Front. Microbiol.* **2016**, *7*, 1020. [CrossRef] [PubMed]

38. Bradford, P.A. Extended-Spectrum β-Lactamases in the 21st Century: Characterization, Epidemiology, and Detection of This Important Resistance Threat. *Clin. Microbiol. Rev.* **2001**, *14*, 933–951. [CrossRef] [PubMed]

39. Peirano, G.; Pitout, J.D. Extended-Spectrum β-Lactamase-Producing Enterobacteriaceae: Update on Molecular Epidemiology and Treatment Options. *Drugs* **2019**, *79*, 1529–1541. [CrossRef]

40. Ahmed, A.M.; Shimamoto, T. Molecular analysis of multidrug resistance in Shiga toxin-producing Escherichia coli O157:H7 isolated from meat and dairy products. *Int. J. Food Microbiol.* **2015**, *193*, 68–73. [CrossRef]

41. Das, L.; Borah, P.; Sharma, R.; Malakar, D.; Saikia, G.K.; Tamuly, S.; Dutta, R.; Sharma, K. Phenotypic and molecular characterization of extended spectrum β-lactamase producing Escherichia coli and Klebsiella pneumoniae isolates from various samples of animal origin from Assam, India. *bioRxiv* **2020**, 1–22. [CrossRef]

42. Geser, N.; Stephan, R.; Hächler, H. Occurrence and characteristics of extended-spectrum β-lactamase (ESBL) producing Enterobacteriaceae in food producing animals, minced meat and raw milk. *BMC Vet. Res.* **2012**, *8*, 21. [CrossRef]

43. Adator, E.H.; Narvaez-Bravo, C.; Zaheer, R.; Cook, S.R.; Tymensen, L.; Hannon, S.J.; Booker, C.W.; Church, D.L.; Read, R.R.; McAllister, T.A. A One Health Comparative Assessment of Antimicrobial Resistance in Generic and Extended-Spectrum Cephalosporin-Resistant *Escherichia coli* from Beef Production, Sewage and Clinical Settings. *Microorganisms* **2020**, *8*, 885. [CrossRef] [PubMed]

44. Vasaikar, S.D.; Obi, L.; Morobe, I.; Bisi-Johnson, M. Molecular Characteristics and Antibiotic Resistance Profiles ofKlebsiellaIsolates in Mthatha, Eastern Cape Province, South Africa. *Int. J. Microbiol.* **2017**, *2017*, 1–7. [CrossRef] [PubMed]

45. Richter, L.; Du Plessis, E.M.; Duvenage, S.; Korsten, L. Occurrence, Phenotypic and Molecular Characterization of Extended-Spectrum- and AmpC- β-Lactamase Producing Enterobacteriaceae Isolated From Selected Commercial Spinach Supply Chains in South Africa. *Front. Microbiol.* **2020**, *11*, 638. [CrossRef] [PubMed]

46. Bauer, M.A.W.; Kirby, M.W.M.M.; Sherris, M.J.C.; Turck, M.M. Antibiotic Susceptibility Testing by a Standardized Single Disk Method. *Am. J. Clin. Pathol.* **1966**, *45*, 493–496. [CrossRef]

47. EUCAST Guidelines for Detection of Resistance Mechanisms and Specific Resistance of Clinical and/or Epidemiological Importance. Available online: https://eucast.org/resistance_mechanisms/ (accessed on 13 November 2020).

48. Amjad, A.; Mirza, I.; Abbasi, S.; Farwa, U.; Malik, N.; Zia, F. Modified Hodge test: A simple and effective test for detection of carbapenemase production. *Iran. J. Microbiol.* **2011**, *3*, 189–193.

49. Korzeniewska, E.; Harnisz, M. Beta-lactamase-producing Enterobacteriaceae in hospital effluents. *J. Environ. Manag.* **2013**, *123*, 1–7. [CrossRef]

50. Anbazhagan, D.; Mui, W.S.; Mansor, M.; Yan, G.O.S.; Yusof, M.Y.; Sekaran, S.D. Development of conventional and real-time multiplex PCR assays for the detection of nosocomial pathogens. *Braz. J. Microbiol.* **2011**, *42*, 448–458. [CrossRef]

51. Anbazhagan, D.; Kathirvalu, G.G.; Mansor, M.; Yan, G.O.S.; Yusof, M.Y.; Sekaran, S.D. Multiplex polymerase chain reaction (PCR) assays for the detection of Enterobacteriaceae in clinical samples. *Afr. J. Microbiol. Res.* **2010**, *4*, 1186–1191.

52. Salehi, T.Z.; Mahzounieh, M.; Saeedzadeh, A. Detection of InvA Gene in Isolated Salmonella from Broilers by PCR Method. *Int. J. Poult. Sci.* **2005**, *4*, 557–559. [CrossRef]

53. Poirel, L.; Walsh, T.R.; Cuvillier, V.; Nordmann, P. Multiplex PCR for detection of acquired carbapenemase genes. *Diagn. Microbiol. Infect. Dis.* **2011**, *70*, 119–123. [CrossRef]

54. Al-Mayahie, S.M.G. Phenotypic and genotypic comparison of ESBL production by Vaginal *Escherichia coli* isolates from pregnant and non-pregnant women. *Ann. Clin. Microbiol. Antimicrob.* **2013**, *12*, 1–7. [CrossRef] [PubMed]

Publisher's Note: MDPI stays neutral with regard to jurisdictional claims in published maps and institutional affiliations.

 *antibiotics*

Article

Comparison of Six Phenotypic Assays with Reference Methods for Assessing Colistin Resistance in Clinical Isolates of Carbapenemase-Producing Enterobacterales: Challenges and Opportunities

Annamária Főldes [1,2,*], Edit Székely [3,4], Septimiu Toader Voidăzan [5] and Minodora Dobreanu [6,7]

1   Department of Microbiology, Laboratory of Medical Analysis, "Dr. Constantin Opriș" County Emergency Hospital, 430031 Baia Mare, Romania
2   Doctoral School of Medicine and Pharmacy, "George Emil Palade" University of Medicine, Pharmacy, Science and Technology, 540142 Targu Mures, Romania
3   Department of Microbiology, Central Clinical Laboratory, County Emergency Clinical Hospital, 540136 Targu Mures, Romania; edit.szekely@umfst.ro
4   Department of Microbiology, "George Emil Palade" University of Medicine, Pharmacy, Science and Technology, 540142 Targu Mures, Romania
5   Department of Epidemiology, "George Emil Palade" University of Medicine, Pharmacy, Science and Technology, 540142 Targu Mures, Romania; septimiu.voidazan@umfst.ro
6   Department of Clinical Biochemistry, Central Clinical Laboratory, County Emergency Clinical Hospital, 540136 Targu Mures, Romania; minodora.dobreanu@umfst.ro
7   Department of Laboratory Medicine, "George Emil Palade" University of Medicine, Pharmacy, Science and Technology, 540142 Targu Mures, Romania
*   Correspondence: anafoldes@gmail.com

Citation: Főldes, A.; Székely, E.; Voidăzan, S.T.; Dobreanu, M. Comparison of Six Phenotypic Assays with Reference Methods for Assessing Colistin Resistance in Clinical Isolates of Carbapenemase-Producing Enterobacterales: Challenges and Opportunities. *Antibiotics* **2022**, *11*, 377. https://doi.org/10.3390/antibiotics11030377

Academic Editor: Francesca Andreoni

Received: 12 February 2022
Accepted: 9 March 2022
Published: 11 March 2022

Publisher's Note: MDPI stays neutral with regard to jurisdictional claims in published maps and institutional affiliations.

Abstract: The global escalation of severe infections due to carbapenemase-producing Enterobacterales (CPE) isolates has prompted increased usage of parenteral colistin. Considering the reported difficulties in assessing their susceptibility to colistin, the purpose of the study was to perform a comparative evaluation of six phenotypic assays—the colistin broth disc elution (CBDE), Vitek 2 Compact (bioMérieux SA, Marcy l'Etoile, France), the Micronaut MIC-Strip Colistin (Merlin Diagnostika GMBH, Bornheim-Hensel, Germany), the gradient diffusion strip Etest (bioMérieux SA, Marcy l'Etoile, France), ChromID Colistin R Agar (COLR) (bioMérieux SA, Marcy l'Etoile, France), and the Rapid Polymyxin NP Test (ELITechGroup, Signes, France)—versus the reference method of broth microdilution (BMD). All false resistance results were further assessed using population analysis profiling (PAP). Ninety-two nonrepetitive clinical CPE strains collected from two hospitals were evaluated. The BMD confirmed 36 (39.13%) isolates susceptible to colistin. According to the BMD, the Micronaut MIC-Strip Colistin, the CBDE, and the COLR medium exhibited category agreement (CA) of 100%. In comparison with the BMD, the highest very major discrepancy (VMD) was noted for Etest ($n = 15$), and the only false resistance results were recorded for the Rapid Polymyxin NP Test ($n = 3$). Only the PAP method and the Rapid Polymyxin NP Test were able to detect heteroresistant isolates ($n = 2$). Thus, there is an urgent need to further optimize the diagnosis strategies for colistin resistance.

Keywords: carbapenemase-producing Enterobacterales; colistin susceptibility testing; broth microdilution; colistin broth disc elution; Vitek 2 compact; rapid polymyxin NP test; Etest; ChromID colistin R agar; micronaut MIC-strip colistin; population analysis profiling

1. Introduction

The emergence and spread of diverse types of carbapenemase producers belonging to the Enterobacterales order (CPE) has been increasingly reported worldwide in recent years [1–4]. These pathogens are involved in different types of human infections, in both community and hospital settings, and frequently coexpress resistance to several classes of antibiotics that are critical in therapy [1,4,5].

Klebsiella pneumoniae is the most common globally mentioned CPE isolate and is mainly associated with nosocomial infections and has devastating effects on patient outcomes [1,3,6,7].

The global fight against the threat of antimicrobial resistance is the best strategy for preventing infections [8]. The potential impact of the current coronavirus disease (COVID-19) pandemic on antimicrobial resistance has not yet been established but was reported to possibly cause an escalation whose severity varies between geographic regions, different hospitals, and even distinct units of the same medical institution [9–11].

The management of infections due to multidrug-resistant (MDR) and extensively drug-resistant (XDR) CPE strains poses a significant challenge for health systems [7,12]. Colistin, tigecycline, aminoglycosides, and fosfomycin are considered second-line antimicrobials, which frequently express in vitro activity against some CPE isolates, but there are concerns regarding their current efficacy, the development of potential toxicity, and the rapid dissemination of resistance [4,7].

Colistin, also known as polymyxin E, is a cationic lipopeptide and bactericidal agent discovered more than half a century ago [13–15], but its systemic administration has largely been fallen out of favor over a substantial period of time, mainly because of its nephro- and neurotoxicity [16]. It recently regained major worldwide clinical importance as a last resort therapy for some severe infections due to MDR and XDR Gram-negative bacilli [17,18].

The complex mechanisms of action of colistin have not been entirely decoded [14,19]. The initial target is the lipopolysaccharide (LPS) of the outer membrane of Gram-negative bacteria, displacing the divalent cations Ca^{2+} and Mg^{2+} and leading to enhanced membrane permeability, loss of membrane integrity, and ultimately cell destruction [13,16,19]. Furthermore, colistin suppresses the endotoxin effect by neutralizing the lipid A component of the LPS and promotes bacterial cell injury through the development of reactive oxygen species and by suppressing enzymes whose functions are indispensable in the bacterial respiratory process [16,20].

Colistin sulfate and colistimethate sodium, which is a less toxic inactive prodrug, are the two commercially available forms of colistin for oral or topical administration and for the systemic route, respectively [13,16,19]. The optimal doses of colistimethate sodium are still a matter of debate [7,17], especially for critically ill patients receiving sustained low-efficiency dialysis and those with acute kidney injury [17].

Recent hospital outbreaks due to colistin-resistant CPE strains have occurred worldwide [15,21–26]. Acquired resistance to colistin is based mainly on diverse chromosomal mutations [14,19,27], but the extensive use of colistin both in human and veterinary health sectors has promoted the emergence and development of the plasmid-encoded mobile colistin resistance (*mcr*) genes starting from *mcr-1* up to *mcr-10*, with multiple variants [14–16,28,29].

Assessing colistin resistance continues to remain a challenge for microbiology laboratories [14,30,31]. Both the European Committee on Antimicrobial Susceptibility Testing (EUCAST) and the Clinical and Laboratory Standards Institute (CLSI) recommend the broth microdilution (BMD) as the reference method for susceptibility testing [32]. This standard technique is laborious, susceptible to errors, difficult to implement into routine practice, and presents limitations in detecting colistin heteroresistance; therefore, an alternative method with satisfactory performance is needed for routine testing [14,30]. Generally, the heteroresistance phenomenon has been described as the emergence of minor subpopulations with higher degrees of antimicrobial drug resistance within the dominant population of the same culture [33]. Population analysis profiling (PAP) is the gold standard method for the evaluation of heteroresistance to an antibiotic [33,34], but this technique is labor- and time-intensive [35], standard guidelines are lacking [33,36], and it requires a high consumption of materials. However, a limited number of studies have investigated the presence of colistin heteroresistance in CPE isolates.

In this context, the aim of this study was to perform a comparative evaluation of six phenotypic methods versus BMD, and all major discrepancies (MDs) (false resistance results) compared to the BMD were further assessed using the PAP method in order to obtain an improved algorithm for reliable and convenient determination of colistin-resistant CPE isolates in daily activity.

2. Results

2.1. Variety of Carbapenemases

The combination disc test was used for phenotypic confirmation of the following carbapenem-hydrolyzing enzymes: *K. pneumoniae* carbapenemase (KPC) ($n = 41$), oxacillinase-48-like (OXA-48-like) ($n = 32$), and metallo-β-lactamase (MBL) ($n = 19$). Of the 22 strains previously tested by polymerase chain reaction (PCR), 12 harbored $bla_{OXA-48-like}$, 6 bla_{KPC}, and 4 bla_{NDM} (New Delhi metallo-β-lactamase), in agreement with the combination disc test results.

2.2. Colistin Testing Results versus BMD

The BMD was used to confirm the colistin susceptibility of 36 (39.13%) strains and 56 (60.86%) isolates resistant to this antimicrobial agent. Distributions of colistin minimum inhibitory concentrations (MICs) obtained with the BMD ranged from 0.0625 to 64 mg/L (Table 1).

Table 1. Distributions of colistin MICs determined by BMD for all isolates.

Species	Carbapenemase Type	Colistin Reference MIC (mg/L)										
		0.0625	0.125	0.25	0.5	1	2	4	8	16	32	≥64
Klebsiella pneumoniae	KPC ($n = 41$)	0	3	4	1	0	0	2	12	6	10	3
	OXA-48-like ($n = 29$)	0	4	5	1	0	0	0	2	5	11	1
	MBL ($n = 8$)	0	0	5	0	0	0	0	0	1	2	0
Citrobacter freundii	MBL ($n = 6$)	0	3	1	1	1	0	0	0	0	0	0
Enterobacter cloacae complex	MBL ($n = 4$)	2	0	1	0	0	0	0	0	0	1	0
Escherichia coli	OXA-48-like ($n = 3$)	1	1	1	0	0	0	0	0	0	0	0
	MBL ($n = 1$)	0	1	0	0	0	0	0	0	0	0	0
Total	$n = 92$	3	12	17	3	1	0	2	14	12	24	4

Legend. MICs: minimum inhibitory concentrations; BMD: reference broth microdilution; KPC: *Klebsiella pneumoniae* carbapenemase; OXA-48-like: oxacillinase-48-like; MBL: metallo-β-lactamase. The red line corresponds to the EUCAST breakpoints (≤ 2 mg/L indicates susceptibility).

The 36 strains with MIC ≤ 0.5 mg/L by Vitek 2 Compact strongly correlated with the BMD results, whereas all isolates with MIC values between 1 and 2 mg/L by Vitek 2 Compact generated false susceptible results ($n = 8$) (Figure 1a).

A perfect linear correlation at the angle of 45° was achieved between the MIC of colistin determined by BMD and by colistin broth disc elution (CBDE) (Figure 1b).

The colistin MIC results obtained with the Micronaut MIC-Strip were strongly correlated with the reference MIC (Figure 1c), whereas a weak correlation was noted for strains determined to have an MIC ≥ 2 mg/L by the Etest method (Figure 1d). Distinct colonies within the ellipse of growth inhibition around the Etest strips were not observed, except in the case of the reference *Enterobacter cloacae* ATCC 13047.

The highest number of false negative results ($n = 15$) was documented for the Etest method, followed by Vitek 2 Compact ($n = 8$) (Table 2), and exclusively for *K. pneumoniae* isolates (Table 3).

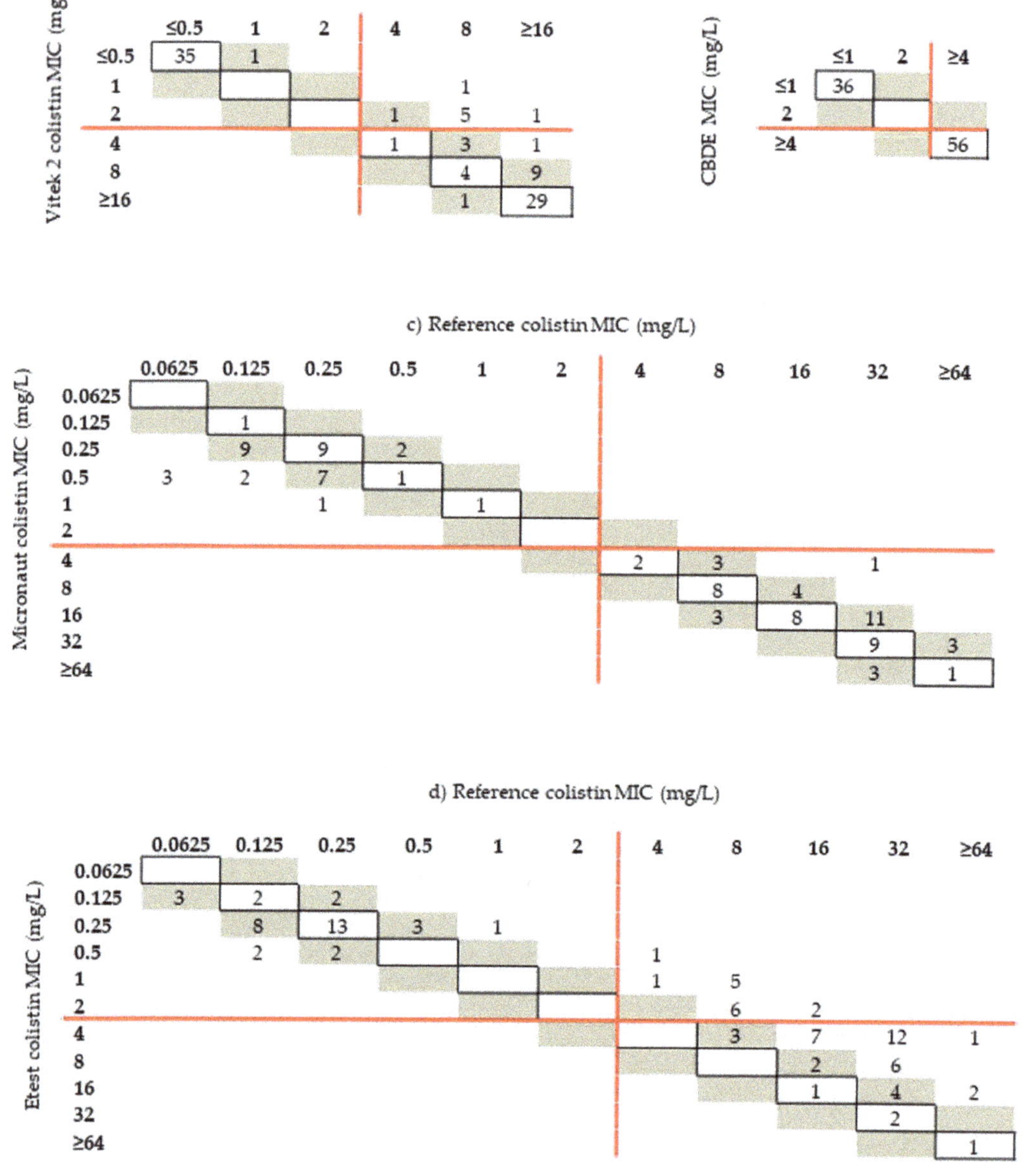

Figure 1. Scattergrams of correlation between reference broth microdilution (BMD) and Vitek 2 Compact (**a**), colistin broth disc elution (CBDE) (**b**), Micronaut MIC-Strip (**c**), and Etest (**d**) for all isolates. Minimum inhibitory concentrations (MICs) identical with those obtained by BMD and the essential agreement (EA) (MICs within ± 1 doubling dilution compared to the BMD) are mentioned as strain numbers within boxes and in shaded gray cells, respectively. The red lines represent the EUCAST breakpoints (≤2 mg/L indicates susceptibility).

Table 2. Performance features of Vitek 2 Compact, Micronaut MIC-Strip, Etest, COLR medium, Rapid Polymyxin NP Test, and CBDE of all isolates according to the BMD.

Parameter	Vitek 2 Compact	Micronaut MIC-Strip	Etest, MHE	COLR Medium	Rapid Polymyxin NP Test	CBDE
True positive (n)	48	56	41	55	56	56
False positive (n)	0	0	0	0	3	0
False negative (n)	8	0	15	0	0	0
True negative (n)	36	36	36	27	33	36
Total (n)	92	92	92	82	92	92
Sensitivity (%)	85.71	100	73.21	100	100	100
Specificity (%)	100	100	100	100	91.67	100
PPV (%)	100	100	100	100	94.92	100
NPV (%)	81.82	100	70.59	100	100	100
EA (%)	91.30	92.39	50	NA	NA	NA
CA (%)	91.30	100	83.69	100	96.73	100
VMD (%)	14.28	0	26.78	0	0	0
MD (%)	0	0	0	0	8.33	0

Legend. BMD: reference broth microdilution; MIC: minimum inhibitory concentration; MHE: Mueller Hinton E agar; COLR: ChromID Colistin R agar; CBDE: colistin broth disc elution; PPV: positive predictive value; NPV: negative predictive value; EA: essential agreement; CA: category agreement; VMD: very major discrepancy; MD: major discrepancy; NA: not applicable.

Table 3. The VMDs and MDs identified in all isolates in comparison with the BMD.

Strain	Carbapenemase Type	Vitek 2 Compact (MIC mg/L)	Micronaut MIC-Strip (MIC mg/L)	Etest, MHE (MIC mg/L)	COLR Medium	Rapid Polymyxin NP Test	CBDE (MIC mg/L)	BMD (MIC mg/L)
Enterobacter cloacae complex ($n = 1$) A	MBL	≤0.5	0.25	0.25	Negative	Positive [1]	≤1	0.25
Klebsiella pneumoniae ($n = 1$) B	KPC	≤0.5	0.5	0.25	Negative	Positive	≤1	0.25
K. pneumoniae ($n = 1$) C	KPC	≤0.5	0.25	0.25	Negative	Positive	≤1	0.5
K. pneumoniae ($n = 1$) D	OXA-48-like	1	16	2	Positive	Positive	≥4	8
K. pneumoniae ($n = 1$) E	KPC	2	4	2	Positive	Positive	≥4	8
K. pneumoniae ($n = 1$) F	KPC	≥16	8	2	Positive	Positive	≥4	16
K. pneumoniae ($n = 1$) G	KPC	≥8	16	2	Positive	Positive	≥4	8
K. pneumoniae ($n = 1$) H	KPC	4	4	0.5	Positive	Positive	≥4	4
K. pneumoniae ($n = 1$) I	KPC	4	8	1	Positive	Positive	≥4	8
K, pneumoniae ($n = 1$) J	KPC	2	8	2	Positive	Positive	≥4	16
K. pneumoniae ($n = 1$) K	KPC	2	4	1	Positive	Positive	≥4	4
K. pneumoniae ($n = 2$) L, M	KPC	2	8	1	Positive	Positive	≥4	8
K. pneumoniae ($n = 1$) N	KPC	4	8	1	Positive	Positive	≥4	8
K. pneumoniae ($n = 1$) O	KPC	2	8	2	Positive	Positive	≥4	8
K. pneumoniae ($n = 1$) P	KPC	2	8	1	Positive	Positive	≥4	8
K. pneumoniae ($n = 1$) Q	KPC	≥8	4	2	Positive	Positive	≥4	8
K. pneumoniae ($n = 1$) R	KPC	8	8	2	Positive	Positive	≥4	8

Legend. VMDs: very major discrepancies (marked in red); MDs: major discrepancies (marked in blue); BMD: reference broth microdilution; MBL: metallo-β-lactamase; KPC: *K. pneumoniae* carbapenemase; OXA-48-like: oxacillinase-48-like; MIC: minimum inhibitory concentration; MHE: Mueller Hinton E agar; COLR: ChromID Colistin R agar; CBDE: colistin broth disc elution. [1] Color change detected at 3 h of incubation but without the same turbidity in comparison to the positive control.

No false positive or false negative results were recorded for any of the 82 strains of *Escherichia coli* and *K. pneumoniae* tested with ChromID Colistin R Agar (COLR). According to the manufacturer, only *E. coli*, *K. pneumoniae*, and *Salmonella* spp. are listed as target microorganisms for qualitative testing. However, one strain of colistin-resistant *E. cloacae* complex determined to have an MIC value of 32 mg/L by BMD developed on this medium, forming blue-green colonies. A positive pattern was also noted for the reference strain *E. cloacae* ATCC 13047.

Three false positive results were observed with the Rapid Polymyxin NP Test for one *E. cloacae* complex MBL strain A and two *K. pneumoniae* KPC isolates B and C (Table 3). All positive results, including these strains, were obvious following 2 h of incubation, with the exception of *E. cloacae* isolate A, where the color change from orange to yellow was clearly noted at 3 h but without the same turbidity in comparison with the positive control well. The test was replicated with the same results.

2.3. Performance of Commercial Methods in Relation to the BMD

The highest essential agreement (EA) was documented for the Micronaut MIC-Strip (85/92, 92.39%, 95% CI 73.21–95.43%), followed by Vitek 2 Compact (84/92, 91.30%, 95% CI 74.81–96.35%), and Etest (46/92, 50%, 95% CI 36.64–66.83%) (Table 2). A category agreement (CA) of >90% was obtained for all methods except the Etest (77/92, 83.69%, 95% CI 66.04–94.86%) (Table 2). Total agreement was noted in the case of the Micronaut MIC-Strip, CBDE, and COLR medium.

Very major discrepancies (VMDs) were strictly limited to Etest (15/56, 26.78%, 95% CI 14.04–44.18%) and Vitek 2 Compact (8/56, 14.28%, 95% CI 7.18–28.24%). Only Rapid Polymyxin NP Test induced MDs (3/36, 8.33%, 95% CI 1.74–24.35) (Table 2). The comparative testing results obtained for all strains in which VMDs and MDs were recorded are summarized in Table 3.

The sensitivity, specificity, positive predictive value (PPV), and negative predictive value (NPV) determined for all methods are presented in Table 2.

2.4. Colistin Testing Results versus PAP

E. cloacae complex strain A and *K. pneumoniae* isolates B and C presented subpopulations with a frequency at 10^8 CFU/mL ranging from 4.0×10^{-7} to 6.6×10^{-4} (Table 4 and Figure 2). *E. cloacae* strain A was interpreted with homogeneous response in the susceptible range with the inhibitory colistin concentration in the PAP assay of 1 mg/L in comparison with the original MIC of 0.25 mg/L by BMD (a fourfold difference) (Tables 3 and 4). The two heteroresistant *K. pneumoniae* isolates B and C had minor resistant subpopulations, and the colistin concentration, which suppressed the entire growth in the PAP technique, was at least 16-fold higher than the native MICs of 0.25 and 0.5 mg/L, respectively, obtained by BMD (Tables 3 and 4). The heteroresistant phenotype of the two mutants belonging to *K. pneumoniae* isolates B and C remained stable after one week of serial passages on colistin-free agar (MIC > 64 mg/L) (Table 4).

Table 4. Clinical aspects and the PAP results of the three isolates with false positive results obtained with the Rapid Polymyxin NP Test.

Strain	Previous Colistin Therapy	Highest Colistin Concentration of Growth in PAP (mg/L)	Inhibitory Colistin Concentration in PAP (mg/L)	Frequency at Highest Colistin Concentration of Growth	MIC by BMD of Colonies before 7 Days Passages (mg/L)	MIC by BMD of Colonies after 7 Days Passages (mg/L)	Strain Classification by PAP
E. cloacae complex A	No	0.5	1	6.6×10^{-4}	0.25	0.25	hO-S
K. pneumoniae B	No	16	≥32	4.0×10^{-7}	>64	>64	hR
K. pneumoniae C	Yes [1]	4	8	8.0×10^{-7}	>64	>64	hR

Legend. PAP: population analysis profiling; MICs: minimum inhibitory concentrations; BMD: reference broth microdilution; hO-S: homogeneous response susceptible; hR: heteroresistant response. [1] Previous parenteral colistin treatment for 2 weeks.

All three strains were isolated from respiratory tract specimens of patients hospitalized in the intensive care unit (ICU) of the same medical institution. They had been treated with colistin according to reported susceptibility results with unfavorable outcomes (Table 4).

2.5. Reproducibility

The expected MIC targets set by EUCAST for BMD were established using reference strains *E. coli* ATCC 25922 and *E. coli* NCTC 13846 for all quantitative methods. The COLR medium, the Rapid Polymyxin NP Test, and the CBDE also showed reproducible results when they were repeatedly assessed with the reference strains. All the qualitative and quantitative methods, including the PAP assay, displayed reproducible results after testing with *E. cloacae* ATCC 13047.

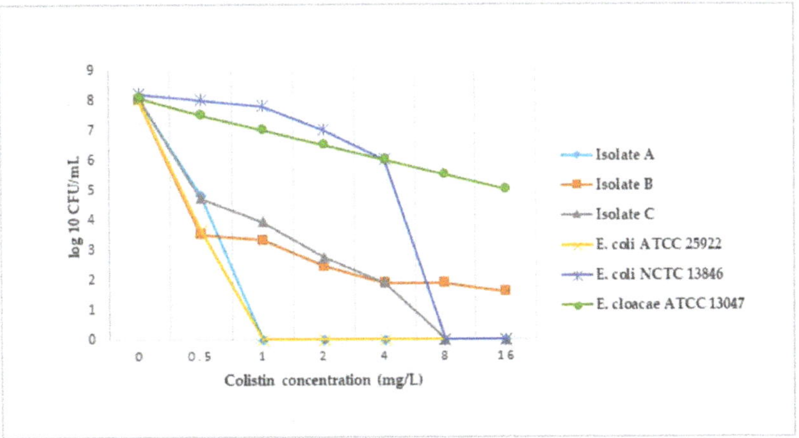

Figure 2. The population analysis profile of isolates A (susceptible), B, C (heteroresistant), and the reference strains at an initial inoculum of 10^8 CFU/mL The data shown are representative of multiple replicates performed in the same experiment and on different working days for each strain.

3. Discussion

Our analysis investigated the performance of six phenotypic methods compared to the reference procedure BMD, and all false positive results were further evaluated using the PAP method in order to assess their appropriateness for susceptibility testing of CPE clinical isolates. To the best of our knowledge, this is the first study to evaluate this combination of tests on CPE strains.

In the clinical arena, colistin is a toxic agent reserved for severe infections due to MDR Gram-negative bacilli [7], and erroneous laboratory results and delays in reporting should be avoided [37]. Finding alternative, accurate, and more practical methods for colistin testing as an alternative to the laborious gold standard BMD method remains a challenge [17,30,37] so long as there are limited laboratories with extensive experience to perform the BMD assay [38].

According to the CLSI, the CBDE and agar dilution MIC methods are also acceptable for colistin susceptibility testing [39]. The CBDE procedure is a simple and affordable method that was obsolete but has recently regained relevance [37,39], and it shows notable concordance with the BMD for Enterobacterales strains, except when *E. coli* isolates harbor *mcr-1* genes [37,40].

In our investigation, the adopted interpretation of the CBDE results was compliant with the EUCAST breakpoints to allow easier comparison with the BMD. Simner et al. suggested that Enterobacterales isolates with a colistin MIC ≥ 2 μg/mL by the CBDE should be validated by BMD, and positive strains should subsequently be genotypically tested for *mcr* genes [40]. None of our isolates exhibited an MIC of exactly 2 mg/L, which is the cutoff value of EUCAST breakpoints. In our analysis, the sensitivity, specificity, and CA were 100% without VMDs or MDs for the CBDE method. These aspects are in agreement with early observations, with the exception of some VMDs reported especially for *mcr-1*-positive *E. coli* strains [40,41].

From a practical perspective, regarding the CBDE method, Humphries et al. mentioned concerns about their limited experience with only one type of cation-adjusted Mueller Hinton broth (CAMHB) (Remel, Lenexa, KS) in pre-aliquoted borosilicate tubes [41]. However, our research showed agreement with the reference method using a different medium (Becton Dickinson) in polypropylene tubes, even though the possibility of colistin adhering to plastic surfaces was mentioned in another study [40].

Previously reported data mentioned inadequate colistin performance testing results obtained with the Vitek 2 Compact according to the standard criteria: CA $\geq$ 90%, EA $\geq$ 90%, VMD $\leq$ 3%, and MD $\leq$ 3% [42–45]. Our results obtained with Vitek 2 Compact confirmed this, but with acceptable EA and CA of 91.30%, and MD of 0%. In contrast to some of the abovementioned publications that reported VMDs for Vitek 2 Compact between 26.3% [42] and 36% [44], our VMD was lower (14.28%), and all eight false negative results were noted exclusively in *K. pneumoniae* strains. Additionally, our collection included only four available isolates of *E. cloacae* complex, a problematic pathogen that was recognized to induce false susceptible results on Vitek 2 Compact and BD Phoenix semiautomated systems [45,46]. None of these isolates had MIC values close to the EUCAST breakpoints.

Colistin susceptibility was correctly identified by Vitek 2 Compact in all 36 of our isolates with MIC $\leq$ 0.5 mg/L. Eight resistant strains with MIC between 4 and 16 mg/L according to BMD were not correctly detected by this instrument, which indicated MICs between 1 and 2 mg/L. In many microbiology laboratories, Vitek 2 Compact represents an important instrument of rapid susceptibility testing [14,30], but with recognized disadvantages regarding underestimation of colistin-resistant Enterobacterales pathogens and impossibility of detection of colistin heteroresistance [45,47]. However, our study reveals that Vitek 2 Compact is a reliable instrument for diagnosing colistin resistance, with no false positive results recorded. Similar observations were previously reported by Pfennigwerth et al. [45].

All our colistin-resistant strains with MIC values close to the EUCAST breakpoints, for which false negative results were obtained by Vitek 2 Compact (n = 8), were accurately assessed using the Micronaut MIC-Strip, the COLR medium, and the Rapid Polymyxin NP Test. Etest failed to detect colistin resistance in any of these cases. The more these particular strains are included in a study, the higher the number of errors, which will contribute to modifying the antimicrobial susceptibility test evaluation results [48].

In our analysis, the Micronaut MIC-Strip was the only reliable MIC technique that fulfilled all requirements with CA and EA > 90%, and no VMD or MD. Furthermore, for this commercial method, Matuschek et al. demonstrated similar scores for EA and CA, acceptable MD, and no VMD on a limited collection of strains (n = 32) of *E. coli* and *K. pneumoniae* [48].

The COLR medium has been particularly designed for detection of some colistin-resistant Enterobacterales species both as a screening method for the detection of colistin-resistant strains directly from biological samples and a qualitative method used directly on bacterial cultures [49]. It contains chromogenic substrates allowing rapid color-based pre-identification of colonies [49]. In our investigation, COLR agar presented excellent performance with both a sensitivity and specificity of 100%, a complete CA with the BMD, and no VMD or MD recorded for *E. coli* and *K. pneumoniae*. Most of the studies assessed the performance of COLR agar using the screening technique [49,50]. However, in a recent report, Bala et al. demonstrated a CA of 94.3% between COLR agar used in the qualitative method and BMD on a collection of 87 characterized Enterobacterales strains [51]. They also remarked that this new chromogenic medium could be a reliable and practical alternative for the taxa recommended by the manufacturer [51].

In our research, the Rapid Polymyxin NP Test was shown to be a simple and rapid assay, with an adequate CA of 96.73%, a sensitivity of 100%, specificity of 91.67%, PPV of 94.92%, and NPV of 100%. The original in-house method, introduced by Nordmann et al. more than 5 years ago, demonstrated a sensitivity and specificity of 99.3% and 95.4%, respectively, when a large group of 200 Enterobacterales strains with different mechanisms of colistin resistance, including one isolate of *K. pneumoniae* previously characterized with colistin heteroresistance, was evaluated [52]. Additionally, in agreement with our findings, a recent study on the commercial test reported a sensitivity, specificity, PPV, and NPV of 100%, 95.9%, 98.3%, and 100%, respectively, on 339 Enterobacterales isolates, including an important proportion of particular strains with colistin MIC close to the cutoff breakpoint values [53].

The Rapid Polymyxin NP Test demonstrated the ability to detect colistin heteroresistance, as well as *mcr-1* and *mcr-2* producers [14,46,54]. On a collection of 70 *mcr-1/mcr-2* producers belonging to the Enterobacterales order with distinct origins, Poirel et al. emphasized that the MICs obtained by the BMD were between 4 and 64 mg/L, and this commercial method showed excellent sensitivity and specificity [54].

Interestingly, our MD of 8.33% for the Rapid Polymyxin NP Test was noted in three strains, A, B, and C, with MIC $\leq$ 0.5 mg/L according to the BMD and applying the reference PAP assay confirmed that two *K. pneumoniae* isolates B and C were colistin-heteroresistant. It has been specified that traditional testing techniques, including the BMD, are not able to identify colistin heteroresistance, which can generate erroneous categorization, and in severe infections these essential findings can explain possible colistin treatment failures [30,34,35]. Moreover, these resistant subpopulations were assumed to contribute through chemical communication, transferring antibiotic resistance to protect more susceptible members [33]. Recently, Band et al., in a multicenter project conducted in the United States, demonstrated that heteroresistance to colistin among 408 CPE strains has largely remained underestimated [35].

In contrast, compared strictly to the BMD, Kon et al. indicated MD, VMD, sensitivity, and specificity of 1.8%, 21.1%, 78.9%, and 98.2%, respectively, for the Rapid Polymyxin NP Test, as well as many inconclusive color changes [55]. In our research, no VMD was observed, and the results of all repeated tests using different size inoculum were reproducible, conducted within the manufacturer's recommended range and applying the same photometric device DensiCHEK, as is the case with the aforementioned authors [55]. However, Jayol et al. demonstrated, on a large collection of 223 Enterobacterales isolates (including 38 *mcr*-like producers and 19 heteroresistant isolates), excellent performance for this commercial kit with MD, VMD, sensitivity, and specificity of 5.1%, 1.9%, 98.1%, and 94.9%, respectively [56].

Our findings revealed some minor difficulties in the interpretation of results obtained with the Rapid Polymyxin NP Test only in the case of *E. cloacae* isolate A, but the PAP method illustrated a homogeneous response with a minor subpopulation (6.6×10^{-4}) that responded to colistin concentrations below the breakpoints. The two heteroresistant *K. pneumoniae* isolates B and C showed a stable phenotype of resistance in the passaging study (MIC > 64 mg/L). Although a previous colistin treatment promoted the emergence of resistance to this antimicrobial agent in *K. pneumoniae* isolates recovered from patients hospitalized in the ICU [57], our study revealed that only the patient with *K. pneumoniae* isolate C had been previously exposed to prolonged therapy to colistin.

In line with several other authors, our Etest results illustrated unacceptably low values of EA and CA of 50% and 83.69%, respectively, and our EA and CA were lower in comparison with others [44,45,48]. The unreliable detection of colistin resistance using the Etest has already been indicated [14,30,32,37,40] and is explained by deficient distribution of the large molecule of colistin into agar media [37,40]. The VMDs outlined for Etest varied between 12% for Enterobacterales strains [44] and 41.5% for *K. pneumoniae* isolates [58]. Our VMD of 26.78% (15 false negative results) for this method was registered exclusively in *K. pneumoniae* strains. However, the main benefit of using Etest remains, which is the possibility to discover colistin-heteroresistant subpopulations, but this aspect is dependent on the type of Mueller Hinton medium used [47]. In the present study, heteroresistant isolates could not be identified using the Etest method.

The current research emphasizes the presence of colistin-heteroresistant subpopulations in *K. pneumoniae* KPC isolates B and C, which were erroneously categorized as colistin-susceptible by the BMD and all applied commercial methods with the exception of the Rapid Polymyxin NP Test.

In this context, clinical microbiology laboratories should select, validate, and implement diverse, accurate, and even combined methods into routine practice [30,38], especially in the case of challenging isolates with an MIC close to the breakpoints, or those with colistin heteroresistance. Moreover, the results of the test performance are influenced by the accessibility and the complex types of isolates included in studies [38,45,59].

Our promising results should be seen in the light of some limitations. This research was strictly limited to phenotypic methods, and the subjacent molecular mechanisms of colistin resistance were not explored despite the fact that it has been highlighted that these methods should be performed concomitantly with phenotypic tests for improved characterization of individual strains [30]. Additionally, colistin heteroresistance has not yet been associated with *mcr* genes [30,60,61]. Although it is fundamental to perform all tests using the same inoculum, this requirement could not be respected in our investigation because of the logistical constraints associated with the vast variety of methods that were performed by a single skilled person. Even when several tests were performed simultaneously, the inoculum was always used within 15 min of preparation. Furthermore, the purpose of our study was not focused strictly on the comparative evaluation of these phenotypic methods but rather an attempt to define an improved algorithm that can be successfully utilized in the routine diagnosis of colistin resistance.

Future Challenges and Perspective

Future studies should provide unequivocal answers to challenges related to colistin susceptibility testing assays and to define an optimal, but more practical, testing strategy. Continuous assessments are essential to confirm the performance of the COLR medium as a qualitative technique and that of the CBDE, which is a recent method approved by the CLSI Guidelines. Furthermore, research should be oriented toward establishing the real significance of colistin heteroresistance and the impact of this phenomenon on phenotypic assays together with its possible involvement in the failure of therapies based on colistin. Definition of a harmonized consensus regarding heteroresistance and a standardized method for detection of this phenomenon is of high importance.

4. Materials and Methods

4.1. Bacterial Strains

A group of 92 nonduplicate clinical CPE isolates, including *K. pneumoniae* ($n = 78$), *Citrobacter freundii* ($n = 6$), *E. cloacae* complex ($n = 4$), and *E. coli* ($n = 4$), were tested. The strains were collected from the Dr. Constantin Opriş County Emergency Hospital Baia Mare, Romania ($n = 60$), from January 2017 to April 2021, and from Targu Mures County Emergency Clinical Hospital, Romania ($n = 32$), from January 2017 to April 2019. The two medical institutions are general acute care public hospitals with 920 and 1089 beds, respectively. The second is a teaching hospital.

Respiratory tract ($n = 37$), urine ($n = 34$), wounds ($n = 13$), blood ($n = 7$), and intravenous catheter tip ($n = 1$) were the sources of isolates. The strains were identified to species level based on conventional methods, Vitek 2 Compact (bioMérieux SA, Marcy l'Etoile, France) or API 20E strip (bioMérieux SA, Marcy l'Etoile, France). At both laboratories, routine testing for colistin susceptibility was performed using Vitek 2 Compact. Strains were selected based on variable levels of MICs as determined by Vitek 2 Compact and interpreted according to the EUCAST breakpoints [62–67]. On this instrument, 44 colistin-susceptible strains presented colistin MICs as follows: $\leq$0.5 mg/L ($n = 36$), 1 mg/L ($n = 1$), and 2 mg/L ($n = 7$). The remaining 48 isolates were colistin-resistant, with MICs between 4 and $\geq$16 mg/L. All isolates were stored at $-70\ ^\circ$C and subcultured twice on solid medium before testing.

4.2. Data Collection

In the case where identification of strains showed discrepancies between the Rapid Polymyxin NP Test and BMD results, the patients' electronic medical records were interrogated for colistin treatments administered before and after specimen collection, the wards of hospitalization, and clinical outcomes.

4.3. Identification of CPE Strains

Screening and phenotypic confirmation of carbapenem-hydrolyzing enzymes were accomplished for all strains using the modified carbapenem inactivation method (mCIM) [68,69] and the combination disc test (KPC, MBL, OXA- 48 Confirm kit, Rosco Diagnostica, Taastrup, Denmark), respectively. Of the total of 92 isolates included in our investigation, 22 strains were characterized by a multiplex PCR assay for the presence of carbapenemase-encoding genes (bla_{KPC}, bla_{NDM}, and $bla_{OXA-48-like}$), as described in two other studies [70,71], according to the method used by Szekely et al. [72].

4.4. Detection of Colistin Resistance

Each CPE isolate was tested using commercial methods according to the manufacturers' instructions and by CBDE versus BMD. In case of false positive results, strains were further examined in detail by application of the PAP method for assessing possible coexistence of residual colistin-heteroresistant subpopulations at baseline.

4.4.1. Colistin MIC Determination

The BMD and the CBDE were carried out simultaneously starting from the same inoculum. The density of direct normal saline bacterial suspension was standardized to 0.5 McFarland using a calibrated photometric device (DensiCHEK, bioMérieux SA).

The reference BMD was accomplished with cation-adjusted BBL Mueller Hinton II Broth (CAMHB, reference 212322, Becton Dickinson, Sparks MD, USA) and colistin sulfate salt powder (reference C4461, Sigma-Aldrich, St. Louis, MO, USA) in 96-well, nontreated, U-bottom, sterile polystyrene plates (reference 734-2782, VWR International, Radnor, PA USA) in accordance with the international guidelines [32,73]. Sterile distilled water was used both as solvent and diluent for preparing stock solutions of colistin [39]. Five batch panels prepared in-house with 100 μL per well using serial twofold dilutions corresponding to a concentration range of 0.125 to 128 mg/L were stored at −70 °C before use. All isolates were tested in duplicate, and all wells contained a targeted final concentration of roughly 5×10^5 CFU/mL microorganisms. A purity plate was prepared as a growth control for each tested isolate. The results were interpreted visually using a mirror after incubation at 35 ± 2 °C for 16–20 h in ambient air.

The CBDE was performed with 4 dilution tubes per isolate and 10 mL cation-adjusted BBL Mueller Hinton II Broth (reference 212322, Becton Dickinson) per polypropylene tube and colistin sulfate discs (10 μg, Thermo Fisher Scientific, Basingstoke, UK) in order to obtain final colistin concentrations of 0 (growth control), 1, 2, and 4 mg/L, respectively, as described previously [39]. The MICs were read by unaided eye after incubation at 33 to 35 °C for 16–20 h in aerobic atmosphere.

Susceptibility testing with Vitek 2 Compact (BioMérieux, Durham, NC) cards AST XN05 and AST N222 allowed the reporting of colistin MICs of ≤0.5 to ≥16 mg/L and was performed for most of the isolates in separate experiments. The susceptibility testing results were available within 18 h.

The Micronaut MIC-Strip Colistin (Merlin Diagnostika GmbH, Borhheim-Hensel, Germany), the gradient diffusion strip Etest (BioMérieux SA, Marcy l'Etoile, France) on Mueller Hinton E agar (MHE) (BioMérieux SA, Marcy l'Etoile, France), and ChromID Colistin R agar (COLR) assays (BioMérieux SA, Marcy l'Etoile, France) were conducted in parallel using the same bacterial suspension in another experimental session.

The design of the Micronaut MIC-Strip allowed testing one isolate per strip, with each strip consisting of 12 plastic wells forming a standard testing panel with dehydrated colistin. Then, 50 μL of each standardized bacterial suspension was homogenized in 11.5 mL CAMHB (Merlin Diagnostika GmbH, Borhheim-Hensel, Germany), followed by inoculation of each well with 100 μL of this prepared suspension. After incubation for 18–22 h at 35–37 °C, Micronaut-MIC strip evaluation was performed visually using a mirror.

Additionally, each standardized bacterial inoculum was swabbed onto MHE, and after a maximum of 15 min of drying of the agar surface, an Etest colistin strip was aseptically applied. After incubation at 36 °C for 16–20 h, the MIC results obtained by Etest were read by naked eye and a magnifying glass in reflected light.

The colistin MICs ranged $\leq$0.0625 to $\geq$64 mg/L for the Micronaut MIC-Strip and from $\leq$0.016 to $\geq$256 mg/L for Etest.

The reference PAP protocol was adapted from Liao et al. and Bergen et al. [60,74]. The preparation of agar plates respected the guidelines for MIC evaluation by the agar dilution method [73], and the prepared plates were refrigerated and used within 4 days of preparation. Isolates were evaluated starting from the standardized inoculum of 0.5 McFarland (approximately 10^8 CFU/mL) prepared from overnight culture. Serial 10-fold saline dilutions ranging from 10^8 to 10^2 CFU/mL were subsequently performed, and 50 µL from each dilution was spread onto solid Mueller Hinton agar (MHA) Thermo Fisher Scientific, Basingstoke, UK) plates supplemented with 0, 0.5, 1, 2, 4, 8, and 16 µg/mL colistin sulfate (reference C4461, Sigma-Aldrich, St. Louis, MO, USA). The inoculation started with a growth control MHA plate containing a similar volume of sterile water instead of colistin, continued from plates with the lowest colistin concentration to plates with the highest drug concentration and, finally, a second growth control plate was used. CFUs on each plate were counted following 48 h of incubation at 36 °C. For each strain, the stability of the phenotype was determined as follows: a single colony selected from the plates with the highest colistin concentrations was passaged onto colistin-free agar for 7 consecutive days, followed by the reassessment of the MICs by BMD [33,75].

4.4.2. Qualitative Phenotypic Assays

A streak plate procedure using 10 µL from each 0.5 McFarland bacterial suspension diluted 1:100 in sterile normal saline to a final concentration of 1×10^6 CFU/mL was applied onto COLR medium. After overnight incubation under aerobic conditions, pink-red and blue-green colonies were suggestive of colistin-resistant *E. coli* and *K. pneumoniae*, respectively.

A standardized inoculum of 3.0–3.5 McFarland (10^9 CFU/mL) for each strain prepared with a calibrated photometric device was homogenized with the Rapid Polymyxin NP Test medium (ELITechGroup, Signes, France). As a result of glucose metabolism, the phenolsulfonphthalein used as a pH indicator changed color from orange to yellow after 2–3 h of incubation, and indicative of a resistant strain in the presence of a defined colistin concentration of 2 mg/L. The interpretations were performed visually.

4.5. Interpretative Criteria

For all MIC determinations, the isolates were considered susceptible to colistin when the MIC $\leq$ 2 mg/L, consistent with the EUCAST breakpoints, including the CBDE [62–67].

All MIC values obtained by Etest were rounded up to the next twofold dilution step for comparative analysis.

A homogeneous response to colistin was designated when the highest inhibitory concentration according to the PAP assay was $\leq$4-fold of the native MIC obtained by BMD, while a difference more than 8-fold higher indicated heteroresistance to this antimicrobial agent [33].

4.6. Quality Controls

Tests using the Micronaut MIC-Strip, the Etest, the COLR medium, the Polymyxin NP Test, and the CBDE were performed at least 3 times with both a colistin-sensitive reference strain *E. coli* ATCC 25922 (0.25–2 mg/L, target value 0.5–1 mg/L) and a colistin-resistant strain *E. coli* NCTC 13846 (*mcr-1*-positive) (2–8 mg/L, target value 4 mg/L).

For Vitek 2 Compact, *E. coli* ATCC 25922 was used in quality control assurance on each new lot of cards, but *E. coli* NCTC 13846 was tested only on two lots of cards.

Quality control tests with *E. coli* ATCC 25922 were carried out for every new lot of API 20E strips.

In the case of the BMD, the two abovementioned reference strains were tested at least 20 times. The BMD quality control scheme was validated by concurrent testing of all these strains on every new lot of panels prepared in-house and on each working day. Additionally, the colony counts on inoculum were performed with *E. coli* ATCC 25922 [73].

All seven aforementioned methods were supplementary tested at least once with the positive heteroresistant strain *E. cloacae* ATCC 13047 (MIC of 256 mg/L).

For the mCIM and the combination disc test, *K. pneumoniae* ATCC BAA-1705 and *E. coli* ATCC 25922 were used to demonstrate positive and negative reactions, respectively. Colistin discs were examined with *Pseudomonas aeruginosa* ATCC 27853 and *E. coli* ATCC 25922.

In the case of PAP method, *E. coli* NCTC 13846 and *E. cloacae* ATCC 13047 were used as positive controls, while *E. coli* ATCC 25922 was the negative control.

A reproducible result was interpreted to be within ±1 dilution for MICs or achieving the same effect for the qualitative tests using the abovementioned reference strains.

4.7. Analysis of the Results

The results acquired with the commercial methods and the CBDE were compared with the reference BMD. In the case of discrepant results for each method, including the BMD, isolates were retested two or three times and the higher MIC value, or the results that appeared most often were accepted. The occurrence of skipped wells was noted only in limited situations, and those pathogens were retested in compliance with the CLSI Guidelines [73].

For the MIC determination methods, the essential agreement (EA = MICs within ±1 doubling dilution from the BMD MICs), the category agreement (CA = the same interpretation category with the BMD), the very major discrepancy (VMD) (false susceptible result), and the major discrepancy (MD) (false resistance result) were evaluated in accordance with the ISO 20776-2 standard [76]. A reliable technique was confirmed when the following criteria were met: CA ≥ 90%, EA ≥ 90%, VMD ≤ 3%, and MD ≤ 3% [76].

Each isolate tested with the PAP method was evaluated using several replicates both on the same working day and in independent experiments, and the most frequent result was recorded. The limit of detection in the PAP assay was one colony per plate (equivalent to 20 CFU/mL) [74]. The PAP was calculated for each strain by dividing the number of colonies obtained on the plates with the highest colistin concentration by the colony counts derived from the growth control plates [75].

4.8. Statistical Analysis

Sensitivity, specificity, PPV, and NPV were calculated for each method using the contingency tables. Microsoft Excel and the GraphPad InStat Demo State Software, version 3.06, San Diego, California, USA, were used for calculations.

4.9. Ethical Approval

Ethical approval for this study was obtained from the ethics committees of the Dr. Constantin Opriş County Emergency Hospital Baia Mare, Romania (reference number 14598/04.06.2019); Târgu Mureş County Emergency Clinical Hospital (reference number Ad. 14925/27.05.2019); and George Emil Palade University of Medicine, Pharmacy, Science, and Technology of Târgu Mureş, Romania (reference numbers 405/11.10.2019, 1024/13.07.2020, and 1217/18.12.2020).

5. Conclusions

The Rapid Polymyxin NP Test is easy to perform and offers rapid results and excellent performance compared to both the BMD and the PAP assay, including in the detection of colistin heteroresistance. The possibility of silent clonal expansion of such heteroresistant mutants in the hospital environment is of great concern, as long as the BMD and the other commercial methods used fail to detect isolates with this particular phenomenon of resistance. There is an urgent need to optimize diagnosis strategies because the reference phenotypic method PAP used for heteroresistance detection cannot feasibly be integrated into routine practice.

Strictly according to the BMD, the Micronaut MIC-Strip, the CBDE, and COLR medium exhibit the best performances in detecting colistin resistance. This report highlights the difficulties of Vitek 2 Compact in detecting isolates with MIC values close to the EUCAST breakpoints. Consequently, in situations when the reference BMD and the PAP technique cannot be performed simultaneously as confirmation, we propose an improved approach of combining all susceptible results obtained with Vitek 2 Compact, the CBDE, the COLR medium, and the Micronaut MIC-Strip with the Rapid Polymyxin NP Test. The performance of Etest gradient strips is unsatisfactory due to an unacceptable number of false negative results.

Author Contributions: Conceptualization, A.F. and E.S.; methodology, A.F. and E.S.; validation, A.F. and E.S.; formal analysis, A.F. and S.T.V.; investigation, A.F.; resources, A.F., E.S. and M.D.; writing—original draft preparation, A.F.; writing—review and editing, A.F., E.S., S.T.V. and M.D.; visualization, A.F.; supervision, E.S. and M.D.; project administration, A.F.; funding acquisition, A.F. and M.D. All authors have read and agreed to the published version of the manuscript.

Funding: This research received partial internal funding from George Emil Palade University of Medicine, Pharmacy, Science, and Technology of Târgu Mureş, Romania, code 25/P3 AA/08.07.2020, ct.2401/21/31.10.2018 BB.

Institutional Review Board Statement: Data are presented in Section 4.9 of this article.

Informed Consent Statement: Not applicable.

Data Availability Statement: Data are contained within the article.

Conflicts of Interest: The authors declare no conflict of interest.

References

1. Nordmann, P.; Poirel, L. The difficult-to-control spread of carbapenemase producers among *Enterobacteriaceae* worldwide. *Clin. Microbiol. Infect.* **2014**, *20*, 821–830. [CrossRef]
2. Albiger, B.; Glasner, C.; Struelens, M.J.; Grundmann, H.; Monnet, D.L. European Survey of Carbapenemase-Producing *Enterobacteriaceae* (EuSCAPE) working group. Carbapenemase-producing *Enterobacteriaceae* in Europe: Assessment by national experts from 38 countries, May 2015. *Eurosurveillance* **2015**, *20*, 30062. [CrossRef]
3. Van Duin, D.; Doi, Y. The global epidemiology of carbapenemase-producing *Enterobacteriaceae*. *Virulence* **2017**, *8*, 460–469. [CrossRef]
4. European Centre for Disease Prevention and Control. Rapid Risk Assessment. In *Carbapenem-Resistant Enterobacteriaceae, Second Update, 26 September 2019*; ECDC: Stockholm, Sweden, 2019. Available online: https://www.ecdc.europa.eu/sites/default/files/documents/carbapenem-resistant-enterobacteriaceae-risk-assessment-rev-2.pdf (accessed on 29 January 2022).
5. Lutgring, J.D.; Limbago, B.M. The problem of carbapenemase-producing-carbapenem-resistant *Enterobacteriaceae* detection. *J. Clin. Microbiol.* **2016**, *54*, 529–534. [CrossRef]
6. Lan, P.; Jiang, Y.; Zhou, J.; Yu, Y. A global perspective on the convergence of hypervirulence and carbapenem resistance in *Klebsiella pneumoniae*. *J. Glob. Antimicrob. Resist.* **2021**, *25*, 26–34. [CrossRef]
7. Rodriguez-Bano, J.; Gutierrez-Gutierrez, B.; Machuca, I.; Pascuala, A. Treatment of infections caused by extended-spectrum-beta-lactamase-, AmpC-, and carbapenemase-producing *Enterobacteriaceae*. *Clin. Microbiol. Rev.* **2018**, *31*, e00079-17. [CrossRef]
8. The United States Centers for Disease Control and Prevention. *Antibiotic Resistance Threats in the United States 2019 (2019 AR Threats Report)*; CDC: Atlanta, GA, USA, 2019. Available online: https://www.cdc.gov/drugresistance/pdf/threats-report/2019-ar-threats-report-508.pdf (accessed on 29 January 2022).
9. Clancy, C.J.; Buehrle, D.J.; Nguyen, M.H. PRO: The COVID-19 pandemic will result in increased antimicrobial resistance rates. *JAC Antimicrob. Resist.* **2020**, *2*, dlaa049. [CrossRef]
10. Rusic, D.; Vilovic, M.; Bukic, J.; Leskur, D.; Perisin, A.S.; Kumric, M.; Martinovic, D.; Petric, A.; Modun, D.; Bozic, J. Implications of COVID-19 pandemic on the emergence of antimicrobial resistance: Adjusting the response to future outbreaks. *Life* **2021**, *11*, 220. [CrossRef]

11. Belvisi, V.; Del Borgo, C.; Vita, S.; Redaelli, P.; Dolce, P.; Pacella, D.; Kertusha, B.; Carraro, A.; Marocco, R.; De Masi, M.; et al. Impact of SARS-CoV-2 pandemic on carbapenemase-producing *Klebsiella pneumoniae* prevention and control programme: Convergent or divergent action? *J. Hosp. Infect.* **2021**, *109*, 29–31. [CrossRef]

12. Magiorakos, A.P.; Srinivasan, A.; Carey, R.B.; Carmeli, Y.; Falagas, M.E.; Giske, C.G.; Harbarth, S.; Hindler, J.F.; Kahlmeter, G.; Olsson-Liljequist, B.; et al. Multidrug-resistant, extensively drug-resistant and pandrug-resistant bacteria: An international expert proposal for interim standard definitions for acquired resistance. *Clin. Microbiol. Infect.* **2012**, *18*, 268–281. [CrossRef]

13. Li, J.; Nation, R.L.; Turnidge, J.D.; Milne, R.W.; Coulthard, K.; Rayner, C.R.; Paterson, D.L. Colistin: The re-emerging antibiotic for multidrug-resistant Gram-negative bacterial infections. *Lancet Infect. Dis.* **2006**, *6*, 589–601. [CrossRef]

14. Poirel, L.; Jayol, A.; Nordmann, P. Polymyxins: Antibacterial activity, susceptibility testing, and resistance mechanisms encoded by plasmids or chromosomes. *Clin. Microbiol. Rev.* **2017**, *30*, 557–596. [CrossRef] [PubMed]

15. European Centre for Disease Prevention and Control. *Expert Consensus Protocol on Colistin Resistance Detection and Characterisation for the Survey of Carbapenem- and/or Colistin-Resistant Enterobacteriaceae-Version 1.0*; ECDC: Stockholm, Sweden, 2019. Available online: https://www.ecdc.europa.eu/sites/default/files/documents/expert-consensus-protocol-colistin-resistance.pdf (accessed on 29 January 2022).

16. El-Sayed, M.A.E.G.; Zhong, L.L.; Shen, C.; Yang, Y.; Doi, Y.; Tian, G.B. Colistin and its role in the era of antibiotic resistance: An extended review (2000–2019). *Emerg. Microbes Infect.* **2020**, *9*, 868–885. [CrossRef] [PubMed]

17. Tsuji, B.T.; Pogue, J.M.; Zavascki, A.P.; Paul, M.; Daikos, G.L.; Forrest, A.; Giacobbe, D.R.; Viscoli, C.; Giamarellou, H.; Karaiskos, I.; et al. International Consensus Guidelines for the optimal use of the polymyxins: Endorsed by the American College of Clinical Pharmacy (ACCP), European Society of Clinical Microbiology and Infectious Diseases (ESCMID), Infectious Diseases Society of America (IDSA), International Society for Anti-infective Pharmacology (ISAP), Society of Critical Care Medicine (SCCM), and Society of Infectious Diseases Pharmacists (SIDP). *Pharmacotherapy* **2019**, *39*, 10–39. [CrossRef] [PubMed]

18. Vasoo, S. Susceptibility Testing for the Polymyxins: Two Steps Back, Three Steps Forward? *J. Clin. Microbiol.* **2017**, *55*, 2573–2582. [CrossRef]

19. World Health Organization. *The Detection and Reporting of Colistin Resistance*; World Health Organization: Geneva, Switzerland, 2018. Available online: https://apps.who.int/iris/rest/bitstreams/1167165/retrieve (accessed on 29 January 2022).

20. Deris, Z.Z.; Akter, J.; Sivanesan, S.; Roberts, K.D.; Thompson, P.E.; Nation, R.L.; Li, J.; Velkov, T. A secondary mode of action of polymyxins against Gram-negative bacteria involves the inhibition of NADH-quinone oxidoreductase activity. *J. Antibiot.* **2014**, *67*, 147–151. [CrossRef]

21. European Centre for Disease Prevention and Control. *Outbreak of Carbapenemase-Producing (NDM-1 and OXA-48) and Colistin-Resistant Klebsiella Pneumoniae ST307, North-East Germany, 2019*; ECDC: Stockholm, Sweden, 2019. Available online: https://www.ecdc.europa.eu/sites/default/files/documents/Klebsiella-pneumoniae-resistance-Germany-risk-assessment.pdf (accessed on 29 September 2021).

22. Bogdanovich, T.; Adams-Haduch, J.M.; Tian, G.B.; Nguyen, M.H.; Kwak, E.J.; Muto, C.A.; Doi, Y. Colistin-resistant, *Klebsiella pneumoniae* carbapenemase (KPC)-producing *Klebsiella pneumoniae* belonging to the international epidemic clone ST258. *Clin. Infect. Dis.* **2011**, *53*, 373–376. [CrossRef]

23. Weterings, V.; Zhou, K.; Rossen, J.W.; van Stenis, D.; Thewessen, E.; Kluytmans, J.; Veenemans, J. An outbreak of colistin-resistant *Klebsiella pneumoniae* carbapenemase-producing *Klebsiella pneumoniae* in the Netherlands (July to December 2013), with inter-institutional spread. *Eur. J. Clin. Microbiol. Infect. Dis.* **2015**, *34*, 1647–1655. [CrossRef]

24. Mezzatesta, M.L.; Gona, F.; Caio, C.; Petrolito, V.; Sciortino, D.; Sciacca, A.; Santangelo, C.; Stefani, S. Outbreak of KPC-3-producing, and colistin-resistant, *Klebsiella pneumoniae* infections in two Sicilian hospitals. *Clin. Microbiol. Infect.* **2011**, *17*, 1444–1447. [CrossRef]

25. Olaitan, A.O.; Diene, S.M.; Kempf, M.; Berrazeg, M.; Bakour, S.; Gupta, S.K.; Thongmalayvong, B.; Akkhavong, K.; Somphavong, S.; Paboriboune, P.; et al. Worldwide emergence of colistin resistance in *Klebsiella pneumoniae* from healthy humans and patients in Lao PDR, Thailand, Israel, Nigeria and France owing to inactivation of the PhoP/PhoQ regulator mgrB: An epidemiological and molecular study. *Int. J. Antimicrob. Agents* **2014**, *44*, 500–507. [CrossRef]

26. Chen, S.; Hu, F.; Zhang, X.; Xu, X.; Liu, Y.; Zhu, D.; Wang, H. Independent emergence of colistin-resistant *Enterobacteriaceae* clinical isolates without colistin treatment. *J. Clin. Microbiol.* **2011**, *49*, 4022–4023. [CrossRef] [PubMed]

27. Li, Z.; Cao, Y.; Yi, L.; Liu, J.H.; Yang, Q. Emergent Polymyxin resistance: End of an era? *Open Forum Infect. Dis.* **2019**, *6*, ofz368. [CrossRef] [PubMed]

28. Gharaibeh, M.H.; Shatnawi, S.Q. An overview of colistin resistance, mobilized colistin resistance genes dissemination, global responses, and the alternatives to colistin: A review. *Vet. World* **2019**, *12*, 1735–1746. [CrossRef] [PubMed]

29. Wang, C.; Feng, Y.; Liu, L.; Wei, L.; Kang, M.; Zong, Z. Identification of novel mobile colistin resistance gene *mcr-10*. *Emerg. Microbes Infect.* **2020**, *9*, 508–516. [CrossRef]

30. Bardet, L.; Rolain, J.M. Development of new tools to detect colistin-resistance among *Enterobacteriaceae* strains. *Can. J. Infect. Dis. Med. Microbiol.* **2018**, 1–25. [CrossRef]

31. European Centre for Disease Prevention and Control. *Antimicrobial Resistance in the EU/EEA (EARS-Net)-Annual Epidemiological Report 2019*; ECDC: Stockholm, Sweden, 2020. Available online: https://www.ecdc.europa.eu/sites/default/files/documents/surveillance-antimicrobial-resistance-Europe-2019.pdf (accessed on 28 January 2022).

32. European Committee on Antimicrobial Susceptibility Testing. Recommendations for MIC Determination of Colistin (Polymyxin E) as Recommended by the Joint CLSI-EUCAST Polymyxin Breakpoints Working Group, EUCAST. 2016. Available online: http://www.bioconnections.co.uk/files/merlin/Recommendations_for_MIC_determination_of_colistin_March_2016.pdf (accessed on 28 January 2022).

33. El-Halfawy, O.M.; Valvano, M.A. Antimicrobial heteroresistance: An emerging field in need of clarity. *Clin. Microbiol. Rev.* **2015**, *28*, 191–207. [CrossRef]

34. Charretier, Y.; Diene, S.M.; Baud, D.; Chatellier, S.; Santiago-Allexant, E.; van Belkum, A.; Guigon, G.; Schrenzela, J. Colistin heteroresistance and involvement of the PmrAB regulatory system in *Acinetobacter baumannii*. *Antimicrob. Agents Chemother.* **2018**, *62*, e00788-18. [CrossRef]

35. Band, V.I.; Satola, S.W.; Smith, R.D.; Hufnagel, D.A.; Bower, C.; Conley, A.B.; Rishishwar, L.; Dale, S.E.; Hardy, D.J.; Vargas, R.L.; et al. Colistin heteroresistance is largely undetected among Carbapenem-Resistant *Enterobacterales* in the United States. *mBio* **2021**, *12*, e02881-20. [CrossRef]

36. Ezadi, F.; Ardebili, A.; Mirnejadc, R. Antimicrobial susceptibility testing for polymyxins: Challenges, issues, and recommendations. *J. Clin. Microbiol.* **2019**, *57*, e01390-18. [CrossRef]

37. Satlin, M.J. The Search for a practical method for colistin susceptibility testing: Have we found it by going back to the future? *J. Clin. Microbiol.* **2019**, *57*, e01608-18. [CrossRef]

38. Humphries, R.M.; Ambler, J.; Mitchell, S.L.; Castanheira, M.; Dingle, T.; Hindler, J.A.; Koeth, L.; Sei, K. CLSI methods development and standardization Working Group best practices for evaluation of antimicrobial susceptibility tests. *J. Clin. Microbiol.* **2018**, *56*, e01934-17. [CrossRef] [PubMed]

39. *M100-S30*; Performance Standards for Antimicrobial Susceptibility Testing, 30th ed. CLSI—Clinical and Laboratory Standards Institute: Wayne, PA, USA, 2020.

40. Simner, P.J.; Bergman, Y.; Trejo, M.; Roberts, A.A.; Marayan, R.; Tekle, T.; Campeau, S.; Kazmi, A.Q.; Bell, D.T.; Lewis, S.; et al. Two-site evaluation of the colistin broth disk elution test to determine colistin in vitro activity against Gram-negative bacilli. *J. Clin. Microbiol.* **2019**, *57*, e01163-18. [CrossRef] [PubMed]

41. Humphries, R.M.; Green, D.A.; Schuetz, A.N.; Bergman, Y.; Lewis, S.; Yee, R.; Stump, S.; Lopez, M.; Macesic, N.; Uhlemann, A.C.; et al. Multicenter evaluation of colistin broth disk elution and colistin agar test: A report from the Clinical and Laboratory Standards Institute. *J. Clin. Microbiol.* **2019**, *57*, e01269-19. [CrossRef] [PubMed]

42. Lellouche, J.; Schwartz, D.; Elmalech, N.; Ben Dalak, M.A.; Temkin, E.; Paul, M.; Geffen, Y.; Yahav, D.; Eliakim-Raz, N.; Durante-Mangoni, E.; et al. Combining VITEK 2 with colistin agar dilution screening assist timely reporting of colistin susceptibility. *Clin. Microbiol. Infect.* **2019**, *25*, 711–716. [CrossRef]

43. Khurana, S.; Malhotra, R.; Mathur, P. Evaluation of Vitek 2 performance for colistin susceptibility testing for Gram-negative isolates. *JAC Antimicrob. Resist.* **2020**, *2*, dlaa101. [CrossRef]

44. Chew, K.L.; La, M.V.; Lin, R.T.P.; Teo, J.W.P. Colistin and polymyxin B susceptibility testing for carbapenem-resistant and mcr-positive *Enterobacteriaceae*: Comparison of Sensititre, MicroScan, Vitek 2, and Etest with broth microdilution. *J. Clin. Microbiol.* **2017**, *55*, 2609–2616. [CrossRef]

45. Pfennigwerth, N.; Kaminski, A.; Korte-Berwanger, M.; Pfeifer, Y.; Simon, M.; Werner, G.; Jantsch, J.; Marlinghaus, L.; Gatermann, S.G. Evaluation of six commercial products for colistin susceptibility testing in *Enterobacterales*. *Clin. Microbiol. Infect.* **2019**, *25*, 1385–1389. [CrossRef]

46. Jayol, A.; Nordmann, P.; Lehours, P.; Poirel, L.; Dubois, V. Comparison of methods for detection of plasmid-mediated and chromosomally encoded colistin resistance in *Enterobacteriaceae*. *Clin. Microbiol. Infect.* **2018**, *24*, 175–179. [CrossRef]

47. Lo-Ten-Foe, J.R.; de Smet, A.M.G.A.; Diederen, B.M.W.; Kluytmans, J.A.J.W.; van Keulen, P.H.J. Comparative evaluation of the Vitek 2, disk diffusion, Etest, broth microdilution, and agar dilution susceptibility testing methods for colistin in clinical isolates, including heteroresistant *Enterobacter cloacae* and *Acinetobacter baumannii* strains. *Antimicrob. Agents Chemother.* **2007**, *51*, 3726–3730. [CrossRef]

48. Matuschek, E.; Ahman, J.; Webster, C.; Kahlmeter, G. Antimicrobial susceptibility testing of colistin-evaluation of seven commercial MIC products against standard broth microdilution for *Escherichia coli*, *Klebsiella pneumoniae*, *Pseudomonas aeruginosa*, and *Acinetobacter* spp. *Clin. Microbiol. Infect.* **2018**, *24*, 865–870. [CrossRef]

49. Girlich, D.; Naas, T.; Dortet, L. Comparison of the Superpolymyxin and ChromID Colistin R screening media for the detection of colistin-resistant *Enterobacteriaceae* from spiked rectal swabs. *Antimicrob. Agents Chemother.* **2019**, *63*, e01618-18. [CrossRef] [PubMed]

50. Garcia-Fernandez, S.; García-Castillo, M.; Ruiz-Garbajosa, P.; Morosini, M.I.; Bala, Y.; Zambardi, G.; Canton, R. Performance of CHROMID® Colistin R agar, a new chromogenic medium for screening of colistin-resistant *Enterobacterales*. *Diagn. Microbiol. Infect. Dis.* **2019**, *93*, 1–4. [CrossRef] [PubMed]

51. Bala, Y.; Fabre, P.; Senot, F.; Loubet, M.; Brossault, L.; Vrignaud, M.; Roche, J.M.; Ghirardi, S.; Zambardi, G. Can CHROMID(R) Colistin R agar (COLR) be used to test colistin susceptibility in Enterobacteriaceae? In Proceedings of the ESCMID Final Programme, 28th ECCMID, Madrid, Spain, 21–24 April 2018; Abstract Number: O0957. Available online: https://www.escmid.org/escmid_publications/escmid_elibrary/material/?mid=63537 (accessed on 24 January 2022).

52. Nordmann, P.; Jayol, A.; Poirel, L. Rapid detection of polymyxin resistance in *Enterobacteriaceae*. *Emerg. Infect. Dis.* **2016**, *22*, 1038–1043. [CrossRef] [PubMed]

53. Yainoy, S.; Hiranphan, M.; Phuadraksa, T.; Eiamphungporn, W.; Tiengrim, S.; Thamlikitkul, V. Evaluation of the Rapid Polymyxin NP test for detection of colistin susceptibility in *Enterobacteriaceae* isolated from Thai patients. *Diagn. Microbiol. Infect. Dis.* **2018**, *92*, 102–106. [CrossRef] [PubMed]

54. Poirel, L.; Larpin, Y.; Dobias, J.; Stephan, R.; Decousser, J.W.; Madec, J.Y.; Nordman, P. Rapid Polymyxin NP test for the detection of polymyxin resistance mediated by the *mcr-1/mcr-2* genes. *Diagn. Microbiol. Infect. Dis.* **2018**, *90*, 7–10. [CrossRef] [PubMed]

55. Kon, H.; Abramov, S.; Dalak, M.A.B.; Elmaliach, N.; Schwartz, D.; Carmeli, Y.; Lellouche, J. Performance of Rapid Polymyxin™ NP and Rapid Polymyxin™ *Acinetobacter* for the detection of polymyxin resistance in carbapenem-resistant *Acinetobacter baumannii* and *Enterobacterales*. *J. Antimicrob. Chemother.* **2020**, *75*, 1484–1490. [CrossRef]

56. Jayol, A.; Kieffer, N.; Poirel, L.; Guerin, F.; Guneser, D.; Cattoir, V.; Nordmann, P. Evaluation of the Rapid Polymyxin NP test and its industrial version for the detection of polymyxin-resistant *Enterobacteriaceae*. *Diagn. Microbiol. Infect. Dis.* **2018**, *92*, 90–94. [CrossRef]

57. Halaby, T.; Al Naiemi, N.; Kluytmans, J.; van der Palen, J.; Vandenbroucke-Graulsb, C.M.J.E. Emergence of colistin resistance in *Enterobacteriaceae* after the introduction of selective digestive tract decontamination in an intensive care unit. *Antimicrob. Agents Chemother.* **2013**, *57*, 3224–3229. [CrossRef]

58. Dafopoulou, K.; Zarkotou, O.; Dimitroulia, E.; Hadjichristodoulou, C.; Gennimata, V.; Pournaras, S.; Tsakris, A. Comparative valuation of colistin susceptibility testing methods among carbapenem-nonsusceptible *Klebsiella pneumoniae* and *Acinetobacter baumannii* clinical isolates. *Antimicrob. Agents Chemother.* **2015**, *59*, 4625–4630. [CrossRef]

59. Giske, C.G.; Kahlmeter, G. Colistin antimicrobial susceptibility testing-can the slow and challenging be replaced by the rapid and convenient? *Clin. Microbiol. Infect.* **2018**, *24*, 93–94. [CrossRef]

60. Liao, W.; Lin, J.; Jia, H.; Zhou, C.; Zhang, Y.; Lin, Y.; Ye, J.; Cao, J.; Zhou, T. Resistance and heteroresistance to colistin in *Escherichia coli* isolates from Wenzhou, China. *Infect. Drug Resist.* **2020**, *13*, 3551–3561. [CrossRef] [PubMed]

61. Cheong, H.S.; Kim, S.Y.; Wi, Y.M.; Peck, K.R.; Ko, K.S. Colistin heteroresistance in *Klebsiella Pneumoniae* isolates and diverse mutations of PmrAB and PhoPQ in resistant subpopulations. *J. Clin. Med.* **2019**, *8*, 1444. [CrossRef] [PubMed]

62. European Committee on Antimicrobial Susceptibility Testing (EUCAST). Breakpoint Tables for Interpretation of MICs and Zone Diameters, Version 7.1. 2017. Available online: https://www.eucast.org/ast_of_bacteria/previous_versions_of_documents/ (accessed on 24 January 2022).

63. European Committee on Antimicrobial Susceptibility Testing (EUCAST). Breakpoint Tables for Interpretation of MICs and Zone Diameters, Version 8.0. 2018. Available online: https://www.eucast.org/ast_of_bacteria/previous_versions_of_documents/ (accessed on 24 January 2022).

64. European Committee on Antimicrobial Susceptibility Testing (EUCAST). Breakpoint Tables for Interpretation of MICs and Zone Diameters, Version 8.1. 2018. Available online: https://www.eucast.org/ast_of_bacteria/previous_versions_of_documents/ (accessed on 24 January 2022).

65. European Committee on Antimicrobial Susceptibility Testing (EUCAST). Breakpoint Tables for Interpretation of MICs and Zone Diameters, Version 9.0. 2019. Available online: https://www.eucast.org/ast_of_bacteria/previous_versions_of_documents/ (accessed on 24 January 2022).

66. European Committee on Antimicrobial Susceptibility Testing (EUCAST). Breakpoint Tables for Interpretation of MICs and Zone Diameters, Version 10.0. 2020. Available online: https://www.eucast.org/ast_of_bacteria/previous_versions_of_documents/ (accessed on 24 January 2022).

67. European Committee on Antimicrobial Susceptibility Testing (EUCAST). Breakpoint Tables for Interpretation of MICs and Zone Diameters, Version 11.0. 2021. Available online: https://www.eucast.org/fileadmin/src/media/PDFs/EUCAST_files/Breakpoint_tables/v_11.0_Breakpoint_Tables.pdf (accessed on 24 January 2022).

68. *M100-S27*; Performance Standards for Antimicrobial Susceptibility Testing, 27th ed. CLSI—Clinical and Laboratory Standards Institute: Wayne, PA, USA, 2017.

69. Van der Zwaluw, K.; de Haan, A.; Pluister, G.N.; Bootsma, H.J.; de Neeling, A.J.; Schouls, L.M. The carbapenem inactivation method (CIM), a simple and low-cost alternative for the Carba NP Test to assess phenotypic carbapenemase activity in Gram-negative rods. *PLoS ONE* **2015**, *10*, e0123690. [CrossRef] [PubMed]

70. Foldes, A.; Molnar, S.; Bilca, D.V.; Voidazan, S.T.; Szekely, E. Molecular epidemiology and the clinical impact of carbapenemase-producing *Enterobacterales* isolates among adult patients: Aspects from a Romanian non-teaching hospital. *Rev. Rom. Med. Lab.* **2020**, *28*, 427–439. [CrossRef]

71. Foldes, A.; Bilca, D.V.; Szekely, E. Phenotypic and molecular identification of carbapenemase-producing *Enterobacteriaceae*-challenges in diagnosis and treatment. *Rev. Rom. Med. Lab.* **2018**, *26*, 221–230. [CrossRef]

72. Szekely, E.; Damjanova, I.; Janvari, L.; Vas, K.E.; Molnar, S.; Bilca, D.V.; Lorinczi, L.K.; Toth, A. First description of *bla*$_{NDM-1}$, *bla*$_{OXA-48}$, *bla*$_{OXA-181}$ producing *Enterobacteriaceae* strains in Romania. *Int. J. Med. Microbiol.* **2013**, *303*, 697–700. [CrossRef]

73. *M07*; Methods for Dilution Antimicrobial Susceptibility Tests for Bacteria That Grow Aerobically, 11th ed. CLSI—Clinical and Laboratory Standards Institute: Wayne, PA, USA, 2018.

74. Bergen, P.J.; Forrest, A.; Bulitta, J.B.; Tsuji, B.T.; Sidjabat, H.E.; Paterson, D.L.; Li, J.; Nation, R.L. Clinically relevant plasma concentrations of colistin in combination with imipenem enhance pharmacodynamic activity against multidrug-resistant *Pseudomonas aeruginosa* at multiple inocula. *Antimicrob. Agents Chemother.* **2011**, *55*, 5134–5142. [CrossRef]

75. Hermes, D.M.; Pormann Pitt, C.; Lutz, L.; Teixeira, A.B.; Ribeiro, V.B.; Netto, B.; Martins, A.F.; Zavascki, A.P.; Barth, A.L. Evaluation of heteroresistance to polymyxin B among carbapenem-susceptible and -resistant *Pseudomonas aeruginosa*. *J. Med. Microbiol.* **2013**, *62*, 1184–1189. [CrossRef]
76. *ISO 20776-2:2007*; Clinical Laboratory Testing and In Vitro Diagnostic Test Systems–Susceptibility Testing of Infectious Agents and Evaluation of Performance of Antimicrobial Susceptibility Test Devices–Part 2: Evaluation of Performance of Antimicrobial Susceptibility Test Devices. International Organization for Standardization: Geneva, Switzerland, 2007. Available online: https://www.iso.org/standard/41631.html(accessed on 24 January 2022).

antibiotics

Article

The Evaluation of Eazyplex® SuperBug CRE Assay Usefulness for the Detection of ESBLs and Carbapenemases Genes Directly from Urine Samples and Positive Blood Cultures

Alicja Sękowska *[ID] and Tomasz Bogiel [ID]

Department of Microbiology, Ludwik Rydygier Collegium Medicum in Bydgoszcz, Nicolaus Copernicus University in Torun, 9 M. Sklodowska-Curie Street, 85-094 Bydgoszcz, Poland; t.bogiel@cm.umk.pl
* Correspondence: asekowska@cm.umk.pl; Tel.: +48-52-585-44-80

Abstract: Increasing antimicrobial resistance of Gram-negative rods is an important diagnostic, clinical and epidemiological problem of modern medicine. Therefore, it is important to detect multi-drug resistant strains as early on as possible. This study aimed to evaluate Eazyplex® SuperBug CRE assay usefulness for beta-lactamase gene detection among Gram-negative rods, directly from urine samples and positive blood cultures. The Eazyplex® SuperBug CRE assay is based on a loop-mediated isothermal amplification of genetic material and allows for the detection of a selection of genes encoding carbapenemases, KPC, NDM, VIM, OXA-48, OXA-181 and extended-spectrum beta-lactamases from the CTX-M-1 and CTX-M-9 groups. A total of 120 clinical specimens were included in the study. The test gave valid results for 58 (96.7%) urine samples and 57 (95.0%) positive blood cultures. ESBL and/or carbapenemase enzymes genes were detected in 56 (93.3%) urine and 55 (91.7%) blood samples, respectively. The Eazyplex® SuperBug CRE assay can be used for a rapid detection of the genes encoding the most important resistance mechanisms to beta-lactams in Gram-negative rods also without the necessity of bacterial culture.

Keywords: carbapenemases; Eazyplex® SuperBug CRE assay; extended-spectrum beta-lactamases; gram-negative rods; LAMP method

Citation: Sękowska, A.; Bogiel, T. The Evaluation of Eazyplex® SuperBug CRE Assay Usefulness for the Detection of ESBLs and Carbapenemases Genes Directly from Urine Samples and Positive Blood Cultures. *Antibiotics* **2022**, *11*, 138. https://doi.org/10.3390/antibiotics11020138

Academic Editor: Francesca Andreoni

Received: 28 December 2021
Accepted: 18 January 2022
Published: 21 January 2022

Publisher's Note: MDPI stays neutral with regard to jurisdictional claims in published maps and institutional affiliations.

1. Introduction

Bacteria of Enterobacterales, especially multi-drug resistant isolates, are one of the most important causes of nosocomial infections. Their significance as a global threat can be explained in several ways; an easy acquisition of antibiotic resistance genes and the capability of these strains to survive in the hospital environment and antimicrobial pressure/bacterial evolution as an answer to common application of a broad-spectrum antibiotic therapy. The problem of multi-drug resistance of Gram-negative rods is mostly associated with the horizontal gene transfer and possible synthesis of several beta-lactamases from the same or different groups of enzymes (e.g., ESBLs-extended spectrum beta-lactamases, carbapenemases). It mainly includes enzymes with a wide range of activity (cephalosporinases or carbapenemases) whose incidences have risen enormously in the last two decades [1–3]. These enzymes have evolved significantly since their discovery and several variants of each enzyme can be currently identified within one enzyme family.

Numerous phenotypic and genotypic methods used in microbiological laboratories allow for the detection of ESBLs and carbapenemase-producing strains. Phenotypic methods always require the culture of the strain, which extends the time needed to obtain a result. On the other hand, these methods are simple to perform and relatively inexpensive, but usually do not allow the identification of a specific enzyme type or variant [4]. An introduction of molecular biology-based methods for routine microbiological diagnostics allows for a simultaneous detection of different antimicrobial resistance genes, also within one strain. Additionally, it enables the identification of a specific enzyme within the bacteria group or family and also directly from a clinical sample or a specimen pre-culture [1,2].

The evaluated Eazyplex® SuperBug CRE assay is based on the Loop-mediated isothermal amplification (LAMP) technique and can be used for the detection of ESBL enzyme genes from CTX-M-1 and CTX-M-9 groups, as well as VIM- (1–37), NDM- (1–7), KPC and OXA-48-like (-48, -162, -204, -244) carbapenemase gene variants.

The aim of this study was to evaluate the usefulness of eazyplex® SuperBug CRE assay for the detection of the genes encoding for the most important beta-lactamases among Gram-negative rods, directly from urine samples and positive blood cultures. The results of the evaluated assay were also compared with the results of conventional methods applied (phenotypic tests for beta-lactamases detection), the results of the standard PCR, and evaluated as an alternative tool for rapid and reliable antimicrobial resistance detection in a large, multidisciplinary university hospital.

2. Results

Of 60 urine samples tested, 40 *K. pneumoniae* strains, 13 strains of *Escherichia coli*, three of *K. oxytoca*, three *Enterobacter cloacae* and one *Klebsiella variicola* strain were recovered. All the strains were isolated in monoculture at a titer of $\geq 10^4$ CFU/mL. The eazyplex® SuperBug CRE assay performed for 58 (96.7%) samples gave valid results. In 51 samples ESBLs genes were detected exclusively, and an additional five samples-ESBLs and carbapenemases genes were present simultaneously (Table 1).

Table 1. Beta-lactamases genes detected in Eazyplex® SuperBug CRE assay and PCR performed directly on urine samples (*n* = 60) and the enzymes activity detected with the application of a particular phenotypic method.

Species Recovered from the Samples	Beta-Lactamases Genes Detected by Eazyplex®	No. of Isolates	Time of Particular Genes Detection by Eazyplex® (min:s)	Double Disc Synergy Test Result	Carba NP Test Result
K. pneumoniae (*n* = 40)	CTX-M-1	32	4:30–10:15	(+)	N/A
	CTX-M-1, CTX-M-9	2	5:30–8:30	(+)	N/A
	CTX-M-1, NDM	3	5:15–15:15	(+)	(+)
	Negative result	2	-	UI	N/A
	Invalid test	1	-	(+)	N/A
E. coli (*n* = 13)	CTX-M-1	9	6:30–11:45	(+)	N/A
	CTX-M-9	3	7:00–8:45	(+)	N/A
	CTX-M1, VIM	1	5:30, 9:15	(+)	(+)
E. cloacae (*n* = 3)	CTX-M-1	2	6:45–7:00	(+)	N/A
	CTX-M-1, VIM	1	6:00, 8:15	(+)	(+)
K. oxytoca (*n* = 3)	CTX-M-1	1	9:45	(+)	N/A
	CTX-M-9	1	7:00	(+)	N/A
	Invalid test	1	-	(+)	N/A
K. variicola (*n* = 1)	CTX-M1	1	7:30	(+)	N/A

UI—uninterpretable-lack of any growth inhibition zone in the applied double disc synergy test (criteria described in the text); (+)—positive results in the applied method, N/A—not applicable.

The detection time ranged from 4:30 min to 11:45 min for the ESBL enzyme genes, and from 8:15 min to 15:15 min for the carbapenemases genes.

For 58 strains derived from urine samples, double-disc synergy test results revealed a characteristic enlargement of the growth inhibition zones between the discs from the side of the beta-lactamase containing disc, which confirmed the presence of ESBLs. For two *K. pneumoniae* strains the results were not interpretable due to the lack of any inhibition zone around the discs applied for ESBLs detection. For five carbapenem-resistant strains with positive results via the Carba NP test and EDTA-supplemented disc method, positive results were obtained in the Eazyplex® SuperBug CRE assay. With the application of standard PCR for the chosen 44 (73.3%) strains (40 *K. pneumoniae*, 3 *K. oxytoca* and 1 *K. variicola*), *bla*CTX-M genes presence was confirmed amongst 39 *K. pneumoniae* strains (including negative and non-valid results of Eazyplex® SuperBug CRE assay), while bla_{TEM} and bla_{SHV} were detected amongst 34 and 26 isolates, respectively (Table 2).

Table 2. Comparison of the results obtained using Eazyplex® SuperBug CRE assay and standard PCR for the chosen urine-derived strains (*n* = 44).

Species Recovered from the Samples	Eazyplex® SuperBug CRE Assay	No. of Isolates	Standard PCR	No. of Isolates
K. pneumoniae (*n* = 40)			CTX-M, TEM, SHV	23
	CTX-M-1	32	CTX-M, TEM	6
	CTX-M-1, CTX-M-9	1	CTX-M	3
	CXT-M1, CTX-M-9	1	CTX-M, TEM, SHV	1
	CTX-M-1, NDM	3	CTX-M, TEM	1
	Negative result	1	CTX-M, NDM	3
	Negative result	1	CTX-M, TEM, SHV	1
	Invalid test	1	TEM	1
			CTX-M, TEM, SHV	1
K. oxytoca (*n* = 3)	CTX-M-1	1	CTX-M	1
	CTX-M-9	1	CTX-M	1
	Invalid test	1	CTX-M, TEM	1
K. variicola (*n* = 1)	CTX-M-1	1	CTX-M	1

For the *K. variicola* strain, the consent results were obtained with the application of the Eazyplex® SuperBug CRE assay and standard PCR. The results for two *K. oxytoca* strains were concordant for the evaluated assay and standard PCR, while one non-valid Eazyplex® SuperBug CRE assay result accompanied with the presence of $bla_{\text{CTX-M}}$ and bla_{TEM} genes in a standard PCR. Comparing the results obtained with the evaluated test based on LAMP method and standard PCR methods, a categorical agreement was obtained for 40 out of 44 ESBL-positive strains and for all three carbapenemase-producing isolates. Comparing the results obtained with the evaluated test and phenotypic methods, a categorical compliance was obtained for 56 out of 60 ESBL-positive strains and for all five carbapenemase producing isolates derived from urine samples.

Of 60 positive blood cultures tested, 46 *K. pneumoniae* strains, three strains of each: *K. oxytoca*, *E. cloacae* and *E. coli*, two strains of *S. marcescens* and *Citrobacter freundii*, and one *Proteus mirabilis* strain were derived. Fifty eight samples (96.7%) gave the strains isolated in monoculture, while from two samples, *K. pneumoniae* and *Enterococcus faecalis*, strains were grown simultaneously. For 59 strains derived from positive blood cultures, ESBLs presence was confirmed with the application of a double disc synergy test.

The Eazyplex® SuperBug CRE assay gave valid results for fifty seven (95.0%) samples of pre-cultured blood; In 50 samples only ESBL enzyme genes were detected exclusively, and an additional four samples-ESBLs and carbapenemases genes were present simultaneously (Table 3).

The detection time for ESBL genes ranged from 4:30 min to 10:15 min, while for carbapenemases genes times ranged from 4:45 min to 11:45 min. Comparing the results obtained for 50 (83.3%) of the chosen blood-derived strains (45 *K. pneumoniae*, 3 *K. oxytoca* and 2 *S. marcescens*) by the application of the evaluated test and standard PCR, a categorical agreement was obtained for 48 out of 50 ESBL-positive strains and for all the carbapenemase-producing strains. Comparing the results obtained by the assay based on the LAMP technique and standard PCR results, a categorical compliance was obtained for 54 out of 60 ESBL-positive strains and for all 5 carbapenemase producing isolates (Table 4).

The presence of $bla_{\text{CTX-M}}$ genes in 44 *K. pneumoniae* strains, bla_{TEM} in 31 and bla_{SHV} in 25 was confirmed by standard PCR (including negative and non-valid results of the Eazyplex® SuperBug CRE assay) (Table 4).

One *K. pneumoniae* strain (negative in the evaluated assay) revealed the bla_{TEM} gene, while for the samples with invalid results in the Eazyplex® SuperBug the CRE assay-$bla_{\text{CTX-M}}$ gene was confirmed with the application of a standard PCR. The results obtained by the evaluated assay and standard PCR methods were also consistent for three *K. oxytoca* strains. For one *S. marcescens* strain, the results obtained with the Eazyplex® SuperBug CRE assay were consistent with the standard PCR method, while for the second *S. marcescens* strain (negative in Eazyplex® SuperBug CRE assay) bla_{CTX-M} and bla_{TEM} genes were detected. For *E. coli* and *P. mirabilis* strains that were negative in Eazyplex® SuperBug CRE assay, ESBLs production was confirmed by the double disc synergy test.

Table 3. Beta-lactamases genes detected with Eazyplex® SuperBug CRE assay and standard PCR while performed on positive blood cultures ($n = 60$) and the enzymes activity detected with the application of a particular phenotypic method.

Species Recovered from the Samples	Beta-Lactamases Genes Detected by Eazyplex®	No. of Isolates	Time of Particular Genes Detection by Eazyplex® (min:s)	Double Disc Synergy Test Result	Carba NP Test Result
K. pneumoniae ($n = 46$)	CTX-M-1	39	4:30–10:15	(+)	N/A
	CTX-M-1, CTX-M-9, NDM	1	4:45–5:15	(+)	(+)
	CTX-M-1, NDM	3	4:45–11:45	(+)	(+)
	KPC	1	9:15	UI	(+)
	Negative result	1	-	(+)	N/A
	Invalid test	1	-	(+)	N/A
E. cloacae ($n = 3$)	CTX-M-1	3	6:15–7:30	(+)	N/A
E. coli ($n = 3$)	CTX-M-1	1	7	(+)	N/A
	CTX-M-9	1	8:15	(+)	N/A
	Invalid test	1	-	(+)	N/A
K. oxytoca ($n = 3$)	CTX-M-1	3	6–7:15	(+)	N/A
S. marcescens ($n = 2$)	CXT-M-1	1	6:30	(+)	N/A
	Negative result	1	-	(+)	N/A
C. freundii ($n = 2$)	CTX-M-1	2	6–7	(+)	N/A
P. mirabilis ($n = 1$)	Invalid test	1	-	(+)	N/A

UI—uninterpretable-lack of any growth inhibition zone in the applied double disc synergy test (criteria described in the text); (+)—positive results in the applied method, N/A—not applicable.

Table 4. Comparison of the results obtained using Eazyplex® SuperBug CRE assay and standard PCR for the chosen positive blood cultures-derived strains ($n = 50$).

Species Recovered from the Samples	Eazyplex® SuperBug CRE Assay	No. of Isolates	Standard PCR	No. of Isolates
K. pneumoniae ($n = 45$)	CTX-M-1	39	CTX-M, TEM, SHV	22
			CTX-M	8
			CTX-M, TEM	7
			CTX-M, SHV	2
	CTX-M-1, CTX-M-9, NDM	1	CTX-M, NDM	1
	CTX-M-1, NDM	3	CTX-M, NDM	2
	Negative result	1	CTX-M, TEM, SHV, NDM	1
	Invalid test	1	TEM	1
			CTX-M	1
K. oxytoca ($n = 3$)	CTX-M-1	3	CTX-M	3
S. marcescens ($n = 2$)	CXT-M-1	1	CTX-M	1
	Negative result	1	CTX-M, TEM	1

Five strains, isolated from positive blood cultures, suspected of producing carbapenemases in the Carba NP test, were also positive with the application of the Eazyplex® SuperBug CRE assay. Among four *K. pneumoniae* strains, production of metallo-beta-lactamases was confirmed by means of EDTA-supplemented discs, and the NDM-1 carbapenemase genes by standard PCR. For one *K. pneumoniae* strain, the synthesis of KPC was confirmed by the test with boronic acid. In one *K. pneumoniae* strain producing KPC-type carbapenemases, no growth inhibition zones were observed around the discs when the phenotypic methods for ESBL detection were applied. Neither the presence of the genes encoding

ESBL from the CTX-M1 and CTX-M9 group was confirmed in the Eazyplex® SuperBug CRE assay, nor were bla_{CTX-M}, bla_{TEM} or bla_{SHV} genes detected by standard PCR.

3. Discussion

In recent years, the frequency of multi-drug-resistant bacterial strains isolation has increased significantly [5–7]. This is mainly due to bacterial spread in hospital environment and unreasonable antibiotic therapy. Antimicrobial pressure on bacterial strains causes the emergence of new mechanisms of antibiotic resistance. Therefore, it is crucial to obtain reliable results of the presence of antimicrobial resistance in the shortest time possible.

It is commonly known that the application of a test which simultaneously detects ESBL and carbapenemase enzymes, or their genes, significantly shortens the time of standard microbiological diagnostics. In addition, results of the phenotypic test performed for carbapenemase-producing strains are often ambiguous or difficult to interpret. For example, in the results of a double-disc synergy test the enlargement of the growth inhibition zones, characteristic for ESBL-positive strains, sometimes does not have a typical shape or does not appear at all. Thus, it requires an application of further methods which sometimes significantly extends the time to give a final result.

The Eazyplex® SuperBug CRE assay is based on isothermal amplification of a genetic material and detects the genes for the following enzymes: KPC, NDM, VIM, OXA-48, CTX-M-1, CTX-M-9 and OXA-181. In the available literature, the first studies on Eazyplex® SuperBug CRE assay application appeared in 2015 [1,8]. The mentioned studies were performed on 94 and 450 carbapenemase positive Gram-negative rods strains, respectively. The results of the mentioned research indicated a high sensitivity and specificity of Eazyplex® SuperBug CRE assay in the detection of resistance mechanisms genes in Gram-negative rods [1,2,8]. The study published by Hinić et al. [9], also in 2015, described the use of the Eazyplex test for the detection of genes of ESBL-positive Gram-negative rods directly in 50 urine samples. In 30 (60.0%) of them the presence of ESBL was confirmed. Thus, the overall sensitivity of the method reached from 95.2% to 100% with a specificity of 97.9%.

In the present study, ESBL and/or carbapenemase enzymes were detected in more than 93% of urine samples. The median detection time of ESBL enzyme genes from urine samples was 7 min 45 s, while it was 9:45 min for carbapenemases genes. For two urine samples with negative results in the Eazyplex® SuperBug CRE assay, the bla_{TEM} gene presence was confirmed by a standard PCR. It is noteworthy that this gene was not detected by the evaluated assay. In the second strain, three different genes encoding ESBL enzymes were detected by the applied confirmatory standard PCR. The false-negative result of the Eazyplex® SuperBug CRE assay for the second strain might have resulted from the low number of gene copies below the assay detection limit. However, the manufacturer assures us that the evaluated test can efficiently detect a small number of gene copies (100% sensitivity).

Recently, Fiori et al. [10] evaluated the usefulness of Eazyplex® SuperBug CRE assay for ESBL and/or carbapenemases genes detection directly in positive blood cultures. The mentioned authors detected the presence of CTX-M and/or KPC and/or VIM-like enzymes genes in 151 of the pre-cultured blood samples among 321 episodes of bloodstream infections. The results obtained by this method allowed for the reduction of time to effective antibiotic therapy introduction in patients with *E. coli* or *K. pneumoniae* bacteraemia. The cited authors also highlight the proposed algorithm for combination of a mass spectrometry identification, directly from a blood sample, and the detection of a resistance mechanism with Eazyplex® SuperBug CRE assay, which significantly reduces diagnostic procedures and the time required to get the final results.

In our study, the median detection time of ESBL enzyme genes from the positive blood cultures was 6 min and for carbapenemases genes it was 7 min 30 s when compared to an overnight incubation of the phenotypic test for a particular antimicrobial resistance mechanism detection. Also, results from Rödel et al. [11] indicate a possible reduction of time and the rationalization of antibiotic therapy in the case of patients with sepsis using the Eazyplex test. However, the mentioned authors used an Eazyplex® MRSA test to detect *mecA* and *mecC* genes among *Staphylococcus aureus* and *Staphylococcus epidermidis* strains.

In turn, Bach et al. [12] assessed the usefulness of the LAMP method-based assay in the identification of selected bacterial species and genes encoding ESBL enzymes from the CTX-M1 and CTX-M9 groups from positive blood cultures. The study included 449 positive blood cultures. In the aforementioned research, with the application of the Eazyplex® BloodScreen GN assay, 100% sensitivity and specificity were obtained for *K. oxytoca* and *E. coli* strains, while 95.7% sensitivity and 100% specificity was obtained for *K. pneumoniae* isolates infections. When assessing the detection of genes encoding CTX-M enzymes, they also obtained 100% sensitivity and specificity.

Slightly over 4% of the tests performed in the present work were not valid. This can be explained by several reasons: sample overload, specimen composition, consistency, the presence of some inhibitory substances in urine or blood samples (patients from whom the samples were collected from often take a number of drugs), the presence of more than one strain in a single sample, the presence of several genes encoding for several resistance mechanisms simultaneously, or particular genes mutations. It may lead to the necessity of repeating the whole procedure, which additionally increases the cost and time of the investigation or leads to the underestimation and oversight of resistance mechanism presence if not consequently repeated.

It is worthy of note that the current the number of recognized beta-lactamases is estimated at around 200 or more. In the double disc synergy tests, it is not possible to identify a specific ESBL family, or whether the strain produces one or more ESBL-like enzymes. In a standard PCR method, it is possible to detect many different ESBL enzymes genes, but this requires a more complex approach, usually a longer time, and often several PCR reactions (or a multiplex approach) to detect specific genes, which also affects the costs. Moreover, whether a strain produces more than one type of ESBLs is of epidemiological significance only and does not clinically affect the antibiotic therapy approach. Hence, the available commercial tests (including the Eazyplex® SuperBug CRE assay) are usually designed to detect the most common antibiotic resistance mechanisms among isolated strains, the selected genes encoding for ESBL and/or carbapenemases.

When comparing the methods used in the study, it should be noted that the phenotypic methods (the double disc synergy test and the Carba NP test) are simple to perform, relatively inexpensive (approximately $2 and $1.5 for one strain, respectively), but require earlier culture (which is related to time delay-16–20 h) and do not allow for the identification of a specific ESBL enzyme or carbapenemase type. On the other hand, the standard PCR method for several genes is more expensive (about $10), requires specialized equipment and laboratories as well as carrying out a multiplexed or multiple separate reactions with different primers (which is associated with a time delay of several hours). The LAMP method, on the other hand, is quite expensive (around $50) but also very fast (less than 20 min). In addition, this method allows for the identification of a specific enzyme type, which is very important in epidemiological studies.

In the present study a great advantage of the Eazyplex® SuperBug CRE assay to obtain a result directly from a clinical sample was confirmed. A short duration of the test and a small sample volume needed to perform are also very important. However, the test has some limitations. It detects only ESBL enzymes from the CTX-M-1 and M-9 groups. On the other hand, CTX-M enzymes are the most common ESBL enzymes in *K. pneumoniae*, and the strains with this resistance mechanism are isolated with the highest frequency, not only in Poland, but also worldwide [13–17]. Moreover, the evaluated assay does not detect all beta-lactamases genes, like ampC-like enzymes. However, the dissemination of ampC-positive *Klebsiella* spp. isolates, at least in our department, does not exceed 1% (data not shown) and thus it was not the issue of the present work. The study also did not compare all the tested strains using the Eazyplex® SuperBug CRE assay and standard PCR, which is a limitation of the study, but it was not possible in this research protocol and was not the key goal of the study.

It is worthy of note that the Eazyplex® SuperBug CRE assay detects only some chosen enzymes genes also among carbapenemases. Interestingly, the strains of Gram-negative rods, derived from the patients in our hospital, that express beta-lactams resistance mainly synthesize class B carbapenemases, mostly the NDM- or VIM-type. Detection of the genes for both mentioned carbapenemases is available in the evaluated assay. Of note, not all types of carbapenemases have been detected among the strains of Gram-negative rods in our hospital so far. Since the first detection of a carbapenemase producing strain in our laboratory, only one KPC-positive and one OXA-48 positive strain have been identified (data not shown). It confirms that the strains expressing this particular resistance mechanism are very rare in our department. Nevertheless, the study was limited to the assessment of the usefulness of the Eazyplex® SuperBug CRE assay only for strains of the Enterobacterales order. On the other hand, the possibility of detecting selected genes in the above assay applies mainly to Enterobacterales rods, and to a lesser extent to bacteria belonging to *Pseudomonas* spp.

The Eazyplex® SuperBug CRE assay is a relatively expensive tool for its use in a routine diagnostic directly for clinical specimens. This approach would significantly increase the cost of a standard microbiological investigation. However, its application might be reasonably taken into account in the diagnostic of some particular cases of infections, such as earlier colonization of the patients with ESBL- or carbapenemase-producing strains, confirmed contact with the infected person, or the presence of the local epidemic outbreak. It seems that in such situations the benefits of using an expensive test are favourable for obtaining the result in a shorter time. Moreover, the possibility of introducing a pre-emptive treatment, while also taking into account the phenotype of the strain, seems to be a very reasonable approach.

4. Materials and Methods

The study included 120 clinical specimens: 60 urine samples and 60 positive blood cultures, in which Gram-negative strains producing beta-lactamases with a broad range of activity: ESBLs and/or carbapenemases were identified. Clinical specimens for the present research were mostly selected based on local epidemiological data. Urine samples are the most common type of clinical material sent for testing from hospital patients, while blood is the main material in the microbiological diagnosis of systemic infections. Each sample included into the study was collected from a specific patient. All the samples included into the study were obtained through a routine diagnostic and clinical microbiology laboratory practice. The identification of the strains was performed by mass spectrometry using the MALDI Biotyper system (Bruker, Karlsruhe, Germany). In the analyzed study period, *K. pneumoniae* strains were the most frequently isolated species from infections in patients in our hospital. Antimicrobial susceptibility testing was determined on a BD Phoenix™ M50 instrument (Becton-Dickinson, Franklin Lakes, NJ, USA) using NMIC-402 panels, performed according to the manufacturer's instructions. The expression of ESBL-type enzymes was assessed simultaneously by the double disc synergy test [18], while carbapenemases activity was assessed by the Carba NP test [19]. Phenotypic tests with the application of boronic acid and EDTA-supplemented discs, as specific carbapenemases inhibitors, were also applied, as recommended by EUCAST documents. For all of the original samples stored from which Gram-negative beta-lactamase-producing rods were cultured, the Eazyplex® SuperBug CRE assay (Amplex Diagnostics, Gießen, Germany) was performed on a Genie II device (OptiGene, Gießen, Germany) according to the manufacturer's instructions [20] (see also Supplementary Material).

Additionally, the chosen 44 strains (all *Klebsiella* spp.) isolated from urine samples and the 50 derived from pre-cultured blood (48 *Klebsiella* spp. and 2 *Serratia marcescens* isolates) were cultured on LB Broth (Biocorp, Issoire, France) to confirm the selected genes presence at the further steps. DNA was extracted from the strains recovered from them (applying an Extractme DNA Bacteria Kit, Blirt, Gdańsk, pomeranian, Poland) and the following genes were detected by standard PCR: bla_{CTX-M}, bla_{TEM}, bla_{SHV}, bla_{VIM} and bla_{NDM-1} in separate reactions, according to the methodology of the previous studies [2,21,22].

5. Conclusions

The Eazyplex® SuperBug CRE assay can be a useful tool for a rapid and reliable identification of resistance mechanism genes in Gram-negative rods, and also directly from urine and pre-cultured blood samples.

Supplementary Materials: The following are available online at https://repod.icm.edu.pl/dataverse/umk-medical-sciences as https://repod.icm.edu.pl/dataset.xhtml?persistentId=doi:10.18150/W6IXRP (accessed on 14 January 2002) Genie®II (OptiGene) User Manual (Instrument Software Version v2.22 (rc1).

Author Contributions: Conceptualization, A.S.; Methodology, A.S.; Data analysis, A.S.; Formal analysis, A.S.; Resources, A.S.; Writing and editing of the original draft, A.S. and T.B. All authors have read and agreed to the published version of the manuscript.

Funding: This research was financially supported by Nicolaus Copernicus University funds for basic research activity of the Department of Microbiology, Ludwik Rydygier Collegium Medicum in Bydgoszcz, Poland (PDB WF 839).

Institutional Review Board Statement: The samples were obtained through standard clinical and diagnostic practice. This study received ethical approval from the Bioethical Commission of Ludwik Rydygier Collegium Medicum in Bydgoszcz Nicolaus Copernicus in Torun, agreement no. 367/2019.

Informed Consent Statement: Not applicable.

Data Availability Statement: The data presented in this study are not publicly available as a matter of confidentiality. However, these data are available upon request from the corresponding author.

Conflicts of Interest: The authors declare that they have no conflict of interest.

References

1. Findlay, J.; Hopkins, K.L.; Meunier, D.; Woodford, N. Evaluation of Three Commercial Assays for Rapid Detection of Genes Encoding Clinically Relevant Carbapenemases in Cultured Bacteria. *J. Antimicrob. Chemother.* **2015**, *70*, 1338–1342. [CrossRef] [PubMed]
2. Sękowska, A.; Bogiel, T.; Gospodarek-Komkowska, E. Evaluation of Eazyplex® SuperBug CRE Test for Beta-Lactamase Genes Detection in *Klebsiella* spp. and *P. Aeruginosa* Strains. *Curr. Microbiol.* **2020**, *77*, 99–103. [CrossRef] [PubMed]
3. Xu, L.; Sun, X.; Ma, X. Systematic Review and Meta-Analysis of Mortality of Patients Infected with Carbapenem-Resistant *Klebsiella pneumoniae*. *Ann. Clin. Microbiol. Antimicrob.* **2017**, *16*, 18. [CrossRef]
4. van der Zwaluw, K.; de Haan, A.; Pluister, G.N.; Bootsma, H.J.; de Neeling, A.J.; Schouls, L.M. The Carbapenem Inactivation Method (CIM), a Simple and Low-Cost Alternative for the Carba NP Test to Assess Phenotypic Carbapenemase Activity in Gram-Negative Rods. *PLoS ONE* **2015**, *10*, e0123690. [CrossRef] [PubMed]
5. Dhesi, Z.; Enne, V.I.; O'Grady, J.; Gant, V.; Livermore, D.M. Rapid and Point-of-Care Testing in Respiratory Tract Infections: An Antibiotic Guardian? *ACS Pharm. Transl. Sci.* **2020**, *3*, 401–417. [CrossRef]
6. Carter, D.; Charlett, A.; Conti, S.; Robotham, J.V.; Johnson, A.P.; Livermore, D.M.; Fowler, T.; Sharland, M.; Hopkins, S.; Woodford, N.; et al. A Risk Assessment of Antibiotic Pan-Drug-Resistance in the UK: Bayesian Analysis of an Expert Elicitation Study. *Antibiotics* **2017**, *6*, 9. [CrossRef]
7. Abdeta, A.; Bitew, A.; Fentaw, S.; Tsige, E.; Assefa, D.; Lejisa, T.; Kefyalew, Y.; Tigabu, E.; Evans, M. Phenotypic Characterization of Carbapenem Non-Susceptible Gram-Negative Bacilli Isolated from Clinical Specimens. *PLoS ONE* **2021**, *16*, e0256556. [CrossRef]
8. García-Fernández, S.; García-Castillo, M.; Melo-Cristino, J.; Pinto, M.F.; Gonçalves, E.; Alves, V.; Vieira, A.R.; Ramalheira, E.; Sancho, L.; Diogo, J.; et al. In Vitro Activity of Ceftolozane-Tazobactam against Enterobacterales and Pseudomonas Aeruginosa Causing Urinary, Intra-Abdominal and Lower Respiratory Tract Infections in Intensive Care Units in Portugal: The STEP Multicenter Study. *Int. J. Antimicrob. Agents* **2020**, *55*, 105887. [CrossRef]
9. Hinić, V.; Ziegler, J.; Straub, C.; Goldenberger, D.; Frei, R. Extended-Spectrum β-Lactamase (ESBL) Detection Directly from Urine Samples with the Rapid Isothermal Amplification-Based Eazyplex® SuperBug CRE Assay: Proof of Concept. *J. Microbiol. Methods* **2015**, *119*, 203–205. [CrossRef]
10. Fiori, B.; D'Inzeo, T.; Posteraro, B.; Menchinelli, G.; Liotti, F.M.; De Angelis, G.; De Maio, F.; Fantoni, M.; Murri, R.; Scoppettuolo, G.; et al. Direct Use of Eazyplex® SuperBug CRE Assay from Positive Blood Cultures in Conjunction with Inpatient Infectious Disease Consulting for Timely Appropriate Antimicrobial Therapy in *Escherichia coli* and *Klebsiella pneumoniae* Bloodstream Infections. *Infect. Drug Resist.* **2019**, *12*, 1055–1062. [CrossRef]

11. Rödel, J.; Bohnert, J.A.; Stoll, S.; Wassill, L.; Edel, B.; Karrasch, M.; Löffler, B.; Pfister, W. Evaluation of Loop-Mediated Isothermal Amplification for the Rapid Identification of Bacteria and Resistance Determinants in Positive Blood Cultures. *Eur. J. Clin. Microbiol. Infect. Dis.* **2017**, *36*, 1033–1040. [CrossRef]

12. Bach, K.; Edel, B.; Höring, S.; Bartoničkova, L.; Glöckner, S.; Löffler, B.; Bahrs, C.; Rödel, J. Performance of the Eazyplex® BloodScreen GN as a Simple and Rapid Molecular Test for Identification of Gram-Negative Bacteria from Positive Blood Cultures. *Eur. J. Clin. Microbiol. Infect. Dis.* **2021**. [CrossRef]

13. Guiral, E.; Pons, M.J.; Vubil, D.; Marí-Almirall, M.; Sigaúque, B.; Soto, S.M.; Alonso, P.L.; Ruiz, J.; Vila, J.; Mandomando, I. Epidemiology and Molecular Characterization of Multidrug-Resistant *Escherichia coli* Isolates Harboring BlaCTX-M Group 1 Extended-Spectrum β-Lactamases Causing Bacteremia and Urinary Tract Infection in Manhiça, Mozambique. *Infect. Drug Resist.* **2018**, *11*, 927–936. [CrossRef]

14. D'Andrea, M.M.; Arena, F.; Pallecchi, L.; Rossolini, G.M. CTX-M-Type β-Lactamases: A Successful Story of Antibiotic Resistance. *Int. J. Med. Microbiol.* **2013**, *303*, 305–317. [CrossRef]

15. Dikoumba, A.-C.; Onanga, R.; Boundenga, L.; Bignoumba, M.; Ngoungou, E.-B.; Godreuil, S. Prevalence and Characterization of Extended-Spectrum Beta-Lactamase-Producing Enterobacteriaceae in Major Hospitals in Gabon. *Microb. Drug Resist.* **2021**, *27*, 1525–1534. [CrossRef] [PubMed]

16. Abrar, S.; Vajeeha, A.; Ul-Ain, N.; Riaz, S. Distribution of CTX-M Group I and Group III β-Lactamases Produced by *Escherichia coli* and *Klebsiella pneumoniae* in Lahore, Pakistan. *Microb. Pathog.* **2017**, *103*, 8–12. [CrossRef] [PubMed]

17. Choudhury, K.; Dhar Chanda, D.; Bhattacharjee, A. Molecular Characterization of Extended Spectrum Beta Lactamases in Clinical Isolates of *Escherichia coli* and Klebsiella Spp from a Tertiary Care Hospital of South Eastern Assam. *Indian J. Med. Microbiol.* **2021**, *in press*. [CrossRef] [PubMed]

18. Jarlier, V.; Nicolas, M.H.; Fournier, G.; Philippon, A. Extended Broad-Spectrum Beta-Lactamases Conferring Transferable Resistance to Newer Beta-Lactam Agents in Enterobacteriaceae: Hospital Prevalence and Susceptibility Patterns. *Rev. Infect. Dis.* **1988**, *10*, 867–878. [CrossRef]

19. Nordmann, P.; Poirel, L.; Dortet, L. Rapid Detection of Carbapenemase-Producing Enterobacteriaceae. *Emerg. Infect. Dis.* **2012**, *18*, 1503–1507. [CrossRef] [PubMed]

20. Genie® II. OptiGene. Available online: http://www.optigene.co.uk/wp-content/uploads/2014/07/Genie%20II%20Mark%20II%20Manual%20version%201.00.pdf (accessed on 12 May 2019).

21. Nordmann, P.; Poirel, L.; Carrër, A.; Toleman, M.A.; Walsh, T.R. How to Detect NDM-1 Producers. *J. Clin. Microbiol.* **2011**, *49*, 718–721. [CrossRef]

22. Pitout, J.D.D.; Gregson, D.B.; Poirel, L.; McClure, J.-A.; Le, P.; Church, D.L. Detection of Pseudomonas Aeruginosa Producing Metallo-Beta-Lactamases in a Large Centralized Laboratory. *J. Clin. Microbiol.* **2005**, *43*, 3129–3135. [CrossRef] [PubMed]

Article

Molecular Detection of Carbapenemases in Enterobacterales: A Comparison of Real-Time Multiplex PCR and Whole-Genome Sequencing

Katja Probst [1,*], Dennis Nurjadi [1], Klaus Heeg [1], Anne-Marie Frede [1], Alexander H. Dalpke [2] and Sébastien Boutin [1,3]

1 Department of Infectious Diseases, Medical Microbiology and Hygiene, University Hospital Heidelberg, 69120 Heidelberg, Germany; dennis.nurjadi@med.uni-heidelberg.de (D.N.); klaus.heeg@med.uni-heidelberg.de (K.H.); anne.frede@gmx.de (A.-M.F.); sebastien.boutin@med.uni-heidelberg.de (S.B.)

2 Institute of Medical Microbiology and Virology, University Hospital Carl Gustav Carus, 01307 Dresden, Germany; alexander.dalpke@uniklinikum-dresden.de

3 Translational Lung Research Center Heidelberg (TLRC), German Center for Lung Research (DZL), Heidelberg University Hospital, 69120 Heidelberg, Germany

* Correspondence: katja.probst@med.uni-heidelberg.de; Tel.: +49-6221-56-36420; Fax: +49-6221-56-5857

Citation: Probst, K.; Nurjadi, D.; Heeg, K.; Frede, A.-M.; Dalpke, A.H.; Boutin, S. Molecular Detection of Carbapenemases in Enterobacterales: A Comparison of Real-Time Multiplex PCR and Whole-Genome Sequencing. *Antibiotics* **2021**, *10*, 726. https://doi.org/10.3390/antibiotics10060726

Academic Editor: Francesca Andreoni

Received: 20 May 2021
Accepted: 15 June 2021
Published: 16 June 2021

Abstract: Carbapenem-resistant Enterobacterales are a growing problem in healthcare systems worldwide. While whole-genome sequencing (WGS) has become a powerful tool for analyzing transmission and possible outbreaks, it remains laborious, and the limitations in diagnostic workflows are not well studied. The aim of this study was to compare the performance of WGS and real-time multiplex PCR (RT-qPCR) for diagnosing carbapenem-resistant Enterobacterales. In this study, we analyzed 92 phenotypically carbapenem-resistant Enterobacterales, sent to the University Hospital Heidelberg in 2019, by the carbapenem inactivation method (CIM) and compared WGS and RT-qPCR as genotypic carbapenemase detection methods. In total, 80.4% of the collected isolates were identified as carbapenemase producers. For six isolates, discordant results were recorded for WGS, PCR and CIM, as the carbapenemase genes were initially not detected by WGS. A reanalysis using raw reads, rather than assembly, highlighted a coverage issue with failure to detect carbapenemases located in contigs with a coverage lower than $10\times$, which were then discarded. Our study shows that multiplex RT-qPCR and CIM can be a simple alternative to WGS for basic surveillance of carbapenemase-producing Enterobacterales. Using WGS in clinical workflow has some limitations, especially regarding coverage and sensitivity. We demonstrate that antimicrobial resistance gene detection should be performed on the raw reads or non-curated draft genome to increase sensitivity.

Keywords: antimicrobial resistance; carbapenem inactivation method; carbapenem-resistant Enterobacterales; real-time multiplex PCR; whole-genome sequencing

1. Introduction

Enterobacterales, including bacterial species such as *Citrobacter freundii*, *Escherichia coli*, *Klebsiella pneumoniae* and the *Enterobacter cloacae* complex, belong to the most common human pathogens and are able to cause a variety of infections [1,2].

In particular, infections with multidrug resistant Enterobacterales lead to high mortality since there are limited treatment options [3]. Carbapenemases are of great concern, as they are able to inactivate the last-resort drug carbapenems in addition to other beta-lactam antibiotics [3,4]. They are mostly plasmid encoded, which facilitates an easy transmission and dissemination through horizontal gene transfer [5]. Worldwide, the most common carbapenemases in Enterobacterales are KPC, NDM, VIM, IMP and OXA-48-like carbapenemases [2,6]. Another less frequent route of carbapenem resistance acquisition is via overexpression of the outer membrane efflux pumps or porin loss combined with the expression of extended-spectrum beta-lactamases or *AmpC* resistance genes [7,8].

Phenotypic screening for carbapenem resistance by Carba-NP test [9], the modified Hodge test [10] or the disc diffusion assay [11] is common in microbiology diagnostics, yet for epidemiological surveillance, high-resolution typing is useful and essential. A few real-time PCR (RT-qPCR)-based assays have been developed to detect carbapenem-resistance genes in Gram-negative bacteria [12–14]. However, these methods are technically limited to a certain number of targets. By contrast, whole-genome sequencing (WGS) provides more comprehensive information and thus has become a powerful tool for surveillance and outbreak investigation [15]. Although there are several studies comparing the performance of phenotypic and commercially available tests for carbapenemase detection [16–18], comparative studies on WGS and RT-qPCR remain scarce. Currently, the application of WGS in the clinical microbiological setting is limited to molecular typing. However, there is still an untapped potential for integrating WGS-based technologies into microbiological diagnostics. Although preparation and turnover time remains a major disadvantage for WGS, the performance and accuracy of WGS compared to those of faster nucleic acid amplification-based and simple phenotypic methods should be investigated.

Our study aimed to retrospectively evaluate the performance of WGS compared to that of RT-qPCR and phenotypic carbapenem-resistant Enterobacterales, identified by antimicrobial susceptibility testing and the carbapenem inactivation method (CIM).

2. Results

A total of 92 phenotypic carbapenem-resistant Enterobacterales were collected in 2019. Carbapenem-hydrolyzing activity could be detected in 74 isolates (80.4%) by CIM. These results were validated by WGS and RT-qPCR. For six isolates, different results occurred between the three methods, as carbapenemases were initially detected by CIM and PCR but not by WGS (Tables 1 and 2). By reanalyzing the raw sequencing data and removing the coverage threshold bla_{NDM-1}, bla_{KPC-2} (2x), bla_{VIM-1} (2x) and bla_{OXA-48} were identified (Table A1). For 18 isolates, all three methods revealed no carbapenemase.

Table 1. Comparison of phenotypic and genotypic carbapenemase detection in Enterobacterales by CIM, RT-qPCR and WGS.

		CIM	
		Positive	**Negative**
RT-qPCR	positive	74	0
	negative	0	18
WGS	positive	70(74) [1]	0
	negative	4(0) [1]	18

[1] After reanalyzing the raw sequencing data.

Table 2. Comparison of genotypic carbapenemase detection in Enterobacterales by WGS and RT-qPCR.

		WGS	
		Positive	**Negative**
RT-qPCR	positive	68(74) [1]	6(0) [1]
	negative	0	18

[1] After reanalyzing the raw sequencing data.

The predominant species of the carbapenemase producers was *E. cloacae* ($n = 30$) followed by *K. pneumoniae* ($n = 17$) and *E. coli* ($n = 15$). *C. freundii* ($n = 7$), *Klebsiella oxytoca* ($n = 3$) and *Serratia marcescens* ($n = 2$) appeared less frequently (Figure 1). OXA-48 (40.5%) was the most prevalent carbapenemase and was detected in all species in this collection. VIM-1 (21.6%) was the second most common enzyme in our study, followed by KPC-2 (12.2%) and NDM-5 (9.5%). Other carbapenemase variants, such as NDM-1, OXA-244, KPC-

3 and OXA-232, were less abundant (<3.0%), and isolates harboring two carbapenemases (8.1%) occurred sporadically (Figure 1, Table A1).

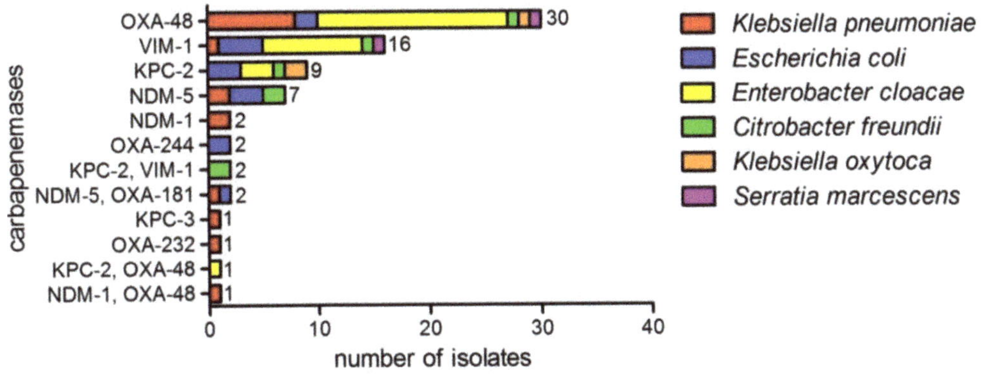

Figure 1. Carbapenemases detected in Enterobacterales by WGS ($n = 74$), showing phenotypic resistance to carbapenem antibiotics. *E. cloacae* (n=30), *K. pneumoniae* ($n = 17$), *E. coli* ($n = 15$), *C. freundii* ($n = 7$), *K. oxytoca* ($n = 3$) and *S. marcescens* ($n = 2$).

3. Discussion

Rapid spreading of carbapenemase-producing Enterobacterales as well as outbreaks of different multidrug resistant bacteria is reported worldwide in clinical settings. For infection control and prevention of further dissemination, monitoring is necessary. Thus, we analyzed 92 phenotypically carbapenem-resistant Enterobacterales by CIM to confirm carbapenem-hydrolyzing activity. We then compared WGS and RT-qPCR to validate performance in detecting carbapenemase genes.

In total, 74 isolates were found to be carbapenemase producers (Figure 1). In six cases, discordant results occurred between WGS and the other two methods, since the carbapenemase was initially not detected by sequencing (Tables 1 and 2 and Table A1). For analyzing WGS data, quality control is crucial, including coverage of the assembly, quality of de novo assembly and detection of potential DNA contamination. The read coverage is of particular importance, as it influences the sensitivity of sequencing [19]. In the initial assembly, we set up a limit of 25× coverage for the full genome, and contigs with a coverage <10× or smaller than 1000 bp were removed because they are potential contaminants or misassemblies. However, our study showed that true signals might be lost during the cleaning of the assembly, since the quality control parameters N50 and the coverage were in the desired range (Table A1). Low-copy number plasmids or plasmid loss during DNA extraction might have led to a low abundance of carbapenemase genes, and, thus, the antimicrobial resistance genes were not detected. While the establishment of such thresholds is crucial for genomic comparison and annotation of a draft genome, our data suggest that antimicrobial resistance gene detection should be performed on the non-curated draft genome to increase sensitivity.

Our findings on carbapenemase variants are in line with the data of the German national reference laboratory (NRL) in the years 2017–2019. In particular, bla_{OXA-48} was detected in all years, followed by bla_{VIM-1}, bla_{KPC-2}, bla_{NDM-1}, bla_{KPC-3}, $bla_{OXA-181}$ and bla_{NDM-5} [20–22], which are detectable with our assay. However, depending on the geographic region, less frequent carbapenemase types, such as GES, GIM and IMI, can occur in Enterobacterales [20–22]. These genes are not included in our assay and, therefore, can lead to false-negative results. In 2019, these carbapenemases were not detected by WGS (Figure 1, Table A1). However, if the epidemiology changes, the PCR should be adapted to the new resistance situation.

The RT-qPCR provides a fast and inexpensive alternative for diagnostic labs without NGS facilities, although the PCR-based assay is limited to known targets. Compared to the

RT-qPCR, WGS is an unbiased method that provides more information, such as genetic relationships and the full resistome. Besides the presence or absence of known resistance genes, novel resistance genes can be identified in phenotypic resistant isolates by WGS [23]. However, the analysis is more complex, and, therefore, bioinformatics expertise is needed.

4. Materials and Methods

4.1. Bacterial Isolates

Clinical samples and rectal swabs were screened for carbapenem-resistant Enterobacterales at the Department of Infectious Diseases, Medical Microbiology, University Hospital Heidelberg in 2019. During routine diagnostics, 92 Enterobacterales showing phenotypic resistance to meropenem and imipenem were collected. Non-duplicate strains were obtained from 79 patients. Multiple isolates ($n = 13$) from the same patient were included in the study due to different bacterial species as determined by MALDI TOF MS (Bruker Daltonics GmbH & Co. KG, Bremen, Germany). The antibiotic susceptibility was tested by the VITEK-2 system (bioMérieux Deutschland GmbH, Nürtingen, Germany) and evaluated according to the valid EUCAST guidelines in the respective year (v 9.0). The isolates were stored at $-20\,°C$ until usage.

4.2. Carbapenem Inactivation Method

CIM was performed, as described elsewhere [24], to examine whether the carbapenem-resistant isolates, identified by antimicrobial susceptibility testing, are able to hydrolyze carbapenem antibiotics.

4.3. DNA Extraction

The isolates were regrown on BD™ Columbia Agar with 5% Sheep Blood (Becton Dickinson GmbH, Heidelberg, Germany) at $37\,°C$. DNA for WGS and RT-qPCR was extracted using the DNeasy Blood and Tissue Kit (Qiagen GmbH, Hilden, Germany) according to the manufacturer's protocol.

4.4. Multiplex Real-Time PCR

The assay based on hydrolysis probes consists of two multiplex PCRs for the detection of bla_{NDM}, bla_{KPC}, bla_{VIM} and bla_{IMP}, and bla_{OXA-23}-like, $bla_{OXA-40/24}$-like, bla_{OXA-58}-like and bla_{OXA-48}-like, respectively. Amplification and detection were performed on the BD MAX™ system, using the protocol for the PCR-only mode, as described elsewhere [25].

4.5. Whole-Genome Sequencing

WGS was performed on the MIseq instrument (2×300 bp), using the Nextera DNA Flex Library Prep Kit (Illumina) for preparing sequencing libraries. Quality control of the raw sequences, assembly and curation (contigs >1000bp and >10× coverage) were performed as described elsewhere [26]. The databases ResFinder 3.0, ARG-ANNOT and CARD-NCBI-BARRGD using ABRIcate (https://git.lumc.nl/bvhhornung/antibiotic-resistancepipeline/tree/master/tools/abricate, accessed on 10 June 2020) were used to determine the resistance genes as previously described [27].

5. Conclusions

Whole-genome sequencing is a powerful tool with high molecular resolution, giving information about bacterial species, plasmid replicon types and the whole resistance pattern, which is needed for surveillance of transmission and outbreak investigation. Real-time PCR is faster but provides less information and cannot detect new carbapenemases that are not included in the panel, which is a general drawback of PCR-based assays. Nevertheless, the additional use of PCR and/or CIM for carbapenemase detection in Enterobacterales was beneficial in our study to ensure high sensitivity, as some carbapenemases would have remained undetected by WGS due to coverage issues.

6. Patents

K.P., K.H. and A.H.D. have a patent (No. 20203612.5) pending.

Author Contributions: Conceptualization, D.N., K.H., A.H.D. and S.B.; methodology, K.P. and A.-M.F.; writing—original draft preparation, K.P.; writing—review and editing, S.B., D.N., A.-M.F., K.H., A.H.D. and K.P.; All authors have read and agreed to the published version of the manuscript.

Funding: This research received no external funding.

Institutional Review Board Statement: Not applicable.

Informed Consent Statement: Not applicable.

Data Availability Statement: Bio project PRJNA634442.

Acknowledgments: We would like to acknowledge the excellent technical support from Delal Sahin, Nicole Henny, Selina Hassel and Suzan Leccese.

Conflicts of Interest: K.P., K.H. and A.H.D. have a patent (No. 20203612.5) pending. The other authors declare no conflicts of interest.

Appendix A

Table A1. Phenotypic carbapenem-resistant Enterobacterales collected in 2019, analyzed by CIM, RT-qPCR and WGS. Quality control parameters for WGS: coverage and N50.

Sample ID	Species	CIM	RT-qPCR	WGS	WGS Reanalyzed	Coverage	N50
KE9539	*E. coli*	positive	bla_{KPC}	bla_{KPC-2}		48	535,993
KE9246	*E. coli*	positive	bla_{KPC}	bla_{KPC-2}		53	135,761
KE9526	*E. cloacae*	positive	bla_{KPC}	bla_{KPC-2}		52	363,822
KE9478	*E. cloacae*	positive	bla_{KPC}	bla_{KPC-2}		96	363,822
BK31926	*E. coli*	positive	bla_{KPC}	bla_{KPC-2}		29	120,862
KE9621	*K. pneumoniae*	positive	bla_{KPC}	bla_{KPC-3}		35	386,401
KE9498	*C. freundii*	positive	bla_{KPC}	bla_{KPC-2}		31	200,582
KE9038	*K. oxytoca*	positive	bla_{KPC}	bla_{KPC-2}		50	285,607
KE9326	*K. oxytoca*	positive	bla_{KPC}	negative	bla_{KPC-2}	42	109,274
KE9511	*C. freundii*	positive	bla_{KPC}, bla_{VIM}	bla_{KPC-2}, bla_{VIM-1}		30	198,406
KE9378	*C. freundii*	positive	bla_{KPC}, bla_{VIM}	bla_{KPC-2}	bla_{KPC-2}, bla_{VIM-1}	39	201,178
KE9132	*E. cloacae*	positive	bla_{KPC}	bla_{KPC-2}		49	363,822
KE9520	*K. pneumoniae*	positive	bla_{NDM}	bla_{NDM-5}		53	186,575
KE9434	*K. pneumoniae*	positive	bla_{NDM}	bla_{NDM-5}		34	292,061
KE9521	*E. coli*	positive	bla_{NDM}, bla_{OXA-48}-like	bla_{NDM-5}, $bla_{OXA-181}$		61	106,471
KE9395	*E. coli*	positive	bla_{NDM}	bla_{NDM-5}		54	94,083
KE9433	*E. coli*	positive	bla_{NDM}	bla_{NDM-5}		36	214,212
KE9636	*K. pneumoniae*	positive	bla_{NDM}, bla_{OXA-48}-like	bla_{NDM-1}, bla_{OXA-48}		27	383,090
KE9616	*C. freundii*	positive	bla_{NDM}	bla_{NDM-5}		50	186,958
KE9522	*E. coli*	positive	bla_{NDM}	bla_{NDM-5}		103	269,697
KE9593	*K. pneumoniae*	positive	bla_{NDM}, bla_{OXA-48}-like	bla_{NDM-5}, $bla_{OXA-181}$		38	296,725
D3014	*C. freundii*	positive	bla_{NDM}	bla_{NDM-5}		36	186,959
KE9449	*K. pneumoniae*	positive	bla_{NDM}	negative	bla_{NDM-1}	25	220,843
KE9500	*K. pneumoniae*	positive	bla_{NDM}	bla_{NDM-1}		33	536,321
KE9382	*E. cloacae*	positive	bla_{OXA-48}-like	bla_{OXA-48}		27	374,725
KE9492	*K. pneumoniae*	positive	bla_{OXA-48}-like	$bla_{OXA-232}$		30	242,997
KE9629	*E. coli*	positive	bla_{OXA-48}-like	$bla_{OXA-244}$		33	238,467
KE9025	*E. cloacae*	positive	bla_{OXA-48}-like	bla_{OXA-48}		49	272,750
KE9469	*E. cloacae*	positive	bla_{OXA-48}-like	bla_{OXA-48}		76	374,315
KE9472	*E. cloacae*	positive	bla_{OXA-48}-like	bla_{OXA-48}		98	382,653
KE9424	*K. pneumoniae*	positive	bla_{OXA-48}-like	bla_{OXA-48}		36	184,292
KE9499	*E. cloacae*	positive	bla_{OXA-48}-like	bla_{OXA-48}		66	486,681
KE9400	*K. pneumoniae*	positive	bla_{OXA-48}-like	bla_{OXA-48}		45	208,351
KE9468	*E. cloacae*	positive	bla_{OXA-48}-like	bla_{OXA-48}		80	383,026

Table A1. *Cont.*

Sample ID	Species	CIM	RT-qPCR	WGS	WGS Reanalyzed	Coverage	N50
KE9638	*E. coli*	positive	$bla_{OXA\text{-}48}$-like	$bla_{OXA\text{-}244}$		37	156,925
KE9493	*E. cloacae*	positive	$bla_{OXA\text{-}48}$-like, bla_{KPC}	$bla_{OXA\text{-}48}$	$bla_{KPC\text{-}2}$, $bla_{OXA\text{-}48}$	44	530,933
KE9456	*K. oxytoca*	positive	$bla_{OXA\text{-}48}$-like	$bla_{OXA\text{-}48}$		28	223,596
KE9443	*K. pneumoniae*	positive	$bla_{OXA\text{-}48}$-like	$bla_{OXA\text{-}48}$		27	225,118
KE9354	*E. cloacae*	positive	$bla_{OXA\text{-}48}$-like	$bla_{OXA\text{-}48}$		66	486,663
BK32270	*E. coli*	positive	$bla_{OXA\text{-}48}$-like	$bla_{OXA\text{-}48}$		35	117,967
KE9626	*E. coli*	positive	$bla_{OXA\text{-}48}$-like	$bla_{OXA\text{-}48}$		53	196,578
KE9208	*S. marcescens*	positive	$bla_{OXA\text{-}48}$-like	$bla_{OXA\text{-}48}$		58	2,797,497
D2902	*E. cloacae*	positive	$bla_{OXA\text{-}48}$-like	$bla_{OXA\text{-}48}$		64	302,960
KE9541	*K. pneumoniae*	positive	$bla_{OXA\text{-}48}$-like	$bla_{OXA\text{-}48}$		47	427,613
KE9554	*C. freundii*	positive	$bla_{OXA\text{-}48}$-like	$bla_{OXA\text{-}48}$		39	165,554
KE9338	*E. cloacae*	positive	$bla_{OXA\text{-}48}$-like	$bla_{OXA\text{-}48}$		47	374,725
KE9355	*E. cloacae*	positive	$bla_{OXA\text{-}48}$-like	$bla_{OXA\text{-}48}$		109	486,681
KE9328	*K. pneumoniae*	positive	$bla_{OXA\text{-}48}$-like	$bla_{OXA\text{-}48}$		43	274,145
KE9510	*E. cloacae*	positive	$bla_{OXA\text{-}48}$-like	$bla_{OXA\text{-}48}$		31	491,022
D3070	*E. cloacae*	positive	$bla_{OXA\text{-}48}$-like	$bla_{OXA\text{-}48}$		44	372,768
KE9428	*E. cloacae*	positive	$bla_{OXA\text{-}48}$-like	$bla_{OXA\text{-}48}$		36	486,663
KE9527	*E. cloacae*	positive	$bla_{OXA\text{-}48}$-like	negative	$bla_{OXA\text{-}48}$	27	339,153
D3018	*K. pneumoniae*	positive	$bla_{OXA\text{-}48}$-like	$bla_{OXA\text{-}48}$		25	473,650
D3082	*E. cloacae*	positive	$bla_{OXA\text{-}48}$-like	$bla_{OXA\text{-}48}$		62	486,663
KE9637	*K. pneumoniae*	positive	$bla_{OXA\text{-}48}$-like	$bla_{OXA\text{-}48}$		36	876,600
D3081	*E. cloacae*	positive	$bla_{OXA\text{-}48}$-like	$bla_{OXA\text{-}48}$		85	383,026
EX1012	*K. pneumoniae*	positive	$bla_{OXA\text{-}48}$-like	$bla_{OXA\text{-}48}$		39	223,327
D3078	*E. cloacae*	positive	$bla_{OXA\text{-}48}$-like	$bla_{OXA\text{-}48}$		54	486,828
KE9366	*E. coli*	positive	bla_{VIM}	$bla_{VIM\text{-}1}$		38	215,473
KE9563	*E. cloacae*	positive	bla_{VIM}	$bla_{VIM\text{-}1}$		35	377,920
KE9409	*E. cloacae*	positive	bla_{VIM}	$bla_{VIM\text{-}1}$		46	486,118
KE9414	*E. cloacae*	positive	bla_{VIM}	$bla_{VIM\text{-}1}$		38	161,463
KE9365	*K. pneumoniae*	positive	bla_{VIM}	$bla_{VIM\text{-}1}$		32	232,474
KE9538	*S. marcescens*	positive	bla_{VIM}	$bla_{VIM\text{-}1}$		40	1,130,420
KE9585	*E. cloacae*	positive	bla_{VIM}	$bla_{VIM\text{-}1}$		25	287,090
KE9559	*C. freundii*	positive	bla_{VIM}	$bla_{VIM\text{-}1}$		41	163,976
KE9549	*E. coli*	positive	bla_{VIM}	$bla_{VIM\text{-}1}$		47	279,067
KE9548	*E. cloacae*	positive	bla_{VIM}	$bla_{VIM\text{-}1}$		46	230,814
KE9579	*E. coli*	positive	bla_{VIM}	$bla_{VIM\text{-}1}$		39	112,495
KE9474	*E. cloacae*	positive	bla_{VIM}	$bla_{VIM\text{-}1}$		38	290,132
KE9462	*E. cloacae*	positive	bla_{VIM}	$bla_{VIM\text{-}1}$		33	502,528
KE9560	*E. cloacae*	positive	bla_{VIM}	$bla_{VIM\text{-}1}$		40	290,117
KE9575	*E. cloacae*	positive	bla_{VIM}	$bla_{VIM\text{-}1}$		38	389,538
KE9536	*E. coli*	positive	bla_{VIM}	negative	$bla_{VIM\text{-}1}$	44	377,920
D2923	*E. cloacae*	negative	negative	negative		58	203,439
KE9347	*E. cloacae*	negative	negative	negative		79	439,426
KE9591	*E. cloacae*	negative	negative	negative		47	279,225
KE9576	*E. coli*	negative	negative	negative		27	228,481
KE9599	*E. coli*	negative	negative	negative		37	281,932
KE9623	*E. coli*	negative	negative	negative		50	93,960
KE9633	*K. aerogenes*	negative	negative	negative		40	495,847
KE9068	*C. freundii*	negative	negative	negative		46	176,242
KE8986	*E. cloacae*	negative	negative	negative		47	230,847
KE9344	*E. cloacae*	negative	negative	negative		48	235,301
KE9475	*E. cloacae*	negative	negative	negative		40	208,042
KE9083	*E. coli*	negative	negative	negative		57	208,544
KE9425	*K. aerogenes*	negative	negative	negative		48	902,223
KE9614	*K. aerogenes*	negative	negative	negative		62	429,809
D3017	*K. pneumoniae*	negative	negative	negative		27	232,937
KE9095	*K. pneumoniae*	negative	negative	negative		66	237,389
KE9171	*K. pneumoniae*	negative	negative	negative		54	481,561
KE9039	*S. marcescens*	negative	negative	negative		40	1,228,444

References

1. Nordmann, P.; Dortet, L.; Poirel, L. Carbapenem resistance in Enterobacteriaceae: Here is the storm! *Trends Mol. Med.* **2012**, *18*, 263–272. [CrossRef] [PubMed]
2. Logan, L.K.; Weinstein, R.A. The Epidemiology of Carbapenem-Resistant Enterobacteriaceae: The Impact and Evolution of a Global Menace. *J. Infect. Dis.* **2017**, *215* (Suppl. 1), S28–S36. [CrossRef]
3. Tzouvelekis, L.S.; Markogiannakis, A.; Psichogiou, M.; Tassios, P.T.; Daikos, G.L. Carbapenemases in *Klebsiella pneumoniae* and Other Enterobacteriaceae: An Evolving Crisis of Global Dimensions. *Clin. Microbiol. Rev.* **2012**, *25*, 682–707. [CrossRef]
4. Delgado-Valverde, M.; Sojo-Dorado, J.; Pascual, Á.; Rodríguez-Baño, J. Clinical management of infections caused by multidrug-resistant Enterobacteriaceae. *Ther. Adv. Infect. Dis.* **2013**, *1*, 49–69. [CrossRef]
5. Van Duin, D.; Doi, Y. The global epidemiology of carbapenemase-producing Enterobacteriaceae. *Virulence* **2017**, *8*, 460–469. [CrossRef]
6. Exner, M.; Bhattacharya, S.; Christiansen, B.; Gebel, J.; Goroncy-Bermes, P.; Hartemann, P.; Heeg, P.; Ilschner, C.; Kramer, A.; Larson, E.; et al. Antibiotic resistance: What is so special about multidrug-resistant Gram-negative bacteria? *GMS Hyg. Infect. Control* **2017**, *12*, 3.
7. Paterson, D.; Doi, Y. Carbapenemase-Producing Enterobacteriaceae. *Semin. Respir. Crit. Care Med.* **2015**, *36*, 74–84. [CrossRef] [PubMed]
8. Nordmann, P.; Cornaglia, G. Carbapenemase-producing Enterobacteriaceae: A call for action! *Clin. Microbiol. Infect.* **2012**, *18*, 411–412. [CrossRef]
9. Tijet, N.; Boyd, D.; Patel, S.N.; Mulvey, M.R.; Melano, R.G. Evaluation of the Carba NP Test for Rapid Detection of Carbapenemase-Producing Enterobacteriaceae and *Pseudomonas aeruginosa*. *Antimicrob. Agents Chemother.* **2013**, *57*, 4578–4580. [CrossRef]
10. Amjad, A.; Mirza, I.; Abbasi, S.; Farwa, U.; Malik, N.; Zia, F. Modified Hodge test: A simple and effective test for detection of carbapenemase production. *Iran. J. Microbiol.* **2011**, *3*, 189–193.
11. Van Dijk, K.; Voets, G.M.; Scharringa, J.; Voskuil, S.; Fluit, A.C.; Rottier, W.C.; Hall, M.A.L.; Stuart, J.W.T.C. A disc diffusion assay for detection of class A, B and OXA-48 carbapenemases in Enterobacteriaceae using phenyl boronic acid, dipicolinic acid and temocillin. *Clin. Microbiol. Infect.* **2014**, *20*, 345–349. [CrossRef] [PubMed]
12. Antonelli, A.; Arena, F.; Giani, T.; Colavecchio, O.L.; Valeva, S.V.; Paule, S.; Boleij, P.; Rossolini, G.M. Performance of the BD MAX™ instrument with Check-Direct CPE real-time PCR for the detection of carbapenemase genes from rectal swabs, in a setting with endemic dissemination of carbapenemase-producing Enterobacteriaceae. *Diagn. Microbiol. Infect. Dis.* **2016**, *86*, 30–34. [CrossRef]
13. Hofko, M.; Mischnik, A.; Kaase, M.; Zimmermann, S.; Dalpke, A.H. Detection of Carbapenemases by Real-Time PCR and Melt Curve Analysis on the BD Max System. *J. Clin. Microbiol.* **2014**, *52*, 1701–1704. [CrossRef] [PubMed]
14. Dallenne, C.; da Costa, A.; Decré, D.; Favier, C.; Arlet, G. Development of a set of multiplex PCR assays for the detection of genes encoding important beta-lactamases in Enterobacteriaceae. *J. Antimicrob. Chemother.* **2010**, *65*, 490–495. [CrossRef]
15. Schurch, A.C.; van Schaik, W. Challenges and opportunities for whole-genome sequencing-based surveillance of antibiotic resistance. *Ann. N. Y. Acad. Sci.* **2017**, *1388*, 108–120. [CrossRef] [PubMed]
16. Baeza, L.L.; Pfennigwerth, N.; Greissl, C.; Göttig, S.; Saleh, A.; Stelzer, Y.; Gatermann, S.; Hamprecht, A. Comparison of five methods for detection of carbapenemases in Enterobacterales with proposal of a new algorithm. *Clin. Microbiol. Infect.* **2019**, *25*, 1286.e9–1286.e15. [CrossRef]
17. Baeza, L.L.; Pfennigwerth, N.; Hamprecht, A. Rapid and Easy Detection of Carbapenemases in Enterobacterales in the Routine Laboratory Using the New Gene POC Carba/Revogene Carba C Assay. *J. Clin. Microbiol.* **2019**, *57*. [CrossRef]
18. Han, R.; Guo, Y.; Peng, M.; Shi, Q.; Wu, S.; Yang, Y.; Zheng, Y.; Yin, D.; Hu, F. Evaluation of the Immunochromatographic NG-Test Carba 5, RESIST-5 O.O.K.N.V., and IMP K-SeT for Rapid Detection of KPC-, NDM-, IMP-, VIM-type, and OXA-48-like Carbapenemase Among Enterobacterales. *Front. Microbiol.* **2020**, *11*, 609856. [CrossRef] [PubMed]
19. Ellington, M.; Ekelund, O.; Aarestrup, F.; Canton, R.; Doumith, M.; Giske, C.; Grundman, H.; Hasman, H.; Holden, M.; Hopkins, K.; et al. The role of whole genome sequencing in antimicrobial susceptibility testing of bacteria: Report from the EUCAST Subcommittee. *Clin. Microbiol. Infect.* **2017**, *23*, 2–22. [CrossRef]
20. Pfennigwerth, N. Bericht des Nationalen Referenzzentrums (NRZ) für gramnegative Krankenhauserreger – Zeitraum 1. Januar 2017 – 31. Dezember 2017. *J. Epidemiol. Bull.* **2018**, *28*, 263–267.
21. Pfennigwerth, N. Bericht des Nationalen Referenzzentrums (NRZ) für gramnegative Krankenhauserreger, 2018. *J. Epidemiol. Bull.* **2019**, *31*, 289–294.
22. Pfennigwerth, N. Bericht des Nationalen Referenzzentrums (NRZ) für gramnegative Krankenhauserreger, 2019. *J. Epidemiol. Bull.* **2020**, *26*, 3–10.
23. Nurjadi, D.; Zizmann, E.; Chanthalangsy, Q.; Heeg, K.; Boutin, S. Integrative Analysis of Whole Genome Sequencing and Phenotypic Resistance Toward Prediction of Trimethoprim-Sulfamethoxazole Resistance in *Staphylococcus aureus*. *Front. Microbiol.* **2020**, *11*, 607842. [CrossRef] [PubMed]
24. Van der Zwaluw, K.; de Haan, A.; Pluister, G.N.; Bootsma, H.J.; de Neeling, A.J.; Schouls, L.M. The carbapenem inactivation method (CIM), a simple and low-cost alternative for the Carba NP test to assess phenotypic carbapenemase activity in gram-negative rods. *PLoS ONE* **2015**, *10*, e0123690. [CrossRef] [PubMed]

25. Probst, K.; Boutin, S.; Bandilla, M.; Heeg, K.; Dalpke, A.H. Fast and automated detection of common carbapenemase genes using multiplex real-time PCR on the BD MAX™ system. *J. Microbiol. Methods* **2021**, *185*, 106224. [CrossRef] [PubMed]
26. Nurjadi, D.; Boutin, S.; Dalpke, A.; Heeg, K.; Zanger, P. Draft Genome Sequence of *Staphylococcus aureus* Strain HD1410, Isolated from a Persistent Nasal Carrier. *Genome Announc.* **2018**, *6*, e00411–e00418. [CrossRef] [PubMed]
27. Eichel, P.C.V.; Boutin, S.; Poschl, J.; Heeg, K.; Nurjadi, D. Altering antibiotic regimen as additional control measure in suspected multi-drug-resistant *Enterobacter cloacae* outbrake in neonatal intensive care unit. *J. Hosp. Infect.* **2019**, *104*, 144–149. [CrossRef]

MDPI

St. Alban-Anlage 66

4052 Basel

Switzerland

Tel. +41 61 683 77 34

Fax +41 61 302 89 18

www.mdpi.com

Antibiotics Editorial Office

E-mail: antibiotics@mdpi.com

www.mdpi.com/journal/antibiotics